Snakes

Snakes

Ecology and Behavior

Richard A. Seigel
Department of Biological Sciences
Southeastern Louisiana University
Hammond, Louisiana

Joseph T. Collins
Museum of Natural History
The University of Kansas
Lawrence, Kansas

THE BLACKBURN PRESS

Reprint of 1993 Edition by McGraw-Hill, Inc.

Snakes
Ecology and Behavior

ISBN-10: 1-930665-14-8
ISBN-13: 978-1-930665-14-9

Library of Congress Control Number: 2001097787

THE BLACKBURN PRESS
P. O. Box 287
Caldwell, New Jersey 07006 U.S.A.
973-228-7077
www.BlackburnPress.com

To Nadia, who has been my main source of encouragement all these years, and without whom I would probably be a lawyer, and to Ben, who I hope I can encourage in the same way.

RAS

To George T. McDuffie, who took me on my first real snake hunt to Shawnee State Forest in southern Ohio; Ralph Dury, who made me at home in the Cincinnati Museum of Natural History; Jack L. Gottschang, who taught my first and only herpetology course at the University of Cincinnati; Karl H. Maslowski, who whetted my interest in wildlife photography; and to the Queen City herpetologists I grew up with: James L. Corrado, Robert Gravenkemper, Douglas E. Haggard, Corson J. Hirschfeld, Martin J. Huelsmann, Frank J. Kramer, Dennis R. Magee, and James B. Murphy.

JTC

Contents

Foreword

In 1987, Joe Collins, Sue Novak, and I put together a series of chapters on snake biology called *Snakes: Ecology and Evolutionary Biology.* This book was successful enough that Joe Collins and I did a follow-up text called *Snakes: Ecology and Behavior*, designed to complement the initial text. Although sales of these two books were quite high, the realities of modern publishing forced the initial publishers to allow them to go out of print in the late 1990s. This was unfortunate, since the demand for the books remained quite high; and many potential readers would ask us, "When will the books be reprinted?"

The answer is "now." Thanks to the interest and cooperation of The Blackburn Press, these texts are again available to our original audiences—graduate students and other professionals seeking to have, at their fingertips, reviews of some of the most topical areas of snake ecology, behavior, and evolutionary biology. We think these original audiences were well-served by the texts, at least judging by their high citation rate in many of the recent papers on snake biology. Indeed, we hope that these books played at least a small role in the virtual explosion of papers and texts on snakes since the late 1980s. We trust that the reprinting of these texts will allow incoming workers in the field to have access to the critical information provided by our authors.

Richard A. Seigel
Joseph T. Collins
Susan S. Novak
2001

Preface

This book is an outgrowth of our 1987 text, *Snakes: Ecology and Evolutionary Biology*. This initial text was based on the idea that an up-to-date volume on snakes that reviewed and summarized the current literature was needed badly. Based on large numbers of citations and on positive reviews, the 1987 text was clearly successful in its goals. However, in the preface to that volume we noted that space requirements precluded us from reviewing all of the areas of snake biology that were of interest, and we tried to emphasize those topics that were in need of review or which were especially exciting. Although we realized that we were leaving many fascinating areas uncovered, we noted at the time that "to adequately cover all aspects of snake biology would require a 'Biology of the Serpentes' series, which neither we nor our publishers were willing to consider."

Okay. So we were wrong.

As time went by, it became apparent that many of the areas we had left uncovered in 1987 were as badly in need of review as were those areas we did cover six years ago. Thus, although we still do not envision a series of texts on snakes, we did feel strongly that a second text was needed. The present text has essentially the same goals as the 1987 volume, specifically (1) to draw together a summary of what is known about the major aspects of snake biology; (2) to summarize the primary literature on snakes, both for the experienced professional who is overwhelmed by the mass of citations available and for the researcher who is just starting to work with these animals; and (3) to stimulate new and innovative research on snakes by drawing attention to those areas of snake biology much in need of additional attention and by making reasonable speculations concerning provocative questions that remain inadequately addressed. Authors were encouraged strongly to interact and cross-reference other chapters in the text. As in 1987, our primary audience is professional researchers and students, but we hoped to include material that would interest other groups (resource managers and amateur naturalists) as well.

In deciding on what topics we would cover this time, we tried to emphasize areas that could not be covered in 1987 due either to lack of space or author availability, or (especially) areas that had developed rapidly over the past few years. Virtually all the topics covered in this text are new; the only chapters that cover the same subject areas are Conservation (Dodd), Foraging (Arnold), and Behavior (Ford and Burghardt), and even for these chapters the orientation and emphasis is very different from chapters in our 1987 text. The authors are new as well; of the 16 authors in the current volume, only four contributed to the 1987 text, and only one of these (Dodd) is writing on the same area. We feel that such a change in authorship provides a healthy new perspective on snake biology, a field which we feel holds considerable promise for future research.

As in 1987, our primary debt goes to the authors of the individual chapters, who continued the tradition of submitting superb reviews on the biology of snakes. We thank each of them for their time and efforts.

We would also like to thank those individuals who served as reviewers for the individual chapters. In addition to those who wished to remain anonymous, we thank David Chiszar, William Cooper, Patrick Gregory, Richard King, Randy Noss, Michael Plummer, and Laurie Vitt. Many of the authors for these chapters reviewed other chapters for the text, and we thank them for their assistance.

RAS would like to thank Gary Childers, the Department Head of Biological Sciences at Southeastern Louisiana University for his generous and willing support of this project. Portions of this text were prepared while RAS was granted summer release time by David Watts, Dean of the College of Arts and Sciences. At the University of Kansas, JTC would like to thank Philip S. Humphrey, Director of the Museum of Natural History, and Edward O. Wiley, Curator of Ichthyology, for their patience and support.

Richard A. Seigel
Joseph T. Collins

Contributors

Editors

Richard A. Seigel *Department of Biological Sciences, SLU 814, Southeastern Louisiana University, Hammond, Louisiana 70402*

Joseph T. Collins *Museum of Natural History, The University of Kansas, Lawrence, Kansas 66045-2454*

Authors

Stevan J. Arnold *Department of Ecology and Evolution, University of Chicago, 940 East 57th Street, Chicago, Illinois 60637*

Edmund D. Brodie III *Department of Integrative Biology, University of California, Berkeley, California 94720*

Gordon M. Burghardt *Department of Psychology, University of Tennessee, Knoxville, Tennessee 37996-0900*

C. Kenneth Dodd, Jr. *National Ecology Center, U.S. Fish and Wildlife Service, 412 NE 16th Avenue, Gainesville, Florida 32601*

Michael E. Dorcas *Department of Biological Sciences, Idaho State University, Pocatello, Idaho 83209-8007*

David Duvall *Life Sciences Program, Arizona State University West, P.O. Box 37100, Phoenix, Arizona 85069*

Neil B. Ford *Department of*

Biology, University of Texas at Tyler, Tyler, Texas 75701-6699

Theodore Garland, Jr. *Department of Zoology, University of Wisconsin, Madison, Wisconsin 53706*

A. Ralph Gibson *Department of Biology, Cleveland State University, Cleveland, Ohio 44115-2403*

Robert W. Henderson *Milwaukee Public Museum, 800 West Wells Street, Milwaukee, Wisconsin 53233-1478*

Harvey B. Lillywhite *Department of Zoology, University of Florida, Gainesville, Florida 32611*

Charles R. Peterson *Department of Biological Sciences, Idaho State University, Pocatello, Idaho 83209-8007*

Howard K. Reinert *Department of Biology, Trenton State College, Trenton, New Jersey, 08650-4700*

Gordon Schuett *Department of Zoology and Physiology, University of Wyoming, Laramie, Wyoming 82701*

Richard A. Seigel *Department of Biological Sciences, Southeastern Louisiana University, Hammond, Louisiana 70402*

Richard Shine *School of Biological Sciences, Zoology Building, University of Sydney, Sydney, New South Wales 2006, Australia*

Chapter

1

Behavioral and Functional Ecology of Arboreal Snakes

Harvey B. Lillywhite

Robert W. Henderson

Introduction

Beginning with fossorial ancestors, snakes have undergone one of the more impressive adaptive radiations in vertebrate evolutionary history. The success of the radiation can be measured by the number of species and their collective persistence in virtually every part of the biosphere except for the deeper oceans and polar regions. The radiation is instructive because of the demonstrated adaptability of form and function coincident with retention of a specialized general body plan. Because of limbless morphology and ectothermy, snakes are closely coupled to both physical forces and the spatial configuration of abiotic features in their environment. The radiation of snakes into marine and arboreal environments was met with particularly challenging problems.

Arboreal snakes pose many fascinating questions, yet numerous aspects of their biology have not been well studied. There is considerable information on these reptiles, derived primarily from case studies of particular species or from scattered data appearing coincident with various analyses of species assemblages. However, few investigators have addressed general questions related to arboreality and the constraints or opportunities of arboreal habitats for limbless reptiles.

Three questions seem of immediate challenge to understanding arboreality. First, in what ways, and to what extent, must snakes evolve to function in arboreal habitats? Second is a similar but more difficult question: what selective forces produce arboreal habits, or is arboreality merely a matter of opportunism? Third, why are there not more arboreal species? In other words, what limits the abundance and diversity of arboreal snakes? Some families of snakes with terrestrial members never show arboreality. Thus, we are curious to understand those factors that favor arboreality in some taxa and those factors that might constrain arboreality in other taxa.

Our present status of information does not allow more than partial or tentative answers to these questions. We hope, however, that the information and ideas that are discussed in this chapter will stimulate imaginative and renewed dedication to investigations of arboreal snakes.

Challenges of Arboreal Environments

Physical environment

Challenges that must be met to achieve evolutionary success in arboreal habitats include both abiotic and biotic factors. The physical environment is particularly important to understanding adaptive successes of arboreal snakes and, of course, provides the underpinning that structures the associated biotic community.

Gravity. The force of gravity is especially relevant to elongate animals that assume vertical postures and utilize three-dimensional axes of their habitat. Physiological processes are affected significantly by gradients of pressure within body fluids. The vascular system is of particular significance in this context, because blood within arteries, veins, and associated vasculature forms long, continuous fluid networks. Hydrostatic pressure increases with depth according to Pascal's laws, so in an elongate animal like a snake, hydrostatic pressures in the lower vessels increase as a sine function of departure from horizontal position. The magnitude of the absolute pressure and the pressure gradient along the vascular column varies directly with the length of the system. Gravitational disturbance to body fluids is therefore potentially greatest in long snakes that assume fully erect posture. Gravity is also important in that it adds an additional vector component of forces that must be dealt with in locomotion. Secondarily, gravity imposes a variety of problems related to acquisition and handling of prey, as well as to social interactions.

Substratum. The nature and configuration of the substratum impose further constraints or requirements for use of arboreal habitats. The structure of these environments varies greatly, depending on latitude, altitude, climatic regime, and floral community. Generally, however, we may regard the substratum as discontinuous in space and potentially unstable. Such features impose the requirement for locomotor skills and, with respect to branches, leaves, or other protruding plant parts, a match between the mass of the snake and the strength and stiffness of the supporting substrate. These factors presumably have played a selective role in the reduction of mass and use of caudal prehension in arboreal species of snakes. Cantilever abilities are important in negotiating gaps in the substrate, while cardiovascular specializations are requirements for counteracting gravity during vertical movement.

Microclimate. Models are available for predicting variations of microclimate in vegetation profiles of varying complexity above the ground (Geiger, 1965; Mitchell et al., 1975; Monteith, 1975; Kira and Yoda, 1989) and for determining limits of body temperatures from heat-exchange characteristics of reptiles including snakes (Tracy, 1982; Peterson, 1987). Although modeling of heat and mass exchanges has not been applied specifically to arboreal snakes, the available literature nonetheless offers insights that are relevant to snakes in above-ground situations.

Radiation and convection are likely to dominate heat-exchange processes in most situations involving snakes in vegetation above the ground. In dense evergreen forests, direct solar radiation and diffuse radiation attenuate rapidly below the upper canopy; therefore, thermal radiation and conduction become dominant modes of heat flux for snakes inhabiting the lower vegetation strata of these forests. Vegetation gaps in such communities may assume increased importance in providing opportunities for snakes to intercept direct solar radiation, especially in closed-canopy forests. In warm and sparsely vegetated habitats such as deserts, shrubs or trees can provide escape from high temperatures that otherwise might be incurred by exposure to heated substrate. Varying degrees of shade are provided, depending on the character of the vegetation. In many situations, temperatures can be moderated above the ground by convection and evapotranspiration of the canopy.

The duration of bright sunshine in regions of tropical rainforests typically ranges from 6 to 7 h day^{-1} (Bazzaz and Pickett, 1980), but even in clearings is effectively reduced to 3 to 4 h day^{-1} because of interception by surrounding trees (Soepadmo and Kira, 1977). If leaf area density (leaf area per unit volume of canopy, LAD) is more or less uniform throughout a canopy space, light flux density inside the canopy is expected to decrease exponentially with increasing depth below the canopy surface (Monsi and Saeki, 1953). The vertical gradient of light extinction varies at different height levels and is steepest in layers with highest LAD. Layers of high LAD typically occur in the main canopy and near the ground, with midlayers of profile structures having relatively low LAD. Extinction of light is also affected by the relative inclination of leaves. Profiles of relative illuminance therefore vary among monolayered herbaceous communities and stratified forest communities. In any event, light flux density decreases significantly within the uppermost levels of vegetation and at ground level beneath typical rainforests is usually less than 1% of that above the canopy surface (Walter, 1971; Leigh, 1975; Alexandre, 1982). The percentage of total light interception tends to decrease slightly in more xerophytic types of tropical evergreen forests or temperate forests, but still exceeds 95%. Considerable light can be intercepted by leafless stems and branches that reduce incident light by 30–70% in some shrub and herb communities (Yim et al., 1969).

In temperate or xerophytic forests the space between the main canopy layer and the ground is relatively open with low LAD. In such forests sunbeams passing through small holes in the canopy can reach the ground layer unintercepted to produce sunflecks. On the other hand, the space beneath the main canopy of typical rainforests is generally filled with tree crowns of various heights, which intercept sunbeams so efficiently that very few sunflecks reach the forest floor. In the undisturbed lowland rainforest of peninsular Malaysia, for example, 50% of the ground surface receives dim light less than 0.2% of that at the canopy surface, and the area of bright spots over 1% of canopy illuminance is only one-hundredth of the total area (Yoda, 1974; Aoki et al., 1975). The brightest spot on the floor of the Malaysian Pasoh forest during fine weather of February and July received only 4–6% of the light at the canopy surface. Thus, there are virtually no sun flecks on the ground of typical equatorial rainforest where no appreciable gaps exist in its canopy. Gaps and disturbed spots are common, however, even in primeval forest, creating very uneven distribution of light flux density over the forest floor. The three-dimensional heterogeneity of light distribution within a forest community has important implications for thermoregulation and resource use by snakes, but these microclimatic interactions have not been studied. It is commonly noted, however, that habitat use by arboreal species of snakes tends to favor edge or disturbed situations (see below).

Air movement approaches zero at the ground surface and increases with height according to velocity profiles that vary with the structure of vegetation (Geiger, 1965; Mitchell et al., 1975; Monteith, 1975; Kira and Yoda, 1989) (Fig. 1.1*A*). Therefore, arboreal snakes are

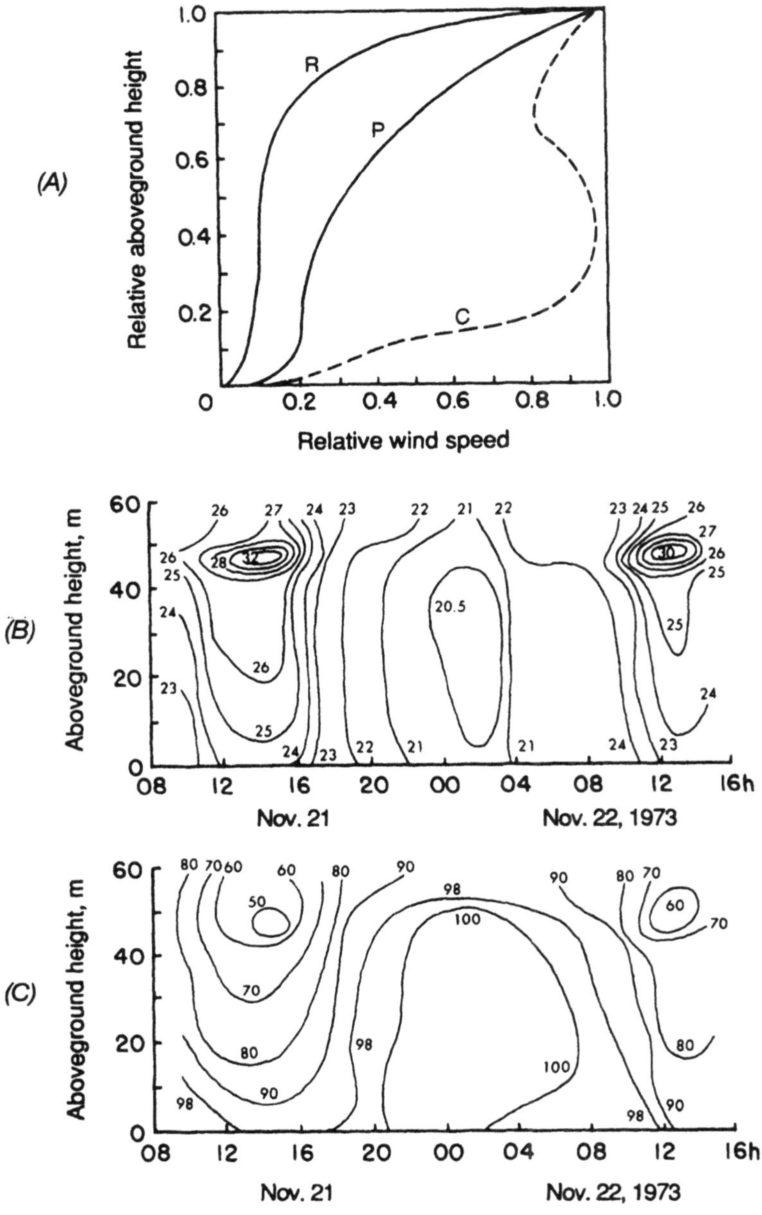

Figure 1.1. (*A*) Comparison of wind-speed profiles in different plant communities. Both wind speed and aboveground height are normalized by taking the values at the canopy surface as the standard. R = paddy field (Aoki et al., 1975); P = Pasoh forest (Aoki et al., 1975); C = Costa Rican rainforest (Allen et al., 1972). (*B*) Diurnal change of air temperature (°C) in Pasoh forest shown as isopleths on height and time coordinates (Aoki et al., 1975). (*C*) Isopleths of relative humidity in Pasoh forest (Aoki et al., 1975). (Modified from Kira and Yoda, 1989.)

potentially exposed to higher levels of convection than are ground-dwelling species, and convective heat exchange and evaporation are significant interactions with the environment. Arboreal snakes require behavioral or morphological protection from evaporative losses of body water, especially in consideration of slender shape and large surface-to-mass ratios. Reductions of mass and increased surface-to-volume ratios increase the coupling of slender animals to convection while reducing the influence of absorbed radiation as determinants of body temperature (Porter and Gates, 1969; Stevenson, 1985). One may hypothesize that selection of microhabitats with respect to vegetative strata might be influenced by physiology related to heat and water exchange as well as biotic or other factors. Wind also affects the mechanical interactions between snakes and substrate by inducing motion of leaves and branches.

The effects of wind are most pronounced in shrub communities or xerophytic temperate forests having comparatively low LADs. In the tropics, however, rainforest regions are generally not windy, and the prevailing wind speed over the surface of such forests generally does not exceed a few meters per second except during local squalls. Moreover, air turbulence is markedly reduced inside dense plant communities, and microclimatic conditions beneath a leaf canopy tend to be less variable than those outside the canopy space.

Much has been published on the vertical distribution of temperature and humidity in tropical rainforests (e.g., Richards, 1952; Walter, 1964, 1971). Relative to ground surface, the canopy crown experiences a wider range of temperatures due to interception of solar radiation during day and radiative cooling at night. As an example, the mean daily range of air temperature in lowland forest of Malaysia studied during November was 10.4°C at the crown surface and 3.9°C at 5 cm above the ground (Aoki et al., 1975). Much of the internal space of the forest remained nearly isothermal, especially at night. Characteristically, vertical thermal gradients are very small in tall forests relative to other types of plant communities, provided there is fairly active eddy diffusion. General patterns of spatial and temporal distribution of relative humidity (RH) are close to the reverse of those in air temperature (Fig. 1.1*B* and *C*). RH values corresponding to afternoon maxima of temperature at the crown surface are as low as 50–60%, while the air near the forest floor remains nearly saturated with water vapor (95–100% RH) throughout the day. The entire forest space may be vapor-saturated during the night.

The persistence of high humidity and a remarkable constancy of microclimatic conditions close to the forest floor are well-known features of tropical rainforests. These microclimatic factors obviously might influence the vertical distribution and activity of arboreal snakes. It should be noted that even the most stable rainforest climates experience long rainless spells that may extend from a few to several months, and the floor of dense forests may become completely dried up. Therefore, "typical" patterns of vegetation use by arboreal species might be altered unpredictably by unusual climatic events. The occurrence or activity of arboreal snakes can be dependent on climatic conditions. As examples, *Thelotornis* (Sweeney, 1971), *Imantodes,* and other species (Henderson et al., 1978) are rare at seasonally dry times of the year.

Heat and water balance. There is little information concerning body temperatures of arboreal snakes and, in particular, their use of thermal "resources." Recently, information on body temperatures and thermal ecology of Rough Green Snakes (*Opheodrys aestivus*) was reported by Plummer (1993). Body temperatures of diurnally active snakes were measured throughout the activity season and compared to operative temperatures measured with copper models. Contrary to expectations, Rough Green Snakes are eurythermal and active over a broad range of body temperatures (17.9–36.8°C). Basking is rare, and body temperatures of snakes conform closely to ambient air temperatures. However, the mean activity temperature (29.3°C) is similar to that reported for a variety of snake species from other habitats (Lillywhite, 1987a). Variation in perch heights selected by snakes in their primary microhabitat (shoreline vegetation overhanging water) does not appear to have significant thermal consequences (Fig. 1.2). Although Rough Green Snakes often forage on the distal portions of branches, they also utilize the less-exposed interior of vegetation during daytime foraging (Plummer, 1981). Snakes often avoid direct solar radiation, and mean body temperatures of Rough Green Snakes are below mean maximum operative temperatures, even on cool mornings. Plummer interpreted body temperatures as passively related to foraging, although basking or foraging in more exposed sites might be a thermoregulatory strategy early in the season when ambient air temperatures are cooler.

Operative temperatures of animal models placed on arboreal perches increase with increasing diameter of the perch (Bakken, 1989). Generally, *Opheodrys aestivus* prefers perches with small diameters (Plummer, 1981; Goldsmith, 1984), and it does not appear that thermal properties of physical perches influence perch selection by Rough Green Snakes. Clearly, this aspect of perch selection deserves further study in this and other species of arboreal snakes.

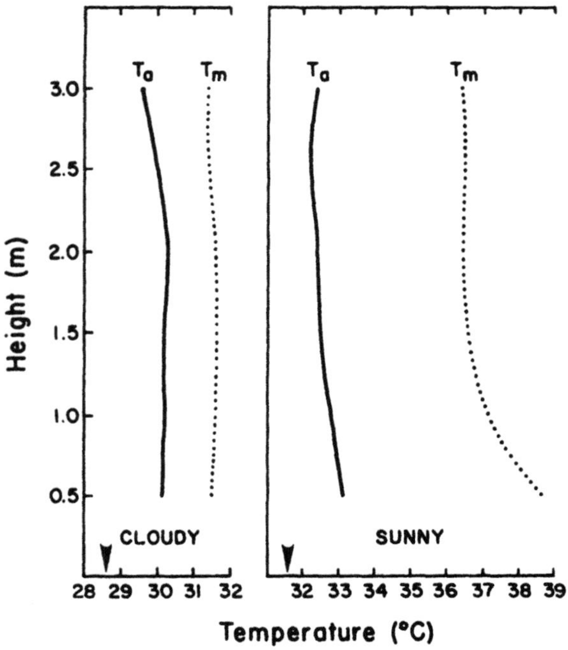

Figure 1.2. Vertical profiles of mean shaded air temperature (T_a) and mean maximum operative temperature (T_m) for a copper model of a Rough Green Snake, *Opheodrys aestivus.* The profile was taken in vegetation at the edge of water during an afternoon with still air on a cloudy day and a sunny day. The arrows indicate water temperatures under each profile. (Redrawn from Plummer, 1993.)

Comparative studies should determine whether imprecise thermoregulation or thermal conformity characterizes other arboreal species, and how the incidence and precision of active thermoregulation might change along a gradient from tropical to cooler temperate climates. Considering the attenuate body shape of arboreal species, it will be important to consider the interplay between thermal ecology and water relations of snakes. See Peterson, Gibson, and Dorcas (Chap. 7, this volume) for additional discussion.

There is little information concerning water exchange and its regulation in arboreal snakes. However, rates of evaporative water loss measured in Rough Green Snakes, *Opheodrys aestivus,* are reported to be lower than those of many terrestrial and fossorial snake species for which data are available (Dove et al., 1982). *Opheodrys aestivus* loses significantly less water than other snakes tested, except vipers and species from extremely xeric habitats. Although measured water fluxes included both respiratory and cutaneous losses, they presumably reflect adjustments in the cutaneous component which accounts for 50–80% of the total and most of the differences among species. These data suggest, therefore, that adaptations to arboreality include reductions of cutaneous water loss, as occurs in arboreal frogs (Wygoda, 1984; Shoemaker, 1988). Further studies of water relations of arboreal snakes should include more comparative data in addition to considerations of microhabitat selection and total water budget.

Biotic environment

We can only speculate about the challenges to snakes presented by co-inhabitants of arboreal habitats relative to those of ground-level terrestrial communities, insofar as there are no studies that have evaluated the relevant issues. Generally, it would seem that snakes inhabiting aboveground vegetation have more limited access to sites offering seclusion and therefore are relatively "more exposed" than in ground-level situations offering holes, burrows, rock crevices, etc. Indirect evidence supporting this supposition can be inferred from the evolution of special defenses, including high incidence of crypsis, immobility, and the relative absence of bright, contrasting color patterns. Whether aerial or avian predation pressures are more intense in arboreal than terrestrial habitats is unknown and probably varies among habitats.

A second consideration of arboreal communities relates to intraspecific social interactions. Because of the heterogeneous and discontinuous nature of the substratum, prey capture and the location of conspecifics for purposes of courtship and mating present special challenges. Selection for crypsis with respect to predation may add to the difficulty. Thus, deposition of pheromonal trails may be particularly important, while gaps in substratum may require enhanced searching on the part of trailing snakes. We must await future careful research to tell us whether "rules" affecting mate acquisition, dispersal, and spatial use of habitat are generally, or necessarily, different between ground-level and aboveground habitats.

Catastrophic events

Snakes living in vegetation above the ground are potentially exposed to impacts of stochastic or periodic physical events such as fire, wind, drought, and cyclonic storms. Therefore, catastrophic factors may play important roles in the population ecology and life history evolution of arboreal snakes.

In xeric or semiarid habitats, fire can be an especially important force causing direct as well as indirect mortality in reptile populations. Whereas snakes on the ground might readily seek refuge in holes or fissures or crevices, those inhabiting elevated vegetation are likely to be consumed by a rapidly advancing fire. In forest habitats, there seems little hope of escape for animals positioned in tall trees that become casualty to a crown fire. The dynamics of survivorship and recovery for snakes in arboreal versus ground microenvironments following fire is a subject in need of investigation.

In drought situations where snakes are especially reclusive or inactive (e.g., Henderson et al., 1978; Lillywhite, 1982), arboreal species can be especially stressed unless terrestrial retreats are sought during these periods. It would be interesting to determine whether habitat shifts occur in specialized arboreal species generally found in tropical or subtropical environments.

Occurrence of Arboreal Habits

Use of arboreal habitat occurs chiefly among colubrids, viperids, and boids. While "arboreal species" may be considered as those species that confine their activity to vegetation (trees or shrubs) above the ground, there is surely a continuum of arboreal behavior. Some snakes alternate activity between terrestrial and arboreal situations. The use of aboveground vegetation may be sporadic and opportunistic or may follow regular diel or seasonal patterns (e.g., terrestrial foraging alternating with arboreal seclusion). Other species are primarily terrestrial but also excellent climbers, ascending on occasion to considerable heights in vegetation. Still other species are predominantly terrestrial or aquatic and do not climb well, but occasionally make brief sojourns into vegetation above the ground.

There is difficulty with attempts to classify snakes according to their degree of arboreal habits, largely because behaviors are not known sufficiently (or quantitatively) to discriminate discrete categories of classification. Therefore, we do not propose at this time a rigorous scheme of terminology with respect to arboreal habits. Throughout this chapter we use *arboreal* in a general sense, usually with reference to those species that spend 50% or more of their time in aboveground vegetation; or, in the case of snakes whose habits are poorly studied, to those species that are known from a limited number of specimens captured in vegetation and referred to as *arboreal* or *partly arboreal* in general guides.

The occurrence of arboreal species, both in absolute numbers and as a percentage of snake fauna, increases from temperate to tropical locations and is highly correlated with mean annual precipitation, presumably reflecting the effect of precipitation on vegetation structure (Shine, 1983). In eastern Kansas, only 1 out of 12 snake species (8.3%) has arboreal tendencies (Fitch, 1982), whereas arboreal species in tropical areas can exceed 50% of the total snake fauna. Snake communities at La Selva, Costa Rica (52 species of boids, colubrids, and viperids), Barro Colorado Island, Panama (43 species), Santa Cecilia, Ecuador (47 species), Manaus, Brazil (52 species), and Manu National Park, Peru (29 species) are represented by a remarkably constant 21.2% (La Selva) to 25.8% (Manu) arboreal species (based on adaptive zones determined by Duellman, 1990), and from 12.9% (Manu) to 28.8% (Manaus) semiarboreal species. Combined, the percentage of arboreal and semiarboreal species ranges from 36.5% (La Selva) to 51.9% (Manaus).

Correlates of arboreal species richness include climate, vegetative structure, productivity, latitude, and historical factors. The combination of ecological factors producing the "driving force" for evolution of arboreal habits in snakes has not been evaluated, although constraints that might limit arboreality can be identified. For example, evolution of cardiovascular specializations appear to be an absolute requirement for climbing behaviors that involve vertical posture (see below). Shine (1983, 1991) has considered the ecology of Australian snakes and suggests that the rare occurrence of arboreality is probably a function both of environment and of the taxonomic composition of the snake fauna. Australia is mainly arid, and its snake fauna consists principally of elapids. The proportion of arboreal species in various geographic regions appears to be consistently lower in elapid faunas than in other families containing arboreal forms (Shine, 1983). It is not readily apparent why arboreal habits are generally rare in the Elapidae, which has many terrestrial members worldwide. However, many elapid species of snakes climb well and are semi- or intermittently arboreal (Heatwole et al., 1973).

Specializations and Convergence in Arboreal Habitats

Arboreal habits have evolved independently in numerous lineages of snakes, many of which exhibit similar, if not identical, specializations of behavior, morphology, and physiology. Despite the recognized difficulties in demonstrating adaptation (e.g., Bock, 1980; Feder et al., 1987), it seems reasonable to infer that reoccurring specializations of form and function are adaptive in arboreal environments and reflect convergent evolution in many cases.

Color

Green or brown color patterns are predominant in arboreal species of snakes, rendering them cryptic against backgrounds of green foliage or branches. Brown coloration provides crypsis against many terrestrial backgrounds and, therefore, is a common component of coloration in nonarboreal as well as arboreal species. The most striking aspect of coloration, therefore, is the occurrence of bright green colors among various colubrid, boid, and viperid snakes from arboreal habitats, particularly in tropical environments where greens are predominant background colors. The prevalence of these patterns suggests that the evolution of coloration was related in major part to selection forces of predation via visually oriented predators. It is also feasible that crypsis is important for snakes as predators.

Body morphology

There is remarkable convergence of body form among arboreal snakes. Most arboreal species are attenuate with light bodies (Fig. 1.3), and, within families that have arboreal members, the tree-

Figure 1.3. An adult *Ahaetulla prasina* from central Thailand.

dwelling species are more attenuate and lighter (e.g., the boids *Chondropython* and *Corallus* and the colubrids *Ahaetulla, Imantodes,* and *Uromacer*) or smaller and lighter (e.g., the viperids *Atheris* and *Bothriechis*) than their ground-dwelling "relatives." Generally, circumference-to-length ratios are as small as 2% in arboreal colubrids and are more than an order of magnitude greater in terrestrial viperids and boids in which ratios may exceed 30%; within families, ratios are nearly uniformly lower in arboreal than in nonarboreal species (Lillywhite, unpublished data).

Guyer and Donnelly (1990) studied length–mass relationships of snakes at La Selva, Costa Rica and described four morphological categories reflecting ecological associations: heavy-bodied, light-bodied, long-tailed, and unextreme. The light-bodied group is comprised of typically arboreal species and corresponds to a similar morphological grouping of arboreal snakes from the Brazilian caatinga (Vitt and Vangilder, 1983). Guyer and Donnelly suggested that length–mass analysis could be used to predict general habitat use, and that it was especially useful in identifying highly arboreal species that also possess other divergent characteristics.

The low mass of arboreal snakes is reinforced by comparatively short intervals between feeding and defecation. In contrast, defecation intervals of considerably stouter terrestrial snakes may be considerably longer (sometimes months in viperids), even when snakes are feeding regularly (Lillywhite, unpublished data). In the heavier forms, storage of fecal matter possibly confers some advantage related to the mere storage of mass, which is disadvantageous for arboreal forms.

There are several possible advantages of the low mass and slender form characteristic of arboreal species. First, a slender body form may enhance crypsis if snakes are situated in or among twigs and branches. Second, the reduction of mass permits snakes to crawl upon lighter branches without weighting them down. The slender, light bodies of these snakes might also enhance the ability to cantilever and thereby span gaps in vegetation. Very slender snakes such as *Imantodes cenchoa* and *Oxybelis aeneus* can easily bridge gaps up to one-half their length. Finally, a small circumference minimizes cardiovascular problems related to gravitational pooling of blood and edema in tissues (see below).

One possible disadvantage of slender body form is that it might limit the reproductive potential of females through limitation of egg or clutch size. The limited data available for reproductive traits of arboreal snakes do not unequivocally confirm this hypothesis, however (Plummer, 1984; Seigel and Ford, 1987).

The very arboreal boas, pythons, vipers, and larger colubrids are laterally compressed. The narrow body is accomplished usually by ribs that are directed downward rather than outward, producing a thin but very deep-bodied animal (Johnson, 1955). Only in the colubrids do narrow vertebrae contribute to the slenderization (see below).

Several analyses have indicated that arboreal snakes tend to possess proportionately long tails, especially colubrid species (e.g., Goldsmith, 1984; Guyer and Donnelly, 1990). This appears to be characteristic of arboreal snakes generally, although ground-dwelling species that show caudal autotomy also have long tails (Guyer and Donnelly, 1990).

Head morphology

Besides elongation of the body and low body mass, many arboreal snakes have evolved narrowed skulls (especially colubrids), elongate snouts, large eyes, and other modifications of facial anatomy. Henderson and Binder (1980) used measurements of such features in comparing four genera of diurnal vine snakes (*Ahaetulla, Oxybelis, Thelotornis, Uromacer*) with other arboreal and ground-dwelling snakes. The vine snakes, even though unrelated, were shown to have a unique combination of head characters and to have longer heads with narrower snouts and smaller eyes than nocturnal arboreal snakes (e.g., *Boiga, Imantodes, Leptodeira, Sibon*). Arboreal predators on lizards (e.g., *Uromacer*) have longer heads, narrower snouts, and smaller eyes than do arboreal predators on frogs (e.g., *Leptodeira, Leptophis*).

The eye. Arboreal snakes may have rounded, vertical or horizontal "keyhole" pupils. Some arboreal species possess a fovea, and some have a double fovea (Underwood, 1970). In snakes that have a keyhole (Fig. 1.4), the slot of the keyhole points forward well beyond the lens, thereby creating an extensive aphakic (lensless) space. The keyhole widens the binocular field without compromising the extent of periscopy (Walls, 1942). During accommodation the lens moves not only forward but also more strongly nasally than in other snakes. In *Ahaetulla,* directionally acute vision is conferred by a line of sight that passes from the temporal fovea (at the outer rim of the retina) through the lens and aphakic portion of the pupil and along a cheek groove in front of the eye, projecting straightforward parallel to the long axis of the head (Fig. 1.4). Visual acuity is also enhanced by slenderized cones (Walls, 1942).

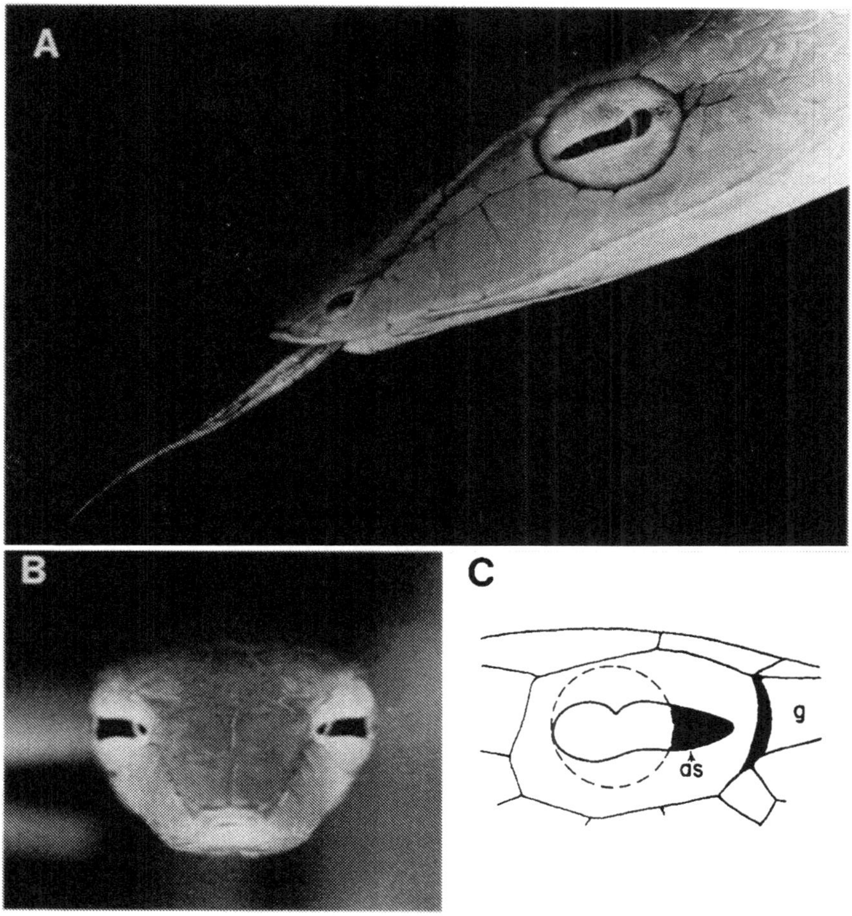

Figure 1.4. (*A*) *Ahaetulla prasina* showing the horizontal "keyhole" pupil and rigid extension of the tongue. Note also the longitudinal groove between eye and snout. (*B*) *Ahaetulla prasina* illustrating forward position of the pupil in relation to anterior and ventral visual fields. (*C*) Details of right eye of *Dryophis* (= *Ahaetulla*) *mycterizans* showing aphakic portion of pupil (as) and cheek groove (g) which permits straightforward vision. Lens is indicated by dashed circular line. (Redrawn from Walls, 1942.)

Snout attenuation in diurnal vine snakes (e.g., *Ahaetulla, Oxybelis*) is likely an aid to binocular vision, conferring considerable anterior as well as ventral overlap of the visual field (Fig. 1.4*B*). In addition, these snakes have longitudinal grooves between the eye and snout that increases straightahead vision. Probably because of these features, the Asian colubrid *Ahaetulla prasina* has the widest binocular field of vision (46°) among those snakes tested for binocular angle (Walls, 1942). A number of arboreal snakes (e.g., *Oxybelis, Uromacer*) have horizontal marks extending from a point posterior of the eye to the tip of the snout—patterns that seem to be less prevalent in terrestrial than arboreal species. Such "eye lines" may mask the eye (e.g., *Thelotornis*) or conceivably act as an aid to vision in species that feed on fast-moving prey (Ficken et al., 1971). Many arboreal snakes that are diurnal ambush predators depend highly on vision for orientation to objects. They can maintain visual contact with a potential prey item by the most subtle of cues: movement of antennae in insects; movement of a toe or respiratory movements of the thorax in lizards. But, if all detectable movement ceases, the snake apparently loses contact with the prey and stalking behavior stops, even while tongue-flicking may continue (Henderson and Binder, 1980; Goldsmith, 1986).

It is of interest that arboreal snakes having wide binocular fields also sway the head back and forth to gain additional parallax. Visual information about the position and prominence of an object is increased by viewing it from more than one angle simultaneously (binocular vision) or in rapid succession (invoked by movements of the head). Head swaying occurs in a number of lateral-eyed vertebrates and is discussed further below.

Structure and function of the cardiovascular system

In terms of physiology, specialized features of the cardiovascular system are especially prominent in arboreal snakes. This is because gravity imposes absolute constraints on blood circulation unless counteracted by adaptation of design and function.

The greatest challenge to blood circulation imposed by gravity is the tendency for blood to pool in dependent vasculature when the posture is inclined from horizontal, the effect becoming greater with increasingly vertical attitude and body length (Fig. 1.5). Blood accumulates in lower tissues due to passive expansion of compliant vessels, especially veins, and filtration of plasma in response to elevated intracapillary pressures, which creates edema. Both processes are favored by weak vascular muscle tone and loose or compliant tissues, especially if coupled to large body girth. Such postural edema and blood pooling reduce venous return to the heart, so the cardiac output falls accordingly and thereby diminishes arterial pressures and central blood flow to vital organs.

The ability of upright snakes to defend central hemodynamics depends on several related features of structure and physiology.

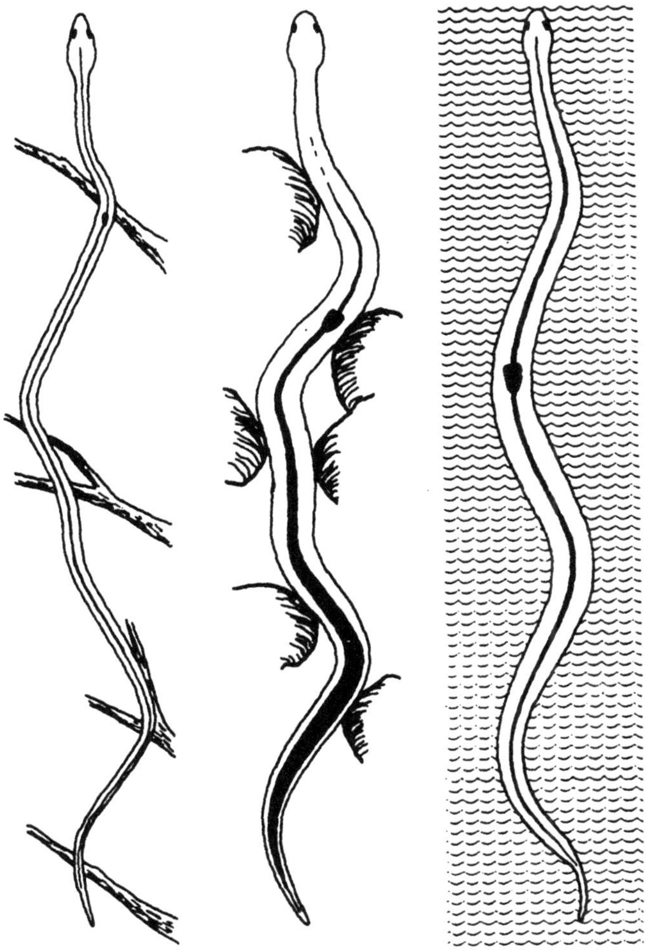

A B C

Figure 1.5. Comparisons of blood pooling in the circulatory systems of (*A*) a generalized arboreal snake, (*B*) a nonclimbing, terrestrial snake, and (*C*) a sea snake. The slender body and tight skin of the arboreal snake act somewhat like a water jacket (external to the sea snake) or "antigravity suit" to counteract intravascular pressures and thereby prevent significant blood pooling. Vascular distension in the arboreal snake also is prevented by a high level of vascular smooth muscle tone mediated by dense adrenergic innervation of blood vessels. Pooling does not occur in the sea snake because the tendency for blood pressure to distend vessels is opposed by external water pressure. Note also the anterior heart location in the arboreal species, which assists blood circulation to the head. The terrestrial snake normally does not climb and lacks the adaptations seen in the arboreal species. Blood pooling is pronounced because the vessels are distensible and expand in response to increasing hydrostatic pressure in the lower body. In this case, the shift of blood volume to the lower body compromises circulation to the head. (Redrawn from Lillywhite, 1988.)

Compared with terrestrial and aquatic species, arboreal snakes pool significantly less blood during vertical posture, and such measurements may approach tenfold differences (Lillywhite, 1985a). Such defense against edema and blood pooling almost certainly reflects low-compliance tissue (Lillywhite, 1987b), slender body shape, and tightly applied integument. In addition, arboreal snakes exhibit extremely dense innervation of posterior arteries and veins (Donald and Lillywhite, 1988), which mediates reflexogenic control of vascular tone and hence arterial resistance and venous capacity. Such control of vascular tone enables arboreal species to regulate arterial pressure and cephalic blood flow during head-up posture (Lillywhite and Gallagher, 1985; Lillywhite, 1987c; 1988; Lillywhite and Donald, 1988). Thus, arboreal species of snakes exhibit superior "postural" regulation of hemodynamics (Seymour and Lillywhite, 1976; Lillywhite, 1987c) in comparison with nonclimbing species (Lillywhite and Pough, 1983; Lillywhite and Smits, 1992). The maintenance of cardiovascular performance may also involve stereotyped body movements that promote venous return to the heart and counteract blood pooling during conditions of postural hypotension (Lillywhite, 1985b).

Arterial pressures are generally higher in arboreal (40–70 mmHg) than in nonclimbing species (20–35 mmHg) (Seymour and Lillywhite, 1976; Lillywhite, 1987c; Seymour, 1987). This "adaptive hypertension" probably reflects comparatively high levels of vascular tone (resistance) and, in any case, helps to ensure that central arterial pressures do not fall below minimum required perfusion pressures when vertical positions cause passive reductions of pressure in the upper regions of vertical blood columns.

Prominent morphological features of arboreal cardiovascular systems include headward position of the heart and reduced length of pulmonary vasculature (Lillywhite, 1987c, 1988; Seymour, 1987). Heart position in arboreal and climbing terrestrial snakes is 15–25% of total body length (TBL) from the head, whereas nonclimbing terrestrial (viperid) and aquatic species have heart distances 25–45% of TBL from the head. The anterior heart helps to ensure adequate perfusion of the head regardless of body posture.

Arboreal species of snakes characteristically have short pulmonary vasculature, either due to short body length (e.g., certain viperids) or a proportionately reduced segment of the lung that is vascular (e.g., many colubrids). In elongate colubrid snakes, the vascular lung may comprise less than 10% of the total body length (Lillywhite, unpublished data). The significance of the short length of vascular parenchyma is that associated blood columns (arteries and veins) do not develop large transmural pressure gradients in response to gravity when the snake is vertical. Thus, excessive intravascular pressures and shifts of blood volume within the lung are minimized or prevented. These aspects of morphometry, in concert with reflexogenic control of the pulmonary circulation (Lillywhite, 1987c; Donald et al., 1990), confer stability of pulmonary function during posture change. In contrast, elongate vascular lungs of some aquatic snakes (>50% of TBL) develop severe pulmonary edema and capillaries actually rupture when these aquatic snakes are positioned vertically in air (Lillywhite, 1987c).

The challenges of gravity to blood circulation and the requirement for adaptive countermeasures are further illustrated by responses of stout, terrestrial viperid snakes (e.g., *Bitis* spp.) to head-up tilt (Lillywhite and Smits, 1992). As an example, head-up tilt of *Bitis* spp. to angles >30° reduces carotid blood flow to zero. Such cardiovascular incompetence is due largely to inability of autonomic reflexes to counteract blood pooling, which presumably is encouraged by compliant skin and girth of the animal. Such lack of adaptation for counteracting gravity, similar to sea snakes, presumably reflects the largely sedentary and horizontal lifestyle of these snakes. Clearly, physiology imposes severe constraints on climbing ability; the effect, however, diminishes with decreasing body length. Physiology is one of a suite of potentially interacting factors that may limit arboreal habits to smaller body size. Thus, one observes that arboreal viperids are generally <1 m in length; longer species might be arboreal as juveniles, but they abandon arboreal habits as longer adults (e.g., March, 1928; Test et al., 1966; Henderson et al., 1976).

Defenses against Predation

We are not aware of any published studies that provide quantitative information concerning predation on arboreal snakes. However, observations suggest that predation can be intense (Plummer, 1990a). Recently, it was reported that Laughing Falcons (*Herpetotheres cachinnans*) inhabiting primary forests in Guatemala prey solely upon snakes, of which a "large proportion" are said to be arboreal species (Parker, 1990).

Cryptic coloration, immobility, and behavior probably constitute the more important means by which arboreal snakes avoid or minimize attacks from potential predators. As already mentioned, green coloration is prevalent among tree snakes, suggesting that convergence in this trait has been an important aspect of adaptation to arboreal life. Pattern as well as color (hue and saturation) is undoubtedly important but has not been analyzed in arboreal species of snakes (see Endler, 1978; Greene, 1988).

Various arboreal species enhance their crypticity by seclusion in epiphytes, beneath bark, or within other elements of arboreal foliage. Slender tree snakes and vine snakes exhibit crypsis in terms of body size and contour matching of twigs and branches, which may be reinforced by behavior. Swaying movements of the neck or body resemble wind-induced oscillatory movements of branches and appear to represent behavioral enhancement of crypsis (e.g., *Ahaetulla* sp., Gans, 1974; Henderson and Binder, 1980; *Opheodrys aestivus,* Goldsmith, 1984; *Oxybelis aeneus,* Fleishman, 1985). Clearly, the behavior may also have visual significance in relation to stalking of prey. However, snakes tend to move when the wind is blowing and in a manner typical of wind-blown vegetation, thereby concealing themselves from potential prey as well as predators (Fleishman, 1985). In addition to visual and protective functions, lateral head movement appears to have been modified and incorporated into the courtship behavior of *Opheodrys aestivus* (Goldsmith, 1981).

Defensive behaviors include displays of head, neck, and body. Gaping is common among arboreal colubrids, often associated with rearing of the head and forebody, dilation to expose interscalar skin, and opening the jaws widely "even so much as to double the lateral expanse of the floor of the mouth and make it shovel-shaped" (*Ahaetulla nasuta,* Wall, 1905). The widely gaped mouth, sometimes accompanied by eversion of the floor and glottis, may expose dark blue tissue that can have a startling effect (Henderson, 1974; Henderson and Binder, 1980). Blue coloration is also associated with interscalar skin and the tongue of Asiatic arboreal colubrids, but its significance is not understood.

Several species of arboreal snakes exhibit specialized tongue-flicking behavior (Gove, 1979), none more unusual than the prolonged (several minutes), rigid tongue protrusions with tines together observed in vine snakes. Prolonged extension of a rigid tongue (Fig. 1.4*A*) occurs typically in defensive situations when a snake is confronted by another animal, but also during prey-stalking (e.g., Kennedy, 1965; Henderson and Binder, 1980). A number of hypotheses have been offered to explain the significance of rigid tongue protrusion. These include (1) luring of prey (Proctor, 1924; Curran and Kauffeld, 1937; Hediger, 1968), (2) disruptive or confusing influence on prey (Keiser, 1975), (3) maintenance of continuous olfactory contact with the environment while promoting crypsis (Keiser, 1975; Gove, 1978), (4) extension of the eye line that is used as a visual reference during prey stalking (Henderson, unpublished data), and (5) defensive act to deter potential predators (e.g., Henderson and Binder, 1980).

Several species of arboreal colubrids flatten and/or extend the anterior part of the body ventrally by inflation (e.g., *Spilotes pullatus,* Rossman and Williams, 1966; *Pseustes poecilonotus,* Greene, 1979). The body may be elevated, extended straight, or thrown into a series of lateral curves. The effect is to increase the size of the snake and to display contrasting markings, as in cobras. However, vertical in contrast to horizontal axis of body presentation seems to be typical of displays in arboreal compared with terrestrial species of snakes. It has been suggested that neck inflation in arboreal *Thelotornis kirtlandi,* in addition to serving as a threat display, possibly mimics a begging, fledgling bird, thereby rendering some species of birds susceptible to predation when harassing these snakes (Goodman and Goodman, 1976).

Other defense mechanisms invoked by arboreal snakes, such as escape movements, striking, envenomation, etc., appear not to differ in any fundamental way from those of other snakes. Vibration of the tail is probably less common among arboreal species (e.g., *Ahaetulla nasuta,* Soderberg, 1971) than terrestrial species of snakes.

Movements

Locomotory specializations

Arboreal locomotion involves various combinations of undulatory, rectilinear, and concertina movements (see Gans, 1974, and Edwards, 1985, for reviews of snake locomotion). The sizes and spacing of supports are key features of the arboreal habitat influencing locomotor behavior, as in other vertebrates (Moermond, 1979; Pounds, 1991). Slender species such as mambas and vine snakes undulate swiftly among branches, executing movements that require capabilities to form variably sized bends and to respond to local instabilities in the substrate. Slower movements (as in foraging) involve both undulatory and concertina movements, sometimes simultaneously, depending on the number, size, and spacing of available supports. Both supportive and propulsive forces are often transmitted at the same site of contact, and factors related to the position, angle, movement, and resilience of supporting structures require numerous adjustments of the animal's curves in relation to the contact sites (Gans, 1974). Snakes may extend the body from branch to branch, either with (boas and vipers) or without (some colubrids) use of caudal prehension as a stabilizing anchor. The body may be fully vertical while ascending or descending a trunk, but surface irregularities are required to secure purchase against slippage and provide points of force application. All of these movements require exceptional neuromotor control and are probably more costly in energetic terms than is locomotion on a level and continuous substrate.

A number of morphological specializations of arboreal snakes likely have significance for locomotory mechanics, although structure–function relationships have not yet been studied explicitly with respect to this context. There is a statistical trend for slender, arboreal snakes to possess elongated, narrow vertebrae with short zygapophyses extending at right angles to the vertebral body (Johnson, 1955; Gasc, 1971). The elongation of vertebrae is apparently related to length of the body, for there is tendency to this condition in terrestrial snakes that are long and whiplike. However, increase in body length can be accomplished either by elongation or by addition of vertebral segments. High numbers of vertebrae tend to be characteristic of constrictors and are not necessarily correlated with body length (Johnson, 1955; Jayne, 1982).

Heavier arboreal boids and viperids tend to have relatively short and broad vertebrae correlated with phylogenetic distributions of this condition (Johnson, 1955). A number of arboreal species (colubrids, boids, and viperids) possess strong and prominent zygapophyseal ridges, which are suggested to prevent downward displacement of epaxial muscles, thereby stabilizing the body while it is extended without support (Johnson, 1955). Apparently, all of the Boidae possess a strong zygapophyseal ridge, although not as pronounced as in the arboreal forms. This observation led Johnson to suggest that the primary radiation of boids might have been arboreal.

Properties of the axial musculature of snakes exhibit considerable interspecific variation that is correlated with habitat specialization as well as locomotion and phylogeny (Mosauer, 1935; Auffenberg, 1958, 1961, 1962, 1966; Gasc, 1967, 1974, 1981; Ruben, 1977; Jayne, 1982). The segmental length of spinalis muscle is generally much greater in arboreal species than that in snakes from other habitats. Differences in the length of these muscles is attributed to disproportionate elongation of the tendinous elements, which correlates well with relative lengths of tendons in the M. longissimus dorsi and M. iliocostalis as well. These three muscles are considered to provide the majority of forces necessary for flexion of the vertebral column and locomotion (Gasc, 1974). Arboreal colubrids exhibit the greatest average segmental length of muscle tissue, while aquatic species appear to have the shortest. A tendon from a single segment of the M. semispinalis spinalis may exceed 30 vertebrae in certain arboreal species, compared with a length of only six vertebrae in some aquatic species (Jayne, 1982).

As discussed by Jayne (1982), the tendinous elongation of axial musculature provides mechanical advantages and preserves locomotory efficiency while conserving mass. The longer muscle segments increase the lever arm through which the muscle acts, thereby adding mechanical efficiency to lateral undulation and improving the ability to support the mass of body that is unsupported by substrate. Additionally, passive tension (due to stretch) in elongated elements on the convex side of lateral flexions may combine with active muscle contraction to increase propulsive thrust from a relatively small amount of contractile tissue. The elongation of muscle segments adds minimal mass to the snake because it is accomplished by increasing the length of relatively lightweight tendon. The use of tendon elongation to minimize mass also occurs in fast-moving, whiplike terrestrial and semiarboreal snakes (Ruben, 1977), but should be especially significant in arboreal species that move about on discontinuous and sometimes fragile supports.

In addition to tendinous elongation, muscle units of arboreal colubrids appear subjectively to be of smaller cross-sectional area than is similar tissue in nonarboreal colubrids (Jayne, 1982). Furthermore, some arboreal species (*Imantodes, Thelotornis*) exhibit marked reduction in the mass of costocutaneous muscles, which presumably limits the use of rectilinear progression (Jayne, 1982).

Other aspects of morphology might contribute to the rigidity of arboreal snakes. Anyone familiar with slender arboreal species is impressed by their "wiry" feel and ability to remain rigid, without drooping, while the anterior body is extended without support. In addition to structural features of the musculoskeletal system, the skin exhibits mechanical features that are relevant to these considerations. The low compliance of integument and its tight apposition to underlying tissue probably confers a component of rigidity to the elongate body. Additionally, arboreal snakes may possess widened ventral and mid-dorsal scales (e.g., dipsadines, *Boiga, Ahaetulla, Dendrelaphis*). The latter have been noted to overlie dense subjacent connective tissue (B. Jayne, personal communication). Such tissue may act mechanically to prevent dorsoventral flexion in the manner of an I-beam. Ventral scutes typically have keeled or sharply angled ventrolateral margins, sometimes formed facultatively, that aid in gripping curved objects or transmitting forces to them. Lateral compression of the body should facilitate prehension of moderate-size objects due to the increased contact area for static friction.

Finally, it seems important to comment on the usefulness of tail length in locomotion, which presently is not understood. One study of terrestrial Common Garter Snakes (*Thamnophis sirtalis*) indicates that tail length has no effect on locomotory performance (Jayne and Bennett, 1989). Tails are used in prehension, but the body can also coil about objects, and the longest relative tail lengths do not appear to be correlated with their use in prehension.

Foraging behavior

Limitations of substrate. There is considerable evidence that arboreal snakes are limited in their use of habitat as a consequence of interactions between their morphology and the physical size of branches. Other factors such as prey distribution and microclimate undoubtedly interact to influence movements as well, but these interactions are not well understood. The mechanics of foraging are critical, for an active snake must be able to span gaps between stems and branches and approach prey without revealing its presence. Perch characteristics as well as foraging height tend to be narrowly selected (Plummer, 1981).

In a field study of diurnal Hispaniolan *Uromacer,* Henderson et al. (1981) found that *U. oxyrhynchus* utilized more slender branches than did *U. catesbyi,* which is heavier and stouter (Fig. 1.6). Both species exploit a range of perch diameters while inactive at night, but during daylight the range of perch diameters used by *U. oxyrhynchus* (4 mm) is less than that used by *U. catesbyi* (13 mm). Selection of perch diameter is correlated with foraging mode and striking behavior. *Uromacer oxyrhynchus* carefully stalks lizard prey and often ambushes them from above. In contrast, *U. catesbyi* often preys on diurnally reclusive tree frogs, strikes more clumsily from a shorter distance, and will chase prey (Henderson et al., 1981; Ulrich and Ford, unpublished data).

The support offered by foliage determines both spatial and ontogenetic foraging patterns. Large (>1.2 m snout–vent length, SVL) tree boas (*Corallus enydris*) on Grenada often forage within 1 m of the ground, whereas smaller tree boas forage anywhere from 1 to 20 m above ground and almost invariably on the distal portions of slender branches at the outer periphery of trees and bushes (Henderson, 1993; see also Chandler and Tolson, 1990). Smaller *C. enydris* (350–700 mm SVL) feed primarily on *Anolis* lizards, which sleep on exposed vegetation at night. Medium-size (700–1000 mm SVL) snakes feed on anoles and small rodents, but larger (>1100 mm SVL) individuals feed exclusively on endothermic prey. A small tree boa (500 mm SVL) weighs only 15 g, a medium-size (850 mm SVL) individual weighs about 90 g, while a large specimen (1000–1500 mm SVL) may weigh anywhere from 150 to 900 g. Larger snakes must (1) use the distal portions of thick branches; (2) use the proximal portions of more slender branches; or (3) forage at or near ground level.

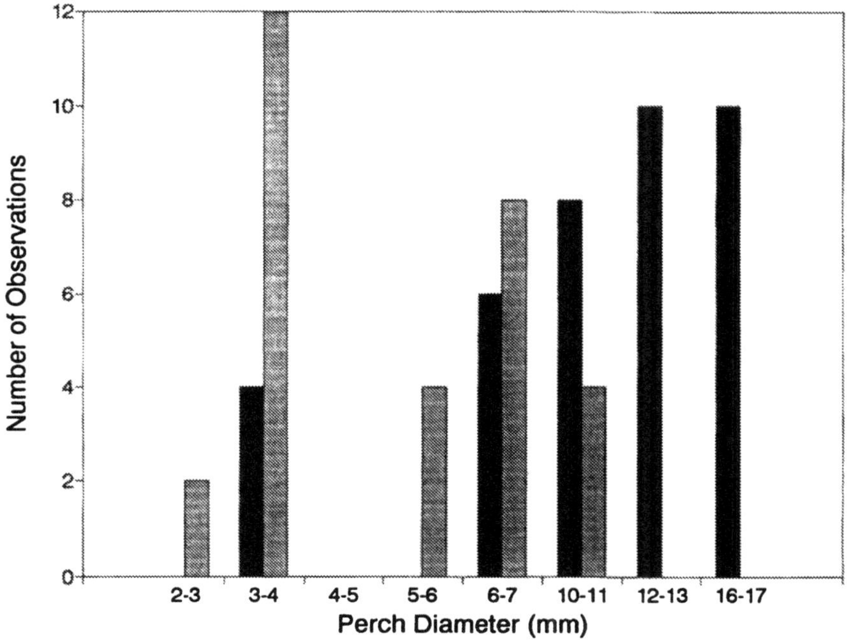

Figure 1.6. Perch discrimination in two species of *Uromacer* on Isla Saona, Dominican Republic. Black bars represent *U. catesbyi* and the gray bars represent *U. oxyrhynchus*. (From Henderson et al., 1981.)

Similarly, on Ile de la Gonâve, Haiti, large *Uromacer frenatus* exhibit a gradual shift in diet from small- to medium-size scansorial *Anolis* to medium to large ground-dwelling *Ameiva*. Henderson and Horn (1983) suggested that this shift in diet might be due to the inability of large (>800 mm SVL) *U. frenatus* to efficiently stalk anoles that use very slender perches.

In summary, available evidence indicates that many species of arboreal snakes must, during the course of their life, make allowances for increases in body mass and their influence on predatory effectiveness on unstable substrate. These accommodations can include shifts in foraging mode, foraging sites, and diet.

Foraging strategies and diet. Snakes generally rely on visual or olfactory cues for locating prey. Less common are heat-sensitive facial or labial pits found in arboreal viperids and in some arboreal boids such as *Chondropython* and *Corallus*. These pits are extremely sensitive and may confer advantages in prey location regardless of whether endothermic or ecothermic prey are targeted (Barret, 1970). Field observations suggest that diurnal sit-and-wait predators (e.g., *Oxybelis* spp.; long-snouted *Uromacer*) are strongly visually oriented (Henderson and Binder, 1980), but diurnal active foragers may be visually oriented (*Opheodrys aestivus,* Goldsmith, 1986) or may use a combination of olfactory and visual stimuli (*Uromacer catesbyi,* Henderson et al., 1981, 1987b).

The use of a combination of visual and olfactory cues by nocturnal species is illustrated from experiments using *Boiga irregularis* (Chiszar et al., 1988a,b). This snake shows a strong response to visual cues of prey, even under conditions of very dim illumination. Nonvolatile chemical cues are used secondarily and opportunistically, i.e., the modality is switched when visual cues are absent or irrelevant (Chiszar, 1990). Because *B. irregularis* exhibits ontogenetic variation in diet (lizards as juveniles, birds and mammals as adults, Savidge, 1988; Greene, 1989), it seems likely that olfactory cues might be important in locating diurnal lizards and birds when they are sleeping at night.

Foraging may account for the majority of a snake's time budget or daily activity pattern. For example, *Uromacer catesbyi* and *U. oxyrhynchus* spend about 95% of daylight hours relatively motionless, presumably looking for active prey (Henderson, unpublished data). The tree boa *Corallus enydris* moves slowly through vegetation from dusk until nocturnal activity ceases (ca. 2330–0300 h), presumably foraging for sleeping anoles. Comparisons of diurnal, arboreal snakes from mainland and insular communities reveal that active foragers (*Leptophis mexicanus, Uromacer catesbyi*) have similar body proportions and feed primarily on diurnally quiescent frogs and bird eggs, whereas sit-and-wait strategists (*Oxybelis aeneus, O. fulgidus, U. oxyrhynchus*) have similar body proportions and feed primarily on diurnally active lizards (Henderson, 1982). Active foragers take prey that is, on average, larger than prey of ambush foragers, but presumably expend more energy in locating prey. Trophic niche breadth is widest in the active foragers and narrowest for the slender snakes. Henderson et al. (1988) hypothesized that in the genus *Uromacer* a sit-and-wait foraging mode and diet specialization evolved from active foraging and a generalized diet.

Recent observations of *Boiga irregularis* on Guam indicate that arboreal snakes can be ecologically very versatile. An area of the island that had been denuded of arboreal lizards and birds sustained a dense population of snakes that were foraging for skinks on the ground (Chiszar, 1990). Conversely, populations of characteristically terrestrial *Boa constrictor* on small cays off the Caribbean coast of Belize are stunted (maximum SVL ca. 1350 mm), completely arboreal, and probably feed on birds (B. Sears, personal communication). *Boa constrictor* and *Boiga irregularis* both display a versatility that is probably unusual in comparison with trophic adaptations of snakes in general.

Chandler and Tolson (1990) examined the relationship between foraging behavior of the nocturnal boid *Epicrates monensis,* the microhabitat of its sleeping prey (*Anolis cristatellus*), and vegetation structure on Cayo Diablo, Puerto Rico. Although anoles occurred throughout the island, boas foraged selectively within groves of *Coccoloba uvifera* that formed closed-canopy woodland. The boas forage and travel along relatively large branches but move out onto smaller branches in attempts to encounter anoles, possibly in response to olfactory cues. Movement through the canopy is limited along the smaller branches, however, so boas return to larger branches to continue foraging. Principal component analysis reveals that female and juvenile anoles sleep on short, thin branches relative to those favored by foraging boas; male anoles occupy sleeping sites of larger diameter and more similar in structure to those used by boas (Fig. 1.7). In contrast to larger boas, and presumably in response to availability of prey, juvenile boas forage in lower vegetation where they are likely to encounter the smaller anoles that they are capable of catching and handling. These relationships suggest that boas are sensitive to the perch heights of their potential prey. On the other hand, anoles seem to respond in their choice of sleeping sites, selecting short and thin branches relative to those that are frequented by boas. Encounters with boas are minimized by moving to the extremities of the smaller branches that will not support the mass of a large snake. If the male anoles are constrained to occupy larger, thicker branches than do females because of territorial behavior, proximity to favorable basking sites, or some other reason, they may compensate by diverging from boas in their use of higher, more sparse vegetation (Fig. 1.7).

Foraging *Boiga irregularis,* regardless of SVL, crawl through foliage supported by many small twigs; there is no correlation between perch diameter and snake SVL (Rodda, 1993). The net travel rate for actively foraging *B. irregularis* is low and varies from 0 to 26 m/h ($\bar{x} = 11.4$) (Rodda, 1992).

Foraging heights. Although most literature does not give specific heights of observation or capture of arboreal snakes, data available for 19 species suggest that foraging heights of arboreal snakes are relatively narrow given the potential for vertical movements in habitats such as rainforest (Table 1.1). Most arboreal snakes studied are

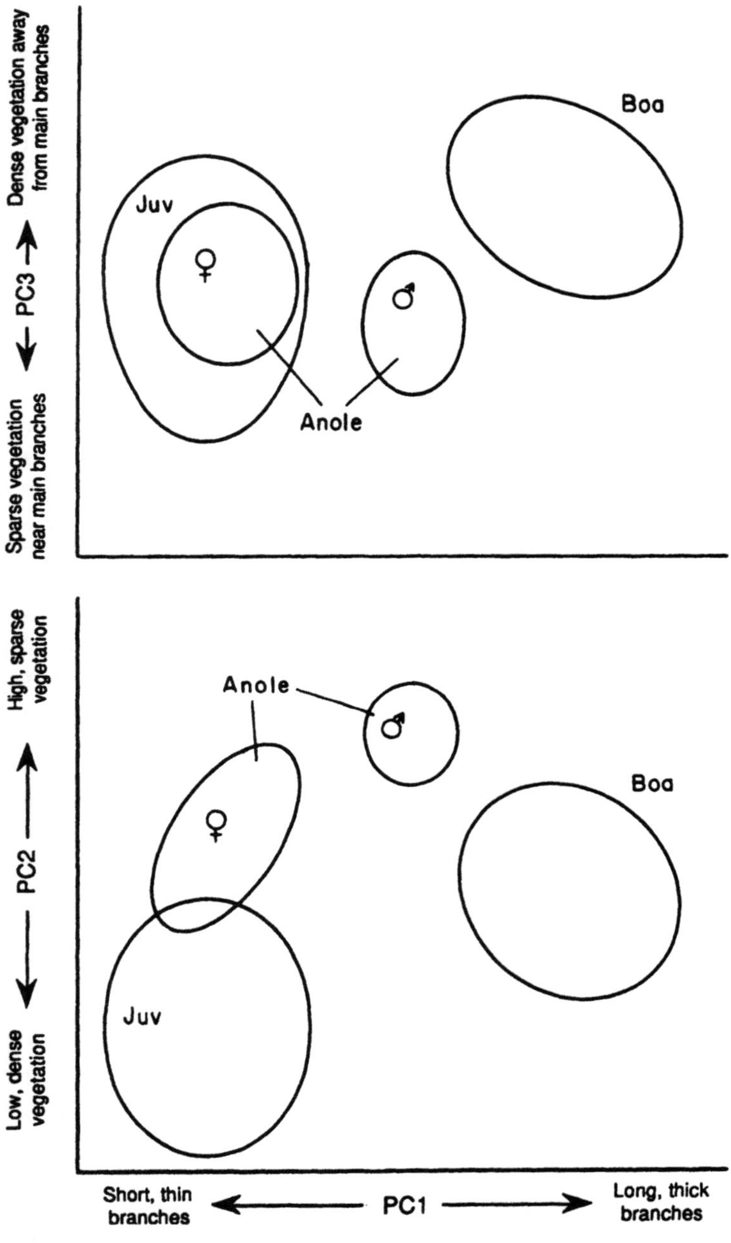

Figure 1.7. Ninety-five percent confidence ellipses for habitat use (as described by principal components analysis) by an arboreal boa (*Epicrates monensis*) and its prey (*Anolis cristatellus*) on Cayo Diablo, Puerto Rico. (Modified from Chandler and Tolson, 1990.)

TABLE 1.1. Foraging Heights of Arboreal Snakes

Species	Observations	Source
Boidae		
Corallus enydris	65% of 230 observations snakes foraging 1–5 m; 40% foraging 1–3 m	Henderson, 1993
Epicrates gracilis	0.5–4.6 m, most 1–2 m	Schwartz and Henderson (1991); R.W.H. (personal observation)
Epicrates monensis	75% of juveniles forage <1.5 m; mean height foraging males 1.62 m, foraging females 1.88 m	Tolson (1988); Chandler and Tolson (1990)
Colubridae		
Ahaetulla nasuta	75% of 12 observations 2.0–2.5 m	Inger et al. (1984)
Boiga dendrophila	Basking 0.5–5.0 m	Minton and Dunson (1978)
Boiga irregularis	77% foraging within 3 m of ground	Rodda (1992)
Chironius fuscus	7 of 11 sleep ing <1.5 m, 4 of 11 at 1.5–4.0 m	Duellman (1978)
Dipsas catesbyi	<4.0 m	Duellman (1978)
Dipsas indica	>75% on vegetation <1.5 m	Duellman (1978)
Imantodes cenchoa	67% of 32 snakes at <1.5 m, remainder 1.5–4.0 m	Duellman (1978)
Imantodes lentiferus	75% of 24 observations <1.5 m	Duellman (1978)
Leptodeira annulata	65% <1.5 m, others up to 6.0 m	Duellman (1978)
Opheodrys aestivus	74% of 100 observations 1.1–4.0 m	Plummer (1981)
Oxybelis aeneus	Mean of 19 observations 1.5 m (0.5–5.0 m)	Henderson (1974)
Oxybelis argenteus	77% of 47 sleeping <1.5 m	Duellman (1978)
Uromacer catesbyi	Mean of 41 observations 2.1 m (0.3–3.5 m)	Henderson et al. (1981)
Uromacer oxyrhynchus	Mean of 28 observations 1.7 m (0.2–3.4 m)	Henderson et al. (1981)
Viperidae		
Bothriechis schlegelii	1–2 m	Alvarez del Toro (1982)
Bothriopsis bilineata	2.0 m	Duellman (1978)

active or sleep at heights less than 1.5–3.0 m (18 of 19 species), even though the range of perch heights is much greater. It is, of course, possible that a collection or observation bias is present inasmuch as it is difficult to locate snakes that are perched much above 3 m. However, *Corallus enydris* has a spectacular eyeshine that can be observed for more than 50 m, yet these snakes are observed foraging at heights comparable to those reported for other arboreal snakes. Only 17 of 230 observations (5.7%) were of snakes foraging at heights greater than 10 m. Some snakes are active in the crowns of tall rainforest trees, but available data suggest that most activity in the majority of species occurs much closer to ground level. Foliage structure, microclimate, body size, and prey availability all probably interact to influence perch and foraging height.

Prey. Arboreal snakes feed on gastropods, arthropods, fish, frogs, lizards, snakes, birds, and mammals, including marsupials, insectivores, bats, carnivores, and artiodactyls. They may be steno- or euryphagous, and there is evidence for naive prey preferences (Henderson et al., 1983). Some arboreal species exhibit ontogenetic or opportunistic variation in diet.

Most arboreal snakes are tropical and prey on frogs or lizards (Leston and Hughes, 1968; Duellman, 1978, 1990; Henderson et al., 1979; Henderson and Crother, 1989). Shine (1983) reported that among arboreal snakes, 53% ate frogs, 65% ate reptiles, 43% ate birds (and their eggs), and 26% ate mammals. His analysis indicated that arboreal snakes and ground-dwelling snakes do not differ significantly in the proportion of species feeding on amphibians or reptiles. Such data can also be interpreted to indicate that, because most arboreal snakes occur in the tropics, vertebrate-eating tropical snakes prey largely on frogs and lizards regardless of their adaptive zone.

Examination of data from five neotropical rainforest sites led Duellman (1990) to recognize three guilds of anuran-eating snakes: (1) diurnal-terrestrial, (2) nocturnal-terrestrial, and (3) nocturnal-arboreal. There should, however, be a diurnal-arboreal guild to accommodate *Leptophis* spp. and semiarboreal *Chironius* spp. Diurnal and nocturnal guilds were recognized for arboreal lizard-eating species and arboreal mammal-eating species. Using the data presented by Duellman (1990: Appendix 1), we determined that strictly arboreal species have narrower trophic niches than do semiarboreal species. For example, 26.9% of 26 species of arboreal snakes preyed exclusively on gastropods, whereas no semiarboreal species exploited that prey group, and 23.1% of the arboreal species ate only frogs in comparison with 11.1% for 18 semiarboreal species (Fig. 1.8). Conversely, semiarboreal species showed a high incidence (55.6% vs. 11.5% in arboreal species) of predation on ectotherms and endotherms, perhaps indicative of ontogenetic variation in diet related to their greater size compared with strictly arboreal species. Also, semiarboreal species are active on the ground, which increases the diversity of available prey.

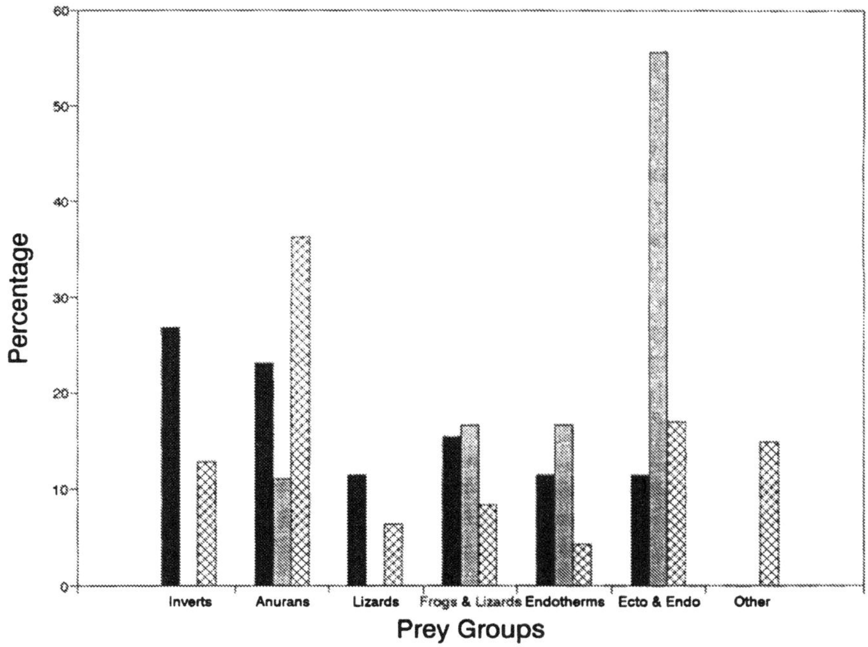

Figure 1.8. Percent occurrence of prey items in arboreal (black bars), semiarboreal (gray bars) and ground-dwelling (cross-hatched bars) snakes at five neotropical rainforest sites. Data derived from Duellman (1990).

There is compelling evidence for partitioning of prey by sympatric species. Using data from Santa Cecilia, Ecuador (Duellman, 1978), we determined that of 37 prey species eaten by colubrid snakes, the arboreal species exploited 18 different prey species, and no two arboreal snakes ate the same prey species. The semiarboreal species exploited eight prey species, and again, there was no overlap. Sixteen prey species were eaten by terrestrial snakes. Looking at pairwise combinations of the three habitat groupings for these snakes, we found that each pair overlapped for only one prey species (in each case a species of *Hyla*).

Diet composition has been studied in relation to morphology and foraging mode in two arboreal snake communities: one in Yucatan, Mexico, and the other on Isla Saona, Dominican Republic (Henderson, 1982). Active foragers (*Leptophis mexicanus, Uromacer catesbyi*) have wider trophic niche breadths than do ambush foragers (*Oxybelis aeneus, O. fulgidus, U. oxyrhynchus*), and snakes with wider heads (*L. mexicanus, O. fulgidus, U. catesbyi*) take larger prey items than do snakes with narrow heads (*O. aeneus, U. oxyrhynchus*). In the Yucatan community, *L. mexicanus* exploits the most prey classes, but *O. fulgidus* exploits more prey taxa within a prey class. *Oxybelis aeneus* is the most specialized species in terms of prey groups and prey taxa exploited and is the most morphologically and behaviorally specialized of the three Mexican species. The two species of *Oxybelis* exhibit the greatest trophic niche overlap value, especially for prey groups exploited (0.958). However, based on prey taxa eaten, *O. fulgidus* has a much wider niche breadth value (5.15) compared to *O. aeneus* (1.64). In addition, mean prey size for *O. fulgidus* is 8.0 ± 2.0 cm^3 compared to 1.4 ± 0.1 cm^3 in *O. aeneus*. Similarly, niche overlap is low between the two species of *Uromacer* on Isla Saona: *U. oxyrhynchus* is an *Anolis* specialist and *U. catesbyi* is a generalist that includes a high percentage of hylid frogs in its diet and takes prey that are larger than that of *U. oxyrhynchus*.

Shine (1983) felt (and we agree) that "the apparently high incidence of bird-eating in arboreal snakes might not be supported by more detailed studies, because authors perhaps suggest that a snake preys on birds simply because the snake is arboreal." Data presented by Duellman (1978) and Dixon and Soini (1986) for snakes from the upper Amazon Basin show that only six of 23 arboreal species are known to exploit birds as food, and none are stenophagous for birds. Among the same 23 species, 13 (56.5%) are known to prey on frogs and/or lizards.

Some arboreal snakes possess highly evolved venom delivery systems, but most do not. Many are rear-fanged, some are constrictors, and others overcome prey with their jaws or loops of their bodies. Some species that can subdue small prey with venom rely on constriction for subduing large prey (Shine and Schwaner, 1985). Captive *Boiga irregularis* invariably rely on constriction to kill rodents, which may be larger than 30% of their own body mass (Chiszar, 1990).

Prey size. Most arboreal snakes exploit a range of prey, although most are small and innocuous. Mean prey sizes (volume) range from less than 1.5 cm^3 in slender species to greater than 66 cm^3 in boids. The diet may vary ontogenetically, and some noteworthy examples are discussed briefly below.

Dietary shifts from lizards to birds or mammals are known in several species of arboreal boids: *Candoia bibroni* (Harlow and Shine, 1992), *Chondropython viridis* (McDowell, 1975; Shine and Slip, 1990), *Corallus enydris* (Henderson, 1993), and *Epicrates striatus* (Henderson et al., 1987a). Observations of boas in the West Indies indicate that foraging sites, and possibly mode, change with increased SVL and body mass. As an example, small- to medium-size *Corallus enydris* actively forage at heights of 1–20 m where they feed primarily on lizards. Very large boas (>120 cm SVL) forage lower (<1.0 m) and employ a sit-and-wait strategy to capture exclusively endotherms, primarily small rodents. On the neotropical mainland, *C. enydris* rarely preys on lizards, but birds are eaten frequently, even by snakes under 50 cm SVL (Henderson, 1991, 1993). Prey size may exceed 100 cm^3, and mass ratios (MR, prey-to-snake ratio) may approach 0.50.

The diets of nocturnal *Boiga* spp. have been studied by examination of stomach contents of 21 species (Greene, 1989). Although some species of *Boiga* are primarily lizard eaters throughout life, Greene described four patterns of ontogenetic variation: (1) lizards in juveniles to birds in adults; (2) lizards to rodents (only *B. irregularis* on New Guinea); (3) lizards to birds and mammals (e.g., *B. blandingi*); (4) lizards to snakes, birds, and mammals (e.g., *B. dendrophila*). The mean MR for eight species of *Boiga* was 0.16 (range 0.004–0.58).

The diet of *B. irregularis* in its native range exclusive of Australia includes frogs (*Rana*), lizards (agamids, scincids, gekkonids), birds (passerines, owlet-nightjars, starlings, quail) and mammals (marsupials, bats, shrews, rodents). Snakes under 60 cm SVL eat only ectotherms; those from 60 to 100 cm SVL consume both ectotherms and endotherms; and large snakes over 100 cm SVL eat only endotherms. Greene (1989) reported MRs of 0.0004–0.241 in *B. irregularis* ($\bar{x}$ = 0.106, n = 13); the largest prey item was a 32-g starling in a 152-g snake. Chiszar et al. (1991) reported instances of MRs >0.50 for *B. irregularis* on Guam.

Boiga irregularis was introduced to Guam where it has devastated the native vertebrate fauna (Savidge, 1987, 1988; Fritts, 1988; Chiszar, 1990; Rodda and Fritts, 1992). Savidge found that these snakes on Guam eat lizards, lizard eggs, a variety of birds (including endemic species), and small mammals. Like Greene, Savidge considered this snake to be somewhat opportunistic and found ontogenetic variation in its diet. In captivity, large *B. irregularis* would not take lizards even when hungry, and the largest snakes in her sample took only birds, bird eggs, and mammals (Savidge, 1988).

The swallowing ability of *B. irregularis,* determined from mandibular protractions and meal size, is comparable to that of broad-headed vipers, and the ability to take a wide range of prey sizes contributes to its successful colonizing ability. This snake exhibits amazing plasticity regarding what it will eat (Savidge, 1988; Chiszar et al., 1992).

Generally, most arboreal snakes take prey that is not particularly large when compared to other snakes, especially ground-dwelling boids, large colubrids, and viperids (Fitch, 1982; Reynolds and Scott, 1982; Greene, 1983; Pough and Groves, 1983). However, arboreal snakes do occasionally ingest prodigious meals. Myers (1982) reported an MR of 0.78 for an *Imantodes cenchoa* that had eaten a large *Anolis,* and Henderson et al. (1987b) calculated an MR of 0.78 for a *Uromacer catesbyi* that had ingested a large hylid frog. Greene (1989) found the remains of a Mouse Deer (*Tragulus javanicus*) in the stomach of a *Boiga dendrophila,* and the rodent was estimated to weigh between 0.7 and 2.0 kg, while the snake weighed no more than 1.0 kg. Small tree boas (*Corallus enydris* under 60 cm SVL) have exceedingly slender necks, and two from South America with neck circumferences of 15.5 and 19.0 mm swallowed birds with maximum circumferences of 70 and 78 mm, respectively (Henderson, unpublished data).

In summary, arboreal snakes exploit a wide variety of prey classes and sizes, but most prey on small, innocuous vertebrates. Strictly arboreal species are more stenophagous than semiarboreal species, and the latter are more likely to exploit both ectotherms and endotherms and to routinely take larger prey (Fig. 1.8).

Habitat Use and Activity Ranges

Although many arboreal snakes sometimes descend to the ground, their movements, activity ranges, and foraging strategies are intimately related to the aboveground environment. Our knowledge of the movement ecology and habitat utilization of arboreal snakes is based on a very limited sample of species: *Corallus enydris, Epicrates monensis, Elaphe obsoleta, Opheodrys aestivus,* and *Oxybelis aeneus.* A critical factor in evaluating movement ecology in arboreal snakes is the three-dimensionality of their activity ranges. Virtually all available analyses of activity ranges in arboreal snakes are concerned with two dimensions, and this presents what may be a very misleading assessment of movements and space requirements of these snakes.

Corallus enydris is a geographically wide-ranging New World boid. It occurs in a wide variety of habitats, but in the West Indies it is almost always an inhabitant of edge situations (road cuts, rivers, agricultural areas). Henderson and Winstel (1992) radio-tracked a single animal in Grenada, and over a 14-d period it always foraged along the boundaries of a road cut or a shallow river. During this time it utilized an area of nearly 900 m^2 (minimum polygon), but each night's foraging usually found the snake within 12 m from where it initiated the night's activity. Although it used the crowns of trees to facilitate movements over a road, circumstantial evidence suggests that it also crossed the road at ground level at least once.

The small arboreal boid *Epicrates monensis granti* is endemic to the Puerto Rico Bank. It is nocturnal and actively forages primarily on *Anolis cristatellus*. Thanks to Tolson (1988) and Chandler and Tolson (1990) we know more about habitat selection in this arboreal snake than in any other. *Epicrates monensis* typically occurs in subtropical dry forest, and factors such as plant composition and structural attributes of the vegetation are critical factors in habitat choice. On Cayo Diablo, foraging *E. monensis* are most commonly associated with *Coccoloba uvifera* (51 of 79 observations). Principal component analysis indicated that *E. monensis* is repeatedly encountered in areas with (1) high compound tree density or high shrub and palm densities; (2) a low or high canopy; and (3) vegetational continuity. The unifying variable is the continuity or interlocking of the branches of shrubs or tree canopy. Tolson (1988) felt that this characteristic decreased the search time between encounters with sleeping *Anolis,* potentially limited encounters with cats and mongooses, and was essential to the foraging success and survival of *E. monensis.* Of 149 inactive *E. monensis* taken from refugia, 43% were in *Cocos* or *Sabal* axils, 36% in termitaria, and 21% were under rocks and debris. Of those snakes taken from termitaria, 51% were females and more than half were gravid. It appears that gravid females use termitaria or sun-baked debris to elevate body temperatures to more than 33°C.

In Arkansas, temperate *Opheodrys aestivus* restrict movements to a particular area, especially within a season (Plummer, 1981). Most movements are parallel to a shoreline at the vegetation edge, and Plummer expressed activity ranges as lengths. The mean for 38 snakes captured four or more times was 62 ± 7.2 m (range 15–102 m), and there was no difference between sexes. Macartney et al. (1988), using Plummer's (1981) data and the convex polygon method, calculated a mean activity range of 0.019 ha for this snake.

An exception to these movement patterns are nesting migrations where *O. aestivus* descend from alder habitat and travel on the ground to nesting sites in hollowed trees away from shoreline habitat (Plummer, 1990b). The mean daily prenesting movement in five gravid females was 2.9 ± 0.4 m; the mean nesting migratory movement was 32.7 ± 5.3 m/day; the mean postnesting migratory movement was 19.4 ± 6.4 m/day; and the mean postnesting movement was 3.1 ± 1.0 m/day.

Movements of *Oxybelis aeneus* were studied at a mangrove edge in Belize (Henderson, 1974). The mean distance moved between recaptures was 42.0 m (0–135 m) in males and 42.5 m (0–103 m) in females. During the day activity was limited to vertical movements from 0.5 to 2.0 m above the ground. Mean convex polygon activity ranges were 0.037 ha in males and 0.05 ha in females. There was no correlation between time and distance moved. Snakes remained in bushes or low trees for prolonged periods of time. Activity ranges were influenced by the distribution of the vegetation, which, in turn, influenced the distribution of suitable prey (primarily *Anolis* lizards).

Movement data are also available for the semiarboreal *Elaphe obsoleta*, which feeds primarily on endothermic prey (Fitch, 1963). Of 47 records of five radiomarked snakes in eastern Kansas (Fitch and Shirer, 1971), 47% were in trees, 23% in burrows, 11% in thick brush, 11% associated with buildings, 6% in tall grass, and 2% in other dense vegetation. Mean activity ranges of *E. obsoleta* were 11.7 ha in males and 9.3 ha in females (circle-radius method). These values did not include movements to and away from hibernacula (Macartney et al., 1988). This snake makes horizontal forays at ground level, but will ascend trees to search for nestling birds. It will then descend, move at ground level, then ascend another tree (Fitch, 1963). There is a strong association of these snakes with edge habitat, especially during the bird-breeding season (Weatherhead and Charland, 1985). In the laboratory, large snakes always perch off the ground immediately after feeding if branches are provided for climbing (Lillywhite, unpublished observation).

The very limited data suggest that large snakes that prey on birds and mammals have larger activity ranges than those species that prey on invertebrates or small ectothermic vertebrates. Generally, measured activity ranges of arboreal snakes are small in comparison with terrestrial species (Macartney et al., 1988). However, these comparisons are probably misleading because they do not take into account the volume of arboreal habitat.

Abundance

Data for estimates of population densities of arboreal snakes are relatively few (Table 1.2). However, the abundance of arboreal snakes in large collections can provide some further indication of their relative abundance. For example, Dunn (1949) reported on a collection of 10,690 specimens collected in Panama between 1933 and 1945. He divided that collection geographically; the most common snake from the Cocle and Herrara areas combined was *Leptodeira annulata* (20.5% of 252 snakes), which is a nocturnal, largely arboreal, frog predator. The most common species in the Sabanas collection was *Oxybelis aeneus* (22.2% of 3914 snakes), a diurnal, arboreal, lizard predator. The Chagres collection contained 10.9% *O. aeneus* (274 of 2500 specimens), again the most common species. Similarly, Dunn's Agua Clara collection (1103 specimens) was represented by 125 *O. aeneus* (11.3%; the most common snake). A collection from Darien included 14.2% *L. annulata* (431 of 3044 specimens) and 11.3% *Corallus enydris* (346 specimens); these were the second and third

TABLE 1.2. Population Densities of Arboreal Snakes

Species	Density (ha^{-1})	Density Method	Locality	Source
Boidae				
Corallus enydris	19–27	Direct counts	Grenada, West Indies	Henderson and Winstel (1992)
Epicrates monensis	>100	Mark–recapture	Cayo Diablo, Puerto Rico Bank	Tolson (1988)
Colubridae				
Boiga irregularis	16	Trapping	Guam	Savidge (1991)
Boiga irregularis	50	Trapping	Guam	Fritts (1988); Fritts et al. (1989)
Elaphe obsoleta	<1	Mark–recapture	Kansas	Fitch (1963, 1982)
Opheodrys aestivus	429	Mark–recapture	Arkansas	Plummer (1985)
Oxybelis aeneus	28–35	Mark–recapture	Belize	Henderson (1974)

most common snakes in the collection, respectively. Finally, the Yavisa collection (2321 specimens) included 354 *L. annulata* (15.3%) and 275 *C. enydris* (7.4%); these were the second and fourth most common snakes in the collection.

On Ile de la Gonâve, off the west coast of Haiti in the West Indies, Henderson (unpublished data) established a snake "market." Over a two-day period, locals brought him 83 arboreal snakes (78 *Uromacer frenatus,* four *U. catesbyi,* one *Epicrates striatus*), and 12 specimens of three species of ground-dwelling colubrids. *Uromacer frenatus* is strictly saurophagous, *U. catesbyi* takes frogs and lizards, and *E. striatus* takes lizards, birds, and mammals.

Duellman (1978) reported the number of specimens of each snake species collected at Santa Cecilia, a lowland site in the Upper Amazon Basin of Ecuador. Of 44 species and 438 specimens of boid, colubrid, and viperid snakes, five of the six most common species are arboreal predators on lizards, frogs, or gastropods, and none feeds on birds or mammals.

Six species of snakes at Cuzco Amazonica, Peru were designated as common (versus uncommon or rare), and these included three arboreal species: *Dipsas catesbyi, Imantodes cenchoa,* and *Oxybelis boulengeri* (Duellman and Salas, 1991). Similarly, a herpetofaunal analysis near Manaus, Brazil, found the five most frequently encountered snake species at night were *Leptodeira annulata, Oxybelis argenteus, Imantodes cenchoa, Dipsas catesbyi,* and *Epicrates cenchria* (Zimmerman and Rodrigues, 1990). The first four are slender arboreal species, and the last is a semiarboreal boid. The most frequently encountered species in daytime was *Dendrophidion dendrophis,* another semiarboreal species.

In summary, arboreal and semiarboreal snake species may be more abundant compared to sympatric or syntopic terrestrial species at neotropical lowland localities. Species that are predators on ectotherms are more abundant than those species that exploit endothermic prey.

Community Structure

Arboreal snakes are spatially segregated from many other snakes in a terrestrial community, at least much of the time. As in other facets of arboreal snake ecology, we are constrained by lack of data for snake communities that include arboreal species, and, again, our attention is focused on tropical rather than temperate communities. There is not a single, thorough study of a tropical snake community, but in the few studies that examine arboreal snakes from a community perspective, it is apparent that resource partitioning is accomplished in one or several dimensions: temporally, spatially, and/or trophically. The broadest generalization is that there are distinctive nocturnal and diurnal herpetofaunas, and these include arboreal snakes.

Duellman (1990) categorized arboreal snakes as either diurnal or nocturnal for five neotropical rainforest sites. The percentage of diurnal species ranged from 12.9% (Manu) to 28.3% (Manaus), and the percentage of nocturnal species ranged from 16.1% (La Selva) to 29.0% (Manu). In general, sites with a higher percentage of arboreal diurnal species had fewer arboreal nocturnal species. In Brazilian caatinga, three of four arboreal colubrids were diurnal, whereas a semiarboreal boid was nocturnal (Vitt and Vangilder, 1983). Of six arboreal or semiarboreal species on Hispaniola, three were diurnal colubrids and three were nocturnal boids. And at a cocoa farm locality in Ghana, at least 50% of the species were diurnal (Leston and Hughes, 1968).

With respect to spatial use of habitat, arboreal snakes can potentially exploit a wide array of shrub and tree species. Many snakes, however, are designated either as shrub or tree dwellers (Duellman, 1990; Duellman and Salas, 1991). Use of habitat can be varied or partitioned on horizontal as well as vertical scales, and in tall forests snakes may occur close to ground level or at heights exceeding 30 m. There are presently no data that allow us to evaluate with certainty why snakes occur at the heights that they do. Generally, lighter-bodied snakes can exploit a wider range of perches and are more likely to use the distal ends of branches than are heavier snakes (Fig. 1.6). Spatial and temporal use of habitat further affects community structure according to interactions with prey and potential predators. Studies of *Epicrates monensis* and *Boiga irregularis* clearly indicate that foraging behavior can affect the abundance of prey and its use of habitat; prey availability, in turn, reflects habitat structure and is likely to be a determinant of snake abundance (Savidge, 1987, 1988; Tolson, 1988; Chandler and Tolson, 1990).

Duellman (1989; Fig. 3.8) presented a three-dimensional graphic interpretation of SVL, microhabitat, diel activity, and diets of snakes at Santa Cecilia, Ecuador. Inspection of the graph indicates that arboreal species that exploit most prey groups are active during day and night, but there is considerable variation in SVL, which may reflect differences in size of the prey exploited. Duellman's presentation illustrates the complexity of a neotropical snake community, the variety of species found in an arboreal adaptive zone, and the multidimensionality of the niche of any of the species.

Conservation

The foregoing information provides a strong inductive base for generalizations applicable to conservation ecology of arboreal snakes. Clearly, arboreal species are highly and convergently specialized in ways that promote functional performance in trees and shrubs. Because of these attributes, arboreal snakes are likely to be especially sensitive to habitat disturbance. Moreover, direct and indirect impact of changes such as deforestation, climatic change, or modification of vegetation can be exacerbated by factors such as high rates of predation, low reproductive potential, foraging limitations, disadvantageous microclimate exposure, and disruption of social interactions. Aspects of vegetation structure and its continuity may be imperative for the survival of some species (e.g., Tolson, 1988), whereas other species might be more versatile or opportunistic in their use of habitat (e.g., Chiszar, 1990). Historical patterns of vegetation change may help to explain present-day distributions and abundance of arboreal snake species and to provide important considerations in formulating management plans where opportunities exist to blunt unchecked development (see Dodd, Chap. 9, this volume, for additional discussion).

Insofar as the majority of highly specialized arboreal snakes are endemic to tropical regions, the continuing destruction of tropical forests is directly relevant to the survival of arboreal snake species. The problem may be especially acute on islands (Tolson, 1988; Henderson, 1992). Considering rates of habitat destruction and modification in many parts of the world (Wilson and Peter, 1988) in relation to ecological investigation of arboreal snakes, our observations and information are woefully inadequate to monitor impending losses of, or changes in, arboreal fauna. Because of difficulties in observing arboreal snakes, the biotic and physical parameters dictating the abundance and distribution of a species are not well-known. Clearly, however, preservation of habitat is the single most important factor relevant to survival of the arboreal herpetofauna. Given an adequate base of habitat, the viability of arboreal snake populations is highly favored by their inconspicuousness in spatially complex environments.

Summary and Future Research

The foregoing discussion is intended to promote interest in arboreal communities and the evolution of arboreal habits among limbless reptiles. Because of the remarkable convergences noted in morphology, behavior, and physiology of unrelated arboreal taxa, intensive studies of arboreal snakes based on general questions related to arboreality are likely to be very fruitful.

Arboreal environments present several interrelated challenges to use by snakes, including discontinuous and potentially unstable substrate, predation risks, rigorous microclimates in some plant communities, and disturbance to blood circulation due to gravitational effects of vertical postures. Evolutionary adaptations to these rigors include reductions of mass and body length, cryptic (often green) coloration, locomotor specializations, visual acuity, and both structural and hemodynamic specializations of the cardiovascular system. Judging from a perspective based on available comparative data, it appears that interactive suites of adaptive characters are indeed prerequisite for snakes to occupy arboreal habitats successfully.

The physical environment is crucially important for understanding adaptive successes and behavioral ecology of arboreal snakes. Yet there is not a single, in-depth study of the biophysical ecology of an arboreal species. Such investigations would contribute to understanding the behaviors, distribution, and physiology of these snakes, and might also provide important insights into the energetics of reproduction and life histories of populations. Reproductive studies are needed to enhance understanding of conservation and management problems such as those related to the ecological impact of *Boiga irregularis* on the island of Guam.

It is evident that in some species the use of arboreal habitat is facultative. For such species the use of resources in either terrestrial or arboreal environments may be opportunistic. Careful studies of facultatively arboreal species might shed important light on the selective forces promoting the use of arboreal resources. Such information, when part of community studies, might also improve understanding the evolution of taxon-specific arboreality.

While arboreal snakes collectively occupy a range of microhabitats, evidence suggests that many species have quite specific requirements for vegetative structure, including heights and diameters of branches used for foraging, dispersal, and perching. Experimental studies using controlled or manipulated substrates, either in the field or laboratory, would be a useful approach to understanding performance measures such as locomotion, foraging, and prey capture. Throughout ontogeny snakes must cope with increasing mass and body size, often reflected in adjustments of foraging behaviors and diet. Preliminary studies suggest that snakes are sensitive to the perch characteristics of their potential prey and, conversely, affect the abundance and use of habitat by prey species. Detailed studies of species interactions are needed to advance our understanding of the diversity, abundance, and spatial distribution of arboreal snakes. Such studies are especially urgent in the tropics where many of the faunistic components of arboreal communities are poorly known.

Data from neotropical rainforests suggest that arboreal snakes may be more trophically specialized than terrestrial counterparts, which may contribute to radiation, diversification, and coexistence of arboreal faunas. Arboreal snakes exploit a wide array of prey classes, but most prey on small, innocuous, ectothermic vertebrates. Prey availability may determine the range of movements by snakes, and competition or other factors may lead to dietary shifts or facultative terrestriality in otherwise stenophagous, strictly arboreal species. Movements and home range size require much further study in arboreal snakes, and must take into consideration the vertical as well as horizontal dimension of habitat. Telemetry should have useful application in these contexts and may reveal quite different patterns of habitat selection than those resulting from conventional mark–recapture techniques (see Reinert, Chap. 6, this volume).

Some of the emergent ecological questions concerning arboreal snakes are directly relevant to long-standing central issues in ecology, such as the relationship between habitat structure and community organization or the role of species interactions in structuring communities (e.g., Strong et al., 1984). Generally, the three-dimensional configuration of arboreal habitats strongly influences locomotion and behaviors of microhabitat specialists, which often reflect patterns of morphological phenotypes (e.g., Pounds, 1988). While it is apparent that coexisting species of arboreal snakes are specialized and partition some resources, little is known as to how structural features of habitat have influenced the morphological and community patterns of these snakes. Understanding such adaptations is likely to be of key importance in understanding the adaptive radiation of arboreal herpetofaunas.

Literature Cited

Alexandre, D. Y., 1982, Etude de l'eclairement du sous-bois d'une foret dense humide sempervirente (Tai, Cote d'Ivoire), *Acta Oecol. Oecol. Generales,* 3:407–447.

Allen, L. H., E. R. Lemon, and L. Muller, 1972, Environment of a Costa Rican forest, *Ecology,* 53:102–111.

Alvarez del Toro, M., 1982, *Los reptiles de Chiapas,* Publ. Instit. Hist. Nat., Tuxtla Gutierrez, Chiapas, Mexico.

Aoki, M., K. Yabuki, and H. Koyama, 1975, Micrometeorology and assessment of primary production of a tropical rain forest in West Malaysia, *J. Agric. Meterol.,* 31:115–124.

Auffenberg, W., 1958, The trunk musculature of *Sanzinia* and its bearing on certain aspects of the myological evolution of snakes, *Breviora,* 82:1–12.

Auffenberg, W., 1961, Additional remarks on the evolution of trunk musculature in snakes, *Amer. Midl. Nat.,* 65:1–19.

Auffenberg, W., 1962, A review of the trunk musculature in limbless land vertebrates, *Amer. Zool.,* 2:183–190.

Auffenberg, W., 1966, The vertebral musculature of *Chersydrus* (Serpentes), *Q. J. Fla. Acad. Sci.,* 29:183–190.

Bakken, G. S., 1989, Arboreal perch properties and the operative temperature experienced by small animals, *Ecology,* 70:922–930.
Barrett, R., 1970, The pit organs of snakes, in C. Gans and T. S. Parsons, eds., *Biology of the Reptilia,* Vol. 2, Academic, New York, pp. 277–300.
Bazzaz, F. A., and S. T. A. Pickett, 1980, Physiological ecology of tropical succession: A comparative review, *Annu. Rev. Ecol. Syst.,* 11:287–310.
Bock, W. J., 1980, The definition and recognition of biological adaptation, *Amer. Zool.,* 20:217–227.
Chandler, C. R., and P. J. Tolson, 1990, Habitat use by a boid snake, *Epicrates monensis,* and its anoline prey, *Anolis cristatellus, J. Herpetol.,* 24:151–157.
Chiszar, D., 1990, The behavior of the brown tree snake: A study in comparative psychology, in D. A. Dewsbury, ed., *Contemporary Issues in Comparative Psychology,* Sinauer Assoc., Sunderland, Massachusetts, pp. 101–123.
Chiszar, D., Drew, D., and H. M. Smith, 1991, Stimulus control of predatory behavior in the brown tree snake (*Boiga irregularis*), III: Mandibular protractions as a function of prey size, *J. Comp. Psychol.,* 105:152–156.
Chiszar, D., K. Fox, and H. M. Smith, 1992, Stimulus control of predatory behavior in the brown tree snake (*Boiga irregularis*), IV. Effect of mammalian blood, *Behav. Neurol. Biol.,* 57:167–169.
Chiszar, D., K. Kandler, and H. M. Smith, 1988a, Stimulus control of predatory attack in the brown tree snake (*Boiga irregularis*), I. Effects of visual cues arising from prey, *The Snake,* 20:151–155.
Chiszar, D., K. Kandler, R. Lee, and H. M. Smith, 1988b, Stimulus control of predatory attack in the brown tree snake (*Boiga irregularis*), II. Use of chemical cues during foraging, *Amphibia-Reptilia,* 9:77–88.
Curran, C. H., and C. Kauffeld, 1937, *Snakes and Their Ways,* Harper, New York.
Dixon, J. R., and P. Soini, 1986, *The Reptiles of the Upper Amazon Basin, Iquitos Region, Peru,* Milwaukee Public Museum, Milwaukee.
Donald, J. A., and H. B. Lillywhite, 1988, Adrenergic innervation of the large arteries and veins of the semi-arboreal ratsnake, *Elaphe obsoleta, J. Morphol.,* 198:25–31.
Donald, J. A., J. E. O'Shea, and H. B. Lillywhite, 1990, Neural regulation of the pulmonary vasculature in a semi-arboreal snake, *Elaphe obsoleta, J. Comp. Physiol., B* 159:677–685.
Dove, L. B., D. A. Baeyens, and M. V. Plummer, 1982, Evaporative water loss in *Opheodrys aestivus* (Colubridae), *Southwest Nat.,* 27:228–230.
Duellman, W. E., 1978, The biology of an equatorial herpetofauna in Amazonian Ecuador, *Univ. Kansas Mus. Nat. Hist. Misc. Publ.,* 65:1–352.
Duellman, W. E., 1989, Tropical herpetofaunal communities: Patterns of community structure in neotropical rainforests, in M. L. Harmelin-Vivien and F. Bourliere, eds., *Vertebrates in Complex Tropical Systems,* Springer-Verlag, New York, pp. 61–88.
Duellman, W. E., 1990, Herpetofaunas in neotropical rainforests: Comparative composition, history, and resource use, in A. H. Gentry, ed., *Four Neotropical Rainforests,* Yale Univ. Press, New Haven, Connecticut, pp. 455–505.
Duellman, W. E., and A. W. Salas, 1991, Annotated checklist of the amphibians and reptiles of Cuzco Amazonico, Peru, *Univ. Kansas Mus. Nat. Hist. Occ. Pap.,* 143:1–13.
Dunn, E. R., 1949, Relative abundance of some Panamanian snakes, *Ecology,* 30:39–57.
Edwards, J. L., 1985, Terrestrial locomotion without appendages, in M. Hildebrand, D. M. Bramble, K. F. Liem, and D. B. Wake, eds., *Functional Vertebrate Morphology,* Belknap Press (Harvard Univ. Press), Cambridge, pp. 159–172.
Endler, J. A., 1978, A predator's view of animal color patterns, in M. K. Hecht, W. C. Steere, and B. Wallace, eds., *Evolutionary Biology,* Vol. 11, Plenum, New York, pp. 319–364.
Feder, M. E., A. F. Bennett, W. W. Burggren, and R. B. Huey (eds.), 1987, *New Directions in Physiological Ecology,* Cambridge Univ. Press, Cambridge.
Ficken, R. W., P. E. Matthiae, and R. Horwich, 1971, Eye marks in vertebrates: Aids to vision, *Science,* 173:936–939.

Fitch, H. S., 1963, Natural history of the black rat snake (*Elaphe o. obsoleta*) in Kansas, *Copeia*, 1963:638–649.
Fitch, H. S., 1982, Resources of a snake community in prairie-woodland habitat of northeastern Kansas, in N. J. Scott, ed., *Herpetological Communities, U.S. Fish Wildl. Serv. Wildl. Res. Rep.* 13, pp. 83–97.
Fitch, H. S., and H. W. Shirer, 1971, A radiotelemetric study of spatial relationships in some common snakes, *Copeia*, 1971:118–128.
Fleishman, L. J., 1985, Cryptic movement in the vine snake *Oxybelis aeneus*, *Copeia*, 1985:242–245.
Fritts, T. H., 1988, The brown tree snake, *Boiga irregularis*, a threat to Pacific islands, *U.S. Fish Wildl. Serv. Biol. Rept.*, 88:1–36.
Fritts, T. H., N. J. Scott, and B. E. Smith, 1989, Trapping *Boiga irregularis* on Guam using bird odors, *J. Herpetol.*, 23:189–192.
Gans, C., 1974, *Biomechanics: An Approach to Vertebrate Biology*, Lippincott, Philadelphia.
Gasc, J. P., 1967, Introduction a l'etude de la musculature axiale des Squamates serpentiformes, *Mem. Mus. Natl. Hist. Nat. Ser. A Zool.*, 48:69–124.
Gasc, J. P., 1971, Les serpents-lianes, *Sci. Nat. Paris*, 105:29–34.
Gasc, J. P., 1974, L'interpretation fonctionelle de l'appareil musculosquelettque de l'axe vertebral chez serpents (Reptilia), *Mem. Mus. Natl. Hist. Nat. Ser. A Zool.*, 83:1–182.
Gasc, J. P., 1981, Axial musculature, in C. Gans and T. S. Parsons, eds., *Biology of the Reptilia*, Vol. 11, Academic, New York, pp. 355–435.
Geiger, R., 1965, *The Climate Near the Ground* (translation of) 3rd German ed., Harvard Univ. Press, Cambridge.
Goldsmith, S. K., 1981, Behavior and ecology of the rough green snake, *Opheodrys aestivus*, Masters Thesis, Univ. Oklahoma, Norman.
Goldsmith, S. K., 1984, Aspects of the natural history of the rough green snake, *Opheodrys aestivus* (Colubridae), *Southwest Nat.*, 29:445–452.
Goldsmith, S. K., 1986, Feeding behavior of an arboreal insectivorous snake (*Opheodrys aestivus*) (Colubridae), *Southwest Nat.*, 31:246–249.
Goodman, J. D., and J. M. Goodman, 1976, Possible mimetic behavior of the twig snake, *Thelotornis kirtlandi* (Hallowell), *Herpetologica*, 32:148–150.
Gove, D., 1978, The form, variation, and evolution of tongue-flicking in reptiles, Doctoral Thesis, Univ. Tennessee, Knoxville.
Gove, D., 1979, A comparative study of snake and lizard tongue-flicking, with an evolutionary hypothesis, *Z. Tierpsychol.*, 51:58–76.
Greene, H. W., 1979, Behavioral convergence in the defensive displays of snakes, *Experientia*, 35:747–748.
Greene, H. W., 1983, Dietary correlates of the origin and radiation of snakes, *Amer. Zool.*, 23:431–441.
Greene, H. W., 1988, Antipredator mechanisms in reptiles, in C. Gans and R. B. Huey, eds., *Biology of the Reptilia*, Vol. 16, Ecology B, Alan R. Liss, New York, pp. 1–152.
Greene, H. W., 1989, Ecological, evolutionary, and conservation implications of feeding biology in Old World cat snakes, genus *Boiga* (Colubridae), *Proc. Calif. Acad. Sci.*, 46:193–207.
Guyer, C., and M. A. Donnelly, 1990, Length-mass relationships among an assemblage of tropical snakes in Costa Rica, *J. Trop. Ecol.*, 6:65–76.
Harlow, P., and R. Shine, 1992, Food habits and reproductive biology of the Pacific island boas (*Candoia*), *J. Herpetol.*, 26:60–66.
Heatwole, H., S. A. Minton, Jr., G. Witten, M. Dick, J. Parmenter, R. Shine, and C. Parmenter, 1973, Arboreal habits in Australian elapid snakes, *Herpetol. Inform. Search Systems* (*HISS*), 1:113.
Hediger, H., 1968, *The Psychology and Behavior of Animals in Zoos and Circuses*, Dover, New York.
Henderson, R. W., 1974, Aspects of the ecology of the neotropical vine snake, *Oxybelis aeneus* (Wagler), *Herpetologica*, 30:19–24.

Henderson, R. W., 1982, Trophic relationships and foraging strategies of some New World tree snakes (*Leptophis, Oxybelis, Uromacer*), *Amphibia-Reptilia,* 3:71–80.
Henderson, R. W., 1991, Distribution and preliminary interpretation of geographic variation in the neotropical tree boa *Corallus enydris:* A progress report, *Bull. Chicago Herpetol. Soc.,* 26:105–110.
Henderson, R. W., 1992, Consequences of predator introductions and habitat destruction on amphibians and reptiles in the post-Columbus West Indies, *Carib. J. Sci.,* 28:1–10.
Henderson, R. W., 1993, Foraging and diet in West Indian *Corallus enydris* (Serpentes: Boidae), *J. Herpetol.,* 27:24–28.
Henderson, R. W., and M. H. Binder, 1980, The ecology and behavior of vine snakes (*Ahaetulla, Oxybelis, Thelotornis, Uromacer*): A review, *Milwaukee Pub. Mus. Contrib. Biol. Geol.,* 37:1–38.
Henderson, R. W., and B. I. Crother, 1989, Biogeographic patterns of predation in West Indian colubrid snakes, in C. A. Woods, ed., *Biogeography of the West Indies: Past, Present and Future,* Sandhill Crane Press, Gainesville, Florida, pp. 479–517.
Henderson, R. W., and H. S. Horn, 1983, The diet of the snake *Uromacer frenatus dorsalis* on Ile de la Gonave, Haiti, *J. Herpetol.,* 17:409–412.
Henderson, R. W., and R. A. Winstel, 1992, Activity patterns, temperature relationships, and habitat utilization in *Corallus enydris* (Serpentes: Boidae) on Grenada, *Carib. J. Sci.,* 28:229–232.
Henderson, R. W., M. A. Nickerson, and S. Ketcham, 1976, Short term movements of the snakes *Chironius carinatus, Helicops angulatus* and *Bothrops atrox* in Amazonian Peru, *Herpetologica,* 32:304–310.
Henderson, R. W., J. R. Dixon, and P. Soini, 1978, On the seasonal incidence of tropical snakes, *Milwaukee Pub. Mus. Contr. Biol. Geol.,* 17:1–15.
Henderson, R. W., J. R. Dixon, and P. Soini, 1979, Resource partitioning in Amazonian snake communities, *Milwaukee Pub. Mus. Contrib. Biol. Geol.,* 22:1–11.
Henderson, R. W., M. H. Binder, and R. A. Sajdak, 1981, Ecological relationships of the tree snakes *Uromacer catesbyi* and *U. oxyrhynchus* (Colubridae) on Isla Saona, República Dominicana, *Amphibia-Reptilia,* 2:153–163.
Henderson, R. W., M. H. Binder, and G. M. Burghardt, 1983, Responses of neonate Hispaniolan vine snakes (*Uromacer frenatus*) to prey extracts, *Herpetologica,* 39:75–77.
Henderson, R. W., T. A. Noeske-Hallin, J. A. Ottenwalder, and A. Schwartz, 1987a, On the diet of the boa *Epicrates striatus* on Hispaniola, with notes on *E. fordi* and *E. gracilis, Amphibia-Reptilia,* 8:251–258.
Henderson, R. W., A. Schwartz, and T. A. Noeske-Hallin, 1987b, Food habits of three colubrid tree snakes (genus *Uromacer*) on Hispaniola, *Herpetologica,* 43:241–248.
Henderson, R. W., T. A. Noeske-Hallin, B. I. Crother, and A. Schwartz, 1988, The diets of Hispaniolan colubrid snakes, II. Prey species, prey size, and phylogeny, *Herpetologica,* 44:55–70.
Inger, R. F., H. B. Shaffer, M. Koshy, and R. Bakde, 1984, A report on a collection of amphibians and reptiles from the Ponmudi, Kerala, south India, *J. Bombay Nat. Hist. Soc.,* 81:551–570.
Jayne, B. C., 1982, Comparative morphology of the semispinalis-spinalis muscle of snakes and correlations with locomotion and constriction, *J. Morph.,* 172:83–96.
Jayne, B. C., and A. F. Bennett, 1989, The effect of tail morphology on locomotor performance in snakes: A comparison of experimental and correlative methods, *J. Exp. Zool.,* 252:126–133.
Johnson, R. G., 1955, The adaptive and phylogenetic significance of vertebral form in snakes, *Evolution,* 9:367–388.
Keiser, E. D., 1975, Observations on tongue extension of vine snakes (genus *Oxybelis*) with suggested behavioral hypotheses, *Herpetologica,* 31:131–133.
Kennedy, J. P., 1965, Notes on the habitat and behavior of a snake, *Oxybelis aeneus* Wagler, in Veracruz, *Southwest Nat.,* 10:136–139.

Kira, T., and K. Yoda, 1989, Vertical stratification in microclimate, in H. Lieth and M. J. A. Werger, eds., *Tropical Rain Forest Ecosystems, Biogeographical and Ecological Studies, Ecosystems of the World,* 14B, Elsevier, New York, pp. 55–71.

Leigh, E. G., 1975, Structure and climate in tropical rain forest, *Annu. Rev. Ecol. Syst.,* 6:67–86.

Leston, D., and B. Hughes, 1968, The snakes of Tafo, a forest cocoa-farm locality in Ghana, *Bull. I.F.A.N. (Inst. Francais Afr. Noir.)* (*Ser. A*), 30:737–770.

Lillywhite, H. B., 1982, Tracking as an aid in ecological studies of snakes, in N. J. Scott, ed., *Herpetological Communities,* U.S. Fish Wildl. Wildl. Res. Rept., 13, pp. 181–191.

Lillywhite, H. B., 1985a, Postural edema and blood pooling in snakes, *Physiol. Zool.,* 58:759–766.

Lillywhite, H. B., 1985b, Behavioral control of arterial pressure in snakes, *Physiol. Zool.,* 58:159–165.

Lillywhite, H. B., 1987a, Temperature, energetics, and physiological ecology, in R. A. Seigel, J. T. Collins, and S. S. Novak, eds., *Snakes: Ecology and Evolutionary Biology,* McGraw-Hill, New York, pp. 422–477.

Lillywhite, H. B., 1987b, Tissue free fluid pressures in relation to behavioral and morphological variation in snakes, *Amer. Zool.,* 27:117A.

Lillywhite, H. B., 1987c, Circulatory adaptations of snakes to gravity, *Amer. Zool.,* 27:81–95.

Lillywhite, H. B., 1988, Snakes, blood circulation and gravity, *Sci. Amer.,* 256:92–98.

Lillywhite, H. B., and J. A. Donald, 1988, Anterograde bias of blood flow in arboreal snakes, *Amer. Zool.,* 28:86A.

Lillywhite, H. B., and K. P. Gallagher, 1985, Hemodynamic adjustments to head-up posture in the partly arboreal snake, *Elaphe obsoleta, J. Exp. Zool.,* 235:325–334.

Lillywhite, H. B., and F. H. Pough, 1983, Control of arterial pressure in aquatic sea snakes, *Amer. J. Physiol.,* 244:R66–73.

Lillywhite, H. B., and A. W. Smits, 1992, Cardiovascular adaptations of viperid snakes, in J. A. Campbell and E. D. Brodie, Jr., eds., *Biology of Pitvipers,* Selva, Tyler, Texas, pp. 143–153.

Macartney, M. J., P. T. Gregory, and K. W. Larsen, 1988, A tabular survey of data on movements and home ranges of snakes, *J. Herpetol.,* 22:61–73.

March, D. D. H., 1928, Field notes on barba amarilla (*Bothrops atrox*), *Bull. Antivenin Inst. Amer.,* 1:92–97.

McDowell, S. B., 1975, A catalogue of the snakes of New Guinea and the Solomons, with special reference to those in the Bernice P. Bishop Museum, Part II. Aniloidea and Pythoninae, *J. Herpetol.,* 9:1–79.

Minton, S. A., and W. A. Dunson, 1978, Observations on the Palawan mangrove snake, *Boiga dendrophila multicincta* (Reptilia, Serpentes, Colubridae), *J. Herpetol.,* 12:107–108.

Mitchell, J., W. Beckman, R. Bailey, and W. Porter, 1975, Microclimatic modeling of the desert, in D. A. deVries and N. H. Afgan, eds., *Heat and Mass Transfer in the Biosphere, Part I. Transfer Processes in the Plant Environment,* Scripta, Washington, D.C., pp. 275–286.

Moermond, T. C., 1979, Habitat constraints on the behavior, morphology, and community structure of *Anolis* lizards, *Ecology,* 60:152–164.

Monsi, M., and T. Saeki, 1953, Uber den Lichtfaktor in den Pflanzengesellschaften und ihre Bedeutung fur die Stoffproduktion, *Jap. J. Bot.,* 14:22–52.

Monteith, J. L. (ed.), 1975, *Vegetation and the Atmosphere, Vol. 1, Principles,* Academic, New York.

Mosauer, W., 1935, The myology of the trunk region of snakes and its significance for ophidian taxonomy and phylogeny, *Publ. Univ. Calif. Los Angeles Biol. Sci.,* 1:81–121.

Myers, C. W., 1982, Blunt-headed vine snakes (*Imantodes*) in Panama, including a new species and other revisionary notes, *Amer. Mus. Novit.,* 2738:1–50.

Parker, M., 1990, *Study of Laughing Falcons (Herpetotheres cachinnans) in Tikal National Park, Guatemala,* Raptor Res. Found., Allentown, Pennsylvania.

Peterson, C. R., 1987, Daily variation in the body temperatures of free-ranging garter snakes, *Ecology,* 68:160–169.
Plummer, M. V., 1981, Habitat utilization, diet and movements of a temperate arboreal snake (*Opheodrys aestivus*), *J. Herpetol.,* 15:425–432.
Plummer, M. V., 1984, Female reproduction in an Arkansas population of rough green snakes (*Opheodrys aestivus*), in R. A. Seigel, L. E. Hunt, J. L. Knight, L. Malaret, and N. L. Zuschlag, eds., *Vertebrate Ecology and Systematics: A Tribute to Henry S. Fitch,* Univ. Kansas Mus. Nat. Hist. Spec. Publ., 10, pp. 105–113.
Plummer, M. V., 1985, Demography of green snakes (*Opheodrys aestivus*), *Herpetologica,* 41:373–381.
Plummer, M. V., 1990a, High predation on green snakes, *Opheodrys aestivus, J. Herpetol.,* 24:327–328.
Plummer, M. V., 1990b, Nesting movements, nesting behavior, and nest sites of green snakes (*Opheodrys aestivus*) revealed by radiotelemetry, *Herpetologica,* 46:190–195.
Plummer, M. V., 1993, Thermal ecology of arboreal green snakes (*Opheodrys aestivus*), *J. Herpetol.,* vol. 27 (in press).
Porter, W. P., and D. M. Gates, 1969, Thermodynamic equilibria of animals with environment, *Ecol. Monogr.,* 39:227–244.
Pough, F. H., and J. D. Groves, 1983, Specializations of the body form and food habits of snakes, *Amer. Zool.,* 23:443–454.
Pounds, J. A., 1988, Ecomorphology, locomotion, and microhabitat structure: Patterns in a tropical mainland *Anolis* community, *Ecol. Monogr.,* 58:299–320.
Pounds, J. A., 1991, Habitat structure and morphological patterns in arboreal vertebrates, in S. S. Bell, E. D. McCoy, and H. R. Mushinsky, eds., *Habitat Structure, The Physical Arrangement of Objects in Space,* Chapman and Hall, New York, pp. 109–119.
Proctor, J. B., 1924, Unrecorded characters seen in living snakes, and a description of a new tree-frog, *Proc. Zool. Soc. Lond.,* 1924:1125–1129.
Reynolds, R. P., and N. J. Scott, Jr., 1982, Use of mammalian resource by a Chihuahuan snake community, in N. J. Scott, Jr., ed., *Herpetological Communities,* U.S. Fish Wildl. Serv. Wildl. Res. Rep. 13, pp. 99–118.
Richards, P. W., 1952, *The Tropical Rain Forest—An Ecological Study,* Cambridge Univ. Press, Cambridge.
Rodda, G. H., 1992, Foraging behaviour of the brown tree snake, *Boiga irregularis, Herpetol. J.,* 2:110–114.
Rodda, G. H., and T. H. Fritts, 1992, The impact of the introduction of the colubrid snake *Boiga irregularis* on Guam's lizards, *J. Herpetol.,* 26:166–174.
Rossman, D. A., and K. L. Williams, 1966, Defensive behavior of the South American colubrid snakes *Pseustes sulphureus* (Wagler) and *Spilotes pullatus* (Linnaeus), *Proc. Louisiana Acad. Sci.,* 29:152–156.
Ruben, J. A., 1977, Morphological correlates of predatory modes in the coachwhip (*Masticophis flagellum*) and rosy boa (*Lichanura roseofusca*), *Herpetologica,* 33:1–6.
Savidge, J. A., 1987, Extinction of an island forest avifauna by an introduced snake, *Ecology,* 68:660–668.
Savidge, J. A., 1988, Food habits of *Boiga irregularis,* an introduced predator on Guam, *J. Herpetol.,* 22:275–282.
Savidge, J. A., 1991, Population characteristics of the introduced brown tree snake (*Boiga irregularis*) on Guam, *Biotropica,* 23:294–300.
Schwartz, A., and R. W. Henderson, 1991, *Amphibians and Reptiles of the West Indies: Descriptions, Distributions and Natural History,* Univ. Florida Press, Gainesville.
Seigel, R. A., and N. B. Ford, 1987, Reproductive ecology, in R. A. Seigel, J. T. Collins, and S. S. Novak, eds., *Snakes: Ecology and Evolutionary Biology,* McGraw-Hill, New York, pp. 210–252.
Seymour, R. S., 1987, Scaling of cardiovascular physiology in snakes, *Amer. Zool.,* 27:97–109.
Seymour, R. S., and H. B. Lillywhite, 1976, Blood pressure in snakes from different habitats, *Nature,* 264:664–666.

Shine, R., 1983, Arboreality in snakes: Ecology of the Australian elapid genus *Hoplocephalus*, *Copeia*, 1983:198–205.
Shine, R., 1991, Strangers in a strange land: Ecology of the Australian colubrid snakes, *Copeia*, 1991:120–131.
Shine, R., and T. Schwaner, 1985, Prey constriction by venomous snakes: A review, and new data on Australian species, *Copeia*, 1985:1067–1071.
Shine, R., and D. J. Slip, 1990, Biological aspects of the radiation of Australasian pythons (Serpentes: Boidae), *Herpetologica*, 46:283–290.
Shoemaker, V. L., 1988, Physiological ecology of amphibians in arid environments, *J. Arid Environments*, 14:145–153.
Soderberg, P. S., 1971, Striking behaviour of the common green whip snake (*Ahaetulla nasutus*), *J. Bombay Nat. Hist. Soc.*, 68:839.
Soepadmo, E., and T. Kira, 1977, Contribution of the IBP-PT Research Project to the understanding of Malaysian forest ecology, in C. B. Sastry, P. B. L. Srivastava, and A. Manap Ahmad, eds., *A New Era in Malaysian Forestry*, Univ. Pertanian Malaysia, Serdang, Selangor, Malaysia, pp. 63–90.
Stevenson, R. D., 1985, Body size and limits to the daily range of body temperature in terrestrial ectotherms, *Amer. Nat.*, 125:102–117.
Strong, D. R., D. Simberloff, L. G. Abele, and A. B. Thistle (eds.), 1984, *Ecological Communities: Conceptual Issues and the Evidence*, Princeton Univ. Press, Princeton, New Jersey.
Sweeney, R. C. H., 1971, *Snakes of Nyasaland*, Asher, Amsterdam.
Test, F. H., O. J. Sexton, and H. Heatwole, 1966, Reptiles of Rancho Grande and vicinity, Estado Aragua, Venezuela, *Misc. Publ. Mus. Zool. Univ. Michigan*, 128:1–63.
Tolson, P. J., 1988, Critical habitat, predator pressures, and the management of *Epicrates monensis* (Serpentes: Boidae) on the Puerto Rico Bank: A multivariate analysis, in *Management of Amphibians, Reptiles, and Small Mammals in North America*, U.S. Dept. Agric. Gen. Tech. Rept., RM-166, pp. 228–238.
Tracy, C. R., 1982, Biophysical modeling in reptilian physiology and ecology, in C. Gans and F. H. Pough, eds., *Biology of the Reptilia*, Vol. 12, Physiology C, Academic, New York, pp. 275–321.
Underwood, G., 1970, The eye, in C. Gans and T. S. Parsons, eds., *Biology of the Reptilia*, Vol. 2, Academic, New York, pp. 1–97.
Vitt, L. J., and L. D. Vangilder, 1983, Ecology of a snake community in northeastern Brazil, *Amphibia-Reptilia*, 4:273–282.
Wall, F., 1905, A popular treatise on the common Indian snakes, Part I, *J. Bombay Nat. Hist. Soc.*, 16:533–554.
Walls, G. L., 1942, The vertebrate eye and its adaptive radiation, *Bull. Cranbrook Inst.*, 19:1–785.
Walter, H., 1964, *Die Vegetation der Erde in Oko-Physiologischer Betrachtung, Band I: Die Tropischen und Subtropischen Zonen*, 2nd ed., Fischer, Jena.
Walter, H., 1971, *Ecology of Tropical and Subtropical Vegetation*, Oliver and Boyd, Edinburgh.
Weatherhead, P. J., and M. B. Charland, 1985, Habitat selection in an Ontario population of the snake, *Elaphe obsoleta*, *J. Herpetol.*, 19:12–19.
Wilson, E. O., and F. M. Peter (eds.), 1988, *Biodiversity*, National Academy Press, Washington, D.C.
Wygoda, M. L., 1984, Low cutaneous evaporative water loss in arboreal frogs, *Physiol. Zool.*, 57:329–337.
Yim, Y.-J., H. Ogawa, and T. Kira, 1969, Light interception by stems in plant communities, *Jap. J. Ecol.*, 19:233–238.
Yoda, K., 1974, Three-dimensional distribution of light intensity in a tropical rain forest in West Malaysia, *Jap. J. Ecol.*, 24:247–254.
Zimmerman, B. L., and M. T. Rodrigues, 1990, Frogs, snakes and lizards of the INPA-WWF reserves near Manaus, Brazil, in A. H. Gentry, ed., *Four Neotropical Rainforests*, Yale Univ. Press, New Haven, Connecticut, pp. 426–454.

Chapter

2

Sexual Dimorphism in Snakes

Richard Shine

Introduction

In many animal species, males and females differ from each other so much that at first sight they may be mistaken for separate species. Although most snakes do not show such extreme dimorphism, sex differences in size, shape, and color are widespread within the suborder. As originally recognized by Charles Darwin (Darwin, 1871), such differences between conspecific males and females offer a unique opportunity for powerful tests of hypotheses about the processes responsible for the origin of biological diversity. Interpretations of sexual dimorphism are not confounded by differing phylogenetic histories or geographic distributions, as would be the case in any attempt to explain the origins of interspecific variation. If males and females evolve to differ, only a limited set of selective forces can be invoked to explain the phenomenon. This tight focus has stimulated a recent resurgence in scientific interest in the evolution of sexual dimorphism (Bradbury and Andersson, 1987).

Differences between the sexes can take many forms, and this chapter will treat only a subset of these. For example, I have not attempted to cover differences in primary sex structures (the gonads and associated genitalia) and have concentrated instead on secondary sexual characteristics. Within this topic, I have concentrated attention on attributes likely to have the greatest biological significance to the animal in terms of its survival or reproductive success. Hence, I have devoted little attention to topics such as sex differences in the sizes and positions of internal organs (Kopstein, 1941; Rossman et al., 1982).

Determinants of Sexual Dimorphism

Theoretical models suggest several ways in which significant sex differences may arise. Most attention has traditionally been focused on *adaptation* and especially on *sexual selection,* whereby differences between males and females in the determinants of reproductive success act as selective pressures for divergence in some characteristic (be it size, morphology, physiology, or behavior). Within the general category of *sexual selection,* work on snakes has emphasized the possible role of intrasexual selection among males (especially, the consequences of male–male combat) rather than epigamic selection (evolution of sex-specific attributes in one sex through active mate selection by the other sex). The extreme scarcity (absence?) of male parental care in squamate reptiles (Shine, 1988b) means that parental investment is negligible in males, and this factor may have reduced the importance of epigamic selection in snakes. Intrasexual selection may, however, be intense, favoring attributes such as mate-finding ability or large body size in males (Gibbons, 1972; Shine, 1978b; Duvall et al., 1992; Madsen et al., 1993).

The evolution of sexual dimorphism in snakes may have been influenced by *natural selection,* as well as sexual selection (note that some authors view sexual selection as a special category of natural selection, whereas others define it to be a separate process). If males and females occupy slightly different ecological niches (perhaps because of sexually selected differences in body sizes or activity patterns), natural selection may then favor independent adaptations in each sex such that they evolve either to be more different or more similar (Slatkin, 1984; Shine, 1989). Although Darwin (1871) proposed this hypothesis at the same time as sexual selection, it has generally been less popular, probably because of the mistaken belief that the hypothesis of ecologically based divergence requires an assumption of resource-based competition between the sexes (Shine, 1989). In fact, both theoretical models and empirical analyses suggest that major sexual dimorphism can evolve because of sex-specific adaptations of foraging biology, without any need to invoke competition between males and females (Slatkin, 1984; Shine, 1989). Sex differences in shape and size might also evolve through *fecundity selection,* whereby selection acts on the female's ability to physically accommodate a large clutch (Pope, 1935; Semlitsch and Gibbons, 1982). Fecundity selection on female body sizes may be particularly intense in aquatic snakes, because less of the body cavity is available to hold the clutch (Shine, 1988a).

Although adaptational hypotheses have attracted the most interest in this field, it is important to recognize that sexual dimorphism may also arise through nonadaptive processes. For example, some sex differences may be incidental consequences of the action of gonadal hormones, without any specific selective advantage or disadvantage. Other patterns in dimorphism might be strongly influenced by absolute body size, such that the degree of development of sexual dimorphism is allometrically constrained. Sexual divergence in some characteristics (such as feeding rates in juveniles) may cause subsequent differences in many other characteristics (such as adult body sizes, relative head sizes, etc.) that are not themselves the target of selection. Hence, caution is needed in interpreting sexually dimorphic attributes as adaptations (Gould and Lewinton, 1979; Lande, 1980).

Hypotheses on the adaptive significance of sexual dimorphism may often be clarified by detailed investigation of the proximate causes of the dimorphism. Unfortunately, little is known about this topic in snakes. On a mechanistic level, sexually dimorphic characters are generally the result of gonadal hormones, and this has been reported to be true for both juvenile growth rates (Crews et al., 1985) and relative head sizes (Shine and Crews, 1988) in Garter Snakes. Reports of sex-biased mortality in Bullsnake eggs incubated at different temperatures (Burger and Zappalorti, 1988) and of long-term effects of incubation temperature on behavior and physiology of the offspring (Burger, 1989) raise the possibility that some sex differences in oviparous snakes may be by-products of different incubation conditions.

Sex Differences in Body Size

Body size is one of the most ecologically important attributes of any organism, and sex differences in this characteristic have attracted considerable scientific attention. On a methodological level, quantifying the degree of sex difference in body size (hereafter abbreviated to sexual size dimorphism, or SSD) is not a trivial problem because of statistical artifacts associated with ratio measures (Atchley et al., 1976). Gibbons and Lovich (1990) suggest a useful measure of the degree of SSD: mean size of larger sex divided by mean size of smaller sex, arbitrarily expressed as positive if females are the larger sex, and negative if males are the larger sex. The difference between this number and a value of 1.0 (or −1.0, in cases where males are larger than females) can be used in statistical analyses, because the resulting ratio is symmetrical around a value of zero. Another obvious difficulty in methodology is in the choice of a variable to describe body size. Body length (usually snout–vent length, or SVL) is the measure most often provided, although mass may be more useful under some circumstances (e.g., when evaluating parental investment into sons versus daughters).

Sexual differences in body size can occur at any stage of the life history, but the most interesting points are at hatching (birth), maturation, and mean and maximum adult body sizes. SSD at hatching seems to be rare among snakes (Shine and Bull, 1977; Fitch, 1981), but has been documented in several taxa including Costa Rican pit vipers (*Bothrops asper,* Solórzano and Cerdas, 1989), European vipers (*Vipera aspis,* Naulleau, 1970), Australian elapids (*Acanthophis antarcticus,* Johnston, 1987), and North American colubrids (*Diadophis punctatus,* Fitch, 1975; *Regina grahamii,* Seigel, 1992). Although theoretical models suggest that differences in the "costs" of producing sons should affect the sex ratio at birth, such adaptive modifications of the sex ratio have apparently not been investigated in snakes (but see Shine and Bull, 1977; Burger and Zappalorti, 1988; Dunlap and Lang, 1990; Madsen and Shine, 1992b). Surprisingly, data on the Australian Death Adder suggest that daughters are produced in greater numbers than are sons, despite the larger size of females at birth (Johnston, 1987).

Although sons and daughters are usually the same size at birth, their growth rates may diverge thereafter. Juvenile growth rates are often similar in males and females (Feaver, 1977), but significant sexual differences in juvenile growth rates have been reported in several species, including *Crotalus* spp. (Klauber, 1956), *Spalerosophis cliffordi* (Dmi'el, 1967), *Heterodon nasicus* and *H. platirhinos* (Platt, 1969), *Carphophis vermis* (Clark, 1970), *Thamnophis sirtalis* (Crews et al., 1985), *Vipera berus* (Madsen and Shine, 1993b), *Crotalus viridis* (Macartney et al., 1990), and *Acrochordus arafurae* (Houston, 1992). Presumably, such differences in growth rates are due either to differences in feeding rates or in metabolic expenditure between the sexes. In *T. sirtalis,* testicular hormones appear to be the actual mechanism responsible for the decreased rate of growth in young males (Crews et al., 1985). In combination with sex differences in age at maturation, growth-rate differences between males and females are major determinants of SSD at maturation and hence SSD at mean adult size (Shine, 1990). Thus, further research on sex differences in juvenile growth rates and ages at maturity is needed before we can understand the evolution of sex differences in adult body size.

Sexual bimaturism is common in snakes and often involves earlier maturation in males than in conspecific females (see review by Parker and Plummer, 1987), probably because males face lower fecundity-independent costs of reproduction (Shine, 1978a; Bell, 1980). Females have less flexibility, in that reproducing with even a small clutch may involve high costs because of the associated behavioral modifications. Thus, females in some populations may be unlikely to survive long enough to breed more than once, and so can maximize their lifetime reproductive success by delaying reproduction until they are able to produce a large litter. More generally, Madsen (1987) showed for Grass Snakes (*Natrix natrix*) that the observed female age at maturation was also the one that maximized lifetime reproductive success, given empirical relationships among female size, clutch size, and survival rates. Madsen and Shine (1993a,b) have pursued this approach with adders and analyzed data on both sexes within this species.

Females grow larger than conspecific males in about two-thirds of snake species for which data are available (Shine, 1978b, 1993; Fitch, 1981). The direction and degree of SSD at mean adult size are set largely at maturity but modified also by sex differences in growth trajectories and survival rates among adults (Shine, 1990, 1993). A detailed analysis of data on Australian snakes showed that the degree of SSD at maturation and at mean body size differed by an average of only 5%, but that the extent of this difference depended on the mating system of the species involved (Shine, 1993). Hence, the proximate mechanisms responsible for the fact that females grow larger than males in most species of snakes are (1) delayed maturation in females compared to males, combined with (2) reduced growth rates after maturation in both sexes. Faster rates of juvenile growth in females may also be important in some taxa.

Analyses of overall patterns in the direction of SSD among snakes have revealed strong patterns. Phylogenetic conservatism in SSD is evident at both the familial and generic levels (Shine, 1978b, 1993b). However, most scientific attention has been focused on correlations between SSD and other factors, notably body size, reproductive mode, and mating system. These patterns are discussed below.

Body size. A strong allometric effect on SSD is evident in comparisons among snakes in general—males tend to be larger, relative to females, in larger species (Shine, 1991b; Shine and Madsen, in preparation). Similar allometric patterns of SSD have been documented in other kinds of animals as well, including other reptiles (Berry and Shine, 1980; Iverson, 1990), and have been attributed to a wide variety of selective forces (Clutton-Brock et al., 1977). Phylogenetically based analysis of the data on snakes indicates that this correlation between absolute size and SSD is not an artifact of taxonomic conservatism—phylogenetic changes in mean body size are consistently associated with changes in SSD (Shine and Madsen, in preparation).

Reproductive mode. Fitch's (1981) review of data suggested a general trend for females to be larger than males in viviparous as opposed to oviparous reptiles, but a more recent, detailed reanalysis shows no significant difference in this respect among snakes (Shine, 1993).

Geographic distribution. Fitch (1981) also detected an overall difference in SSD between tropical and temperate-zone reptiles. More detailed analysis of his data on snakes confirmed this correlation but showed that it was due primarily to phylogenetic conservatism—that is, lineages that occur mainly at higher latitudes tend to be more dimorphic, with larger females relative to males (Shine, 1993). Phylogenetic analysis did not show any consistent tendency *within lineages* for species invading cooler areas to evolve to be more dimorphic.

Male combat. The influence of the mating system on SSD was first emphasized by Darwin (1871). Because of their elongate shape, male snakes are apparently unable to forcibly inseminate females (Devine, 1975, 1984), so that selection for large body size to facilitate forcible insemination (a possible selection pressure on male lizards) does not apply to snakes. Instead, selection should favor male abilities to locate receptive females and induce them to copulate. Although these abilities might depend on body size (e.g., if larger males are more mobile, face greater energetic costs in locomotion, or are less vulnerable to predators during mate-searching movements), male body size is likely to be especially important in taxa that show male–male combat during the mating season. This kind of behavior has now been recorded in more than 100 species of snakes (23 boid species from 10 genera, 31 colubrids from 17 genera, 29 elapids from 14 genera, and 40 viperids from 10 genera, Shine, 1993).

Large males are more likely to engage in combat (Madsen et al., 1993) and more likely to win combat bouts (Andrén and Nilson, 1981; Andrén, 1986; Schuett and Gillingham, 1989; Madsen et al., 1993) and thereby obtain matings (Madsen et al., 1993). Hence, straightforward Darwinian theory predicts that male–male combat should impose a selection pressure for large body size in adult males (Gibbons, 1972). The difficulty with testing this idea is that no prediction is possible about the absolute size of males or of male size relative to female size. The only prediction is that males should evolve to be larger than they would have been in the absence of male–male combat (Greenwood and Adams, 1987; Shine, 1987). Other selective forces may act to reduce male size or increase female size, so that the direction and degree of SSD cannot be predicted simply from the mating system.

Two solutions to this difficulty have been proposed. One is to measure the reproductive success of individuals directly, so that the intensity of selection on male and female body sizes can be compared. This technique requires information on success rates of males in combat and courtship in the field. Although such information is very difficult to obtain for most snakes because of their secretive habits and low population densities, one population of European Adders (*Vipera berus*) in Sweden has been studied in this way. As predicted, selection for large body size is intense both in males (because of combat) and in females (because of fecundity selection) (Madsen and Shine, 1992a; Fig. 2.1). Interestingly, the intensity of selection for larger body size in males varies considerably from year to year, depending on the operational sex ratio. In years when only a few reproductive females are available for a large group of males, attempted courtships and matings are often interrupted by the arrival of other males, and hence success in combat is an important determinant of male reproductive success. In years when more females reproduce, many matings occur without prior interruption or combat, and hence smaller males may be almost as successful as larger animals (Madsen and Shine, 1992c; Fig. 2.2).

These data suggest that the operational sex ratio may be of general importance in determining the intensity of selection for larger body sizes in males. Thus, for example, habitats (such as trees or water) or locomotory modes (such as swimming or sidewinding) that make it difficult for a male to track a female using chemosensory cues may reduce the probability that two males will encounter each other near a receptive female. Low population densities may enhance this effect. If these factors act to reduce the importance of

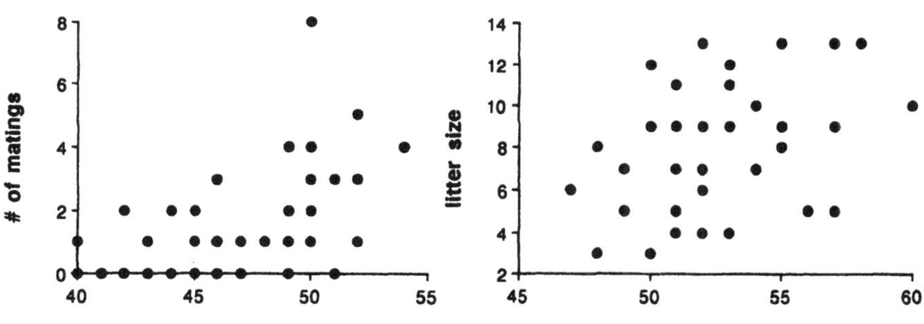

Figure 2.1 Body size influences reproductive success in European Adders (*Vipera berus*) of both sexes, but more so for males than for females. Nonetheless, females grow larger than males, probably because of delayed maturation due to high fecundity-independent costs of reproduction in this sex. (Redrawn from Madsen and Shine, 1992a.)

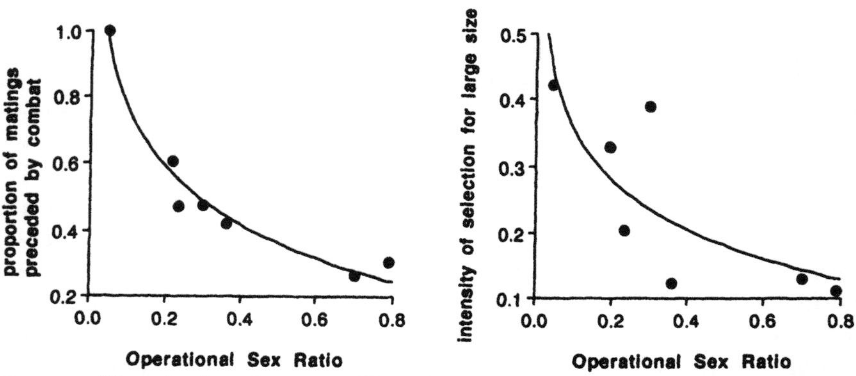

Figure 2.2 The selective advantage of large body size in male European Adders (*Vipera berus*) depends upon the sex ratio of reproductive adults in the local population (ratio of fertilizable females to sexually active males, or operational sex ratio, OSR). In years when there are many more males than females (low OSR), most matings are preceded by male–male combat and thus only the largest males obtain matings. (Redrawn from Madsen and Shine, 1992c.)

male–male combat in the mating system, the intensity of selection for large body size in males may also be lower. Given the advantages of earlier maturation to lifetime reproductive success in males, smaller male size is then likely to enhance male fitness. In keeping with this hypothesis, the only rattlesnake that moves about mainly by sidewinding (*Crotalus cerastes*) is the only one in which males average smaller than females, despite the existence of male–male combat (Klauber, 1956).

There are surprisingly few data on determinants of male reproductive success in any snakes other than European Adders. The available data suggest that larger body size probably does not enhance male reproductive success in File Snakes (Shine, 1986) or Garter Snakes (Joy and Crews, 1988), but a laboratory study of "mating balls" in Grass Snakes (*Natrix natrix*) showed that larger males were more successful because of their greater ability to wrestle tails of rival males out of the way (Madsen and Shine, 1993c). Hence, male–male competition in male snakes can take many forms, and the highly ritualized "combat dance" of viperids may be only one of a series of strategies employed by males. Some of the attributes thereby favored by sexual selection (e.g., mate-searching ability, copulatory plugs, and pheromonal confusion of other males, Devine, 1975, 1984; Mason and Crews, 1985; Duvall et al., 1992) may be unaffected by male body size, but physical battles between rival males are likely to favor larger size, and such battles may be more widespread and more diverse in form than is currently appreciated.

The idea that male–male combat has favored the evolution of larger body size in males has also been tested by another method—one that is less direct but has the virtue of using information from a wide variety of phylogenetic lineages. This method attempts to infer the action of sexual selection from associations between male–male combat and SSD in interspecific comparisons. If other selective forces on both male and female body sizes are independent of the mating system, then we might expect that males would tend to be larger *relative to conspecific females* in species with male–male combat than in species where such combat does not occur. This prediction is strongly supported—males tend to grow at least as large as females in taxa with male–male combat, whereas females tend to be significantly larger than males in species where such combat is not known to occur (Shine, 1978b). Exceptions to this generalization, especially species with male–male combat but males smaller than females, may be due to high fecundity-independent costs of reproduction favoring delayed maturation in females (Madsen and Shine, 1993b).

One problem with earlier comparative tests of the association between male–male combat and SSD in snakes is that such tests ignored the effects of phylogenetic conservatism. More recently developed techniques of analysis (Pagel and Harvey, 1989) take this factor into account by superimposing the values of SSD for each species onto a phylogeny so that phylogenetic *changes* in SSD can be calculated, and compared to concurrent shifts in the mating system. When this is done, the evolution (or loss) of male–male combat in snakes is significantly associated with shifts in SSD, and thus the correlation between these two variables among snakes is not simply an artifact of phylogenetic conservatism (Shine, 1993).

Other hypotheses. Many other hypotheses have also been invoked to explain the evolution of sexual size dimorphism in snakes. For example, Schwaner and Sarre (1988) discuss a number of alternative ideas and interpret the fact that male Black Tiger Snakes (*Notechis ater*) are larger than females as an adaptation to the highly seasonal environment of this species. Because female Black Tiger Snakes do not feed while they are gravid (in midsummer, when food is most abundant), they must remain active during cool weather in spring and autumn. At these times, a smaller body size may be advantageous because it enables the snakes to reach operating temperatures more rapidly. Similarly, Forsman (1991b) suggested that variations in the degree of SSD among island European Adders (*Vipera berus*) were related to prey availability, with females growing much larger than the "optimum size for survival" (and hence, male size) in areas where this optimum size was small, but remaining closer to this optimum size when it was larger. This hypothesis offers a possible explanation for the general allometry of SSD in snakes (see above), because it predicts an increase in male size (relative to female size) in areas where the mean adult body size is greater.

As well as varying among species—often, congeneric species (Shine, 1978, 1993b; Fitch, 1981)—SSD in snakes may vary significantly through the geographic range of a species. The best examples of geographic variation in SSD come from island populations of Australian Black Tiger Snakes (*Notechis ater,* Schwaner, 1985), European Adders (*Vipera berus,* Forsman, 1991) and Grass Snakes (*Natrix natrix,* Madsen and Shine, 1993d). In all three cases, statistical analysis of the data on body lengths (two-factor analysis of variance with gender and locality as the factors) shows a significant interaction (gender X locality) term, testifying to significant differences among populations in male size relative to female size (I thank T. D. Schwaner for raw data on Tiger Snake sizes for analysis). All three cases involve significant geographic variation in food availability. In the case of the Grass Snakes, growth rates of captive offspring from the two populations showed that the body-size differences in the field were entirely attributable to phenotypic plasticity (Madsen and Shine, 1993d). However, other studies have documented microevolutionary changes in SSD through time within a single population (Madsen and Shine, 1992a), and many factors other than food availability clearly influence the degree of SSD in a population.

Sex Differences in Body Shape

Because of their simplified external morphology, snakes display relatively few attributes of body shape with the potential to differ between conspecific males and females. Nonetheless, significant dimorphism has been documented in relative head size, head shape, relative tail size, and in the relationships between (1) body length and mass and (2) body mass and abdominal volume.

Sexual dimorphism in head length relative to body length is widespread among snakes and may be related to sex differences in diets (Shine and Crews, 1988; Shine, 1991a). The most obvious alternative interpretation—that larger head size in males evolves because of its advantages in male–male combat—is untenable for most snakes because (1) it is usually females, not males, that have the larger heads, and (2) even where male–male combat occurs in snakes, it only rarely involves biting (Carpenter and Ferguson, 1977). The proximate mechanism responsible for the sex difference in relative head sizes is known for at least one species. In the Red-sided Garter Snake

(*Thamnophis sirtalis parietalis*), the dimorphism is due to gonadal hormones. Newborn male Red-sided Garter Snakes castrated soon after birth developed heads almost as large as their sisters, whereas castrated males with interperitoneal androgen implants developed "normal" male-sized heads (Shine and Crews, 1988).

Sexual dimorphism in head length relative to body length has been found in newborn offspring of several taxa and so is not a result of sex differences in prey types or feeding rates of young snakes. I found statistically significant differences in head size relative to body length (using analysis of covariance) in 47% of 114 species that I examined from seven families (Shine, 1991a). The direction and degree of sex differences in relative head size show great phylogenetic lability among snakes and vary considerably even among widespread populations of single species. In several species that have been studied in detail, the head-size dimorphism is correlated with dietary differences between the sexes (see Shine, 1991a for a review).

If one sex evolves a larger head primarily as an adaptation to ingesting larger prey items, then one might predict that male and female heads would differ in shape as well as size, because the primary difference between the sexes should be in the size of feeding structures rather than in other parts of the head. Detailed analysis of head shape in four species of snakes supported this prediction—feeding structures were disproportionately larger in the sex with the larger head size in three of the four species examined (Camilleri and Shine, 1990). Male Grass Snakes (*Natrix natrix*) have more teeth than do females on the maxilla, palatine, pterygoid, and dentary (Thorpe, 1973). Bizarre sexually dimorphic rostral protuberances influence head shape in the Madagascan colubrid genus *Langaha* (see below). These data, together with the recent documentation of significant intraspecific geographic variation in relative head size related to variation in prey sizes (Forsman, 1991a), suggest that snakes may provide ideal model systems in which to investigate the ways in which natural selection produces niche divergence between closely related animals.

Other, quite bizarre differences in head shape between the sexes have also been reported in snakes. Males have larger eyes (relative to head size) than do females in the Malayan tree snake *Ahaetulla picta* (Kopstein, 1941) and the South American viperid *Bothrops moojeni* (Leloup, 1975). In the Central American tree snakes *Imantodes cenchoa* and *I. inornatus*, males have longer tongues (relative to head length) than do conspecific females (Myers, 1982). Male cobras of one species (*Naja naja*) but not another (*N. melanoleuca*) have longer fangs than do females of the same body length (Bogert, 1943), although it remains possible that this difference may be a consequence of dimorphism in head size. Higher venom yields in male than in female Eastern Tiger Snakes (*Notechis scutatus*) of the same body size (Weiner, 1960) could have many explanations, including differences in head size (and hence, volume of the venom glands), feeding frequency, or willingness to expel venom.

Tail length relative to snout–vent length also shows significant sexual dimorphism in many types of snakes, with males typically having longer tails (Klauber, 1943, 1956; Clark, 1967). Tails of males and females also differ in shape, as well as size, in some species (especially sea snakes, Wall, 1921; Pernetta, 1977). King (1989) has recently identified three alternative hypotheses for the evolution of sexual dimorphism in tail length among snakes: (1) morphological constraint (male tails must contain hemipenes and retractor muscles); (2) female reproductive output (fecundity selection favoring more posterior placement of the vent); and (3) male mating ability (sexual selection for courtship success). King's review of data on 56 colubrid genera showed that tail-length dimorphism was more extreme in taxa with relatively short tails, that hemipenes and retractor muscles occupied a greater proportion of the tail in taxa with relatively short tails, and that the dimorphism in tail length was more male-biased in taxa in which body-size dimorphism was more female-biased. These results support both the morphological constraint and female reproductive output hypotheses. Although King's analysis is open to the criticism that it may be confounded by phylogenetic biases (for example, natricine colubrids tend to have highly female-biased body-size dimorphism and highly male-biased tail-length dimorphism), phylogenetically based analysis of these data confirms that the significant correlations reported by King (1989) reflect functional relationship rather than common inheritance. In particular, evolutionary shifts in the degree of dimorphism in tail length have been accompanied by concurrent shifts in body-size dimorphism and in tail length relative to body length (Shine, unpublished analyses).

In some but not all snake species, females are more heavy-bodied than are males (Boettger, 1888; Blanchard, 1931; Bergmann, 1942; Kaufman and Gibbons, 1975; Feaver, 1977; Semlitsch and Moran, 1984; Macartney et al., 1990). Such data are difficult to interpret because of profound seasonal shifts in mass–length relationships of females due to reproduction. Female snakes often produce a mass of eggs equaling one-third of their own body mass (Seigel and Ford, 1987) and hence may be grossly swollen prior to parturition. Such a modification in shape may be biologically significant. For example, gravid Checkered Garter Snakes (*Thamnophis marcianus*) crawl more slowly than do nongravid animals, and the decrement in crawling speed is proportional to the relative clutch mass (Seigel et al., 1987). Other differences in body shape between male and female snakes may be maintained outside of the reproductive season. For example, female Arafura File Snakes (*Acrochordus arafurae*) are much more heavy-bodied than their brothers at birth and maintain this difference throughout their lives (Shine, 1986). Many records suggest that a similar but less pronounced difference in body shape occurs in a wide array of both terrestrial and aquatic snake taxa (see above references). The less gracile shape of females in such species may be a result of fecundity selection—their stouter shape may provide more space for reproductive materials within the body cavity. In keeping with this interpretation, measurements of abdominal volume relative to body mass show that female snakes consistently exceed conspecific males in this respect (Shine, 1992).

Many other, usually minor, shape differences between males and females have been reported. For example, female *Natrix natrix* have larger cloacal glands (Thorpe, 1973). Sensory pits on the postoculars and temporals also differ between the sexes in this species (Thorpe, 1973). Other shape differences between the sexes may be influenced by dimorphism in the size and position of internal organs, many of which appear to be displaced either anteriorly or posteriorly in females in such a way as to leave more clear space for developing eggs or embryos in the midbody region (Rossman et al., 1982). Body musculature may also differ between the sexes. Schwaner and Sarre (1988) used an ingenious device to measure muscular strength of Black Tiger Snakes (*Notechis ater*). Although adult males showed similar mass–length relationships as did nongravid females, the males pulled on a rope with about twice the force of females. As Schwaner and Sarre (1988) point out, there are several possible explanations for this difference, and it merits further investigation.

Sex Differences in Scalation

The taxonomic literature on snakes abounds with examples of minor differences in scalation between conspecific males and females, but the biological significance of such differences is difficult to evaluate. Many are probably related closely to the shape dimorphisms discussed above: for example, relative tail length will affect the numbers of ventral versus subcaudal scales, and overall shape (heavy-bodied or gracile) may influence the number of midbody scale rows or the relative position of scale row reductions (Boettger, 1888; Thorpe, 1975; Thomas and Dixon, 1976; Kminiak and Kalúz, 1983; Ota et al., 1986; Dmi'el et al., 1990). The general trend toward more ventrals in females and more subcaudals in males (Pope, 1935; de Silva, 1969) has been attributed to a simple anterior displacement in the position of the vent (Boulenger, 1913). However, Pope (1935) has argued against such simplistic interpretations, suggesting instead that fecundity selection has favored an increase in maternal abdominal volume and that this has been achieved by an increase in the number of ventrals. Such anatomical modifications may also be responsible for sex differences in the position of the umbilical scar (Edgren, 1958; Platt, 1969; Houston, 1992). Rather than consider the myriad examples of slight differences of this kind, I choose to focus on some more extreme developments of dimorphism in scalation. These take three main forms: keeled or tubercular scales (usually on the chins or cloacal regions of males), cloacal spurs in boids, and rostral protuberances.

Two main types of sexually dimorphic keeled scales occur in males of a number of otherwise smooth-scaled snake species. The first type consists of tubercles or knobs under the male's chin (Table 2.1). Noble's (1937) experiments suggested that these tubercles have a sensory function, but this conclusion has been challenged (Kubie et al., 1978). Male snakes typically adpress their chins firmly to the female's dorsal surface during courtship, and the tubercles may function to stimulate the female during this activity or have some sensory function to facilitate pheromonal uptake (Gillingham, 1987).

The other main type of keeled scale occurs on the male's body rather than the head and is generally concentrated in the area surrounding the cloaca. Such scales (also seen in many lizards) tend to develop only in mature males and are seen in a variety of taxa (Table 2.1). Like the chin keels, they may function either in mate stimulation or in sensory reception. Another possibility is that the increase in friction due to this roughened surface enhances a courting male's ability to push the female's tail into position for successful intromission or push his rival's tail out of the way. Such rugosities are common in sea snakes and may be evident even prior to birth (Wall, 1921; Deraniyagala, 1955). In some hydrophiid sea snakes (*Lapemis hardwickei, L. curtus*), the roughened scales of the male extend over much of his body (Wall, 1921; Cogger, 1975). Surprisingly, they may be concentrated primarily in the anterior rather than posterior regions. Similar rugosities are present, although less well-developed, on some larger females (R. Shine, personal observation). The function of these keels remains unknown, although they might serve to maintain friction (and thus enhance continued contact) between intertwined combative males or courting pairs (Boettger, 1888). The carinate anterior mid-dorsal scales of males in the terrestrial colubrid *Drymarchon corais* are less easy to explain.

Most species of boid snakes possess small thornlike spurs on either

Table 2.1. Snake Species Reported to Show Sex Differences in Scalation

Taxon	Form of dimorphism	Authority
Colubridae		
Ahaetulla prasina	Supra-anal keels in males, not females	Pope (1935)
Aspidura, 5 spp.	Supra-anal keels in males, not females	Smith (1943)
Carphophis amoenus	Supra-anal keels in males, not females	Blanchard (1931)
Cerberus rynchops	Supra-anal keels in males, not females	Kopstein (1941)
Chironius carinatus	Vertebral scales more strongly carinate in males	Guibe (1953); do Amaral (1978)
Clonophis kirtlandii	Supra-anal keels in males, not females	Blanchard (1931)
Diadophis punctatus	Supra-anal keels in males, not females	Blanchard (1931)
Drymarchon corais	Supra-anal keels in males, not females	Layne and Steiner (1984)
Enhydris chinensis	Tubercles on chins of adult males, not on females	Pope (1935)
Enhydris chinensis	Supra-anal keels in males, not females	Pope (1935)
Enhydris plumbea	Tubercles on chins of adult males, not on females	Pope (1935)
Enhydris plumbea	Supra-anal keels in males, not females	Pope (1935)
Farancia abacura and *F. erytrogramma*	Supra-anal keels in males, not females	Blanchard (1931)
Fordonia leucobalia	Supra-anal keels in males, not females	Kopstein (1941)
Homalopsis buccata	Number of scale rows differ in some parts of body	Bergmann (1951)
Homalopsis buccata	Supra-anal keels in males, not females	Kopstein (1941)
Hypsiglena torquata	Supra-anal keels in males, not females	Blanchard (1931)
Langaha intermedia and *L. nasuta*	Nasal protuberance conical in male, broad and leaflike in female	Guibe (1953); Schmidt and Inger (1957)
Leimadophis flavilatus	Supra-anal keels in males, not females	Blanchard (1931)
Macrelaps microlepidotus	Tubercles on chins of adult males, not on females	Broadley (1982)
Macrelaps microlepidotus	Supra-anal keels in males, not females	Broadley (1982)
Macrelaps plumbicolor	Supra-anal keels in males, not females	Smith (1943)
Natrix chrysarga	Supra-anal keels in males, not females	Kopstein (1941)
Natrix natrix	Supra-anal keels in males, not females	Thorpe (1975)
Nerodia erythrogaster	Supra-anal keels in males, not females	Blanchard (1931)
Nerodia fasciata	Supra-anal keels in males, not females	Blanchard (1931)
Nerodia rhombifer	Tubercles on chins of adult males, not on females	Blanchard (1931)

(*Continued*)

Table 2.1. Snake Species Reported to Show Sex Differences in Scalation (Continued)

Taxon	Form of dimorphism	Authority
Colubridae (*cont.*):		
Nerodia rhombifer	Supra-anal keels in males, not females	Blanchard (1931)
Nerodia sipedon	Supra-anal keels in males, not females	Blanchard (1931)
Nerodia taxispilota	Supra-anal keels in males, not females	Blanchard (1931)
Nerodia spp.	Tubercles on chins of adult males, not on females	Blanchard (1931)
Opisthotropis katunensis	Supra-anal keels in males, not females	Pope (1935)
Opisthotropis kuaturensis	Tubercles on chins of adult males, not on females	Pope (1935)
Opisthotropis latouchii	Tubercles on chins of adult males, not on females	Pope (1935)
Opisthotropis spp.	Supra-anal keels in males, not females	Smith (1943)
Pseudoxenodon fukiensis	Supra-anal keels in males, not females	Pope (1935)
Pseudoxenodon karlschmidti	Supra-anal keels in males, not females	Pope (1935)
Pseudoxenodon macrops	Supra-anal keels in males, not females	Smith (1943)
Pseudoxenodon striaticaudatus	Supra-anal keels in males, not females	Pope (1935)
Regina grahamii	Supra-anal keels in males, not females	Blanchard (1931)
Rhabdophis subminiatus	Supra-anal keels in males, not females	Kopstein (1941)
Rhabdophis tigrinus	Supra-anal keels in males, not females	Pope (1935)
Rhadinaea decorata	Supra-anal keels in males, not females	Blanchard (1931)
Rhadinaea flavilata	Supra-anal keels in males, not females	Myers (1967)
Seminatrix pygaea	Supra-anal keels in males, not females	Blanchard (1931)
Sinonatrix aequifasciata	Tubercles on chins of adult males, not on females	Pope (1929)
Sinonatrix annularis	Tubercles on chins of adult males, not on females	Pope (1935)
Sinonatrix percarinata	Tubercles on chins of adult males, not on females	Pope (1929)
Thamnophis butleri	Supra-anal keels in males, not females	Blanchard (1931)
Thamnophis elegans	Supra-anal keels in males, not females	Blanchard (1931)
Thamnophis ordinoides	Supra-anal keels in males, not females	Blanchard (1931)
Thamnophis radix	Tubercles on chins of adult males, not on females	Smith (1943)
Thamnophis sauritus	Supra-anal keels in males, not females	Blanchard (1931)
Thamnophis sirtalis	Supra-anal keels in males, not females	Blanchard (1931); Harrison (1933)
Tretanorhinus nigroluteus	Tubercles on chins of adult males, not on females	Henderson and Hoevers (1979)
Tretanorhinus variabilis	Tubercles on chins of adult males, not on females	Henderson and Hoevers (1979)

Trirhinopholis styani	Supra-anal keels in males, not females	Pope (1935)
Tropidoclonion lineatum	Males develop supra-anal keels at smaller size than females	Force (1936)
Tropidoclonion lineatum	Supra-anal keels in males, not females	Blanchard (1931)
Elapidae		
Micrurus, 12 spp.	Supra-anal keels in males (absent in 25 other spp.)	Campbell and Lamar (1989)
Micrurus spp.	Supra-anal keels in males, not females	Blanchard (1931)
Hydrophiidae		
General	Males have fewer ventrals, fewer scale rows around neck, longer tails, and thinner, spinier scales on venter	Boetgger (1888)
Enhydrina schistosa	Scale tubercles are larger in males	Wall (1921)
Hydrophis cyanocinctus	Scale tubercles are larger in males	Wall (1921)
Hydrophis fasciatus	Scale tubercles are larger in males	Wall (1921)
Hydrophis spiralis	Scale tubercles are larger in males	Wall (1921)
Lapemis curtus	Scale tubercles are larger in males	Wall (1921)
Lapemis hardwickei	Scale tubercles are larger in males	Cogger (1975)
Microcephalophis cantorus	Scale tubercles are larger in males	Wall (1921)
Microcephalophis gracilis	Scale tubercles are larger in males	Wall (1921)
Pelamis platurus	Scale tubercles are larger in males	Wall (1921)
Viperidae		
Amblycephalus moellendorffi	Males with tubercular scales on chin	Pope (1935)
Bitis caudalis	Subcaudals keeled in females, not in males	Branch (1988)
Bitis cornuta	Males have larger, more prominent horns over eyes	Mehrtens (1987)
Bitis cornuta	All subcaudals keeled in females, only at tip of tail in males	Branch (1988)
Bitis schneideri	All subcaudals keeled in females, only at tip of tail in males	Branch (1988)
Hypnale hypnale	Terminal scute longer, turned slightly upward in male	Wall (1921); Smith (1943)
Trimeresurus wiroti	Males more rugose (upturned scales)	Mehrtens (1987)

side of the cloacal aperture, and these spurs are typically larger in males than in conspecific females (McDowell, 1979; Shine and Slip, 1990). Indeed, they are completely lacking in females of some species, and in a high proportion of females in some others (McDowell, 1979; Harlow and Shine, 1992). In the Pacific boa *Candoia carinata,* females on some islands have small spurs whereas females on other islands lack them entirely (McDowell, 1979). The male's spurs are raked against the female's body and tail base during courtship and may serve to stimulate receptivity on her part (i.e., induce cloacal gaping) or actually maneuver her tail out of the way to facilitate intromission (Gillingham 1987; Slip and Shine, 1988). Combative males may also use their spurs to gouge away at their rivals (Carpenter et al., 1978; Barker et al., 1979). Interestingly, males also retain spurs in another snakelike squamate lineage, the pygopodid lizards (Greer, 1989).

Perhaps the most remarkable sexual dimorphisms in scalation are the rostral appendages of the Madagascan colubrid genus *Langaha* and the hydrophiid sea snake *Emydocephalus* (Table 2.1). Both sexes of *Langaha* have nasal protuberances, but the male's is straight whereas the female's is foliose (Guibe, 1948). The viperid *Hypnale hypnale* shows a similar but less extreme dimorphism, whereby the terminal rostral scale of the male is longer and turned slightly upward (Wall, 1921). The function of the nasal protuberance of *Langaha* and *Hypnale* is unknown (perhaps it breaks up the snake's outline as the snake lies in ambush for prey?). Recent studies show that male *Emydocephalus annulatus* use their rostral spines to repeatedly prod the female's lateral surface during courtship (Michael Guinea, personal communication), so that sexual behavior rather than sex-specific foraging (Heatwole et al., 1978) seems to be the likely selective force for their bizarre sexual dimorphism.

Sex Differences in Coloration

Although snakes do not show colors as vivid as those seen in many sexually dichromatic lizard species, sex differences in color are more common among snakes than has generally been appreciated (Table 2.2). Many of the sex differences are relatively subtle and involve minor differences in background color or pattern. However, some are more obvious. For example, males of the New World colubrid *Manolepis putnami* are tan in color with a dark brown mid-dorsal stripe and an immaculate venter, whereas females are gray with a light brown stripe and a heavily stippled venter. Male *Hydrodynastes gigas* (Brazilian False Water Cobras) are strongly blotched, whereas females are not. Male European Adders (*Vipera berus*) are usually

TABLE 2.1. Snake Species Reported to Show Sex Differences in Color

Species	Male coloration	Female coloration	Authority
Boidae			
Chondropython viridis	Blue phase uncommon	Blue phase more common	S. Hammack (personal communication)
Eryx miliaris	Darker color	Lighter color	Trutnau (1986)
Colubridae			
Atretium schistosum	Brighter red dorsal line	Less bright dorsal line	Wall (1921)
Cerberus rynchops	Yellow knobbed keels on para-anal scales	Not present	Kopstein (1941)
Coronella austriaca	Brownish red belly, dorsally brown or red brown	Greyish brown belly, dorsally gray or brownish black	Noble (1937); Hellmich (1962); Trutnau (1986)
Dasypeltis scabra	Sexes differ in number or dorsal blotches	(see Male coloration)	Stevens (1973)
Dispholidus typus	Variable colors (black, brown, green, blue)	Usually brown	Broadley (1982); Mehrtens (1987); Branch (1988)
Drymarchon corais	Sexes differ in extent of red or creeam color about chin, throat, and cheeks (geographically variable)		Mount (1975); Moulis (1976)
Enhydris enhydris	Dark median stripe on tail extends to ventrals	Stripe does not extend to ventrals	Kopstein (1941)
Farancia erytrogamma	Hatchlings have smaller, less distinct, more variable spots under tail	Spots more distinct, less variable	Richmond (1954); Neill (1954, 1964); Gibbons et al. (1977)
Fordonia leucobalia	Keels on para-anal scales are yellow	Keels absent, no yellow pigmentation	Kopstein (1941)
Heterodon nasicus	Fewer dorsal blotches	More dorsal blotches	Platt (1969)
Holarchus chinensis	Less reddish	More reddish	Pope (1929)
Hydrodynastes gigas	Yellowish with dark brown or black blotches, venter yellow	Light brown with indistinct or no blotches, venter brown	Rheinhard and Vogel (1972); Trutnau (1986); Mehrtens (1987)
Lampropeltis getula	Lineate color phase more common	Lineate phase less common	Bartlett (1988)
Langaha intermedia	Back, abdomen uniform brown; sides and upper lips bright yellow	Gray-brown with transverse dark bands; abdomen and sides gray	Guibe (1953)
Liopeltis balioderus	Uniform gray or greenish brown	Brighter pattern, gray-brown with whitish spots	Kopstein (1941); Angel (1950)
Malpolon monspessulanus	Uniform coloration	More or less spotted	de Haan (1984)
Manolepis putnami	Tan, dark-brown mid-dorsal stripe, venter immaculate	Grayish; mid-dorsal stripe light brown in center; stippled venter	Werler and Smith (1952)
Natrix natrix	More lateral blotches	Fewer lateral blotches	Thorpe (1973)
Philodryas nattereri	Light orange on side of neck	Lacks orange on neck	L. Vitt (personal communication)

(Continued)

TABLE 2.2. Snake Species Reported to Show Sex Differences in Color (Continued)

Species	Male coloration	Female coloration	Authority
Colubridae (*cont.*):			
Philothamnus semivariegatus	Underside lighter than back	Underside same color as back	Loveridge (1958)
Psammodynastes pulverulentus	Generally shades of brown, often banded or striped	Darker	Kopstein (1941); Angel (1950); Rasmussen (1975); Leviton (1983)
Pseudoxenodon bambusciola	Uniform color	More vivid, contrasting color	Pope (1929)
Pseudoxenodon nothus	Yellow with no trace of red	Reddish neck	Maslin (1950)
Ptyas korros	Ventral scales yellow-green	Ventral scales white	Kopstein (1941)
Seminatrix pygaea	Four color patterns, two of which occur mostly in females	(see Male coloration)	Dowling (1950)
Sinonatrix annularis	More bands	Fewer bands	Pope (1929)
Telescopus dhara	Reddish, especially on cross bands and head	Gray or gray-brown; brown cross bands; head yellow-brown	Zinner (1985)
Thamnophis melanogaster	Red phase less common	Red phase more common	Gregory et al. (1983)
Trimorphodon vandenburghi	More body blotches	Fewer body blotches	Klauber (1940); Wright and Wright (1957)
Uromacer frenatus	Mid-dorsal stripe less clearly defined	Mid-dorsal stripe more clearly defined	Schwartz (1976)
Uromacer oxyrhynchus	Three color phases, one of which occurs only in females	(see Male coloration)	Schwartz and Henderson (1984)
Elapidae			
Micrurus fulvius	More bands on tail	Fewer bands on tail	Clark (1967); Quinn (1977)
Naja nivea	Spotted or speckled	Brown	Hewitt (1937); Broadley (1982)
Naja haje	Banded phase more common	Banded phase less common	Broadley (1968); Branch (1988)
Vermicella annulata	Sexes differ in number and relative width of rings		Cogger (1986)
Hydrophiidae			
Aipysurus laevis	Light brown	Blue-gray	Burns (1984)
Aipysurus pooleorum	Brownish	Purplish-brown	Storr et al. (1986)
Astrotia stokesii	Black cross bands	Gray dorsally	Krefft (1869)
Laticaudidae			
Laticauda semifaciata	More tail rings	Fewer tail rings	Tu et al. (1990)
Tropidophiidae			
Tropidophis melanurus	Black tails more common	Black tails less common	Grant (1957)

Viperidae			
Agkistrodon bilineatus taylori	Dark	Gray, more ornate pattern	Burchfield (1982)
Agkistrodon contortrix	More bands on tail	Fewer bands on tail	Quinn (1979)
Amblycephalus margartophorus	No nuchal markings	White or yellow nuchal collar or spot	Campden-Main (1970)
Hypnale hypnale	Markings less intense	Light with dark markings more intense and contrasting	Taylor (1950)
Bitis arietans	More brightly colored	Less brightly colored	Branch (1988)
Bitis caudalis	Variable, strongly patterned	Sandy or reddish-orange	Mehrtens (1987)
B. caudalis	More brightly colored	Less brightly colored	Branch (1988)
Bothrops asper	Neonates have yellow tail tips	Tail tip not yellow	Tryon (1985); Solórzano and Cerdas (1989)
Bothrops atrox	Juveniles have yellow tail tips	Tail tip not yellow	Burger and Smith (1950); Neill (1960); Hoge and Federsoni (1977)
Bothrops moojeni	Less defined markings dorsally; spotted labials; yellow tail tip persists for 18 months from birth	Clear markings dorsally; yellow-white labials; yellow tail tip fades within six months	Leloup (1975)
Bothrops ophryomegas	Darker background dorsally	Lighter	Solórzano et al. (1988)
Cerastes vipera	Tail same as body, terminal scute black	Black tail	Marx (1958); Schnurrenberger (1959)
Crotalus adamanteus	More bands on tail	Fewer bands on tail	Klauber (1956)
Crotalus atrox	More bands on tail	Fewer bands on tail	Quinn (1979)
Crotalus horridus	Usually yellow	Usually black	Klauber (1956); Mehrtens (1987); but see Schaeffer (1969)
Crotalus lepidus klauberi	Green dorsally	Gray dorsally	Jacob and Altenbach (1977)
Crotalus viridis	More bands on tail	Fewer bands on tail	Quinn (1979)
Crotalus viridis lutosus	Chocolate-brown blotches on body	Light-colored blotches on body	L. Vitt (personal communication)
Porthidium picadoi	Neonates with a few points on top of head	Neonates with two lines on top of head	Solórzano (1990)
Sistrurus catenatus	More bands on tail	Fewer bands on tail	Quinn (1979)
Sistrurus miliarius	Lighter background color	Darker background color	Clark (1963)
Trimeresurus albolabris	Pale lateral stripes in some	Pale lateral stripe absent	Campden-Main (1970)
Trimeresurus puniceus	Darker with more distinct transverse bands	Lighter, less distinctly banded	Kopstein (1941); Angel (1950)
Trimeresurus stejnegeri	Stripes white to yellow to brick red; additional red stripe	Stripes white to yellow	Pope (1929); Mao (1962); Campden-Main (1970)
Trimeresurus wagleri	Retain juvenile (green) color past maturity	Disruptive pattern with bright yellow, black, blue	Vogel (1991)
Trimeresurus wiroti	Dichromatic at hatching	(see Male coloration)	Mehrtens (1987)
Vipera ammodytes	Pale gray background, zigzag dorsal stripe dark gray or black	Grayish or reddish brown, stripe brown	Stewart (1971); Street (1979); Mehrtens (1987)
Vipera aspis	Lighter, more conspicuously colored	Darker	Noble (1937); Angel (1950); Street (1979); Naulleau (1973b)

(Continued)

TABLE 2.2. Snake Species Reported to Show Sex Differences in Color (Continued)

Species	Male coloration	Female coloration	Authority
Viperidae (*Cont.*)			
Vipera berus	Dark eyes, black subcaudals; dorsum lighter and more brightly patterned	Yellow eyes, tip of tail gray-brown; dorsum darker and less contrasting pattern	Noble (1937); Angel (1950); Stewart (1971); Kheruvimov et al. (1977); Andrén and Nilson (1981); Belova (1982); Madsen and Shine (1992c)
Vipera kaznakovi	Bright colors as juveniles, becoming melanic with white or yellow stripes	Drab as juvenile, eventually melanic with yellow or orange stripes	Kretz (1971)
Vipera latastei	Gray or brown	Often reddish-brown	Stewart (1971)
Vipera palastinae	Venter usually pale gray	Venter usually pale yellow or speckled	Mehrtens (1987)
Vipera xanthina	Lighter color, less distinct zigzag line on dorsum	Darker color and marked white between zigzag	Nilsen and Andrén (1986)

more brightly colored than their mates, with more strongly contrasting dorsal zigzag markings—especially during the mating season. Some of the more subtle differences include sex biases in the frequency of different morphotypes (e.g., *Naja haje, Vipera berus, Lampropeltis getula, Seminatrix pygaea, Thamnophis melanogaster*). The African Boomslang (*Dispholidus typus*) offers perhaps the most remarkable example—females are generally brown throughout the range of this species, whereas males occur in a wide array of color morphs (e.g., green, blue, black) that differ among areas. African Boomslangs from rainforest habitats tend to show little dichromatism (both sexes are brown), whereas males from savanna populations are often vivid green dorsally (T. Madsen, personal communication).

Many other examples of snake dichromatism are much less obvious and their ecological significance (if any) difficult to discern. For example, why should newly hatched male *Farancia erytrogramma* have smaller, less distinct spots under their tails than do their sisters, and why should this dichromatism occur in some populations but not others? Why should eye color differ between male and female vipers (Naulleau, 1973a)? Sex differences in the number of bands on the tail (as in *Laticauda semifasciata, Crotalus atrox,* etc.) are presumably a simple consequence of the longer tail in males (see above), but many examples of dichromatism cannot be easily explained. Sex-specific retention of juvenile (green) coloration in the arboreal viper *Trimeresurus wagleri* may reflect an ontogenetic shift in habitat use and a size-related change from mimicry of a noxious caterpillar by small animals (juveniles and adult males) to crypsis through a disruptive color pattern in the larger adult females (Vogel, 1991). There are few consistent patterns among the published examples of sexual dichromatism in snakes (Table 2.2). Many different phylogenetic lineages and adaptive "types" (arboreal, terrestrial, aquatic, fossorial) are represented. There seems to be no consistent tendency for either sex to be more brightly colored than the other. Indeed, males of one species (*Vipera kaznakovi*) are reported to be more brightly colored than females as juveniles, but less brightly colored than females as adults (Kretz, 1971). Another problem with interpretation is the reliability of the data in Table 2.2. Most of these records refer to nonquantified statements by authors, and some may be based on only a few captive animals.

One of the most consistent results from my survey of sexual dichromatism in snakes was that this characteristic shows a high degree of phylogenetic and geographic lability. In several cases, the dichromatism is restricted to only one part of a species' range, with monomorphism the rule in other areas (e.g., *Crotalus lepidus, Vipera berus, V.*

xanthina, Dispholidus typus, Farancia erytrogramma, Drymarchon corais, Malpolon monspessulanus, Agkistrodon bilineatus). Similarly, I found many cases where a strongly dichromatic taxon was closely related to a monomorphic species (e.g., within the genera *Vipera, Bothrops, Cerastes*). Seasonal changes in coloration are also common in snakes (Banks, 1981) and may influence the degree of sexual dichromatism. For example, European Adders (*Vipera berus*) court and mate in springtime, soon after the males have sloughed. Hence, males are brightly colored at this time, and the sexual dichromatism is at its most intense. In contrast, no dichromatism is evident in the related *V. aspis,* a species that commences mating in autumn, without a preliminary slough by males. The bright colors of courting male *V. berus* may function to reduce the vulnerability of these active snakes to predation, by inducing "flicker fusion" in the visual systems of vertebrate predators through the rapid movement of black-and-white bands (Shine, 1980; Shine and Madsen, in preparation).

The adaptive significance of sexual dichromatism in snakes remains obscure. The hypothesis that these colors have evolved as sexual signals is unconvincing, because (1) most snakes seem to rely on chemoreception rather than vision in locating mates and recognizing rivals, so that elaborate visually oriented challenge or courtship displays are unknown (Carpenter and Ferguson, 1977); (2) anatomical studies on the eye suggest that many kinds of snakes are probably unable to distinguish between colors (Engelmann and Obst, 1981); and (3) only rarely does the dichromatism take a form where it could be displayed in social interactions. Of course, sexual dichromatism in snakes may have evolved for many reasons, and epigamic or intrasexual selection may well have been involved in some lineages. Nonetheless, the relatively subtle sexually dichromatic features of snakes offer a striking contrast to the vivid patches of sex-specific color seen in many lizard species. Lizards probably show more subtle dichromatisms as well, similar to those seen in snakes (e.g., *Pedioplanis breviceps,* Branch, 1988; *Uta stansburiana,* V. deMarco, personal communication), but these may have generally been overlooked because of the prevalence of intense sexually selected dichromatisms in this group. Because some snake genera (notably *Vipera*) incorporate both dichromatic and nondichromatic species (and sometimes subspecies), they offer ideal material for a comparative phylogenetically based analysis of the evolution and loss of dichromatism.

Sex Differences in Ecology

In many species of snakes, males and females differ in important ecological characteristics. Many of these differences are related to the disparity in body sizes between the sexes. For example, adult male Arafura File Snakes (*Acrochordus arafurae*) are small (mean = 105 cm SVL, 660 g) and forage in relatively shallow water for small fishes, whereas adult females are much larger (mean = 135 cm, 1.4 kg) and tend to feed on larger fishes in deeper water (Shine, 1986; Houston and Shine, 1993). I recently reviewed published cases of sex-based dietary differences in snakes, including records from acrochordids, colubrids, elapids, laticaudids, and viperids (Shine, 1991a). Many of these cases presumably reflect a simple tendency for larger predators to eat larger prey items, as is common in snakes (Mushinsky et al., 1982; Shine, 1991a; Arnold, Chap. 3, this volume), and the sex difference in diets is an inevitable consequence of the difference in body sizes. However, the existence of significant dimorphism in relative head sizes and shapes in many of these taxa (see above) suggests that the disparity in body sizes has been reinforced by natural selection for differential niche use. Analogous geographic variation in relative head size (Shine, 1991a) and shape (Forsman, 1991a) has also been documented recently.

Differences in habitat use, as noted above for *Acrochordus arafurae,* also seem to be important in many cases of sex-based ecological divergence in snakes. For example, the larger sex may forage in deeper water (as in *A. arafurae* and *Laticauda colubrina*) or on the ground rather than in the trees (as suggested by reports on *Telescopus fallax* and *Trimeresurus wiroti,* Shine, 1991a; see also Manjarrez and Macias Garcia, 1991). In general, differences in body size brought about by fecundity selection (for larger females) or sexual selection (for larger males) may then be reinforced (or opposed) by natural selection working on foraging abilities of the two sexes. The same pattern of reinforcement of an original sexually selected difference may apply to other sex-specific foraging morphologies. For example, the bright yellow tail tips of newborn males of some *Bothrops* species may function to enhance the effectiveness of caudal luring, and may have evolved because existing differences in feeding rates, tail lengths, or caudal mobility between the sexes made bright caudal coloration advantageous for males but not females. Hence, many of the sexually dimorphic attributes of snakes are likely to reflect a combination of selective forces.

Thermal relationships with the environment may also differ between male and female snakes. The larger sex will heat and cool more slowly because of its higher thermal time constant, and may have greater physiological control over rates of temperature change (Hillman, 1969). Also, reproduction may profoundly affect feeding rates, movement patterns, habitat selection, and thermal preferenda in both sexes (Naulleau, 1979; Gibson and Falls, 1979; Reinert and

Kodrich, 1982). In many viviparous species, females reduce feeding rates during the latter part of gestation and spend much of their time basking (Shine, 1979; Schwaner and Sarre, 1988; but see Shine and Lambeck, 1990). Laboratory studies on lizards and snakes confirm that higher basking rates accelerate embryogenesis and permit earlier parturition (Naulleau, 1986; Schwarzkopf and Shine, 1992). Some of the color differences between male and female snakes may reflect adaptations of thermal reflectance to facilitate the maintenance of high and relatively constant body temperatures. Darker colors enhance heating rates (Gibson and Falls, 1979), and melanistic morphs are more common in females than in males in several snake species (Table 2.2).

Another ecologically relevant sex difference, possibly widespread although rarely described, is a consistent difference in general behavior and in response to stress. For example, Leloup (1975) noted that male *Bothrops moojeni* were more "nervous" than females and less prepared to feed in captivity. This kind of difference may influence activity patterns and feeding rates in the field as well as in the laboratory (Gibbons and Semlitsch, 1987). Feaver (1977) noted that male *Nerodia sipedon* stopped feeding in midsummer, despite an abundant food supply. Differences in activity levels may also explain some of the consistent differences between males and females in the size of the home range, especially during the breeding season (see review by Gregory et al., 1987). Movement patterns, activity levels, and dichromatism may also influence vulnerability to predation. Tail-break frequencies (and hence, it has been inferred, the incidence of unsuccessful predation attempts) differ between the sexes in some snake populations (Bergmann, 1958) but not in others (Schwaner and Sarre, 1988).

Overview

Sexual dimorphism occurs in snakes from a wide variety of phylogenetic lineages and a considerable diversity of ecological "types," and involves a broad range of characteristics including morphology (size, shape, scalation), coloration, and ecology (habitat use, thermal selection, diet). Some of these differences can be attributed to specific selective pressures—such as sexual selection for large male body size in species with male–male combat—but interpretation is difficult for most characteristics. Undoubtedly multiple selective forces are at work, and nonadaptive processes such as genetic correlation between the sexes, or pleiotropic effects, may also exert substantial effects. Nonetheless, the available data are sufficient to dismiss any simple argument that invokes sexual selection as the sole agent for the evolution of sexual dimorphism in these animals. From what we know of the sensory abilities and mating systems of snakes, it is difficult to interpret most cases of sexual dichromatism or larger relative head size in females as sexually selected adaptations. Instead, sexual dimorphism in snakes appears to result from a complex interplay of intrasexual selection, fecundity selection, and natural selection on ecologically relevant attributes of morphology, color, and behavior.

Indeed, one major inference from my review is that male and female snakes often differ quite profoundly in their day-to-day activities and in their general ecological relationships. I predict that detailed investigations will show cases in which males of related sympatric species resemble each other more (in terms of body sizes, body shapes, activity patterns, thermoregulatory strategies, and foraging tactics) than they resemble females of their own species. These kinds of subtle but important ecological differences between the sexes may well have acted as significant selective forces for the evolution of further sex differences in size, shape, scalation, coloration, temperature selection, and so forth. Hence, natural selection may have acted to adapt each sex to a slightly different ecological niche and thus may have constrained or amplified sex differences that evolved initially through fecundity selection or sexual selection. This hypothesis suggests that features varying in a complex way at the level of the local environment (prey availability, thermal opportunities, predation pressures, etc.) should influence the evolution of dimorphic characteristics, and hence that the type and degree of dimorphism should vary considerably even among closely related populations or species. This prediction is strongly supported: perhaps the strongest and most consistent result from my survey of sexually dimorphic characters in snakes is that most of these characters display extreme phylogenetic lability and geographic (interpopulational) variation. Such geographic variation is difficult to reconcile with any explanations for the evolution of sexual dimorphism that do not invoke a role for local environmental features.

Summary and Future Research

Many authors have bemoaned the difficulties of studying snakes, especially in the field, and some have even suggested that the difficulties are so great that in many cases the attempt may not be worthwhile (Turner, 1977). My review suggests a much more optimistic prognosis—snakes may be ideally suited to studies on sexual dimorphism. They have several advantages, such as the following:

1. A diversity of mating systems exists, imposing different intensities of intrasexual selection for large body size in males. Importantly, male body size seems to be almost irrelevant to reproductive success in many abundant and easily studied taxa (e.g., natricine colubrids). This situation removes an otherwise confounding factor.

2. Even in snake species where males fight with each other during the mating season, this fighting often does not involve biting—and hence is not likely to favor disproportionate enlargement of the jaws in males (as is commonly seen in lizards, Vitt and Cooper, 1985). Thus, significant sexual dimorphism in relative head size (which is widespread and common in snakes) cannot be attributed to sexual selection on males.

3. Snakes are gape-limited predators, generally relying upon occasional ingestion of relatively large prey items. In many taxa, the snake's head size may limit maximum ingestible prey size, so that sex-based divergence in prey sizes (exploitation of larger prey by one sex) necessarily involves evolutionary modifications to the trophic apparatus. Hence, sex-based dietary divergence is likely to result in measurable morphological divergence, a condition not satisfied in other types of animals that are not gape-limited predators. For example, male and female lizards could differ profoundly in diets but have identical trophic structures.

4. Sexually dimorphic characters in snakes show a great degree of variation among related species and even among conspecific populations, providing an ideal opportunity for tightly controlled comparisons and phylogenetically focused tests of explanatory hypotheses.

5. Because of their continued growth after maturation, both sexes include a wide range of adult body sizes. Although this factor introduces a complicating variable (compared to the relatively invariant body sizes of some other types of animals, such as birds, cephalopods, and terrestrial arthropods), it also means that there is generally a wide overlap in body sizes of males and females. This overlap facilitates direct comparison by factoring out any allometric effects.

Although snakes are thus well-suited to analyses of sexual dimorphism, this potential has rarely been exploited. Comparative analyses have attempted to infer past evolutionary processes from present-day patterns and have provided convincing evidence of the influence of the mating system on patterns of SSD. However, there are remarkably few data on the determinants of reproductive success in free-ranging snakes of either sex. Radiotelemetry has eased the considerable logistical impediments to such studies, and Thomas Madsen's pioneering studies on European Adders in Sweden have shown what can be achieved. His studies have been the first on snakes to measure the actual intensities of selection on body size and to quantify the dependency of reproductive tactics on the size of the reproducing animal. Such studies will not be feasible on most types of snakes, but surely there are other cases where analogous work would be both feasible and productive.

Ecological differences between the sexes have also attracted little attention. Most published compilations of dietary data (including my own!) do not separate data for the two sexes. In at least one case, a subsequent reanalysis revealed a hitherto-unsuspected sex difference in diets (Shine, 1991a). North American Garter Snakes (*Thamnophis*) are likely to be ideal for work of this kind, because they are abundant and often highly dimorphic both in body sizes and in relative head sizes. Some geographically widespread Garter Snakes show significant intraspecific variation in the degree of dimorphism in both of these variables (Shine, 1991a), and hence would provide excellent study systems.

Research to date has focused mostly on adult snakes, and much more remains to be learned about juveniles and about the determinants of sexual maturation. Sexual bimaturism and differential growth rates in juveniles are clearly major influences on adult SSD but have attracted little attention. Similarly, performance measures have rarely been attempted but have great potential. Studies such as those of Schwaner and Sarre (1988) on muscle strength, Burger (1989) on behavior of neonates, and Seigel et al. (1987) on crawling speeds of gravid female snakes warrant replication and more detailed analysis. The proximate basis of sexual dimorphism also warrants attention, because mechanistic models for the origin of a characteristic can often greatly illuminate (and sometimes, falsify) adaptationist models. For example, to what degree are the observed sex differences in relative head sizes among snakes a consequence of different growth rates (Forsman, 1991a) rather than direct hormonal effects (Shine and Crews, 1988)? Much remains to be learned, but we can say confidently that snakes are well-suited to studies of this kind and that most of these questions can be resolved by careful study.

In summary, female snakes are generally longer than conspecific males, and this trend is particularly evident in smaller species. The direction (and to a lesser extent, the degree) of sexual size dimorphism is basically set at maturation and is due mainly to sex differences in age at maturation, although sex differences in size at birth and in juvenile growth rates may also play a role. Males tend to be larger relative to females in species showing male–male combat than in species where no such combat occurs. In many taxa, females tend to be more heavy-bodied than males and to have shorter tails (relative to their snout–vent length), with corresponding minor differences in scalation.

Head size relative to body length also often differs between males and females with the larger sex often having a disproportionately larger trophic apparatus. In at least one species, the dimorphism in relative head sizes and in juvenile growth rates appears to result from the action of testicular hormones. In many different types of snakes, males have keeled scales in some parts of their bodies whereas females do not. These keels tend to be either under the male's chin, or in the supra-anal area. Male boid snakes tend to have larger cloacal spurs than do females. In a few species, the scale dimorphisms are much more extreme, sometimes involving rostral appendages that differ in form between the sexes. Color differences are also widespread but are generally relatively subtle. Sex differences in activity patterns, habitat use, and diets have been documented in several species. Most sexually dimorphic characters in snakes show great phylogenetic lability and often intraspecific geographic variation.

The available data suggest that fecundity selection, together with higher fecundity-independent costs of reproduction in females, is responsible for the fact that female snakes are generally larger than conspecific males. Fecundity selection may also explain the trend for females to be more heavy-bodied. Intrasexual selection, operating mainly through the advantages of larger body size in male–male combat, has resulted in the evolution of large male size in some taxa. Ecological divergence between the sexes may also have been an important selective pressure for the evolution of sexual dimorphism in snakes, especially with respect to characters such as coloration and relative head size.

Acknowledgments

I thank Richard Seigel and Joseph T. Collins for asking me to contribute to this volume, Geordie Torr for data entry, Michael Guinea for allowing me to cite his observations on sea snake courtship, and Monika Ostercamp for translating relevant papers. Thomas Madsen deserves special thanks. Our collaboration on the analysis of his remarkable data set on European Adders stimulated many hours of vigorous discussion on sexual dimorphism in snakes and greatly clarified my thinking on this topic. I stand in awe of Tom's enthusiasm, dedication, and ability. Lastly, I thank the Australian Research Council for financial support during the preparation of this manuscript.

Literature Cited

Andrén, C., 1986, Courtship, mating and agonistic behavior in a free-living population of adders, *Vipera berus*, *Amphibia-Reptilia*, 7:353–383.

Andrén, C., and G. Nilson, 1981, Reproductive success and risk of predation in normal and melanistic color morphs of the adder, *Vipera berus*, *Biol. J. Linn. Soc.*, 15:235–246.
Angel, F., 1950, *Vie et Mouers des Serpents*, Payot, Paris.
Atchley, W. R., C. T. Gaskins, and D. Anderson, 1976, Statistical properties of ratios, I. Empirical results, *Syst. Zool.*, 25:137–148.
Banks, C. B., 1981, Notes on seasonal color change in a western brownsnake, *Herpetofauna*, 13:29–30.
Barker, D. G., J. B. Murphy, and K. W. Smith, 1979, Social behavior in a captive group of Indian pythons, *Python molurus* (Serpentes, Boidae) with formation of a linear hierarchy, *Copeia*, 1979:466–477.
Bartlett, R. D., 1988, *In Search of Reptiles and Amphibians*, E. J. Brill, Leiden.
Bell, G., 1980, The costs of reproduction and their consequences, *Amer. Nat.*, 116:45–76.
Belova, Z. V., 1982, Color variations in the common viper *Vipera berus* L., *Moskov. Obshchet. Ispyt. Prior. Biull. Nov. Ser.*, 87:42–45.
Bergmann, R. A. M., 1951, The anatomy of *Homalopsis buccata*, *Proc. Koninkl. Neder. Akad. Weten. Ser. C*, 54:511–524.
Bergmann, R. A. M., 1958, The anatomy of *Natrix piscator*, *Biol. Jarb. (Koninkl. Natuur. Gen.)*, 26:77–99.
Bergmann, R. A. M., 1942, *Enhydrina schistosa*, *Natuur. Tijd. Nederl. Indie*, 102:9–12.
Berry, J. F., and R. Shine, 1980, Sexual size dimorphism and sexual selection in turtles (Order Chelonia), *Oecologia (Berlin)*, 44:185–191.
Blanchard, F. N., 1931, Secondary sexual characters of certain snakes, *Bull. Antivenin Inst. Amer.*, 4:95–104.
Boettger, O., 1888, Über äussere Geschlechtscharactere bei den Seeschlangen, *Zool. Anz.*, 284:395–398.
Bogert, C. M., 1943, Dentitional phenomena in cobras and other elapids with notes on adaptive modifications of fangs, *Bull. Amer. Mus. Nat. Hist.*, 81:285–360.
Boulenger, G. A., 1913, *The Snakes of Europe*, Methuen, London.
Bradbury, J. W., and M. B. Andersson, 1987, *Sexual Selection: Testing the Alternatives*, Wiley, Chichester.
Branch, W. R., 1988, *A Field Guide to the Snakes and Other Reptiles of Southern Africa*, Struik, Cape Town.
Broadley, D. G., 1968, A review of African cobras of the genus *Naja* (Serpentes, Elapidae), *Arnoldia*, 29:1–14.
Broadley, D. G., 1982, *FitzSimons' Snakes of Southern Africa*, Delta Books, Cape Town.
Burchfield, P. M., 1982, Additions to the natural history of the crotaline snake *Agkistrodon bilineatus taylori*, *J. Herpetol.*, 16:376–382.
Burger, J., 1989, Incubation temperature has long-term effects on behavior of young pine snakes (*Pituophis melanoleucus*), *Behav. Ecol. Sociobiol.*, 24:201–207.
Burger, J., and R. T. Zappalorti, 1988, Effects of incubation temperature on sex ratios in pine snakes: Differential vulnerability of males and females, *Amer. Nat.*, 132:492–505.
Burger, W. L., and P. W. Smith, 1950, The coloration of the tail tip of young fer-de-lance: Sexual dimorphism or adaptive coloration, *Science*, 112:431–433.
Burns, G. W., 1984, Aspects of population movements and reproductive biology of *Aipysurus laevis*, the olive sea snake, Doctoral thesis, Univ. New England, Armidale, New South Wales.
Camilleri, C., and R. Shine, 1990, Sexual dimorphism and dietary divergence: Differences in trophic morphology between male and female snakes, *Copeia*, 1990:649–658.
Campbell, J. A., and W. W. Lamar, 1989, *The Venomous Reptiles of Latin America*, Cornell Univ. Press, Ithaca.
Campden-Main, S. M., 1970, *A Field Guide to the Snakes of South Vietnam*, U.S. Natl. Museum, Washington, D.C.
Carpenter, C. C., and G. W. Ferguson, 1977, Variation and evolution of stereotyped

behavior in reptiles, in C. Gans and D. W. Tinkle, eds., *Biology of the Reptilia,* Academic, New York, pp. 335–554.
Carpenter, C. C., J. B. Murphy, and L. A. Mitchell, 1978, Combat bouts with spur use in the Madagascan boa (*Sanzinia madagascariensis*), *Herpetologica,* 34:207–212.
Clark, D. R. Jr., 1963, Variation and sexual dimorphism in a brood of the western pygmy rattlesnake (*Sistrurus*), *Copeia,* 1963:157–159.
Clark, D. R. Jr., 1967, Notes on sexual dimorphism in tail length in American snakes, *Trans. Kansas Acad. Sci.,* 69:226–232.
Clark, D. R. Jr., 1970, Ecological study of the worm snake *Carphophis vermis* (Kenicott), *Univ. Kansas Publ. Mus. Nat. Hist.,* 19:89–194.
Clutton-Brock, T. H., P. H. Harvey, and B. Rudder, 1977, Sexual dimorphism, socionomic sex ratio and body weight in primates, *Nature,* 269:797–800.
Cogger, H. G., 1975, Sea snakes of Australia and New Guinea, in W. A. Dunson, ed., *The Biology of Sea Snakes,* Univ. Park Press, Baltimore, pp. 59–140.
Cogger, H. G., 1986, *Reptiles and Amphibians of Australia,* 3rd ed., Reed, Sydney.
Crews, D., M. Diamond, J. Whittier, and R. Mason, 1985, Small male body size in garter snakes depends on testes, *Amer. J. Physiol.,* 249:R62–R66.
Darwin, C., 1871, *The Descent of Man and Selection in Relation to Sex,* 2nd ed., John Murray, London.
de Haan, C. C., 1984, Dimorphisme et comportement sexuel chez *Malpolon monspessulanus,* considerations sur la denomination subspecifique *insignitus, Bull. Soc. Herpetol. Fr.,* 30:19–26.
de Silva, P. H. D. H., 1969, Taxonomic studies on Ceylon snakes of the family Colubridae, *Spolia Zeylanica,* 31:431–546.
Deraniyagala, P. E. P., 1955, *A Colored Atlas of Some Vertebrates from Ceylon, Serpentoid Reptilia,* Vol. 3, National Museum, Colombo.
Devine, M. C., 1975, Copulatory plugs, restricted mating opportunities and reproductive competition among male garter snakes, *Nature,* 267:345–346.
Devine, M. C., 1984, Potential for sperm competition in reptiles: Behavioral and physiological consequences, in R. L. Smith, ed., *Sperm Competition and the Evolution of Animal Mating Systems,* Academic, New York, pp. 509–521.
Dmi'el, R., 1967, Studies on reproduction, growth, and feeding in the snake *Spalerosophis cliffordi, Copeia,* 1967:332–346.
Dmi'el, R., G. Perry, and H. Mendelssohn, 1990, Sexual dimorphism in *Walterinnesia aegyptia* (Reptilia: Ophidia: Elapidae), *The Snake,* 22:33–35.
do Amaral, A., 1978, *Serpentes do Brasil: Iconografía colorida,* 2nd ed., Univ. São Paulo, São Paulo.
Dowling, H. G., 1950, Studies of the black swamp snake, *Seminatrix pygaea* (Cope), with descriptions of two new subspecies, *Misc. Publ. Mus. Zool. Univ. Michigan,* 76:1–38.
Dunlap, K. D., and J. W. Lang, 1990, Offspring sex ratio varies with maternal size in the common garter snake, *Thamnophis sirtalis, Copeia,* 1990:568–570.
Duvall, D., S. J. Arnold, and G. Schuett, 1992, Pitviper mating systems: Ecological potential, sexual selection, and microevolution, in J. A. Campbell and E. D. Brodie, Jr., eds., *Biology of the Pitvipers,* Selva, Tyler, Texas, pp. 321–336.
Edgren, R. A., 1958, Umbilical scar position and sexual dimorphism in hog-nosed snakes, genus *Heterodon*: A review, *Nat. Hist. Misc.,* 163:1–6.
Engelmann, W., and F. J. Obst, 1981, *Snakes: Biology, Behavior and Relationship to Man,* Exeter, New York.
Feaver, P. E., 1977, The demography of a Michigan population of *Natrix sipedon* with discussions of ophidian growth and reproduction, Doctoral Thesis, Univ. Michigan, Ann Arbor.
Fitch, H. S., 1975, A demographic study of the ringneck snake (*Diadophis punctatus*) in Kansas, *Univ. Kansas Mus. Nat. Hist. Misc. Publ.,* 62:1–53.
Fitch, H. S., 1981, Sexual size differences in reptiles, *Univ. Kansas Mus. Nat. Hist. Misc. Publ.,* 70:1–72.
Force, E. R., 1936, The relation of the knobbed anal keels to age and sex in the lined snake *Tropidoclonion lineatum* (Hallowell), *Pap. Mich. Acad. Sci., Arts Lett.,* 21:613–617.

Forsman, A., 1991a, Adaptive variation in head size in *Vipera berus* L. populations, *Biol. J. Linn. Soc.*, 43:281–296.
Forsman, A., 1991b, Variation in sexual size dimorphism and maximum body size among adder populations: Effects of prey size, *J. Anim. Ecol.*, 60:253–267.
Gibbons, J. W., 1972, Reproduction, growth and sexual dimorphism in the canebrake rattlesnake (*Crotalus horridus atricaudatus*), *Copeia*, 1972:222–226.
Gibbons, J. W., J. W. Coker, and T. M. Murphy, Jr., 1977, Selected aspects of the life history of the rainbow snake (*Farancia erytrogramma*), *Herpetologica*, 33:276–281.
Gibbons, J. W., and J. E. Lovich, 1990, Sexual dimorphism in turtles with emphasis on the slider turtle (*Trachemys scripta*), *Herpetol. Monogr.*, 4:1–29.
Gibbons, J. W., and R. D. Semlitsch, 1987, Activity patterns, in R. A. Seigel, J. T. Collins, and S. S. Novak, eds., *Snakes: Ecology and Evolutionary Biology*, McGraw-Hill, New York, pp. 396–421.
Gibson, R., and J. B. Falls, 1979, Thermal biology of the common garter snake *Thamnophis sirtalis* (L.), 1. Temporal variation, environmental effects and sex differences, *Oecologia (Berlin)*, 43:79–97.
Gillingham, J. C., 1987, Social behavior, in R. A. Seigel, J. T. Collins, and S. S. Novak, eds., *Snakes: Ecology and Evolutionary Biology*, McGraw-Hill, New York, pp. 184–209.
Gould, S. J., and R. C. Lewinton, 1979, The spandrels of San Marco and the panglossian paradigm: A critique of the adaptationist programme, *Proc. Roy. Soc. Lond., B*, 205:581–598.
Grant, C., 1957, The black tailed *Tropidophis* (Reptilia: Serpentes), *Herpetologica*, 13:154.
Greenwood, P. J., and J. Adams, 1987, Sexual selection, sexual dimorphism and a fallacy, *Oikos*, 48:106–108.
Greer, A. E., 1989, *The Biology and Evolution of Australian Lizards*, Surrey Beatty, Australia.
Gregory, P. T., L. A. Gregory, and J. M. Macartney, 1983, Color-pattern variation in *Thamnophis melanogaster*, *Copeia*, 1983:530–534.
Gregory, P. T., J. M. Macartney, and K. W. Larsen, 1987, Spatial patterns and movements, in R. A. Seigel, J. T. Collins, and S. S. Novak, eds., *Snakes: Ecology and Evolutionary Biology*, McGraw-Hill, New York, pp. 366–395.
Guibe, J., 1948, Sur le dimorphisme sexuel des especes du genre *Langaha* (Ophidia), *C. R. Seances l'Acad. Sciences*, 226:1219–1220.
Guibe, J., 1953, The sexual dimorphism of the reptiles, *La Nature (Paris)*, 3217:129–133.
Harlow, P., and R. Shine, 1992, Food habits and reproductive biology of the Pacific island boas (*Candoia*), *J. Herpetol.*, 26:60–66.
Harrison, M.B., 1933 The significance of knobbed anal keels in the garter snake, *Thamnophis sirtalis sirtalis* (Linnaeus), *Copeia*, 1933:1–3.
Heatwole, H. F., S. A. Minton, Jr., R. Taylor, and V. Taylor, 1978, Underwater observations of sea snake behavior, *Rec. Aust. Mus.*, 31:737–761.
Hellmich, W., 1962, *Reptiles and Amphibians of Europe*, Blandford, London.
Henderson, R. W., and L. G. Hoevers, 1979, Variation in the snake *Tretanorhinus nigroluteus lateralis* in Belize with notes on breeding tubercles, *Herpetologica*, 35:245–248.
Hewitt, J., 1937, *A Guide to the Vertebrate Fauna of the Eastern Cape Province, South Africa, Part II: Reptiles, Amphibians and Freshwater Fishes*, Govt. Printers, Grahamstown.
Hillman, P. E., 1969, Habitat specificity in three sympatric species of *Ameiva* (Reptilia: Teiidae), *Ecology*, 50:476–481.
Hoge, A. R., and P. A. Federsoni, Jr., 1977, Observations on a brood of *Bothrops atrox* (Linneaux, 1758): (Serpentes: Viperidae: Crotalinae), *Mem. Inst. Butantan (São Paulo)*, 40/41:19–36.
Houston, D., 1992, Ecology of the filesnake, *Acrochordus arafurae*, Doctoral Thesis, Univ. Sydney, New South Wales.
Houston, D.L., and R. Shine, 1993, Sexual dimorphism and niche divergence: feeding habits of the Arafura filesnake, *J. Anim. Ecol.*, in press.

Iverson, J. B., 1990, Phylogenetic hypotheses for the evolution of modern kinosternine turtles, *Herpetol. Monogr.*, 4:1–27.
Jacob, J. S., and J. S. Altenbach, 1977, Sexual color dimorphism in *Crotalus lepidus klauberi* Gloyd (Reptilia, Serpentes, Viperidae), *J. Herpetol.*, 11:81–84.
Johnston, G. R., 1987, Reproduction and growth in captive death adders *Acanthophis antarcticus* (Squamata: Elapidae), *Trans. Roy. Soc. South Aust.*, 111:123–125.
Joy, J. E., and D. Crews, 1988, Male mating success in red-sided gartersnakes: Size is not important, *Anim. Behav.*, 36:1839–1841.
Kaufman, G. A., and J. W. Gibbons, 1975, Weight-length relationships in thirteen species of snakes in the southeastern United States, *Herpetologica*, 31:31–37.
Kheruvimov, V. D., A. S. Sokolov, and L. A. Sokolova, 1977, Sex and age determination of the common adder, *Vestnik Zoologii*, 6:39–44.
King, R. B., 1989, Sexual dimorphism in snake tail length: Sexual selection, natural selection, or morphological constraint?, *Biol. J. Linn. Soc.*, 38:133–154.
Klauber, L. M., 1940, The lyre snakes (*Trimorphodon*) of the U.S., *Trans. San Diego Soc. Nat. Hist.*, 9:163–194.
Klauber, L. M., 1943, Tail-length differences in snakes with notes on sexual dimorphism and the coefficient of divergence, *Bull. Zool. Soc. San Diego*, 18:1–60.
Klauber, L. M., 1956, *Rattlesnakes: Their Habits, Life Histories and Influence on Mankind*, Univ. Calif. Press, Berkeley, California.
Kminiak, M., and S. Kalúz, 1983, Evaluation of sexual dimorphism in snakes (Ophidia, Squamata) based on external morphological characters, *Folia Zool.*, 32:259–270.
Kopstein, F., 1941, Sexual dimorphism in Malaysian snakes, *Temminckia*, 6:101–185.
Krefft, G., 1869, *The Snakes of Australia: An Illustrated and Descriptive Catalogue of All the Known Species*, T. Richards, Sydney.
Kretz, J., 1971, Uber *Vipera kaznakovi* Nikolskij 1909 aus Nordostanatolien (Reptilia, Viperidae), *Bern Jarbusch Naturhist. Museum*, 4:125–134.
Kubie, J., A. Vagvolgyi, and M., H., 1978, Roles of vomeronasal and olfactory systems in courtship behavior of male garter snakes, *J. Comp. Physiol. Psychol.*, 92:627–641.
Lande, R., 1980, Sexual dimorphism, sexual selection, and adaptation in polygenic characters, *Evolution*, 34:292–305.
Layne, J. N., and T. M. Steiner, 1984, Sexual dimorphism in occurrence of keeled dorsal scales in the eastern indigo snake (*Drymarchon corais couperi*), *Copeia*, 1984:776–778.
Leloup, P., 1975, Observations sur la reproduction de *Bothrops moojeni* Hoge en captivite, *Acta. Zool. Pathol. Antverp.*, 62:173–201.
Leviton, A. E., 1983, Contributions to a review of Philippine snakes, XIV. The snakes of the genera *Xenopeltis, Zaocys, Psammodynastes* and *Myersophis, Philippine J. Sci.*, 112:195–223.
Loveridge, A., 1958, Revision of five African snake genera, *Bull. Mus. Comp. Zool.*, 118:1–198.
Macartney, J. M., P. T. Gregory, and M. B. Charland, 1990, Growth and sexual maturity of the western rattlesnake, *Crotalus viridis*, in British Columbia, *Copeia*, 1990:528–542.
Madsen, T., 1987, Cost of reproduction and female life-history tactics in a population of grass snakes, *Natrix natrix*, in southern Sweden, *Oikos*, 49:129–132.
Madsen, T., and R. Shine, 1992a, A rapid, sexually-selected shift in mean body size in a population of snakes, *Evolution*, 46:1220–1224.
Madsen, T., and R. Shine, 1992b, Sexual competition among brothers may influence offspring sex ratio in snakes, *Evolution*, 46:1549–1552.
Madsen, T., and R. Shine, 1992c, Temporal variability in sexual selection on reproductive tactics and body size in male snakes, *Amer. Nat.* 141:167–171.
Madsen, T., and R. Shine, 1993a, Costs of reproduction in a population of European adders, *Oecologia*, in press.
Madsen, T., and R. Shine, 1993b, Costs of reproduction influence the evolution of sexual size dimorphism in snakes, *Evolution*, in review.
Madsen, T., and R. Shine, 1993c, Male mating success and body size in European grass snakes, *Copeia*, 1993:529–532.

Madsen, T., and R. Shine, 1993d, Phenotypic plasticity in body sizes and sexual size dimorphism in European grass snakes, *Evolution,* in press.
Madsen, T., R. Shine, J. Loman, and T. Hakansson, 1993, Determinants of mating success in male adders, *Vipera berus, Anim. Behav.,* in press.
Manjarrez, J., and C. Macias Garcia, 1991, Feeding ecology of *Nerodia rhombifera* in a Veracruz swamp, *J. Herpetol.,* 25:499–502.
Mao, S.-H., 1962, Sexual dimorphism of Taiwan bamboo vipers, *Bull. Inst. Zool. Acad. Sinica,* 1:41–46.
Marx, H., 1958, Sexual dimorphism in coloration in the viper *Cerastes vipera* L., *Nat. Hist. Misc.,* 164:1–2.
Maslin, T. P., 1950, Snakes of the Kiukiang-Lushan area, Kiangsi, China, *Proc. Cal. Acad. Sci.,* 26:419–466.
Mason, R. T., and D. Crews, 1985, Female mimicry in garter snakes, *Nature,* 316:59–60.
McDowell, S. B., 1979, A catalogue of the snakes of New Guinea and the Solomons, with special reference to those in the Bernice P. Bishop Museum, Part III, Boinae and Acrochordoidea, *J. Herpetol.,* 13:1–92.
Mehrtens, J. M., 1987, *Living Snakes of the World in Color,* Sterling, New York.
Moulis, R., 1976, Autecology of the eastern indigo snake, *Drymarchon corais couperi, Herp (Bull. New York Herpetol. Soc.),* 12:14–23.
Mount, R. H., 1975, *Reptiles and Amphibians of Alabama,* Auburn Univ. Agric. Exp. Stat., Auburn, Alabama.
Mushinsky, H. R., J. J. Hebrard, and D. S. Vodopich, 1982, Ontogeny of water snake foraging ecology, *Ecology,* 63:1624–1629.
Myers, C. W., 1967, The pine woods snake, *Rhadinea flavilata* (Cope), *Bull. Fla. State Mus. Biol. Sci.,* 11:47–97.
Myers, C. W., 1982, Blunt-headed vine snakes (*Imantodes*) in Panama, including a new species and other revisionary notes, *Amer. Mus. Novit.,* 2738:1–50.
Naulleau, G., 1970, La reproduction de *Vipera aspis* en captivité dans des conditions artificielles, *J. Herpetol.,* 4:113–121.
Naulleau, G., 1973a, Le mélanisme chez *Vipera aspis* et chez *Vipera berus, Bull. Soc. Zool. Fr.,* 98:595–596.
Naulleau, G., 1973b, *Les Serpents de France,* Rev. Francaise d'Aquariologie, Nancy, France.
Naulleau, G., 1979, Etude biotelemetrique de la thermoregulation chez *Vipera aspis* (L.) elevee on conditions artificielles, *J. Herpetol.,* 13:203–208.
Naulleau, G., 1986, Effects of temperature on 'gestation' in *Vipera aspis* and *V. berus* (Reptilia: Serpentes), in Z. Röcek, ed., *Studies in Herpetology,* Charles Univ., Prague, pp. 489–494.
Neill, W. T., 1954, Variation and sexual dimorphism in hatchlings of the rainbow snake, *Farancia erythrogramma, Copeia,* 1954:87–92.
Neill, W. T., 1960, The caudal lure of various juvenile snakes, *Quart. J. Fla. Acad. Sci.,* 23:173–200.
Neill, W. T., 1964, Taxonomy, natural history, and zoogeography of the rainbow snake, *Farancia erythrogramma* (Palisot de Beauvois), *Amer. Midl. Nat.,* 71:257–295.
Nilson, G., and Andren, C. 1986, A review of the *Vipera xanthina* complex, in Z. Röcek, ed., *Studies in Herpetology,* Charles Univ., Prague, pp. 223–226.
Noble, G. K., 1937, The sense organs involved in the courtship of *Storeria, Thamnophis,* and other snakes, *Bull. Amer. Mus. Nat. Hist.,* 73:673–725.
Ota, H., M. Toriba, and H. Takahashi, 1986, The alteration pattern of dorsal scale rows in the yellow-lipped sea krait *Laticauda colubrina,* with special reference to sexual dimorphism, *Jap. J. Herpetol.,* 11:145–151.
Pagel, M. D., and P. H. Harvey, 1989, Comparative methods for examining adaptation depend on evolutionary models, *Folia Primatol.,* 53:203–220.
Parker, W. S., and M. V. Plummer, 1987, Population ecology, in R. A. Seigel, J. T. Collins, and S. S. Novak, eds., *Snakes: Ecology and Evolutionary Biology,* McGraw-Hill, New York, pp. 253–301.
Pernetta, J. C., 1977, Observations on the habits and morphology of the sea snake *Laticauda colubrina* (Schneider) in Fiji, *Can. J. Zool.,* 55:1612–1619.

Platt, D. R., 1969, Natural history of the hognose snakes *Heterodon platyrhinos* and *Heterodon nasicus, Univ. Kansas Publ. Mus. Nat. Hist.,* 18:253–420.

Pope, C. H., 1929, Notes on reptiles from Fukien and other Chinese provinces, *Bull. Amer. Mus. Nat. Hist.,* 58:335–487.

Pope, C. H., 1935, *The Reptiles of China,* Amer. Mus. Nat. Hist., New York.

Quinn, H. R., 1977, A method to dorsally determine the sex of the coral snake, *Micrurus fulvius tenere, Bull. Okla. Herpetol. Soc.,* 2:13.

Quinn, H. R., 1979, Sexual dimorphism in tail pattern of Oklahoma snakes, *Texas J. Sci.,* 31:157–160.

Rasmussen, J. B., 1975, Geographical variation, including an evolutionary trend, in *Psammodynastes pulverulentus* (Boie, 1827) (Boiginae, Homalopsidae, Serpentes), *Vidensk. Meddr dansk naturh. Foren.,* 138:39–64.

Reinert, H. K., and W. R. Kodrich, 1982, Movements and habitat utilization by the massasauga, *Sistrurus catenatus catenatus, J. Herpetol.,* 16:162–171.

Rheinhard, W., and Z. Vogel, 1972, Colubrid snakes, in B. Grzimek, ed., *Animal Life Encyclopedia,* Van Nostrand Reinhold , New York, pp. 381–414.

Richmond, N. D., 1954, Variation and sexual dimorphism in hatchlings of the rainbow snake, *Abastor erythrogrammus, Copeia,* 1954:87–92.

Rossman, N. J., D. A. Rossman, and N. K. Keith, 1982, Visceral topography of the New World snake tribe Thamnophiini (Colubridae, Natricinae), *Tulane Stud. Zool. Bot.,* 23:123–164.

Schaeffer, G. C., 1969, Sex-independent ground color in the timber rattlesnake, *Crotalus horridus horridus, Herpetologica,* 25:65–66.

Schmidt, K. P., and R. F. Inger, 1957, *Living Reptiles of the World,* Doubleday, New York.

Schnurrenberger, H., 1959, Observations on behavior in two Libyan species of viperine snakes, *Herpetologica,* 15:70–72.

Schuett, G. W., and J. C. Gillingham, 1989, Male-male agonistic behavior of the copperhead, *Agkistrodon contortrix, Amphibia-Reptilia,* 10:243–266.

Schwaner, T. D., 1985, Population structure of black tiger snakes, *Notechis ater niger,* on offshore islands of South Australia, in G. C. Grigg, R. Shine, and H. Ehmann, eds., *The Biology of Australasian Reptiles and Frogs,* Royal Zool. Soc. New South Wales, Sydney, pp. 35–46.

Schwaner, T. D., and S. D. Sarre, 1988, Body size of tiger snakes in southern Australia, with particular reference to *Notechis ater serventyi* (Elapidae) on Chappell Island, *J. Herpetol.,* 22:24–33.

Schwartz, A., 1976, Variation in the Hispaniolan colubrid snake *Uromacer frenatus* (Reptilia, Serpentes, Colubridae), *J. Herpetol.,* 10:319–327.

Schwartz, A., and R. W. Henderson, 1984, *Uromacer oxyrhynchus, SSAR Cat. Amer. Amphib. Rept.,* 358:1.

Schwarzkopf, L., and R. Shine, 1992, Thermal biology of reproduction in viviparous skinks, *Eulamprus tympanum:* Why do gravid females bask more? *Oecologia (Berlin),* in press.

Seigel, R. A., 1992, Ecology of a specialized predator: *Regina grahami* in Missouri, *J. Herpetol.,* 26:32–37.

Seigel, R. A., and N. B. Ford, 1987, Reproductive ecology, in R. A. Seigel, J. T. Collins, and S. S. Novak, eds., *Snakes: Ecology and Evolutionary Biology,* McGraw-Hill, New York, pp. 210–252.

Seigel, R. A., M. M. Huggins, and N. B. Ford, 1987, Reduction in locomotor ability as a cost of reproduction in snakes, *Oecologia (Berlin),* 73:481–465.

Semlitsch, R. D., and J. W. Gibbons, 1982, Body size dimorphism and sexual selection in two species of water snakes, *Copeia,* 1982:974–976.

Semlitsch, R. D., and G. B. Moran, 1984, Ecology of the redbelly snake (*Storeria occipitomaculata*) using mesic habitats in South Carolina, *Amer. Midl. Nat.,* 111:33–40.

Shine, R., 1978a, Growth rates and sexual maturation in six species of Australian elapid snakes, *Herpetologica,* 34:73–79.

Shine, R., 1978b, Sexual size dimorphism and male combat in snakes, *Oecologia (Berlin),* 33:269–278.

Shine, R., 1979, Activity patterns in Australian elapid snakes (Squamata: Serpentes: Elapidae), *Herpetologica,* 35:1–11.
Shine, R., 1980, Reproduction, feeding and growth in the Australian burrowing snake *Vermicella annulata, J. Herpetol.*, 14:71–77.
Shine, R., 1986, Sexual differences in morphology and niche utilization in an aquatic snake, *Acrochordus arafurae, Oecologia (Berlin)*, 69:260–267.
Shine, R., 1987, Sexual selection in amphibians: A reply to Halliday and Verrell, *Herpet. J.*, 1:202–203.
Shine, R., 1988a, Constraints on reproductive investment: A comparison between aquatic and terrestrial snakes, *Evolution,* 42:17–27.
Shine, R., 1988b, Parental care in reptiles, in C. Gans and R. B. Huey, eds., *Biology of the Reptilia,* Vol. 16, Alan R. Liss, New York, pp. 275–330.
Shine, R., 1989, Ecological causes for the evolution of sexual dimorphism: A review of the evidence, *Quart. Rev. Biol.*, 64:419–464.
Shine, R., 1990, Proximate determinants of sexual differences in adult body size, *Amer. Nat.*, 135:278–283.
Shine, R., 1991a, Intersexual dietary divergence and the evolution of sexual dimorphism in snakes, *Amer. Nat.*, 138:103–122.
Shine, R., 1991b, *Australian Snakes, A Natural History,* Reed, Sydney.
Shine, R., 1992, Relative clutch mass and body shape in lizards and snakes: Is reproductive investment constrained or optimized? *Evolution,* 46:828–833.
Shine, R., 1993, Sexual size dimorphism in snakes revisited, in preparation.
Shine, R., and J. J. Bull, 1977, Skewed sex ratios in snakes, *Copeia,* 1977:228–234.
Shine, R., and D. Crews, 1988, Why male garter snakes have small heads: The evolution and endocrine control of sexual dimorphism, *Evolution,* 42:1105–1110.
Shine, R., and R. Lambeck, 1990, Seasonal shifts in the thermoregulatory behavior of Australian blacksnakes, *Pseudechis porphyriacus, J. Therm. Biol.*, 15:301–305.
Shine, R., and D. J. Slip, 1990, Biological aspects of the adaptive radiation of Australasian pythons (Serpentes: Boidae), *Herpetologica,* 46:283–290.
Slatkin, M., 1984, Ecological causes of sexual dimorphism, *Evolution,* 38:622–630.
Slip, D. J., and R. Shine, 1988, The reproductive biology and mating system of diamond pythons, *Morelia spilota* (Serpentes, Boidae), *Herpetologica,* 44:396–404.
Smith, M. A., 1943, *The Fauna of British India, Ceylon and Burma Including the Whole of the Indo-Chinese Sub-region, Reptilia and Amphibia, Serpentes,* Vol. 3, Taylor and Francis, London.
Solórzano, A., 1990, Reproduction in the pit viper *Porthidium picadoi* Dunn (Serpentes: Viperidae) in Costa Rica, *Copeia,* 1990:1154–1157.
Solórzano, A., and L. Cerdas, 1989, Reproductive biology and distribution of the terciopelo, *Bothrops asper* Garman (Serpentes: Viperidae) in Costa Rica, *Herpetologica,* 45:444–450.
Solórzano, A., J. M. Gutíerrez, and L. Cerdas, 1988, *Bothrops ophryomegas* Bocourt (Serpentes: Viperidae) en Costa Rica: Distribución, lepidosis, variación sexual y cariotipo, *Rev. Biol. Trop.*, 36:187–190.
Stevens, R. A., 1973, A report on the lowland viper, *Atheris superciliaris* (Peters), from the Lake Chilwa floodplain of Malawi, *Arnoldia,* 22:1–22.
Stewart, J. W., 1971, *The Snakes of Europe,* Fairleigh Dickinson Univ. Press, Rutherford.
Storr, G. M., L. A. Smith, and R. E. Johnstone, 1986, *Snakes of Western Australia,* West. Aust. Museum, Perth.
Street, D., 1979, *The Reptiles of Northern and Central Europe,* B. T. Batsford, London.
Taylor, E. H., 1950, A brief review of Ceylonese snakes, *Univ. Kansas Sci. Bull.*, 33:519–603.
Thomas, R. A., and J. R. Dixon, 1976, Scale row formulae in *Elaphe guttata* (Linnaeus) and notes on their interpretation, *Nat. Hist. Misc.*, 195:1–5.
Thorpe, R. S., 1973, Intraspecific variation of the ringed snake *Natrix natrix* (L.), Doctoral Thesis, London Polytechnic, London.
Thorpe, R. S., 1975, Quantitative handling of characters useful in snake systematics

with particular reference to intraspecific variation in the ringed snake *Natrix natrix* (L.), *Biol. J. Linn. Soc.*, 7:27–43.

Trutnau, L., 1986, *Nonvenomous Snakes*, Barron's, Toronto.

Tryon, B., 1985, *Bothrops asper* (terciopelo), Reproduction, *Herpetol. Rev.*, 16:28.

Tu, M. C., S. C. Fong, and K. Y. Lue, 1990, Reproductive biology of the sea snake, *Laticauda semifasciata*, in Taiwan, *J. Herpetol.*, 24:119–126.

Turner, F. B., 1977, The dynamics of populations of squamates, crocodilians and rhyncocephalians, in C. Gans and D. W. Tinkle, eds., *Biology of the Reptilia*, Vol. 7, Academic, New York, pp. 157–264.

Vitt, L. J., and W. E. Cooper, Jr., 1985, The evolution of sexual dimorphism in the skink *Eumeces laticeps:* An example of sexual selection, *Can. J. Zool.*, 63:995–1002.

Vogel, C. M., 1991, Captive maintenance and reproduction of the temple viper (*Tropidolaemus wagleri*), *Vivarium*, 3:19–22.

Wall, W. F., 1921, *Ophidia Taprobanica* or *The Snakes of Ceylon*, H. R. Cottle, Colombo.

Weiner, S., 1960, Venom yields and toxicity of the venoms of male and female tiger snakes, *Med. J. Aust.*, 2:740–741.

Werler, J. E., and H. M. Smith, 1952, Notes on a collection of reptiles and amphibians from Mexico, 1951–1952, *Texas J. Sci.*, 4:551–573.

Wright, A. H., and A. A. Wright, 1957, Handbook of Snakes of the United States and Canada, Cornell Univ. Press, Ithaca.

Zinner, H., 1985, On behavioral and sexual dimorphism of *Telescopus dhara* Forscal 1776 (Reptilia: Serpentes, Colubridae), *J. Herpetol. Assoc. Africa*, 31:5–6.

Chapter

3

Foraging Theory and Prey-Size–Predator-Size Relations in Snakes

Stevan J. Arnold

Introduction

The aim of this chapter is to encourage additional studies of prey-size–snake-size relations. The fascinating evolutionary vistas sketched by Greene (1983), Mushinsky (1987), Pough and Groves (1983), and Voris and Voris (1983) have not been explored as assiduously as they might. In this chapter I focus on one tantalizing result from the recent literature on snake diets in the hope of encouraging more work. The result is that in many of the snake species studied so far larger snakes drop small prey items out of their diet. The implication is that snakes pass over, perhaps even avoid, some of the smallest prey items they encounter. Foraging theory is used to devise some hypotheses to explain this apparently enigmatic result. In trying to use foraging theory to this end, I was constantly plagued by the lack of relevant data. However, the lack of data is undoubtedly a reflection of our failure to adapt theory to snake biology. Perhaps even a provisional application of theory to the problem will help break the logjam.

Trend in the snake diet literature

A progressive trend is apparent in the literature on snake diets, with later papers providing ever more detailed information. The earliest reports (and some later ones) provide only simple lists of prey with no indication of relative importance (Surface, 1906; Wright and Wright,

1957; Mori et al., 1989). Beginning in the 1940s investigators reported the relative importance of prey types in the diet. Relative importance has been variously based on number, weight, volume, or energy content of prey; on the number of snakes that have eaten particular prey types; or on indices that are functions of two or more of these variables (Fitch and Twining, 1946; Hamilton, 1951; Carpenter, 1952; Hamilton and Pollack, 1956; Fitch, 1960; Catling and Freedman, 1980; Arnold, 1981; Shine, 1986; Hasegawa and Moriguchi, 1989). Surprisingly, the uses, virtues, and limitations of these different modes of representing relative importance seem not to have been comprehensively discussed in the snake literature. Such a discussion is, however, beyond the scope of the present chapter. In the 1960s investigators began reporting the relative importance of different prey types as a function of snake size (Fitch, 1963, 1965; Godley, 1980; Saint Girons, 1980). In the simplest cases, the diets of juveniles and adults are compared. In the late 1970s workers began reporting prey size, as well as relative importance, as a function of snake size (Shine, 1977; Voris and Moffett, 1981; Mushinsky et al., 1982; Henderson and Horn, 1983; Plummer and Goy, 1984; Seib, 1984; Henderson et al., 1987; Shine, 1987; Henderson et al., 1988; Slip and Shine, 1988; Cobb, 1989). Finally, in the last few years workers have used field studies of prey-size–snake-size relations to motivate laboratory studies of prey handling and swallowing ability (Jayne et al., 1988; Shine, 1991).

How can we make foraging theory work for us?

Foraging theory (MacArthur and Pianka, 1966; Schoener, 1971; Stephens and Krebs, 1986) apparently did not promote the progressive trend in the snake diet literature. Optimal foraging theory is occasionally mentioned in snake diet studies, but it seems to have been used post hoc rather than to inspire (but see Shine, 1991). The progressive trend appears to have been propelled as much by herpetological craft and tradition as by theoretical developments outside the discipline. Within the discipline, Greene (1983) argued persuasively that prey diameter in relation to snake head size (ingestion ratio) and prey mass in relation to snake mass (weight ratio) are the keys to understanding snake evolution and ecology. The connection between Greene's key variables and foraging economics needs to be drawn out so that we can evaluate his propositions. More generally, how do we make foraging theory work for us?

Foraging theory could motivate studies of snake diet in three ways. We might test the theory's assumptions or its predictions, or we might use the theory to highlight certain data trends and generate new hypotheses. I will concentrate on the third approach, but first I will review some important assumptions of the theory.

An Economic Perspective on Prey-Size–Predator-Size Relations in Snakes

Crucial assumptions of optimal foraging theory

Optimality is one of the most hazardous assumptions of foraging theory. Important lessons can be drawn from the theory, however, while avoiding the pitfalls of the optimality assumption. The goal of foraging theory is to predict how animals should feed (Schoener, 1969, 1971; Pyke et al., 1977; Krebs and McCleery, 1984; Stephens and Krebs, 1986). To make predictions we need to evaluate foraging alternatives. The use of optimality to evaluate alternatives has three aspects: (1) the choice of a currency; (2) the choice of a cost–benefit function that specifies the relationship between foraging and the currency; and 3) the solution of the function for the foraging traits that maximizes the currency (Schoener, 1971). The third aspect is a straightforward mathematical problem, but the first two aspects involve potentially hazardous assumptions and implications. From a practical standpoint the best choice for a currency is amount of energy gained, time spent foraging, or some combination of the two. Time and energy are both readily measured and natural currencies for the various activities or stages of foraging. From an evolutionary standpoint, however, fitness (the number of progeny produced in a lifetime) is the natural currency (Crow and Kimura, 1970). To use energy and time as a currency for evolutionary predictions we must know or specify the relationship between our currency (based on time and/or energy) and fitness. The standard assumption is that the relationship is linear or monotonically increasing. Either version is equivalent to saying that the currency is under perpetual directional selection with no intermediate optimum. The second hazardous assumption is that the population will (and has) evolved to the optimum specified by the cost–benefit function. Frequency-dependent selection and various kinds of genetic constraint can cause violations of this assumption (Lewontin, 1978; Gould and Lewontin, 1979; Lande, 1979). One posture is to argue that these two assumptions are not hazardous. Another posture is to acknowledge the potential pitfalls of the two assumptions and use foraging theory in a way that avoids those pitfalls.

The pitfalls of optimality can be avoided by using foraging theory to characterize selection rather than to predict evolutionary outcome. We can think of selection as the statistical relationship between a phenotypic trait (such as a foraging tendency) and fitness (Lande, 1979; Lande and Arnold, 1983). If we visualize the relationship as a pathway, we can recognize two parts: the path from our foraging trait to our energy and time currency and the path from the currency to fitness (Arnold, 1983, 1988). Foraging theory gives prescriptions for measuring the first part of the pathway, the energy gradient (Arnold, 1988). In other words, foraging theory guides measurement of part of the phenotypic selection that acts on foraging. In principle, we could also measure the second part of the path (the fitness gradient), which is often and conveniently assumed to be linear, and so characterize the total selection that acts on foraging. If we have not accomplished this second task (i.e., the fitness gradient remains unknown) then we must qualify our conclusions by recognizing that we have measured only a part of the selection. If we had measured total selection, we still would need to measure genetic constraints to predict the outcome of selection. But even if we use foraging theory just to make statements about selection, it can be an informative tool.

Prey size versus snake size

Ontogenetic shift in lower size limit for prey. In this section I will review studies in which prey size is plotted as a function of snake size. In all the studies it is apparent that larger snakes tend to eat larger prey. Furthermore, regression line intercepts for plots relating prey dimensions to snake dimensions are often appreciably different from zero (Voris and Moffett, 1981; Plummer and Goy, 1984; Seib, 1984; Cobb, 1989). Such nonzero intercepts mean that prey-size–snake-size ratios will vary within a species as a function of snake size. In the two studies in which the regression slope was estimated for log prey mass as a function of log snake mass, the allometric slope was less than one (Voris and Moffett, 1981; Jayne et al., 1988), indicating that prey mass does not increase in proportion to snake mass (Schmidt-Nielsen, 1984). The range of prey is also larger for larger snakes. These important trends may be universal in snakes and have been discussed by Shine (1991). I will not remark on these trends as I survey published studies. Instead, I will focus on the lower size limit for prey as a function of snake size. Two patterns are apparent in the snake literature and are shown diagramatically in Fig. 3.1. In some species the lower limit does not increase, so that the prey of small snakes is a subset of the prey sizes eaten by large snakes (Fig. 3.1*A*). The range of prey sizes eaten by smaller snakes telescopes within the range of larger snakes. In most species, however, the lower limit increases with snake size (Fig. 3.1*B*). In other words, in most species larger snakes delete small prey from their diets. In Figs. 3.2–3.10 N refers to the number of snakes in the sample and n refers to the number of prey.

Shine (1977, 1987) used museum specimens to study the diets of three genera of terrestrial elapids in Australia. *Notechis scutatus* is a

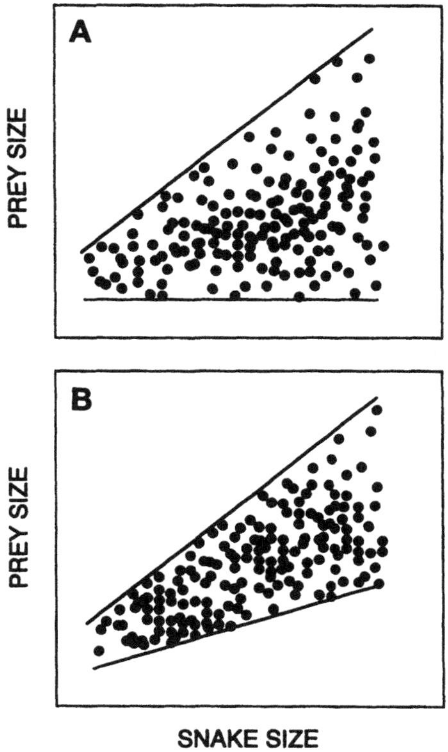

Figure 3.1 Patterns of prey-size–snake-size relationships. (*A*) Ontogenetic telescope. (*B*) Ontogenetic shift in lower size limit.

swamp dweller that feeds on frogs and nestling birds (Shine, 1977). Frog length is plotted against snake length in Fig. 3.2*A*. *Pseudechis porphyriacus* is a riparian forager that feeds on frogs and lizards (Shine, 1977). Frog length is plotted against snake length in Fig. 3.2*B*. Australian Copperheads (*Austrelaps*) are active searchers that feed on lizards and frogs (Shine, 1987). Lizard (mainly scincids) length is plotted against snake length in Fig. 3.2*C*. The data in Fig. 3.2*C* are pooled from three species of *Austrelaps* (*A. labialis, A. ramsayi,* and *A. superbus*). In all three genera of Australian elapids there is no indication that larger snakes delete small prey from their diets. Shine (1977) argues that large elapids continue to eat small prey because capture and ingestion costs are small in relation to energy content of prey and because envenomation of prey eliminates risks.

Seib (1984) used special collections and museum specimens to study the diets of three colubrid snakes in the genera *Drymobius* and *Mastigodryas*. All three species are frog and lizards predators. Seib

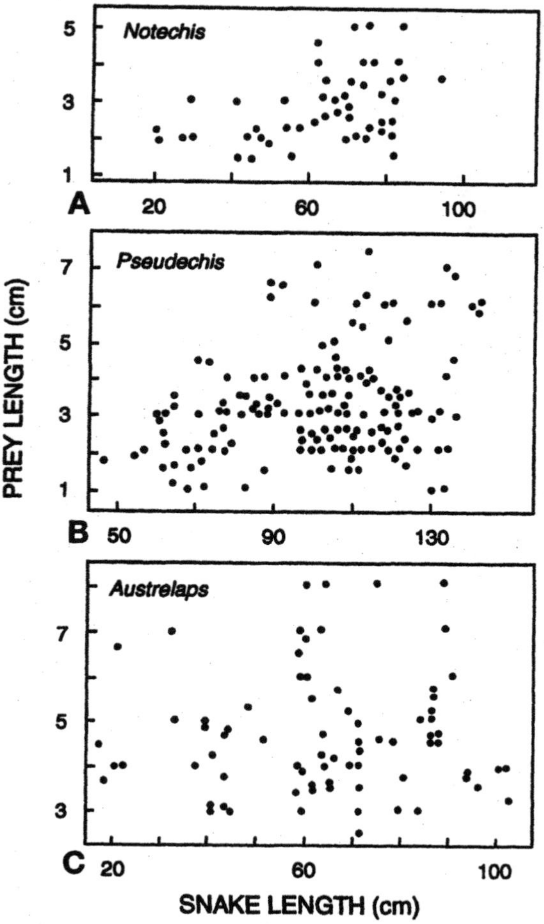

Figure 3.2 Prey-size–snake-size relationships in Australian elapids. (*A*) *Notechis scutatus* feeding on frogs (n = 83) (data from Shine, 1977). (*B*) *Pseudechis porphyriacus* feeding on frogs (n = 198) (data from Shine, 1977). (*C*) *Austrelaps* (three species) feeding on lizards (n = 147) (data from Shine, 1987).

plotted the cube root of prey mass against the cube root of snake mass for all three species (Fig. 3.3). This transformation should make variance about regression more homogeneous than on the original scale, but variances are still heterogeneous on the cube-root scale. It appears that larger specimens of *D. chloroticus* and *M. melanolomus* continue to eat small frogs (Fig. 3.3*A*,*B*). Larger specimens of the third species (*D. margaritiferus*), however, appear to drop small anurans out of the diet (Fig. 3.3*C*). That trend is especially clear when prey length is plotted against snake head length (Fig. 3.3*D*).

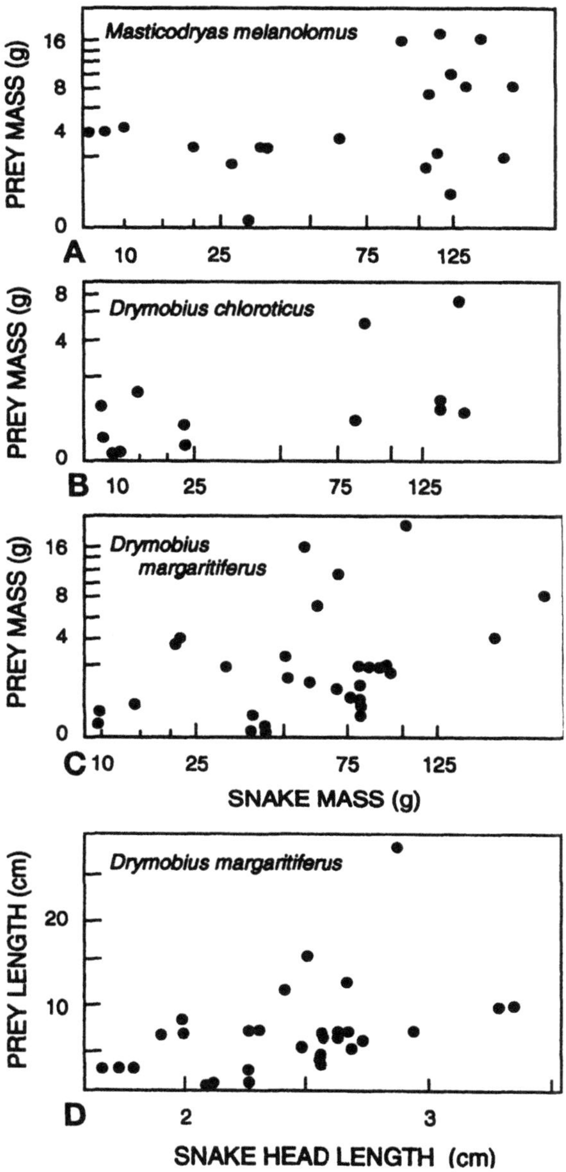

Figure 3.3 Prey-size–snake-size relationships in neotropical racers (data from Seib, 1984). (*A*) *Mastigodryas melanolomus* feeding on frogs, lizards, snakes, and mammals ($n = 20$). Cube-root scales on both axes. (*B*) *Drymobius chloroticus* feeding on frogs ($n = 13$). Cube-root scales on both axes. (*C*) *Drymobius margaritiferus* feeding on frogs and lizards ($n = 30$). Cube-root scales on both axes. (*D*) Same specimens as *C*. Linear scales on both axes.

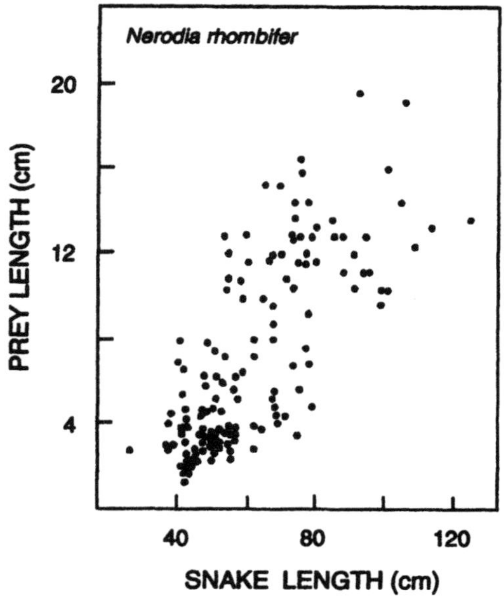

Figure 3.4 The prey-size–snake-size relationship in the Diamondback Water Snake (*Nerodia rhombifer*) (N = 134) feeding on fish (n = 194) (data from Plummer and Goy, 1984).

Plummer and Goy (1984) studied the diet of a large, freshwater natricine (*Nerodia rhombifer*) at a catfish hatchery. Channel Catfish (*Ictalurus punctatus*) dominated the diet, but Fathead Minnows (*Pimephales promelas*) and centrarchids were also eaten. A plot of fish length versus snake length (snout–vent length, SVL) shows a striking tendency for larger snakes to drop small fish from their diet (Fig. 3.4).

Godley et al. (1984) studied the diets of two crayfish-eating natricines (*Regina grahamii* and *R. septemvittata*). They found that crayfish gastroliths were a good predictor of crayfish size and used radiography of museum specimens of *Regina* to determine the gastrolith sizes of partially digested prey. Larger specimens of both species of *Regina* showed no tendency to drop small crayfish from the diet (Fig. 3.5).

Henderson et al. (1987, 1988) used museum specimens to study the diets of three species of semiarboreal colubrids (*Uromacer*) in Hispaniola. *Uromacer frenatus* is a slender-bodied, sit-and-wait forager that feeds only on lizards. Larger individuals of *U. frenatus* add *Leiocephalus* and *Ameiva* to their diets but continue to eat small *Anolis* (Fig. 3.6).

Cobb (1989) studied the diet of the Flathead Snake (*Tantilla gracilis*), a small, secretive colubrid, in Texas. The diet was dominated by

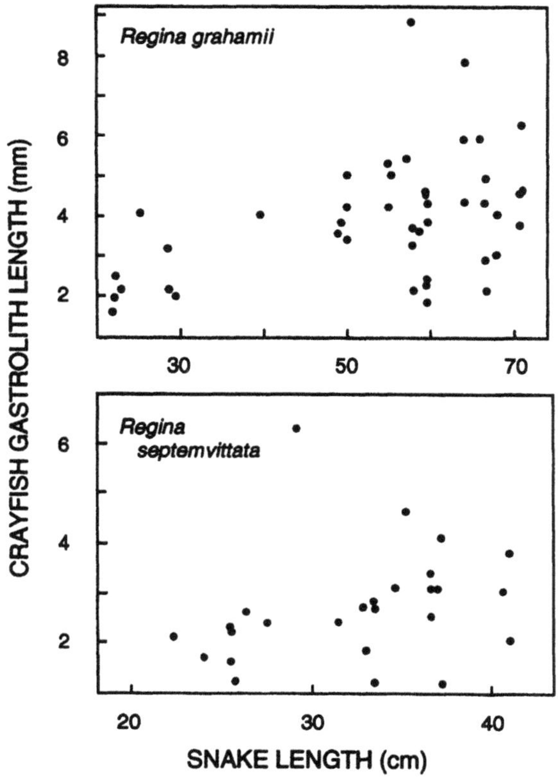

Figure 3.5 Prey-size–snake-size relationships in the crayfish specialists *Regina grahamii* (N = 25, n = 44) and *R. septemvittata* (N not reported, n = 26) (data from Godley et al., 1984). Crayfish body length is highly correlated with gastrolith length (r = 0.94).

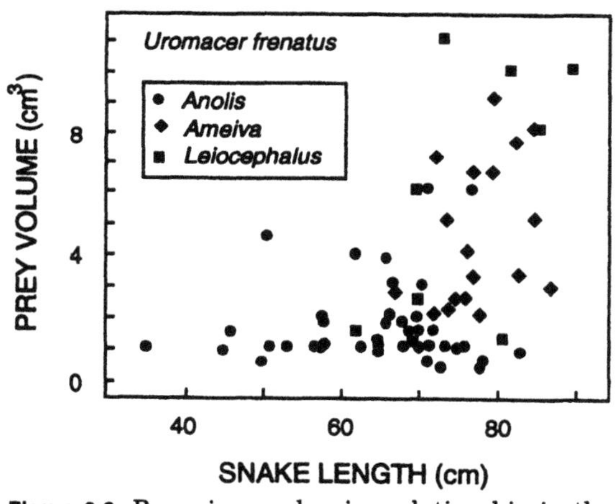

Figure 3.6 Prey-size–snake-size relationship in the Tree Snake *Uromacer frenatus* feeding on lizards (n = 108) (data from Henderson et al., 1987).

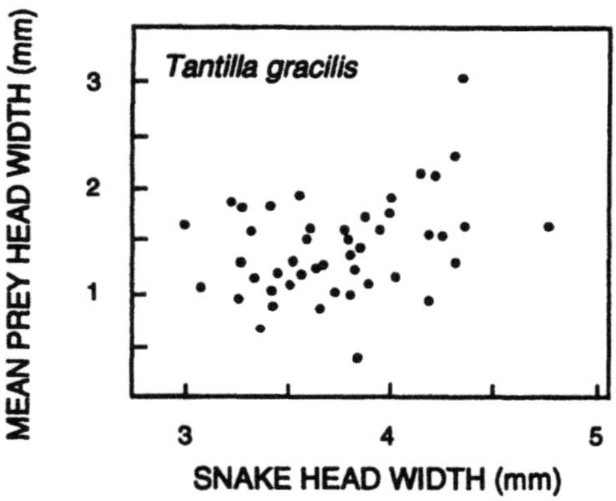

Figure 3.7 Prey-size–snake-size relationship in the colubrid snake, *Tantilla gracilis* ($N = 65$) feeding on arthropods and molluscs ($n = 158$) (data from Cobb, 1989).

beetle larvae (82% of items) and centipedes (11% of items). Diameters of snakes and their prey are compared in Fig. 3.7. Larger snakes take larger prey and appear to drop small prey from their diet.

Voris and Moffett (1981) studied the diet of Beaked Sea Snakes (*Enhydrina schistosa*) in Malaysia and produced an exemplary analysis of snake-size–prey-size relationships. One species of ariid catfish (*Tachysurus maculatus*) dominated the diet of the sea snakes. The authors reconstructed the original size of partially digested fish using measurements of intact fins and regression equations. Three measures of fish size (diameter, length, and mass) were plotted against three comparable measures of snake size (Fig. 3.8). In all three plots it is apparent that small fish are missing from the diet of larger snakes. Notice that a log transform yields uniform variance about regression in the fish mass versus snake mass plot.

Jayne et al. (1988) studied the diet of the homalopsine snake *Cerberus rynchops* in mangrove habitat in a river mouth in Malaysia. The diet was exclusively fish and was dominated by four species of oxydercine gobies. The authors analyzed total mass of prey for each snake as a function of snake mass. A log-log transformation was used to make variance about regression uniform. The resulting plot (Fig. 3.9) indicates that larger snakes have a larger mass of prey in their stomachs. The plot also suggests that larger snakes drop small prey from their diets, but this needs to be confirmed by plotting the sizes of individual fish against snake size. Because Jayne et al. (1988) found that larger snakes showed no tendency to have more fish in their stomachs, a plot of individual fish as a function of snake size should look very much like Fig. 3.9.

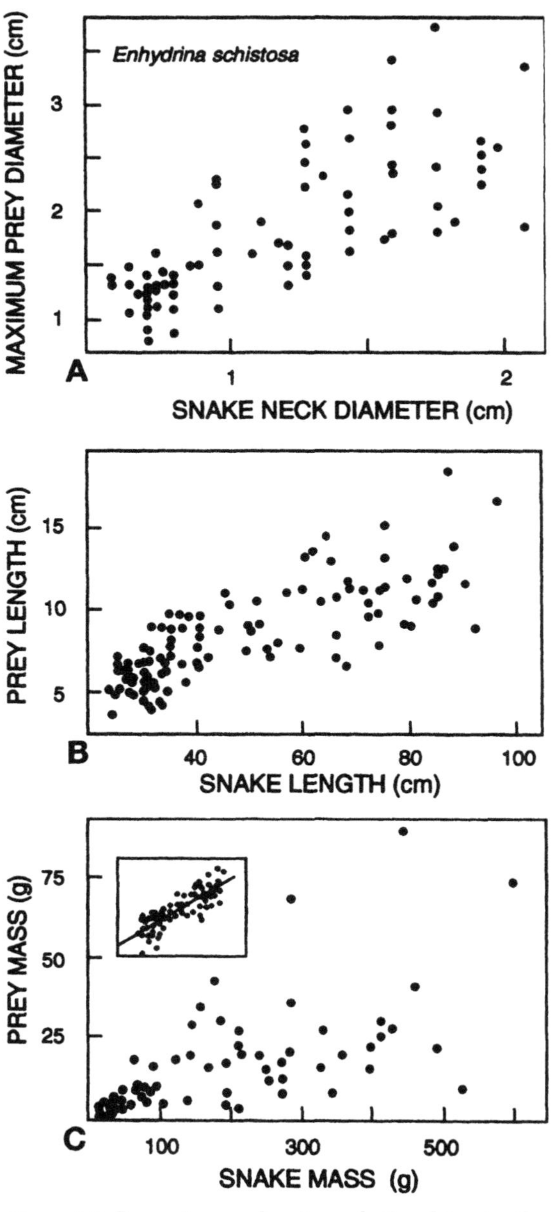

Figure 3.8 Prey-size–snake-size relationships in the Sea Snake *Enhydrina schistosa* feeding on catfish (data from Voris and Moffett, 1981). (*A*) Maximum prey diameter as a function of snake neck diameter (*N* = 94). (*B*) Prey length as a function of snake length (*N* = 106). (*C*) Prey mass as a function of snake mass (*N* = 104). Inset shows data after log-log transformation.

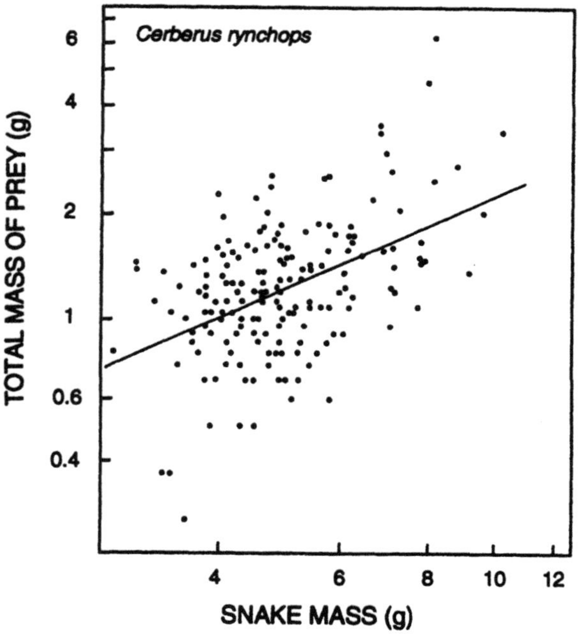

Figure 3.9 Prey-size–snake-size relationship in the homalopsine snake *Cerberus rynchops* ($N = 181$) feeding on fish (data from Jayne et al., 1988). Logarithmic scales on both axes.

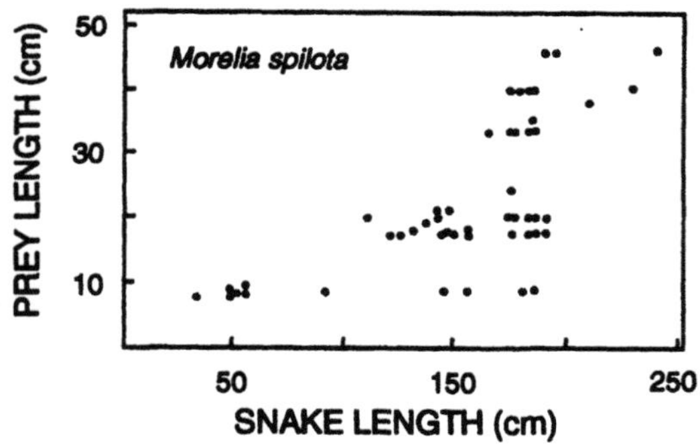

Figure 3.10 Prey-size–snake-size relationship in the Diamond Python *Morelia spilota* ($N = 49$, $n = 57$) feeding on rodents (data from Slip and Shine, 1988).

Slip and Shine (1988) found that Diamond Pythons (*Morelia spilota*) in Australia fed predominantly on mammals (mainly *Rattus*). A companion study of radiotelemetered snakes suggested that *M. spilota* is an ambush predator. Stomach analyses (Fig. 3.10) indicated that large snakes continued to eat small prey.

In summary, as most snake species grow larger they drop small prey from their diets. This deletion trend is apparent in all the studies just reviewed with the exception of Australian elapids (Shine, 1977, 1987), a boid (Slip and Shine, 1988), crayfish-eating natricines (Godley et al., 1984), and possibly some neotropical racers (Seib, 1984). A tendency for larger snakes to drop small kinds of prey from the diet has been noticed in a large number of studies in which snake size was measured but prey size was not (Fitch, 1963, 1965; Godley, 1980; Saint Girons, 1980; Mushinsky et al., 1982). In the next section we shall consider possible explanations for the prey-deletion phenomenon and how they might be tested. But before leaving empirical studies of snake-size–prey-size relations, let us consider some methodological issues.

It could be argued that larger snakes continue to eat small prey but the lower limit for prey size increases as a statistical artifact. William Magnusson (personal communication) has pointed out that if we plot the average prey size in each snake's stomach as a function of snake size, then we expect the lower limit for prey size to increase with snake size even if snakes continue to eat small prey. The reason for the illusion of prey deletion is that the variance of the mean decreases with sample size (Sokal and Rohlf, 1981, p. 183). If larger snakes have larger samples of prey in their stomachs, the lower limit of prey size might increase as a simple statistical consequence. This explanation, however, does not appear to apply to any of the studies reviewed above (with the possible exception of Cobb, 1989). Apparently, the data points in all the plots (except Cobb, 1989) represent individual prey items, not the averages of items in individual stomachs. In the study by Jayne et al. (1988) the total mass of stomach content was plotted against snake mass, but the authors also report that prey number is unrelated to snake size. Magnusson's interpretation should be considered, however, in future studies of prey-size–snake-size relations. Although visual inspection of a plot based on individual prey items is a good first start at determining whether the lower prey-size limit increases with snake size, such a plot is not the best form of data for describing the average relationship between prey size and snake size.

Perhaps the best procedure for estimating the regression slope relating prey size to snake size is to use the average prey size for each snake. This approach takes account of the likely possibility that the sizes of prey within each snake are correlated. In other words, the individual snake stomachs can be treated as independent data points even though all the individual prey items may not be independent data. One can account for the variation in number of items per stomach by using a weighted regression. Since the variance in mean prey size is inversely related to the number of items per stomach, a natural weighting scheme is to weight each snake stomach by the number of items it contains (Neter et al., 1990). Weighted regression is available in some computer packages (e.g., SAS™).

In all of the studies we have reviewed, variance in prey size increases with snake size. While biologically interesting, such inconstancy of variance violates a modeling assumption that is made for the purposes of testing the statistical significance of the regression slope. Transformations such as the log-log transform used by Voris and Moffett (1981) may make the variance about regression uniform and so render tests of significance more trustworthy. Another advantage of such log-log plots (known as allometric plots) is that scaling relationships can be easily interpreted from the regression coefficients (Schmidt-Nielsen, 1984).

The quantitative characterization of the lower or upper prey-size limit as a function of snake size is a difficult statistical problem. Voris and Moffett (1981) and Seib (1984) quantified the lower limit by using only the 10 or 11 prey items along the lower limit in a regression of prey size on snake size. The problem with this approach is that the decision of which points to include is arbitrary. Maller et al. (1983) discuss two other approaches that might be applicable to the problem. (I am grateful to W. Magnusson for this reference.)

Quantifying the prey-size–snake-size relationship from a field collection (Figs. 3.2 to 3.10) can tell us whether large snakes delete small prey from their diet but it cannot tell us why the deletion takes place. Once the field study has performed the important function of identifying which prey types of what size are deleted from the diets of which-sized snakes, we can use insights from foraging theory to formulate hypotheses that might answer the *why* question.

Why are small prey deleted from the diet of large snakes?—The conversion of worthy prey to worthless prey with increasing predator size. Schoener (1971) gives a useful model for the costs and benefits of foraging that can be modified to reflect snake biology. Let us recognize three possible outcomes for a snake that has encountered a prey item: (1) no capture; (2) capture but no ingestion (e.g., the prey escapes during the ingestion process); and (3) capture, ingestion, and digestion. These three outcomes are mutually exclusive and hold for each of the various prey types that the snake might encounter. To evaluate the economics of feeding on various types of prey we need to know the probabilities of the three outcomes for each prey type, as well as their costs and benefits. The three probabilities can be expressed usefully as functions of the probabilities of transition between the various possible events (Fig. 3.11). Thus, the probability of an outcome from a given encounter is the product of the various transition probabilities that lie along the relevant path in Fig. 3.11. For example, the probability that the snake

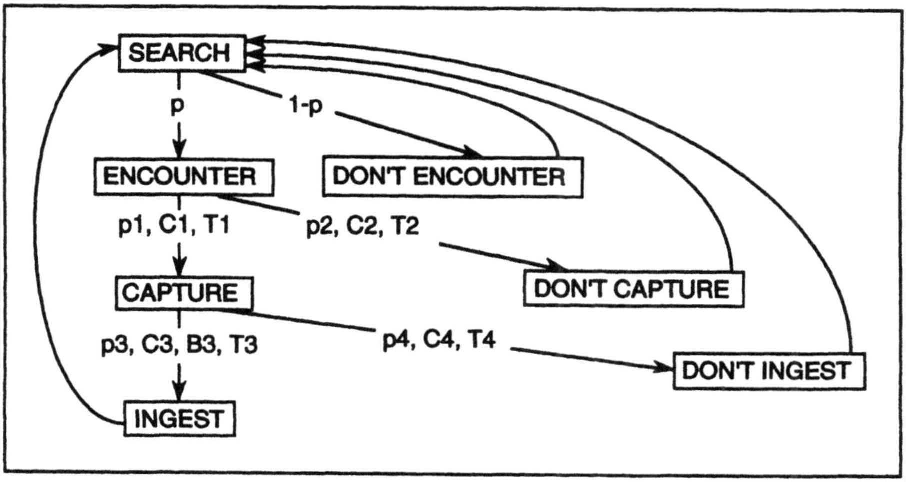

Figure 3.11 Foraging activities in the model and possible transitions between them. p and p1–p4 are transition probabilities. C1–C5 are energetic costs. B3 is benefit or energetic gain. T1–T5 are times spent in various activities.

will not capture the prey it has encountered is p2. The probability of encounter, capture, but failed ingestion is p1 p4. The probability of capture, ingestion, and digestion is p1 p3. These three probabilities correspond to the three possible outcomes given an encounter and so sum to one. Symbols for the benefits and costs of various transitions are also shown in Fig. 3.11. Only successful prey ingestion benefits the snake. Benefit is denoted B3 after capture in Fig. 3.11. All of the various events could incur costs in the form of energy expenditure and are denoted C2 for pursuit costs, etc. The various events could also incur costs in terms of time expenditure and these are denoted T1, T2, etc.

The ingredients just listed can be used to express the net energy expected from a particular prey type,

$$
\begin{aligned}
e_i = [&\text{expected energy gain from successful ingestion} = \text{p1 p3 B3} \\
&- \text{expected cost of successful prey pursuit} = \text{p1 p3 C1} \\
&- \text{expected cost of successful prey ingestion} = \text{p1 p3 C3} \\
&- \text{expected cost of prey digestion} = \text{p1 p3 C5} \\
&- \text{expected cost of failed capture attempts} = \text{p2 C2} \\
&- \text{expected cost of failed ingestion attempts} = \text{p1 p4 C4}].
\end{aligned} \tag{1}
$$

The first four terms (involving B3, C1, C3, and C5) represent the net benefit of a successful prey encounter that goes all the way to digestion. Similarly we can express the expected time devoted to pursuing, ingesting, and digesting the *i*th prey type as

t_i = [expected time spent in successful pursuit = p1 p3 T1

+ expected time spent in successful ingestion = p1 p3 T3

+ expected time spent in prey digestion = p1 p3 T5 (2)

+ expected time spent in failed capture attempts = p2 T2

+ expected time spent in failed ingestion attempts = p1 p4 T4].

All of the variables in the preceding expressions should carry an *i* subscript to indicate that they pertain to the *i*th prey type, but those subscripts have been left off for simplicity.

The ratio e_i/t_i represents the net energy expected per unit time from the *i*th prey item given encounter. This ratio can be used to rank prey times on the basis of their profitability. Even with a ranking of all prey items, however, it would be hard to predict where the cutoff point should be. What is the least profitable prey type that should nevertheless be included in the diet? A useful mathematical approach is to cast the issue as a standard maximization problem. As Schoener (1971) put it, we need to choose a currency and an appropriate cost–benefit function and then solve for the maximum. We need a function that includes more than e_i and t_i for each prey type. In particular, we need to take the relative availability of prey into account and also the costs of prey search. The simplest way to incorporate search costs is to assume that the snake searches for all prey types simultaneously (i.e., that there are no special search costs associated with different prey types). Schoener (1971) proposed maximizing the following function:

$$\frac{\sum_{i=N_1}^{N_2} p_i(e_i) - Cs\ Ts}{\sum_{i=N_1}^{N_2} p_i(t_i) + Ts}, \qquad (3)$$

where p_i is the probability of encountering the *i*th prey type, *Cs* is the cost per unit time of searching for prey, *Ts* is the expected time between prey encounters (whether or not they result in prey capture). Thus, *Cs Ts* is the expected energy expended in prey search for each prey encountered. The numerator gives the expected net energy gain from search, pursuing, ingesting, and digesting all the prey types that are included in the diet. The denominator gives the expected time spent in these activities. We wish to maximize their ratio, which is the rate of energy gain for a particular diet. N_1 denotes the most profitable prey type (i.e., the prey with the highest e_i/t_i). The maximization problem is to find the least profitable prey type (denoted N_2) that should be included in the diet.

Some insights can be gained without amassing all the relevant data needed to solve the diet optimization problem for a snake population. Many models of optimal foraging, including the present one, predict that a greater range of prey types should be taken when the overall food abundance is lower (longer Ts) (Schoener, 1971). Also, larger predators should take larger prey when one assumes exponential forms for handling and ingestion times (Schoener, 1971). Finally, some prey should be dropped from any diet because they are absolutely worthless or net the snake too small an energy gain. We can recognize a number of varieties of worthless prey using Eqs. (1)–(3).

Varieties of worthless prey

The first four varieties that we will consider are absolutely worthless. That is to say, the expected energy gain from these prey is zero or negative. On the average the snake gains nothing or actually loses energy in the process of hunting, pursuing, and trying to ingest these prey. In other words, Eq. (1) or the term that occurs in the numerator of Eq. (3) and pertains to a particular prey type is zero or negative.

Hard-to-find prey. Finding such prey involves a special and possibly costly search (large prey-specific $Cs\ Ts$ term). This search is special in the sense that the snake cannot search for or encounter other prey while looking or waiting for the prey in question (see Shine, 1991). Prey that occur only in localized habitats fall into this category. For this aspect of prey to raise the lower prey limit as snake size increases, larger snakes must incur larger search costs than small snakes.

For many fish-eating snakes an ontogenetic shift to larger fish may mean deleting small fish from the diet because large and small fish occupy different habitats. At Eagle Lake in northern California, for example, small minnows that are preyed on by juvenile Western Terrestrial Garter Snakes (*Thamnophis elegans*) are found most abundantly in shallow, warm water at the lake's edge. In contrast, larger minnows are found only in deeper water. Because adult snakes ambush and pursue large minnows at a depth of 1–2 m, they necessarily miss encounters with schools of small minnows.

Hard-to-catch prey. Such prey are unprofitable because they are likely to escape once encountered or because the cost of a failed capture attempt is high (large p3 C1 or p2 C2). Elusive prey such as fish fall into the first category. Racers and other active pursuers (e.g., *Coluber, Masticophis, Psammophis*) may incur substantial costs in pursuing prey but we need actual data to test this possibility. An ontogenetic shift away from elusive prey could occur if large snakes are more inept at catching small prey than are small snakes or if large snakes incur greater costs.

Hard-to-eat prey. Such prey are unprofitable because ingestion is likely to be unsuccessful or costly (large p3 C3 or p4 C4). Anguid lizards, for example, can thwart snake predation by grasping some part of the snake's anatomy or even their own tail in their jaws (Fig. 3.12; Fitch, 1935). Slugs and salamanders can impede or even thwart ingestion with sticky secretions (Figs. 3.13, 3.14; Arnold, 1982). The most likely circumstance that could cause an ontogenetic shift away from such prey is if large snakes are more inept at prey handling than small snakes. Such a circumstance is not inconceivable, but it seems more likely that large snakes would be more capable of overwhelming the defenses of hard-to-eat prey than small snakes.

The analysis of trials involving hard-to-eat prey could be accomplished in the laboratory. Events leading up to prey encounter depend on a large number of special ecological circumstances, but predatory outcome after prey capture can be accurately scored and measured in the laboratory if a little care is taken. Special ecological factors that need to be controlled are the body temperatures of the snake and its prey, their psychological states, and perhaps the substrate (which might adhere to sticky prey). Arnold (1982) discusses scoring predatory outcome and statistical analysis. The anaerobic costs of prey handling can be assessed by dropping the snake into liquid nitrogen at the end of the encounter and later measuring its lactic acid content (Feder and Arnold, 1982). The aerobic costs of prey handling could be assessed by staging encounters in sealed containers and measuring oxygen consumption, as in the study by Pough and Andrews (1985) of the lizard *Chalcides ocellatus.*

Feder and Arnold (1982) used the liquid-nitrogen technique to stop the metabolic action after predatory encounters between Western Terrestrial Garter Snakes (*Thamnophis elegans*) and Woodland Salamanders (*Plethodon jordani*). Anaerobic costs were measured directly and other costs were estimated. The average time to ingest a salamander was 14.4 min. A typical 16-g snake that ingested a 2.2-g salamander with an energy content of about 2000 cal might expect to pay as much as 6.7 cal in aerobic metabolism and 2.4 cal in anaerobic metabolism during capture and ingestion plus another 6.1 cal to pay off the incurred oxygen debt and 260 cal in digestion costs. Thus, the

Figure 3.12 A Racer (*Coluber constrictor*) during an unsuccessful attempt to swallow an anguid lizard (*Gerrhonotus multicarinatus*) that thwarts ingestion by holding its tail in its mouth.

Figure 3.13 Predation by the Western Terrestrial Garter Snake (*Thamnophis elegans*) on the slug (*Ariolimax columbianus*). (*A*) The snake during attack on the slug. (*B*) After ingestion the snake's mouth is filled with slug mucus, which also causes the snake's head to adhere to the substrate.

total energetic costs of the encounter are no more than 15.2 cal, while the energetic content of the prey (subtracting digestion costs) is at least 1740 cal. Thus, the energetic cost of the encounter is less than 1% of the energy assimilated from the prey (Feder and Arnold, 1982).

We can use the preceding results to estimate how elusive salamanders would have to be for them to be deleted from the diet on energetic grounds. From our discussion so far we have B3 = 2000, C1 + C3

Figure 3.14 A plethodontid salamander (*Ensatina eschscholtzii*) thwarts ingestion by a Western Terrestrial Garter Snake (*Thamnophis elegans*) by wrapping its tail around the snake's neck. Sticky secretions have been released from the salamander's tail, causing it to adhere to the snake's neck.

= 15.2, and C5 = 260. Feder and Arnold (1982) found that salamanders escaped from snakes in 32% of encounters, so p4 = 0.32 and p3 = 0.68. Let us assume that the cost of a failed capture attempt is about 1% of the costs of successful capture and ingestion (C2 = 2) and that the cost of a failed ingestion attempt is about the same as the costs of successful capture and ingestion (C4 = 15.2). Using Eq. (1) we can solve for probability of successful capture given an encounter (p1) that will make net energy gain positive (e_i>0). We find that even if only 1 in 1000 capture attempts is successful, salamanders should still be retained in the diet. In other words, salamanders would have to be so elusive that they escaped capture in more than 99.915% of encounters to be deleted from the diet. While it seems unlikely that salamanders could ever be this elusive, fish may very well have such high escape rates from snakes.

Low-energy prey. Such prey are unprofitable because even when ingestion is successful, net energy gain is small. In particular, the net gain may be small or negative when the usable energy content of the prey (B3) is low and/or the costs of capture, ingestion, and digestion (C1–C5) are high. Low-energy prey are candidates for ontogenetic shifts if large snakes experience larger capture, ingestion, or digestion costs, because those costs may overwhelm the possible energy gain from such prey.

Some snake prey have a relatively low-energy content. A slug with the same live weight as a mouse has only about one-quarter the energy content (Table 3.1). Adult *Thamnophis elegans* in at least some populations appear to drop slugs from their diet while adding

TABLE 3.1. Average Energy Content of Major Prey Categories (Data from Cummins and Wuycheck, 1971). Values Flagged with Asterisks Are Based on Small Samples

Prey	Ash-free dry weight, cal/g	Wet weight, cal/g
Mollusks	5492	480
Annelids	5628	782*
Arthropods	5673	2315*
Fishes	5296	1493*
Tetrapods	6542	1853

mice (C. R. Peterson, personal communication). An energetic comparison of slug and mouse predation in this snake species might be an informative exercise.

Marginal prey. The snake turns a profit from these prey but the gain rate (e_i/t_i) is so small relative to other prey types that the snake's overall energy budget is improved by dropping marginal prey from the diet. In contrast to the first four varieties of unprofitable prey, marginal prey are not absolutely worthless. The two important causes of marginal status are low net energy gain and large time expenditure. In the preceding categories we discussed the energetic aspect of predation but not time expenditure. When search, capture, or ingestion are energetically costly they are likely to be time consuming as well. Time spent in these activities cannot be spent in predation on more profitable prey, in thermoregulation, in mating, or in other activities. Perhaps more importantly, time spent in predatory activities exposes the snake to predation. At present we do not know whether exposure to predation constitutes a sizable hazard. Brodie's (1993) recent work with snake models, however, provides a field protocol for measuring at least part of the hazard. It may be that predatory costs are so trivial [as Feder and Arnold's (1982) results suggest] that time is the informative currency for evaluating the economics of snake predation.

Pough and Groves (1983) measured swallowing time in a series of trials in which snake size and prey size were varied systematically. These authors used number of maxillary protractions as an index of handling time because that index was not sensitive to disturbance of the snake. During feeding trials the authors noted that snakes would sometimes cease all swallowing movements for several minutes. If such bouts of inactivity are not induced by human interference, they should be included in swallowing time if the aim is to pursue an energetic model of the type specified in Eq. (3). Nevertheless, it is interesting to note that the number of maxillary protractions shows a linear relationship to prey size over a more than fivefold range in prey diameter (Fig. 3.15*A*).

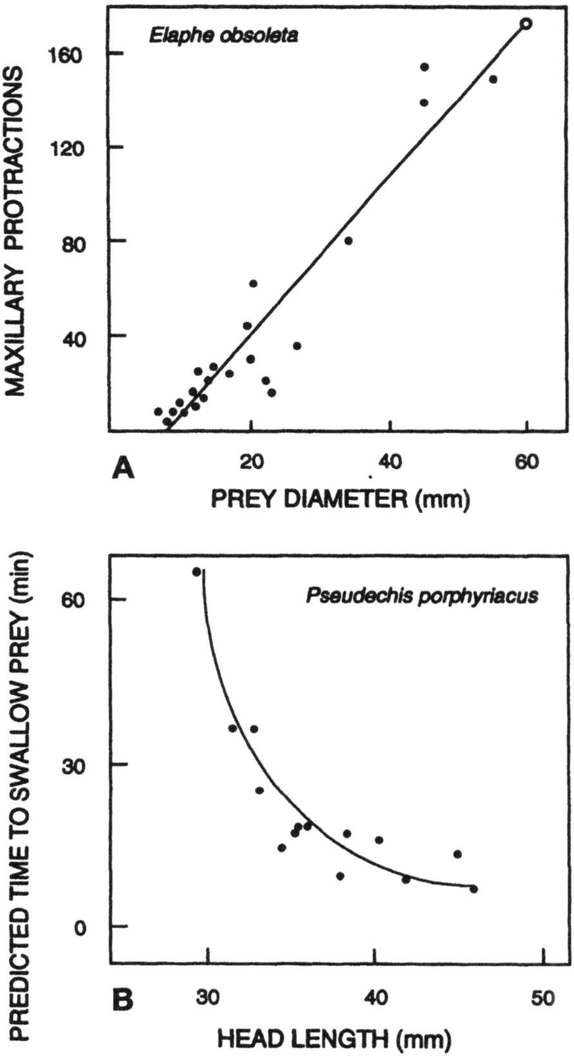

Figure 3.15 Swallowing time as a function of prey size and snake size. (*A*) An individual Rat Snake (*Elaphe obsoleta*) feeding on rodents (data from Pough and Groves, 1983). Swallowing time is measured as the number of maxillary protractions during ingestion. Solid circles denote successful ingestion of prey. The single unsuccessful ingestion attempt is shown with an open circle. (*B*) Predicted time to swallow a 50-g mouse as a function of snake head length in Black Snakes (*Pseudechis porphyriacus*) (data from Shine, 1991). Curve was fitted by eye.

Jayne et al. (1988) measured handling times for the homalopsine snake *Cerberus rynchops* feeding on live fish (*Periophthalmus chrysospilos*). These workers recognized three phases of prey handling: (1) an initial quiescent phase (during which the motionless snake holds the fish in its mouth); (2) positioning time (the time from the onset of jaw-walking until the snake positions the fish's snout in its mouth); and (3) swallowing time. Fish size and snake size were significant predictors of the durations of all three phases of prey handling as well as total handling time (in all cases larger fish increased handling durations and larger snakes decreased handling times). The duration of fish struggling affected only the duration of the initial quiescent phase, while the initial position where the fish was seized affected only the duration of the positioning phase.

Shine (1991) conducted a particularly illuminating study of ingestion times in Black Snakes (*Pseudechis porphyriacus*) and Diamond

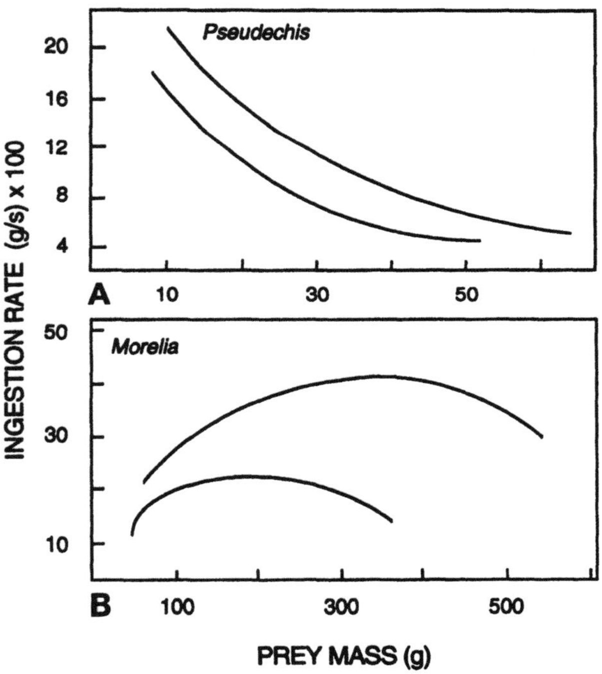

Figure 3.16 Prey mass ingestion rates as a function of prey size for individual Black Snakes (*Pseudechis porphyriacus*) and Diamond Pythons (*Morelia spilota*). Curves were fitted by eye to the data of Shine (1991). (*A*) Ingestion curves for two individual Black Snakes. (B) Ingestion curves for two individual pythons.

Pythons (*Morelia spilota*) handling dead mice and rats. In both species ingestion times increased with prey size and decreased with snake size (Fig. 3.15*B*). Shine's most surprising result was that the ratio of prey mass to ingestion time for individual snakes declined steadily with increasing prey mass in *Pseudechis* but showed an intermediate peak in *Morelia* (Fig. 3.16). Thus, individual Diamond Pythons were relatively inefficient when ingesting both very large and very small prey.

One of the many unresolved issues involving swallowing time is snake performance in the vicinity of the breaking point. Arnold (1982) defined the breaking point as the smallest size of prey that the snake is incapable of ingesting. For example, Pough and Grove's (1983) results for a Rat Snake (Fig. 3.15*A*) suggest that its breaking point is in the vicinity of 55–60 mm prey diameter, although more trials with prey 55 mm and larger would be required to establish this. Does swallowing time increase disproportionally fast as the breaking point is approached? Shine's (1991) results for Black Snakes (*Pseudechis porphyriacus*) suggest that swallowing time increases exponentially as the breaking point is approached (Fig. 3.15*B*).

Behavioral Basis of Ontogenetic Shifts in Diet

We can imagine that an ontogenetic diet shift simply reflects the experience and changing fortunes of the growing snake. Such a purely ecological perspective on diet shifts is challenged by an important study conducted by Mushinsky and Lotz (1980). By rearing Water Snakes (*Nerodia*) on specified diets and periodically testing their chemoreceptive responses to various prey, Mushinsky and Lotz were able to show that the intrinsic perceptual program of at least one species shows an interesting maturational change. Young *N. erythrogaster* show a strong tongue-flicking reaction to fish odor that persists for several months whether the snakes are reared on fish or frogs. At 8–9 months of age *N. erythrogaster* show a strong reaction to frog odor, regardless of their rearing diet. In particular, even snakes reared on fish show a strong reaction to frog odor at 8–9 months. Analysis of stomach contents indicates that *N. erythrogaster* show an ontogenetic diet shift from fish to frogs (Mushinsky et al., 1982). Mushinsky and Lotz's (1980) results indicate that an intrinsic maturational change is the proximate factor determining the diet change in the field. Of the many species that show ontogenetic shifts in diet, *N. erythrogaster* and *N. fasciata* are apparently the only species in which maturational reactions to prey have been studied.

Summary and Future Research

In the last decade a number of snake ecologists have succeeded in quantifying the relationship between prey size and snake size in a number of snake species. Although the number of studies is too few to reveal broad ecological or phylogenetic patterns, several trends stand out. Larger snakes tend to eat larger prey species. More specifically, larger snakes take larger individual prey items, and they often add larger prey species to their diet. The range and variance in prey size also increase with snake size. An enigmatic result is that in many snake species larger individuals drop small prey items from the diet. To date this trend seems characteristic of fish-eating snakes.

Foraging theory can be used to erect hypotheses to explain why the lower prey-size limit increases as a function of snake size. To test these hypotheses we will need to measure the costs and benefit of prey pursuit, capture, and ingestion, while varying both prey size and snake size. Various experimental studies with lizards (DeMarco et al., 1985; Pough and Andrews, 1985) and snakes illustrate how particular aspects of such a program could be tackled. What is particularly needed is an allometric approach to snake-feeding performance and predatory costs that is firmly rooted in field results. In particular, field studies of diet could be used to identify the prey types and sizes to manipulate in the laboratory (Jayne et al., 1988; Shine, 1991).

The behavioral basis of ontogenetic diet shift is another potentially important but neglected topic. One pathbreaking study illustrates how to combine experimental and observational approaches.

Acknowledgments

An abbreviated version of this manuscript was presented at the First World Congress of Herpetology at Canterbury, England, in September 1989, as part of a symposium on snake ecology and behavior organized by R. Shine and H. Saint Girons. The preparation of this manuscript was supported by National Science Foundation Grant BSR 91-19588.

Literature Cited

Arnold, S. J., 1981, The microevolution of feeding behavior, in A. Kamil and T. Sargent, eds., *Foraging Behavior: Ecological, Ethological and Psychological Approaches*, Garland Press, New York, pp. 409–453.

Arnold, S. J., 1982, A quantitative approach to antipredator performance: Salamander defense against snake attack, *Copeia*, 1982:247–253.

Arnold, S. J., 1983, Morphology, performance and fitness, *Amer. Zool.*, 23:347–361.

Arnold, S. J., 1988, Behavior, energy and fitness, *Amer. Zool.*, 28:815–827.

Brodie, E. D., III, 1993, Differential avoidance of coral snake banded patterns by free-ranging avian predators in Costa Rica, *Evolution*, 47:227–235.

Carpenter, C. C., 1952, Comparative ecology of the common garter snake (*Thamnophis s. sirtalis*), the ribbon snake (*Thamnophis s. sauritus*), and Butler's garter snake (*Thamnophis butleri*) in mixed populations, *Ecol. Monogr.*, 22:235–258.
Catling, P. M., and B. Freedman, 1980, Food and feeding behavior of sympatric snakes at Amherstburg, Ontario, *Can. Field-Nat.*, 94:28–33.
Cobb, V. A., 1989, The foraging ecology and prey relationships of the flathead snake, *Tantilla gracilis*, Master's Thesis, Univ. Texas, Tyler.
Crow, J. F., and M. Kimura, 1970, *An Introduction to Population Genetics Theory*, Burgess, Minneapolis.
Cummins, K. W., and J. C. Wuycheck, 1971, Caloric equivalents for investigations in ecological energetics, *Int. Vereinigung für Theor. u. Angewand, Limnologie, Mitteilung No.*, 18:1–147.
DeMarco, V. G., R. W. Drenner, and G. W. Ferguson, 1985, Maximum prey size of an insectivorous lizard, *Sceloporus undulatus garmani*, *Copeia*, 1985:1077–1080.
Feder, M. E., and S. J. Arnold, 1982, Anaerobic metabolism and behavior during predatory encounters between snakes (*Thamnophis elegans*) and salamanders (*Plethodon jordani*), *Oecologia (Berlin)*, 53:93–97.
Fitch, H. S., 1935, Natural history of the alligator lizards, *Trans. St. Louis Acad. Sci.*, 29:1–38.
Fitch, H. S., 1960, Autecology of the copperhead, *Univ. Kansas Publ. Mus. Nat. Hist.*, 13:85–288.
Fitch, H. S., 1963, Natural history of the racer *Coluber constrictor*, *Univ. Kansas Publ. Mus. Nat. Hist.*, 15:351–468.
Fitch, H. S., 1965, An ecological study of the garter snake, *Thamnophis sirtalis*, *Univ. Kansas Publ. Mus. Nat. Hist.*, 15:493–564.
Fitch, H. S., and H. Twining, 1946, Feeding habits of the Pacific rattlesnake, *Copeia*, 1946:64–71.
Godley, J. S., 1980, Foraging ecology of the striped swamp snake, *Regina alleni*, in southern Florida, *Ecol. Monogr.*, 50:411–436.
Godley, J. S., R. W. McDiarmid, and N. N. Rojas, 1984, Estimating prey size and number in crayfish-eating snakes, genus *Regina*, *Herpetologica*, 40:82–88.
Gould, S. J., and R. C. Lewontin, 1979, The spandrels of San Marco and the Panglossian paradigm: A critique of the adaptationists programme, *Proc. Roy. Soc. Lond. B*, 205:581–598.
Greene, H. W., 1983, Dietary correlates of the origin and radiation of snakes, *Amer. Zool.*, 23:431–441.
Hamilton, W. J., Jr., 1951, The food and feeding behavior of the garter snake in New York State, *Amer. Midl. Nat.*, 46:385–390.
Hamilton, W. J., Jr., and J. A. Pollack, 1956, The food of some colubrid snakes from Fort Benning, Georgia, *Ecology*, 37:519–526.
Hasegawa, M., and H. Moriguchi, 1989, Geographic variation in food habits, body size and life history traits of the snakes on the Izu Islands, in M. Matui, T. Hikida, and R. C. Goris, *Current Herpet. in East Asia*, The Herpetological Society of Japan, pp. 414–432.
Henderson, R. W., and H. S. Horn, 1983, The diet of the snake *Uromacer frenatus dorsalis* on Ile de la Gonave, Haiti, *J. Herpetol.*, 17:409–412.
Henderson, R. W., A. Schwartz and T. A. Noeske-Hallin, 1987, Food habits of three colubrid tree snakes (genus *Uromacer*) on Hispaniola, *Herpetologica*, 43:241–248.
Henderson, R. W., T. A. Noeske-Hallin, B. I. Crother, and A. Schwartz, 1988, The diets of hispaniolan colubrid snakes. II. Prey species, prey size, and phylogeny, *Herpetologica*, 44:55–70.
Jayne, B. C., H. K. Voris and K. B. Heang, 1988, Diet, feeding behavior, growth, and numbers of a population of *Cerberus rynchops* (Serpentes: Homalopsinae) in Malaysia, *Fieldiana*, 50:1–15.
Krebs, J. R., and R. H. McCleery, 1984, Optimization in behavioural ecology, in J. R. Krebs and N. B. Davies, eds., *Behavioural Ecology: An Evolutionary Approach*, Sinauer, Sunderland, Massachusetts, pp. 91–121.
Lande, R., 1979, Quantitative genetic analysis of multivariate evolution, applied to brain: Body size allometry, *Evolution*, 33:402–416.

Lande, R., and S. J. Arnold, 1983, The measurement of selection on correlated characters, *Evolution,* 37:1210–1226.
Lewontin, R. C., 1978, Fitness, survival, and optimality, in D. H. Horn, R. Mitchell and G. R. Stairs, eds., *Analysis of Ecological Systems,* Ohio State Univ. Press, Columbus, pp. 3–21.
MacArthur, R. H., and E. R. Pianka, 1966, On the optimal use of a patchy environment, *Amer. Natur.,* 100:603–609.
Maller, R. A., E. S. de Boer, L. M. Joll, D. A. Anderson and J. P. Hinde, 1983, Determination of the maximum foregut volume of western rock lobsters (*Panulirus cygnus*) from field data, *Biometrics,* 39:543–551.
Mori, A., J. Daming, H. Moriguchi and M. Hasegawa, 1989, Food habits of snakes in east Asia: A biogeographical approach to resource partitioning, in M. Matui, T. Hikida, and R. C. Goris, *Current Herpet. in East Asia,* The Herpetological Society of Japan, pp. 433–436.
Mushinsky, H. R., 1987, Foraging ecology, in R. A. Seigel, J. T. Collins and S. S. Novak, eds., *Snakes: Ecology and Evolutionary Biology,* McGraw-Hill, New York, pp. 302–334.
Mushinsky, H. R., J. J. Hebrard and D. S. Vodopich, 1982, Ontogeny of water snake foraging ecology, *Ecology,* 63:1624–1629.
Mushinsky, H. R., and K. H. Lotz, 1980, Chemoreceptive responses of two sympatric water snakes to extracts of commonly ingested prey species, *J. Chem. Ecol.,* 6:523–535.
Neter, J., W. Wasserman, and M. H. Kutner, 1990, *Applied Linear Statistical Models, Regression, Analysis of Variance, and Experimental Designs,* 3rd ed., Irwin, Homewood.
Plummer, M. V., and J. M. Goy, 1984, Ontogenetic dietary shift of water snakes (*Nerodia rhombifera*) in a fish hatchery, *Copeia,* 1984:550–552.
Pough, F. H., and J. D. Groves, 1983, Specializations of the body form and food habits of snakes, *Amer. Zool.,* 23:443–454.
Pough, F. H., and R. M. Andrews, 1985, Energy costs of subduing and swallowing prey for a lizard, *Ecology,* 66:1525–1533.
Pyke, G. H., H. R. Puliam, and E. L. Charnov, 1977, Optimal foraging: A selective review of theory and tests, *Q. Rev. Biol.,* 52:137–154.
Saint Girons, H., 1980, Modifications sélectives du régime des Vipères (Reptilia: Viperidae) lors de la croissance, *Amphibia-Reptilia,* 1:127–136.
Schmidt-Nielsen, K., 1984, *Scaling, Why is Animal Size So Important?* Cambridge Univ. Press, Cambridge.
Schoener, T. W., 1969, Models of optimal size for solitary predators, *Amer. Nat.,* 103:277–313.
Schoener, T. W., 1971, Theory of feeding strategies, *Ann. Rev. Ecol. Syst.,* 2:369–404.
Seib, R. L., 1984, Prey use in three syntopic neotropical racers, *J. Herpetol.,* 18:412–420.
Shine, R., 1977, Habitats, diets, and sympatry in snakes: A study from Australia, *Can. J. Zool.,* 55:1118–1128.
Shine, R., 1986, Ecology of a low-energy specialist: Food habits and reproductive biology of the Arafura filesnake (Acrochordidae), *Copeia,* 1986:424–437.
Shine, R., 1987, Ecological ramifications of prey size: Food habits and reproductive biology of Australian copperhead snakes (*Austrelaps,* Elapidae), *J. Herpetol.,* 21:21–28.
Shine, R., 1991, Why do larger snakes eat larger prey items? *Funct. Ecol.,* 5:493–502.
Slip, D. J., and R. Shine, 1988, Feeding habits of the diamond python, *Morelia s. spilota:* Ambush predation by a boid snake, *J. Herpetol.,* 22:323–330.
Sokal, R. R., and F. J. Rohlf, 1981, *Biometry,* Freeman, New York.
Stephens, D. W., and J. R. Krebs, 1986, *Foraging Theory,* Princeton Univ. Press, Princeton.
Surface, H. A., 1906, The serpents of Pennsylvania, *Bull. Div. Zool., Penn. State Dept. Agricul.,* 4 (4–5):113–208.
Voris, H. K., and M. W. Moffett, 1981, Size and proportion relationship between the beaked sea snake and its prey, *Biotropica,* 13:15–19.

Voris, H. K., and H. H. Voris, 1983, Feeding strategies in marine snakes: An analysis of evolutionary, morphological, behavioral and ecological relationships, *Amer. Zool.*, 23:411–425.
Wright, A. H., and A. A. Wright, 1957, *Handbook of Snakes of the United States and Canada,* Cornell Univ. Press, Ithaca.

Chapter

4

Perceptual Mechanisms and the Behavioral Ecology of Snakes

Neil B. Ford

Gordon M. Burghardt

Introduction

Snakes as models in behavior studies

To study behavior in animals we typically test questions of how the behavior works (physiology), how it develops (genetics, maturation, and experience), its function (social interactions and ecology), and its phylogenetic history (evolution). Snakes as a group lend themselves well to testing of certain questions, such as the role of experience and genetics in prey preferences. However, they are less amenable for other studies, e.g., territorial behavior. Because of their limblessness, snakes have often unique solutions to various life-history problems, such as how to find mates, forage, and avoid predation. Just as wings and flying influence most aspects of the physiology, social behavior, and ecology of birds, so does the lack of legs affect nearly all aspects of the life of snakes. In particular, their perceptual capabilities and orientation mechanisms are influenced by their limblessness. For example, the increase in environmental interference at ground level (i.e., vegetation; Gillingham, 1987) may constrain the use of vision in many snakes and increase their dependence on chemical cues. In addition, the cylindrical morphology of snakes and the increase in tactile receptors needed for locomotion predispose them to communicate by touch (Gillingham, 1987).

Snakes are a diverse group, with over 2300 species in some 11 or 12 families (McDowell, 1987). Within the limitations of their physical characteristics they have solved life-history problems in a number of ways. In relatively well-studied genera such as *Thamnophis, Nerodia, Elaphe, Vipera,* and *Crotalus,* the primary behavioral hypotheses explored have revolved around locomotion, prey selection and handling, antipredator behaviors, reproduction, and intraspecific communication. Obviously, perceptual mechanisms are critical features of each of these suites of behaviors. Therefore, an understanding of snake sensory capabilities and roles in particular behaviors can give insight into the areas mentioned in the first paragraph: how the behavior works, how it develops, how it functions, and how it may have evolved. The last is the most problematic, because a given aspect of behavioral ecology has rarely been examined across an array of snake species.

Specializations and constraints

The dominant perceptual modalities used by snakes are vision (including infrared), chemoreception (vomerolfaction and nasolfaction) and touch. These senses are involved in varying amounts and in different aspects of social behavior, foraging, habitat selection, and predator defense. The auditory and gustatory senses are morphologically constrained in snakes. Although the capacity for hearing and taste buds are both present, these modalities, especially the former, play minimal roles in ophidian natural history. In addition, sensory modalities have been well-studied in only a limited number of snakes, but there is accumulating evidence that there are phylogenetic differences in the importance of particular senses.

Overview of Sensory Systems

Eyes

The eyes of snakes are distinctive among reptiles (Underwood, 1970), not so much because the capabilities of their eyes differ, but because their anatomy and physiology are unique. In particular, the organization of the retina and the mechanism of accommodation are singular to snakes. Underwood (1970) suggested that new adaptations evolved when snakes reinstated vision as an important sense during their return to terrestrial life from a fossorial existence, during which the usefulness of eyes had been reduced (see also Repérant, 1992).

Retina. The outer segments of snake photoreceptors may be all conical or cylindrical (simplex) or of both types (duplex pattern) (Peterson, 1992). Walls (1942) and Underwood (1970) both suggest that the duplex pattern is the primitive condition for snakes. As snakes returned to surface life, the number of cones increased among the rods of the retina and extended closer to the sclera, which produced a two-tiered retina (Underwood, 1970). The inner tier of mostly oligosynaptic (ordinary) rods has maximum sensitivity in low light but provides only poor acuity. The outer tier of cones has high acuity but low sensitivity. The dipsadinae, boiginae, and some lycodontinae have increased cone sensitivity by enlargement of these outer segments (Underwood, 1970). Other snakes (Underwood's viperine pattern) evolved with more cells involved in visual acuity but having less sensitivity, and the rods and cones are no longer segregated in tiers. This change resulted in the duplex polysynaptic retina found in some colubridae, viperidae, and elapidae. The final commitment to diurnal activity came in some species with the loss of scotopic pigment-containing rods, resulting in a pure-cone retina. Most colubrids, some elapidae, and some lycodontinae have this pattern. However, snakes that have returned to fossorial or nocturnal activity typically have some cones containing scotopic pigments (polysynaptic rods).

Accommodation. Accommodation mechanisms also had to be reelaborated after the snakes returned from a burrowing existence. Because the lens is hard in most snakes (*Natrix* is the only known exception, Beer, 1898) deformation of the lens is apparently not used to achieve accommodation. Instead, using muscles at the root of the iris, snakes increase the pressure in the posterior chamber, which pushes the lens forward. Pupillary contraction can also assist in producing better acuity. Sea Snakes contract their pupils when on land (Walls, 1942) and the semiaquatic *Thamnophis couchii* does this in water (Schaeffel and de Queiroz, 1990). The usefulness of accommodation in nocturnal snakes with large pupil apertures would likely be reduced. *T. melanogaster*, an aquatic foraging specialist (Drummond, 1983), exhibits a marked ability to accommodate. Schaeffel and de Queiroz (1990) suggest that it must use some as yet unknown mechanism to deform the lens.

Polarized light reception. A third structure in the retina was redeveloped by snakes; the refringent organelles of the cones (Underwood, 1970). These structures refract incident light and may allow detection of the plane of polarization.

Binocular vision. Most snakes have laterally placed eyes that produce a binocular overlap of between 30° and 40° (Porter, 1972). Several arboreal snakes (*Ahaetulla, Oxybelis, Uromacer, Thelotornis*) have attenuated snouts and more forward-oriented eyes, which allow binocular fields up to 46° (Henderson and Binder, 1980; see Lillywhite and Henderson, Chap. 1, this volume).

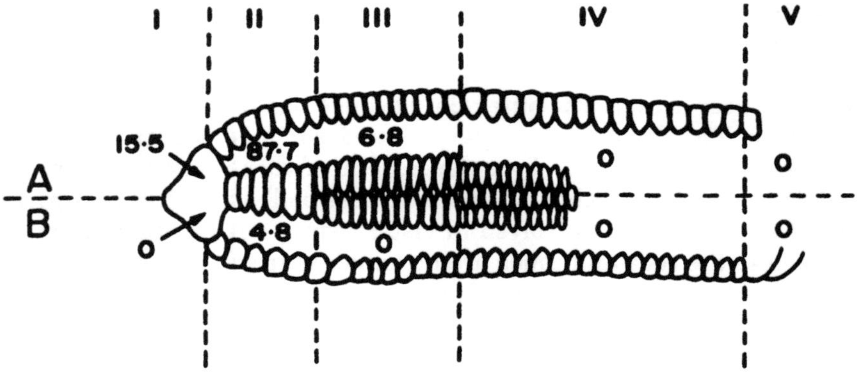

Figure 4.1 Regions of the tail of the sea snake, *Aipysurus laevis* (letters and Roman numerals) and percent of stimulations by light resulting in tail movements (numbers). (Zimmerman and Heatwole, 1990; reprinted with permission.)

Extraoptic photoreception. Many reptiles are known to have photoreceptive organs in the parietal eye (Quay, 1979; Underwood, 1992), but this is not known to occur in snakes, although light may affect the pineal gland through the skull (V. Hutchison, personal communication). Interestingly, cutaneous photoreceptors have been documented in the tail of the sea snake, *Aipysurus laevis,* and apparently assist in concealing the animal in clumps of coral (Fig. 4.1; Zimmerman and Heatwole, 1990).

Thermal pits

Two families of snakes have evolved thermal pits on the head that detect radiant heat of intermediate to long infrared wavelengths (see reviews by Barrett, 1970 and Molenaar, 1992). The crotalines of the viperidae have facial pits (one on each side), whereas viperines of the same family do not have pits but may still have the ability to detect heat (Breidenbach, 1990). The boidae have variable numbers of labial pits (also supralabial, infralabial, and rostral pits; Bullock and Barrett, 1968; Warren and Proske, 1968). In both groups the pits are enervated by phasic (fast adapting) fibers of the trigeminal nerve. Integration of the information occurs in the optic tectum, which also receives visual input (Hartline et al., 1978; Newman et al., 1980; Newman and Hartline, 1981; Schroeder, 1981). Both visual and infrared information are then transmitted to the forebrain (Benson and Hartline, 1988). Although these receptors respond slightly to touch, they are primarily responsive to radiant heat above the background level, for which they are extremely sensitive (Bullock and Diecke, 1956; Molenaar, 1992). Both facial and labial pits provide directional information as well (Barrett, 1970). These receptors could function in detecting prey or predators (Cock Bunning, 1983) as well as any environmental substrate warmer than the background (Goris and Nomoto, 1967). The usefulness of thermal pits in thermoregulation is likely limited by the fact that these sense organs adapt quickly (Barrett, 1970). Kardong and MacKessy (1991) suggest that a switch to thermal sensing would occur whenever snakes hunt at night or in burrows. Greene (1992) recently presented evidence that the ancestral role for pit organs in snakes was in predator detection. Molenaar (1992) discusses in some detail the oriententional maneuvers involved in using the pit organs to perform a successful strike.

Tactile, kinesthetic, auditory, and vibrational receptors

Very little is known about the cutaneous receptors of snakes, although they surely can detect temperature, touch, and pain (Porter, 1972). For an elongated animal like a snake in which tactile cues are very important for locomotion, the number of tangoreceptors and sensitivity to touch should be high. Indeed, a snake's sensitivity to even light contact by a conspecific appears to be acute (*Trimeresurus*, Nishimura, personal communication), which would suggest that this sense could be useful in conveying information during any behavior involving contact with other organisms. In Noble and Clausen's (1936) seminal work on the cues involved in nonsexual aggregations of *Storeria dekayi*, they found that after the snakes' chemical senses were eliminated the snakes would stay in contact with models of snakes. During courtship and male–male interaction of snakes, conspecifics also are typically tightly intertwined, but again the role of tactile cues in snake communication has not specifically been examined. Wall- and cover-seeking behavior (thigmotaxis) are common antipredator mechanisms and are well-known phenomena to researchers designing experiments for snakes, but only a few studies have specifically tested the effects of tactile cues (Chiszar et al., 1987).

Auditory capability is present even in the extreme modification of the inner ear of snakes (Baird, 1970). Although there is meager behavioral evidence for hearing of airborne sounds (Weaver, 1978), there is physiological evidence from several species in six families (Weaver and Vernon, 1960; Hartline and Campbell, 1969). In addition, attachment of the quadrate to the inner ear suggests that vibrational stimuli could be transmitted by the jaw. However, vibrational cues could just as likely be detected by tangoreceptors on the venter of the snake.

Chemosensory organs

Snakes, like other terrestrial vertebrates, possess several sensory organs that perceive chemicals: the nasal olfactory system, the vomeronasal organ, and taste buds. These senses are termed nasolfaction, vomerolfaction, and gustation, respectively. A current comprehensive review of the anatomy of both nasal senses is available (Halpern, 1992) and so they will only be touched on here. Alberts (1992) discusses the constraints on the chemical communication systems in vertebrates that also apply to snakes.

The olfactory system originates with unconvoluted but ciliated sensory epithelia in the posterior and dorsal nasal cavity (Halpern, 1983; Halpern and Kubie, 1984). The vomeronasal organ consists of two dome-shaped pits of thick unciliated epithelia encased in bone in the roof of the mouth (Halpern, 1983). Ducts from the oral cavity lead to each vomeronasal lumen. The tongue mediates the transfer of chemicals to this system by way of the vomeronasal ducts (Halpern and Kubie, 1980; Cooper and Burghardt, 1990b). The tongue can deliver nonvolatile chemicals, whereas odorants must be volatile to reach the olfactory mucosa.

These two systems send information directly to the telencephalon but are separate in both their neural pathways and their termination points in the limbic system and cortex (Halpern, 1980; Halpern and Kubie, 1984). The bipolar neurons of the olfactory system synapse in the main olfactory bulb, whereas the abundant neurons of the vomeronasal nerves pass between the olfactory bulbs to the accessory olfactory bulbs of the telencephalon (Halpern, 1980). The areas receiving neural input from the accessory olfactory bulbs are all known to concentrate gonadal hormones (Halpern et al., 1982), and the vomeronasal system is especially important in mediating social behavior in snakes (see reviews by Halpern and Kubie, 1984; Halpern, 1987; Halpern, 1992).

The anatomy of snake taste buds are poorly known, but they are present in papillae of the oral epithelium of at least some snakes although apparently not on the tongue (Kroll, 1973; Schwenk, 1985). Taste buds are enervated by the trigeminal nerve (Burns, 1969). It has been suggested that gustation is involved when snakes bite and then reject unpalatable prey (Burghardt et al., 1973; Garton and Mushinsky, 1979).

Methods of Studying Sensory System Function

One of the primary concerns of early twentieth-century ethologists was the role of sensory mechanisms in behavior. Although vision and audition were emphasized in these studies, tactile, chemical, electrical, proprioceptive, thermal, and pain sensation were also investigated using methods as varied as inferences from neuroanatomy to tests of discrimination learning. Hess (1973) provides a useful review of methods and findings from this early work.

This traditional emphasis on the senses is no longer found in ethological studies, which is unfortunate because advances in neuroscience, molecular biology, and computers have brought forth a wealth of new techniques. Indeed, before employing sophisticated new approaches with snakes, knowledge of and familiarity with more traditional methods are necessary. Given the paucity of our knowledge of snake perception, methods that are relatively "low tech" are more likely to be useful.

The sensory and perceptual systems of snakes can be studied in many ways, and various methods are suited for uncovering different types of information. While it might seem obvious that studies in naturalistic settings are most useful to behavioral ecology, this is not necessarily true. Unfortunately, workers on various aspects of snake perception appear to be unaware of the benefits of other techniques and may even be hostile toward followers of a different drummer. This lack of communication is not conducive to effective progress in the field. Certainly all approaches or published works are not equally useful or valid. Indeed, many legitimate criticisms can be leveled against specific studies. The point is that workers must have both a healthy respect for alternative approaches and enough understanding of the findings they have generated to evaluate and use them perceptively. It is quite clear to us that many of the published studies on snakes could have been improved if their authors had a larger methodological toolbox to draw from.

Here we give a brief overview of the methods employed and a sampler of the findings available for snakes. An occasional example will be drawn from lizards or other groups. However, we used examples from snakes wherever possible to illustrate a point, even if a study on another animal was more thorough or elegant. We also admit a bias toward species and studies with which we are most familiar. Subsequent sections will cite studies that employ these methods.

Setting

The first decision that must be made is whether to study a process or organism in its natural habitat, in the laboratory, or in some intermediate captive setting with some naturalistic components such as a zoo or large enclosure. For many students of biology and psychology, tradition or convenience has too often been decisive and alternatives have not been considered impartially. "Only studies in a species' nat-

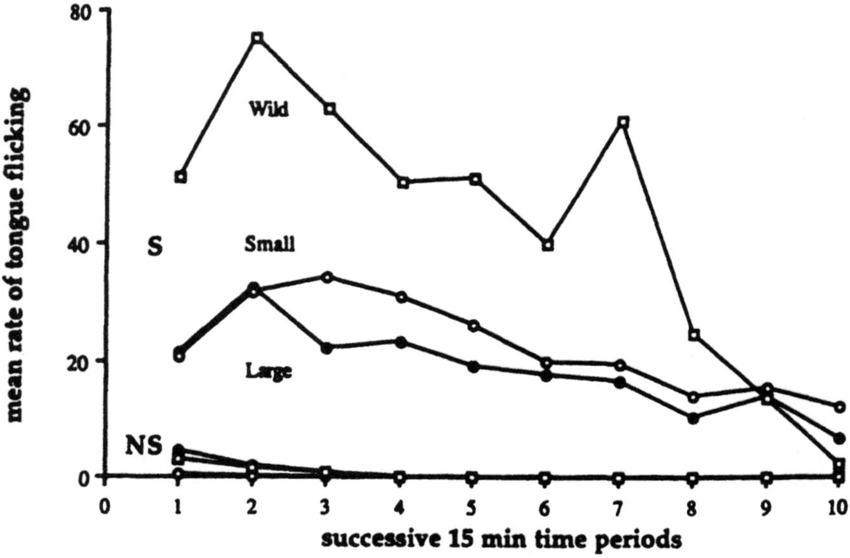

Figure 4.2 A SICS experiment involving three groups of the rattlesnake, *Crotalus enyo*. Two groups were born in captivity and were raised in large and small cages, respectively. The third group consisted of wild-caught adults (Marmie et al., 1990; reprinted with permission.)

ural environment have any validity" and "Only in well-controlled laboratory settings can the role of specific factors be determined" set the outer extremes of the debate. This dichotomy is particularly relevant in the study of perception, where one needs to know both the function and mechanism of specific cues and sensory processes in understanding the natural behavior of snakes. Only rarely have the setting's effects on the behavior been carefully evaluated (for example, cage size—Fig. 4.2; Marmie et al., 1990).

Description and natural history. In later sections we review the role of perceptual mechanisms in the following categories of behavior: orientation and migration, habitat selection, foraging, defense, and intraspecific communication. Our knowledge about all these phenomena in most species is limited. Thus descriptive and nonexperimental information is essential.

Field observations range from casual or opportunistic notes to highly systematic long-term descriptive studies. Unfortunately there are few of the latter for snakes. Thus the reporting of instances of predation, courtship, nesting, parturition, and fighting can be valuable even if based on a limited sample. Reports of predation or defense often give some hints as to the sensory cues involved as well as the sequence of behavioral events. For example, Ota (1986) reported on field observations of predatory episodes by the Japanese Striped Snake, *Elaphe quadrivirgata,* that suggest an important role for vision.

Obviously, extensive observations involving large sample sizes and quantitative recording of events along with durations and sequential organization are best. Workers studying a species for some time are likely to be familiar with its basic repertoire (ethogram), to be able to identify the species, sex, and perhaps even individual, and also to recognize the significance of details that might elude even a skilled general naturalist. Nonetheless, perceptual studies oriented toward natural events need to consider all available information. The same is true of captive observations made incidentally or haphazardly. Older work of many noted herpetologists (see Wall, 1921; Mell, 1929; Ditmars, 1936) contains a wealth of information that has generally been ignored. Often this is because the relevant observations are buried in monographs, field guides, or autobiographical writings and frequently are not in English.

Field experiments. Although field experiments are often difficult to plan and carry out with snakes, the potential of such studies is great, since they address the role of perception in its evolved context. For example, Duvall et al. (1985) studied prey localization in radiotagged Western Rattlesnakes, *Crotalus viridis,* in Wyoming, and then Duvall and Chiszar (1990) designed field experiments utilizing cages containing mice to look at the role of chemical cues from prey in influencing the spring migration of Western Rattlesnakes. They found that females stayed in the vicinity of cages containing mice or mice-derived stimuli longer than they did to control stimuli.

Enclosure experiments. One method that is particularly suitable for snakes is to transfer and test them in large arenas in outdoor natural habitats. A recent example is a study by Hampton and Gillingham (1989), who evaluated the approach direction of a simulated predator on antipredator responses in Common Garter Snakes, *T. sirtalis.* They found that the same stimulus (a gray piece of cardboard mounted on a stick) presented overhead elicited a greater response than when presented at eye level. They also found evidence of both short-term and long-term habituation.

Laboratory experiments. The laboratory is often the essential setting to understand the specific cues involved in behaviors first observed in the field, in particular the details of responses and especially neurophysiological control mechanisms. Halpern and her colleagues, for example, have used several devices to assess the role of the various chemical senses used by Garter Snakes in locating their earthworm prey (Kubie and Halpern, 1979).

Environmental manipulation

After a decision is reached on the experimental setting, the method of manipulating the snake or its environment must be addressed.

General. First one can manipulate the context in which the animal responds to the normal cue. For example, in prey-trailing studies in pit vipers that release prey after striking, trails laid down by nonenvenomated mice or those killed by other means and dragged along the ground can be presented to see if they confuse the snake. In actuality, such studies have shown that a snake can recognize an envenomated mouse against a nonenvenomated one, even if it did not do the envenomation (Duvall et al., 1978; Chiszar et al., 1992).

Another approach is to modify the background environment. For example, Czaplicki and Porter (1974) found that Water Snakes, *Nerodia sipedon* and *N. rhombifer,* preferentially captured black goldfish over gold goldfish when presented in tanks with gold backgrounds and vice versa when the background was black. This was a simple but clear-cut demonstration of the role of visual contrast that did not involve manipulating either the snake's sensory system or the normal stimulus object (see also Fig. 4.3; Teather, 1991).

Temperature, a critical variable in work with poikilothermic vertebrates, can best be studied experimentally in controlled settings. For example, tongue-flicking in Skinks, *Eumeces* (Cooper and Vitt, 1986), and various components of defensive behavior in *Thamnophis* (Schieffelin and de Queiroz, 1991) are under strong temperature influence. However, the direction and amount of influence are not always obvious (Fig. 4.4; cf. Arnold and Bennett, 1984, Stevenson et al., 1985). On the other hand, correlations between ambient or snake body temperature and behavior can be assessed in the field over naturally occurring ranges. For example, Layne and Ford (1984) found a strong relationship between flight distance and body temperature in Queen Snakes (*Regina septemvittata*) but no sex or body-size differences. See Peterson, Gibson, and Dorcas (Chap. 7, this volume) for a review of the thermal ecology of snakes.

Other environmental factors could be discussed here including ambient light levels, humidity, barometric pressure, structural complexity (influence of perches, escape routes, etc.), and background chemical stimuli such as odors. These factors do not operate independently. Gregory (1984) found that temperature could affect habitat preference in the field and occupation of sunny versus shady spots by *Thamnophis.*

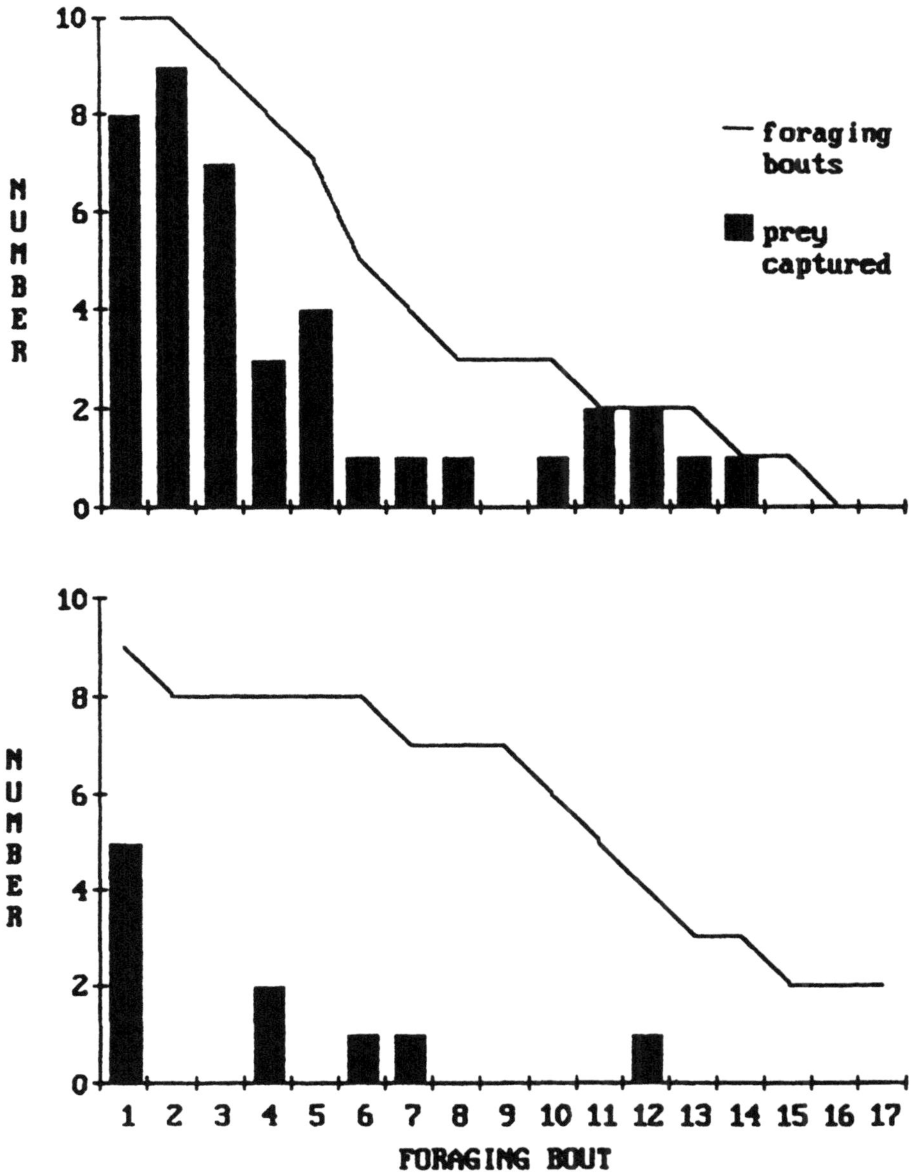

Figure 4.3 Total numbers of fish captured by Common Garter Snakes, *T. sirtalis,* in each feeding bout by snakes foraging on fish against a white background (top) and a mottled background (bottom). (Teather, 1991; reprinted with permission.)

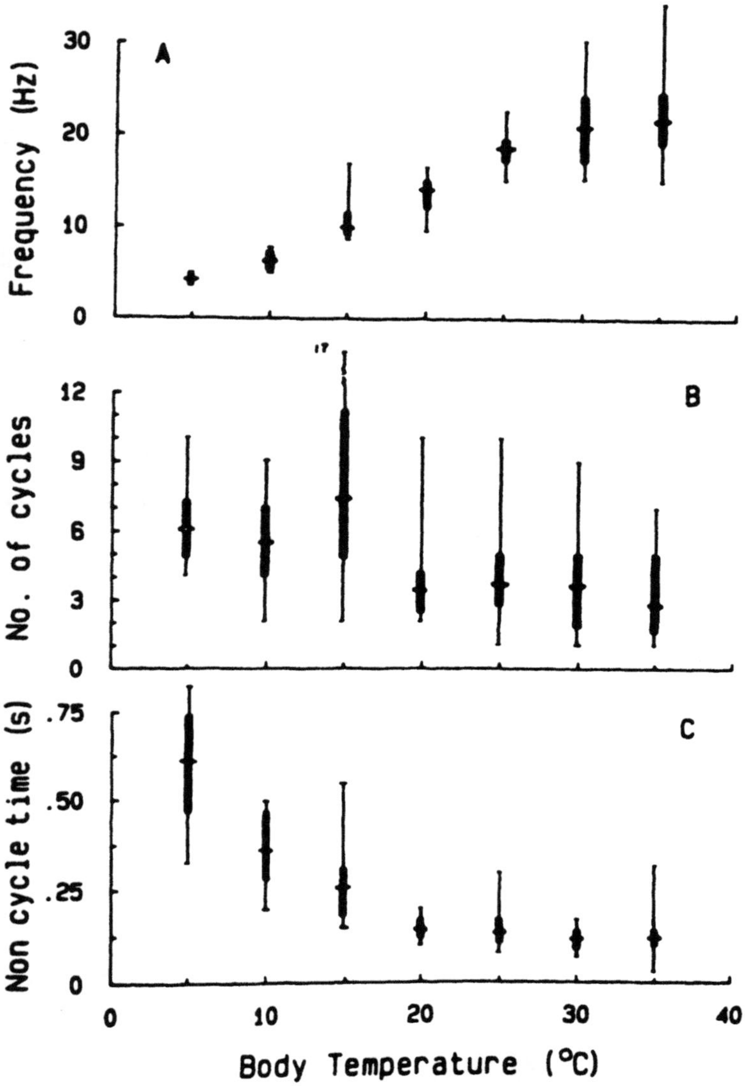

Figure 4.4 Three components of tongue protraction in Western Terrestrial Garter Snakes, *T. elegans,* as a function of body temperature, including: (*A*) frequency of up-and-down motions of the tongue; (*B*) number of up-and-down motions or cycles per tongue flick (at 15°C); and (*C*) the time the tongue was outside of the mouth in which it was not cycling up and down. The horizontal line is the population mean. The thick vertical bars indicate the range in individual mean values, and the vertical line indicates the range of individual observations. (From Stevenson et al., 1985; reprinted with permission.)

Temporal. Although little studied, seasonal and daily rhythms may influence the response of snakes to identical experimental stimuli, even when kept under constant laboratory conditions from birth. For example, newly hatched Corn Snakes (*Elaphe guttata*) showed locomotor cyclicity even when under constant light and temperature from hatching (Burghardt, 1978). Animals captured in the wild or those reared in captivity may develop anticipatory responses at certain times of day or after set intervals. Thus, if snakes are used in prey-recognition experiments, the time of day when they are tested should be standardized. In many published experiments it is difficult to determine if such was the case. Similarly, time of year may also be a factor. Snakes may decrease food intake in winter even when kept in standard lab conditions year round (R. Gibson, personal communication). Andrén (1982) has reported differences in activities of *Vipera berus* over years as a function of prey density. Modification of light, temperature, and their daily cycles, as well as hormones, can alter the responses of snakes to various stimuli. Although these manipulations are most often employed in studying reproduction, other responses and sensory systems may be affected as well. Reversal of day–night cycles is less problematic and can be a real convenience to those studying nocturnal snakes in captivity (Chiszar et al., 1988a,b).

Cue manipulation

The manipulation of cues from the specific objects of interest is more intrusive in some respects than modifying the environment but is a more common method, and the results are often more easily interpreted. Most behavioral ecology work with snakes has been based on cues associated with another animal: prey, predator, or conspecific.

Object modification. This was the earliest method, often used in informal experiments to assess the role of a given cue. Was movement by prey or mate important in feeding or courtship? Simply test by offering the snake a dead animal. However, it did not take long to realize that a dead animal may vary from a live one in more than just visual movement cues. Perhaps odors, sounds, and other cues are different also. An early experiment by Noble (1937) investigating the cues used by male Garter Snakes in recognizing females tried to solve this problem by presenting female snakes with broken backs so they could not move. We do not recommend such methods.

There are other ways to eliminate sensory cues. One method is to block certain stimuli from the subject. Building on the pioneering work of Baumann (1929), Chiszar and his colleagues (Chiszar et al., 1988a,b) separated the role of chemical and visual cues in the predatory attack of the Brown Tree Snake (*Boiga irregularis*). They used Plexiglas cubes sequentially containing no cues, rat-soiled bedding, a live rat pup (no odor escaping), and a live rat pup in a tube with holes. For this species vision was extremely important in foraging. Using a similar technique of presenting live fish in clear sealed vials, Burghardt (1966) found that visual cues would lead neonatal Common Garter Snakes (*T. sirtalis*) to approach and orient to prey, but were insufficient to elicit attacks. Teather (1991) used fish in plastic bags and also concluded chemical cues were necessary to elicit foraging in *T. sirtalis*. Drummond (1985) showed that the same species would approach water in which fish had been housed ("fishy water") in preference to clean water. Gillingham and Clark (1981) demonstrated that visual stimulation could "turn on" tongue-flicking and chemosensory searching in Western Diamondback Rattlesnakes (*Crotalus atrox*) without a strike by using tethered prey that were moved in and out of the snakes' sight. In short, there are many ways in which the importance of various cues can be established.

Models. The use of live predators, prey, or conspecifics has the advantage of at least using ecologically valid stimuli, even if the experimental context is artificial. There are several problems, however: (a) live animals may vary in size and condition, leading to difficulties in equating stimuli across experiments and, if the stimulus animal is changed for each trial, even within tests; (b) the stimulus animal is alive and behaves; moving, changing posture, vocalizing, emitting chemicals, etc., and these cues can themselves vary within the same animal. In some sophisticated experiments where potential confounding stimuli can be identified in advance, it is possible to measure aspects of the animal's behavior such as amount of movement or apparent size. These values can then be used as covariates in analyses of variance (ANOVAs) or figured into regression analyses. However, such detailed data are often impractical to gather or still not as statistically powerful as experimental control of such variables.

The success of early ethologists in using models to study the theoretical concept of innate releasing mechanisms (Tinbergen, 1951), exemplifies the advantages models have over live stimuli. The many studies of the perceptual responses of fish, butterflies, and birds using crude two-dimensional cutout figures gave astonishing results. Today, more controlled and sophisticated experimental designs are necessary, but the basic approach retains its validity. First, the features of the specific cue can be manipulated, such as size of a prey item, color of a body part, or whether a predator has eyes or not. Both qualitative and quantitative variation can be used. The latter is especially useful in determining sensitivity or thresholds. A second advantage is that the "behavior" of the model can be controlled precisely. For example, Drummond (1985) suspended realistic model fish on wires and, placing them in water, could control their movement in evaluating visual cues in the foraging of Garter Snakes. Similarly, Burghardt and Denny (1983) used adjustable motors to rotate artificial earthworm sections behind glass to evaluate the role of movement in prey selection by *Thamnophis.*

Chemical cues can be isolated from natural prey, predators, and conspecifics and used to assess responses (Halpern, 1992; Weldon et al., 1992; Weldon et al., in press). Such chemicals can be identified (Burghardt et al., 1988; Wang et al., 1988; Mason et al., 1989; Mason, 1992) and through the use of synthesized compounds (as models) the specific chemical cues involved in the response can be demonstrated.

Less recognized formally as a method, but one frequently employed, is combining a stimulus derived from the natural object with an artificial vehicle; for example, presenting prey odors on cotton swabs or artificial worm bits (Burghardt, 1970a; Wang et al., 1988). Another method, already alluded to, is presenting stimuli derived from the object such as laying trails with conspecifics or prey, testing mouse bedding in boxes, or evaluating responses to feces. The assumption, which should not be treated gratuitously, is that the derived stimulus actually represents that produced by the real thing. Gillingham et al. (1990), for example, showed that the casings deposited by earthworms attract foraging *T. sirtalis* but that the effect disappears after 48 h.

Snake manipulation

One of the oldest and most powerful methods to study perception is intervention with the sensory system itself. There are ethical considerations to such methods, however. Snakes must be treated with respect, and anesthesia and stress-reducing procedures employed if necessary. To reduce the number of animals used, refined rather than crude procedures need to be developed, although this itself may mean several studies.

Sensory blocking. Blocking studies attempt to evaluate the modality being used in a response by eliminating a sensory channel. One can do this by blocking signal reception at the level of the sense organ or by intervening more centrally in the nervous system.

Peripheral blocking. An example of a peripheral blocking study would be blinding or putting a blindfold on an animal, plugging its nares, sealing the vomeronasal organ ducts, or cutting off its tongue. These methods were used in the classic studies by Noble on feeding and courtship in snakes (Noble and Clausen, 1936; Noble, 1937).

The results from these studies can be clear-cut and dramatic. However, they also have problems that must be addressed: (a) abnormal responses can occur, such as blindfolded snakes preoccupied with rubbing off the blindfold, or restricted breathing in animals with blocked nares; (b) blocking the modality prevents determination of the aspects of the normal stimulus involved in the response. That is, a blindfolded snake cannot be used to determine the relative roles of prey movement, color, shape, contrast, or size.

Central blocking. More sophisticated methods are used to enter into the sensory system pathway. Classic work showing the importance of the vomeronasal system rather than olfaction in prey recognition of Garter Snakes was carried out by Wilde (1938) and later extended by others (reviewed in Halpern, 1992). In these studies the nerves leading from the olfactory or vomeronasal epithelium to the main or accessory olfactory bulbs were severed, and neurophysiological responses to prey and conspecific cues measured. Central blocking studies have the advantage that only cues specific to that system will cause a response. As Graves and Halpern (1990) point out, behaviors always assumed to be associated with a particular sensory system may occasionally occur in other contexts (i.e., snake tongue-flicking may occur as part of a social display).

Stimulation studies. After one has identified the part of the snakes' nervous system that receives input from cues from an object, then stimulation of the receptor, sensory nerve, target brain area, or the effector system can be used to map out the internal system. For example, Meredith and Burghardt (1978) recorded from the accessory olfactory bulb of *Thamnophis* and documented that cells fired at the moment the snake retracted its tongue after flicking prey-derived chemicals. Andry and Luttges (1972) used a comparable method to assess habituation to stimuli. A problem with neurophysiological studies is that the scientists involved are often remote from behavioral ecology concerns and typically use artificial stimuli and nonbiological responses. In addition, those interested in understanding naturalistic behavior have often neglected these powerful methods.

Increasingly important will be studies of the interaction between the physiological status of the animal and its perceptions. For example, hormonal states influence response to sexual stimuli and aggression toward same-sex conspecifics. Nutritional states can shift the probability of devoting time to mating or foraging in *Vipera berus* (Andrén, 1982), and this could operate through shifting the salience of sensory cues. Short-term satiation on food can also affect responsivity to chemical prey cues (Burghardt, 1970a) and simulated predatory stimuli (Herzog and Bailey, 1987). In the future we see direct stimulation of the nervous system with neurotransmitters, imaging, brain scanning, and other developing neuroscience technology as fostering significant new understanding of reptile perception.

Evaluating relative importance of cues

Most methods currently used vary one sensory channel or component at a time with the aim of seeing how that cue limits performance of the normal response. Unfortunately, animals receive information from several stimuli at once. Attempts to rank the importance of stimuli from a series of univariate experiments have several problems. One is that they work best when one modality is primary, as in chemoreception in many snakes. Another difficulty is that they do not consider possible synergistic effects of more than one modality or cues within a modality (i.e., size and movement)

Integration of Sensory Information

The sensory control of behavior patterns in snakes, particularly those involved in foraging, defense, and courtship, has been studied more in the laboratory than in the field. The methods outlined above have been used to gather information on a wide range of snake behaviors. Unfortunately, comparative diversity is limited and what we can conclude about even the most well-studied species is also less than satisfactory. Most sustained long-term studies have been carried out on vipers and thamnophines. Early studies such as those by Baumann, Wiedemann, Noble, Naulleau, and Dullemeirer have been reviewed in Burghardt (1970a) and more recent work in Burghardt (1980, 1990), Halpern (1992), Mason (1992), and Weldon et al. (1992). As we cannot be comprehensive on even the more recent work on snakes we will summarize what is known about snake perception in natural behavior using information from where it is most well-established. We hope from this review that researchers will be encouraged to study these phenomena in other families, genera, and species.

Habitat selection

Although the nature of habitat utilization by snakes is not well-understood (see Reinert, Chap. 6, this volume), numerous studies suggest that snakes occupy specific habitats that provide resources important to the survival of individuals of that particular species

(Reinert, 1984a,b; Gregory et al., 1987). Indeed, many studies of snake movements deal with the nature of the habitats chosen by the different sexes during particular seasons. For example, in northern North America it is common for viviparous snakes to overwinter in one area and mate there. Males and nonpregnant females then move to a foraging area in late spring. Pregnant females move to a different place where they can safely and effectively thermoregulate as their embryos develop (Duvall et al., 1985; Lawson, 1991). Presumably, the physiological needs of each sex and condition require the particular habitat selected.

Similar conclusions have been made for seasonal shifts in arboreality, i.e., thermal requirements vary in midsummer (Mushinsky et al., 1980; Lillywhite and Henderson, Chap. 1, this volume). In addition, foraging and pregnant animals are known to select microhabitats that relate to their particular requirements, for example, open sunny sites for gravid females or fallen logs for snakes ambushing prey (Reinert and Bushar, 1984; Reinert and Zappalorti, 1988). Although deposition of pheromones and landmark orientation have been suggested as mechanisms snakes use to locate preferred sites, few studies have dealt with the nature of perceptual cues in habitat selection.

Movement to and from mating and foraging sites may involve celestial orientation or the use of landmarks (see the section "Orientation and navigation," p. 140). However, because trail pheromones are nonvolatile (Ford and Low, 1984), routes marked by long-lasting pheromones are also a possibility (Gregory et al., 1987). Tests of how internal changes stimulate shifts in cue preferences would be a mechanism to determine the nature of habitat selection. For example, foraging success influences habitat preferences in European Adders (Andrén, 1982), suggesting that fat stores must regulate a migratory "restlessness" in these snakes. Comparison of the orientation mechanisms of gravid, nongravid, and male Garter Snakes (Lawson,1991) therefore gives additional insight into the stimuli involved in their movements.

The sensory cues involved in microhabitat selection have been even less studied (Gibbons and Semlitsch, 1987). It is known that the thermal nature of the microhabitat is important in regulating the position and behavior of snakes in their habitat (Gibson and Falls, 1979; Charland and Gregory, 1990). The mechanisms used by snakes to evaluate environmental variation in temperature have been much less studied (Huey et al., 1989). The infrared receptors have been implicated in thermal selection but the phasic nature of these organs suggest that they might be ineffective (Barrett, 1970), and of course, most snakes do not have such receptors and still thermoregulate effectively. Visual cues, i.e., sunny spots versus dark areas, are also likely stimuli for local orientation but have not been tested (Gregory, 1984). Indeed, simple laboratory experiments adding and removing cues, such as those performed by Chiszar et al. (1987) to study the roles of light and tactile cues in the cover-seeking behavior of Red Spitting Cobras, could substantially add to our knowledge of habitat preference. The interaction of humidity and desiccation, shelters and tactile cues, and temperature and light all need to be examined in a variety of snakes.

Intraspecific communication

Although social behavior has been well-studied in at least some species of snakes, experiments on the role of the sensory cues involved have been limited to descriptive analysis or have concentrated on the role of chemical communication. The vomeronasal system, in particular, has been found to be critical for the performance of several snake social behaviors. Although early work by Noble and his colleagues suggested an important role for olfaction in social behavior in snakes (reviewed in Burghardt, 1970a), more recent studies give little evidence of its necessity (Kubie et al., 1978). Vision also is not likely to be as important for intraspecific communication in most snakes as it is in other squamates. Structural modifications for visual displays, such as the dewlap of anoline lizards, are probably constrained in snakes by their locomotory requirements. Although visible behavior acts have been noted in some snake social activity, i.e., swaying during combat dances, the information conveyed by such cues has not been determined. Tactile cues may also transmit information during contact between conspecific snakes, i.e., courtship, combat, and aggregation, but specific experiments to elucidate the information conveyed by touch have not been conducted in snakes.

Reproductive behavior. The neural connection of the vomeronasal system to the hypothalamus supports the evidence that pheromones picked up through tongue flicks and deposited in Jacobson's organ mediate reproductive and aggressive behavior in snakes (Andrén, 1982; Halpern and Kubie, 1984; Halpern, 1992). For species that have been examined (primarily *Thamnophis*), it appears that chemicals released from the dorsal and lateral skin of females are species specific (Ford, 1986) and indicate receptivity (Gartska and Crews, 1981). Pheromones deposited as trails allow males to locate the female (Ford and Low, 1984), and contact pheromones stimulate his courtship, which includes mounting the female's dorsum and pressing his chin and snout along her back. The courtship pheromones have been characterized for *T. sirtalis* (Mason et al., 1989). Whether the pheromones in the skin are the same as those deposited as the trail is unknown. Other courtship behavior patterns such as biting (in the nonthamnophine colubrids *Lampropeltis, Elaphe, Coluber,* and *Pituophis*; reviewed in Secor, 1987), spurring (in boids, Barker et al., 1979; Gillingham and Chambers, 1982), or the male searching with his tail for the female's vent are also likely to be initiated by vomerolfaction.

Visual cues may be involved in reproduction in the initial locating of females, at least after pheromone trailing has placed the male in the general vicinity of the female. Many colubrids raise the head from a stationary position during trailing and apparently visually search for the female (Lillywhite, 1985; Larson, 1987; N. Ford, personal observation). Vision is also involved during chase sequences of courtship in which the male has lost physical contact with the female (Perry-Richardson, 1987). Actual movement of the female appears to be the stimulus that draws the male into contact. Other functions for visual cues in reproduction, i.e., species identification, have not been documented.

Male snakes are not known to produce species identification pheromones and visual cues that females might recognize are not readily apparent. However, male behavior in several species of colubrids, some viperids, and boids indicates that a diversity of tactile signals occurs (Gillingham and Chambers, 1982; Secor, 1987; Schuett and Gillingham, 1988). Such activities as body-jerking, caudal-cephalic waves, chin-pressing, bouncing, biting, spurring, and tail-searching are performed in contact with the female. Different species of related snakes show specific male courtship patterns (Gillingham, 1987; Secor, 1987). The types of activities, their sequence, and the rate of performance of each motor pattern may contain cues to the male's species identity (Secor, 1987) and serve as premating reproductive isolating mechanisms.

In snakes in which male combat occurs, subordinate animals show decreased courtship activity (Gillingham, 1987). Therefore, male fitness may be indicated to females in these species by the willingness of the male to court. However, in species in which males do not combat, tactile aspects of courtship could signal information concerning the male's age (size) or energetic capabilities, both possible measures of fitness (Perry-Richardson, 1987) and therefore serve as releasers of the female's receptivity.

Thamnophines do not exhibit male combat, and apparently females do not select larger males (Joy and Crews, 1988; Perry-Richardson et al., 1990). However, there is evidence that females choose more vigorous ones. In tests with pairs of males courting a single female Checkered Garter Snake, *T. marcianus,* successful males show more tail-searches per minute and are faster at remounting females after chases than are unsuccessful males (Perry-Richardson, 1987).

Although females of species in which males exhibit combat may be less selective of mates, tactile cues are likely involved in more than just species identification. Male *Elaphe* often switch to a more aggressive courtship pattern, including biting, to induce less receptive females to mate (Gillingham, 1979; Lewke, 1979). This resistance on the part of the female may also help ensure that the male is fit. In those species where the male must follow the female for several weeks before mating (southern temperate regions, Ford and Cobb, 1992) the same principle could be involved, as weaker males might not be able to keep up with the female.

Agonistic behavior. Both epigamic and intrasexual sexual selection may occur in snakes (Ford and Holland, 1990; see Duvall et al., 1992 for mating systems in pit vipers). Intrasexual competition, involving ritualized combat between male snakes for access to females (other resources such as food may also stimulate combat), has been reported in 25 viperids, 11 elapids, 21 colubrids, and 4 boids (Gillingham, 1987; S. Secor, in Ford and Holland, 1990; see Shine, Chap. 2, this volume). Vision, the vomeronasal system, and tactile senses are all likely to be involved in this agonistic behavior (Carpenter, 1977, 1984). Both chemical trails and vision could guide males together. Shine et al. (1981) described chemosensory searching (tongue-flicking and head-weaving) by the elapid *Pseudechis porphyriacus* prior to male combat, and tongue flicks have been described during male–male approach in the Western Diamondback Rattlesnake, *Crotalus atrox* (Gillingham et al., 1983). Andrén (1982) suggested that male European Adders, *Vipera berus,* used trailing behavior to locate sloughed males when the vomeronasal organ was intact. However, because females were in the area in both Andrén's and Shine's studies and no one has documented the presence of trail pheromones in male snakes, the female may have actually been the source of attraction. Adders could approach males even with the vomeronasal organ blocked, appearing to visually locate moving animals. However, blocking the vomeronasal system prevented aggressive behavior in these animals. Therefore, it is likely that contact pheromones function as releasers during an encounter (possibly by identifying both species and sex of the snakes). Because male Adders begin combat as soon as even a small area of old skin is sloughed from a competitor, Andrén's (1982) suggestion of a male pheromone is likely valid. In several laboratory studies with *Thamnophis,* R. Mason (personal communication) also found that some chemical from male skin could inhibit the normal courtship response to female pheromones.

Shine et al. (1981) described spreading of hoods in the initial encounter of male *Pseudechis* and others have described short-duration, noncontact head-raising in viperids as a solicitation posture (Carpenter, 1977). Visual components of such elevated postures may achieve dominance without combat, particularly if the animals have been in contact previously (Carpenter, 1977). However, in all species of snakes, physical contact appears to be involved in the determination of dominance during a combat bout. These displays include cues that can indicate size and strength of the males. Aggressive behaviors typically involve either raising the head high above the combatant or attempting to push the other snake down. North American colubrids tend to combat while stretched out and intertwined horizontally (Carpenter, 1984; Secor, 1990). Various pushing and pinning movements of the heads of the participants is normal. Old world colubrids (*Elaphe longissima, Coronella austriaca,* and *Ptyas mucosa*) show a pattern of elevation of the anterior body (Stemmler-Morath, 1935; Shaw, 1951; Andrén, 1976). Elapids also tend to combat in a horizontal posture with intertwining of the mid- and posterior body and some elevation of the anterior trunk and neck occurring (Carpenter, 1977; Shine et al., 1981). Both old- and new-world vipers show dramatic elevation of the anterior body during combat (*Bitis arietans* is an exception, Thomas, 1972, in Carpenter, 1977). While vertical, members of this family sway back and forth, hook bodies and necks, and attempt to push the opponents' head down. The boidae are different in that much of the tactile stimulation during combat occurs in the posterior region of the body (Carpenter et al., 1978; Barker et al., 1979). Pelvic spurs are erected and are used to scratch during their entwining of tails (Gillingham and Chambers, 1982). Heads of the opponents are often some distance from each other. Biting is seen in some colubrids (Secor, 1990), but is not common in combat rituals of elapids or viperids (Carpenter, 1977; Shine et al., 1981; Burchfield, 1982). Striking with the mouth closed has been described in the combat behavior of Gaboon Vipers, *Bitis gabonica* (Akester, 1979).

Determination of dominant animals appears therefore to involve both visual and tactile signals. In viperids, visual signals relating to which animal is more elevated are important (Carpenter, 1977), although it has also been suggested that fatigue from the raising of their heavy bodies may be involved (Schuett and Gillingham, 1989). In viperids and in colubrids such as *Lampropeltis* and *Elaphe,* pushing down the subordinate may signal the strength of the dominant animal. Schuett and Gillingham (1989) documented that swaying with up to one-half of the anterior body raised and hooking and forcing the opponent to the substrate are both activities during combat that lead to winning in the Copperhead, *Agkistrodon contortrix*.

Nonsexual social behavior. Tactile and chemosensory information is likely to be involved in various nonsexual social activities of snakes such as early neonate movements and aggregations of adult animals (i.e., den site location and formation of aggregations of gravid females). However, only the role of the vomeronasal system has been well-studied in this context. Although work has been conducted on several colubrids (*Elaphe, Thamnophis, Nerodia,* and *Pituophis*) and some species of vipers (*Crotalus*), the variety of methodologies used and their probable influence on the results precludes any general conclusions.

If, for example, shelters are important in predator avoidance, movements in a testing arena may be more influenced by the need to be in a secure area rather than by cues that might be important in the wild. A common testing apparatus for aggregation experiments consists of a small rectangular aquarium with soiled bedding on one side; measurements of the time a test animal spends on each side of the aquarium are used as evidence of avoidance or preference of an odor. Such an experiment may work for neonates of "calm" individuals and species, but results with adults or nervous snakes very likely are influenced by stress factors. For example, when Southern Copperheads, *Agkistrodon contortrix contortrix,* were tested in a large arena with shelters, they would only move out of those areas at night (Ford and Holland, 1990).

Experiments with adult snakes are also difficult to assess because of the influence of sexual pheromones. Tests during nonbreeding seasons do not necessarily eliminate all sexual attraction pheromones, as is sometimes assumed in experimental designs (Allen et al., 1984). Therefore, our discussion of cues involved in these behaviors will be general, with an emphasis on various hypotheses generated by research to this point.

Aggregations of neonates occur in the field and may slow down desiccation (Cohen, 1975; Graves et al., 1986) or decrease predation rates (Graves et al., 1986). The cues involved in this phenomenon have been studied primarily in the laboratory beginning with the influential paper by Noble and Clausen (1936). Neonates can detect and follow trails produced by conspecifics (Constanzo, 1989) apparently from lipids derived from the skin (Graves and Halpern, 1988). However, there are indications of species differences in how neonates respond to conspecific chemicals (Burghardt, 1983; Allen et al., 1984). Porter and Czaplicki (1974) found that *Nerodia rhombifer* avoid conspecific odors, whereas several *Thamnophis* species, *Nerodia sipedon,* and

Crotalus viridis have been documented to prefer areas soiled by conspecifics (Scudder et al., 1980; Heller and Halpern, 1982; Allen et al., 1984; Graves et al., 1986; Halpin, 1990) as have hatchling *Pituophis* (Burger, 1990b). In addition, recent studies have indicated that diet of the subjects can affect aggregation tendencies in somewhat unexpected ways. *Thamnophis butleri* and *T. radix* neonates select sites marked with feces from conspecifics but often avoid feces from conspecifics eating the same diet (Burghardt, 1990; Lyman, 1990). These neonates also tend not to aggregate with individuals with which they have competed for food (Yeager and Burghardt, 1991). Although suggestions of ecological functions for these avoidance patterns (i.e., avoiding competition) have been made (Yeager and Burghardt, 1991), conclusions as to the adaptive nature of neonate aggregations are premature.

There is some evidence that aggregations of gravid animals occur in the field (Gordon, 1980; Reichenbach, 1983), which may reduce predation of newly parturient young (although limited suitable refuges for the females could also produce such aggregations). Female snakes may locate other females by pheromones (Ford and O'Bleness, 1986; Constanzo, 1989), and chemical cues sensed by vomerolfaction seem involved in most nonsexual aggregations that have been studied (Dundee and Miller, 1968; Burghardt, 1980, 1983; Heller and Halpern, 1982; Graves et al., 1986). However, the maintenance of the aggregation may be dependent on tactile cues, as snakes in aggregations tend to be in physical contact (Burghardt, 1983; Lawson, 1991).

Orientation and navigation

The orientation mechanisms involved in locating dens in the fall and movements to foraging areas in the spring have been examined in *Thamnophis, Pituophis,* and *Crotalus.* Both chemosensory and celestial cues can be important.

Pheromone trails, at least for neonatal snakes, are used to locate traditional dens (Brown and MacLean, 1983; Graves et al., 1986; Reinert and Zappalorti, 1988; Burger, 1989a). Experiments examining nonsexual trailing in adult snakes have been confounded by our lack of understanding of the seasonality of sexual pheromones. Constanzo (1989) gave some evidence that adult *Thamnophis* follow conspecific trails in the fall. However, King et al. (1983) indicated that adult *Crotalus viridis* do not trail, although neonates are known to do so (Graves et al., 1986). Older, experienced snakes may locate traditional dens or foraging sites by sun compass or landmark orientation.

Landmarks were suggested to be important for orientation by *Pituophis melanoleucus* (Parker and Brown, 1980); solar orientation has been demonstrated in *Crotalus atrox* (Landreth, 1973) and *T. radix* (Lawson, 1985), and phase-shift tests of true navigation have been conducted for *Nerodia sipedon, Regina septemvittata* (Newcomer et al., 1974), *T. sirtalis,* and *T. ordinoides* (Lawson, 1989, 1991). Lawson's experiments are particularly interesting because she examined both migratory and nonmigratory populations. In tests in a large arena, she found that solar cues, including polarized light, affected the orienting ability of both migratory and nonmigratory *T. sirtalis.* Pheromone trails, however, could override the sun compass of migratory snakes but were less likely to do so in nonmigratory populations.

Orientation mechanisms used by snakes in their core areas on a daily basis have not been specifically examined, although some evidence indicates that pheromone trails are involved. Snakes show typical chemosensory searching both when disturbed in their natural habitat and when observed undisturbed (Gillingham et al., 1990). Refuge and feeding locations both could be identified in this manner. Adult snakes often show tendencies to return to refuges they previously occupied (Heller and Halpern, 1982). In the field, sites could be localized by landmark orientation. Homing to a foraging area has been documented in the Rat Snake, *Elaphe obsoleta,* and likely includes both navigation and landmark orientation (Weatherhead and Robertson, 1990).

Foraging

Snakes can be classified by foraging strategy (sit-and-wait or active searchers), habitat used (arboreal, aquatic, terrestrial), and so forth. Such distinctions are useful, even if they oversimplify and stereotype. What is apparent is that snakes use a variety of cues and methods to locate, recognize, and capture prey. The foraging sequence can be divided into appetitive and consummatory phases, and different stimuli can be involved in each stage.

Chemosensory and visual cues are most important in the appetitive phase of foraging, although tactile cues may also alert snakes to the presence of prey. Volatile chemical stimuli from prey detected by nasolfaction often lead to increased tongue-flicking and searching behavior that help localize prey and aid in detecting more proximal nonvolatile molecules (Burghardt, 1969, 1977; Drummond, 1979; Teather, 1991). Visual cues are also important in orienting snakes to specific objects in the environment, especially if the prey are moving (Herzog and Burghardt, 1974; Burghardt and Denny, 1983; Drummond, 1985; Chiszar et al., 1988a; Teather, 1991). In addition, prey that contrast visually with their background are more frequently attacked than the same prey against a noncontrasting background

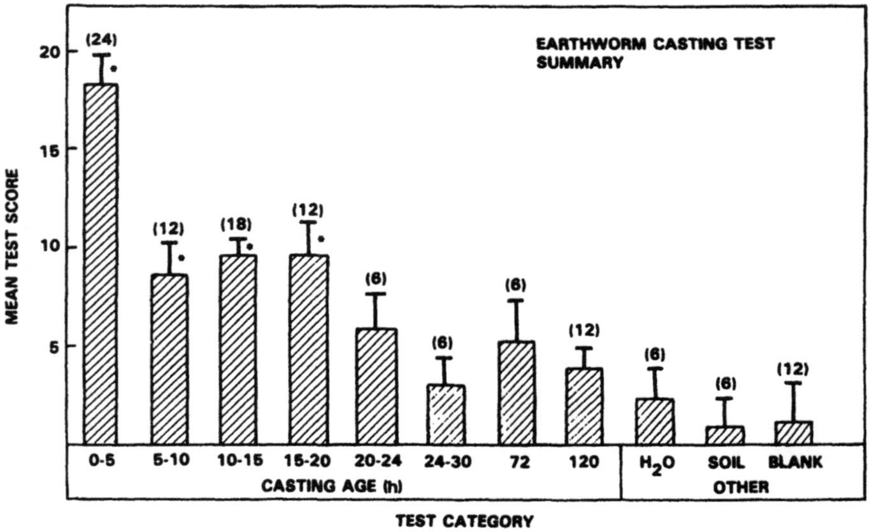

Figure 4.5 Mean test scores for Common Garter Snakes, *T. sirtalis,* tested with variously aged earthworm castings and control substances. An asterisk indicates that the mean score was significantly different from controls at the 0.05 confidence level using the Mann–Whitney U-test. Sample size is indicated parenthetically. (Gillingham et al., 1990; reprinted with permission.)

(Porter and Czaplicki, 1974). Although most of these studies employed laboratory-reared thamnophiines, Gillingham et al. (1990) have shown that wild-caught *T. sirtalis* can recognize recent versus old (48 h) fecal deposits of earthworms at burrow entrances (Fig. 4.5).

Snakes respond to chemical cues derived from species-typical prey with increased tongue-flicking and even open-mouthed attack (the consummatory act). Colubrid snakes, particularly *Thamnophis,* have been used in much of this work (reviews in Burghardt, 1970a, 1980, 1990; Arnold, 1981c; von Achen and Rakestraw, 1984; Cooper, 1990; Halpern, 1992). Several different methods and scoring techniques have been used, some being more discriminating than others (Cooper and Burghardt, 1990a). The ecological validity of such responses is strong, but not complete. For example, *T. butleri* will attack fish and fish chemical cues although they have never been reported eating fish in the field. Perhaps fish-eating is a characteristic retained from its presumed prey-generalist ancestral species, *T. radix* (Burghardt, 1970a, in press; Ford, 1986).

Responses of other colubrids to aqueous extracts of surface substances of prey are often not as marked as in *Thamnophis* (Burghardt, 1970a). While species differences may simply be due to presentation of inappropriate prey chemicals, they may also indicate that other stimuli are needed to facilitate or elicit prey attack. Multiple cues may especially be required for snakes that specialize on rapidly moving prey such as fish, birds, and mammals (Chiszar et al., 1988a,b). Even within *Thamnophis* large differences exist in the tendency to respond to prey extracts and models (Drummond, 1985). The aquatic *T. melanogaster* is less responsive to chemical cues, even from highly preferred prey such as fish, than is the generalist *T. sirtalis,* while the earthworm specialist *T. butleri* is more sensitive to its preferred prey, earthworms, but less sensitive to fish (Burghardt, in press).

Snakes may become more responsive to other cues after exposure to odors from prey. Thus fish-eating *Nerodia* and *Thamnophis* are more likely to attack fish models in the presence of "fishy" water (Drummond, 1979,1985). Likewise, many species of snakes will rapidly respond to touch with predatory attacks after exposure to appropriate prey odors (N. Ford, personal observation). However, the importance of chemical stimulation may differ with species and experience. While neonatal *T. sirtalis* do seem to need chemical cues to elicit prey attack (Burghardt, 1966; Teather, 1991), visual cues alone appear sufficient in neonates of the aquatic specialist *T. melanogaster* (Drummond, 1985).

The Brown Tree Snake (*Boiga irregularis*) has been the focus of experiments that elegantly document the role of context in prey recognition. An initial study showed that adult snakes did not attack cotton swabs rubbed on prey (mice) or investigate cues from prey bedding. A more formal experiment using both visual and chemical cues showed that attacks and increased tongue-flicking only occurred if visual cues were present (Chiszar et al., 1988a). Subsequently, Chiszar et al. (1988b) demonstrated that chemical cues were used in darkness, when visual cues could not confirm prey identification, as when a snake was in a nest or burrow. Nonvolatile stimuli presumably received by the vomeronasal organ, rather than airborne chemicals, were necessary and this was further confirmed with prey-trailing experiments (see also Kardong and Smith, 1991).

For snakes prey size has a major influence on caloric intake, handling time, risk of predation while feeding, and risk of being injured by the prey. While there are several studies on the size of prey eaten by snakes (Miller and Mushinsky, 1990; Shine, 1991; reviewed in Arnold, Chapter 3, this volume), the perceptual basis of prey-size selection has been rarely explored. Visual cues may well mediate approach and attack, but tactile and other cues encountered during succeeding events (constriction, swallowing, etc.) may play major roles. A general finding is that large individuals in a species take a greater range of prey sizes than do smaller individuals. However, it is hard to know if this is based on prey availability, preference, or handling constraints (Shine, 1991). Clearly, the stimuli underlying prey-size selection merit attention.

Another area of major study in snake foraging is strike-induced chemosensory searching (SICS). This phenomenon, the activation of tongue-flicking and trailing behavior after a venomous strike, was first described in vipers (Baumann, 1929), but has been shown to occur in a wide variety of vipers, elapids, and colubrids (O'Connell et al., 1982, 1985; Cooper et al., 1989) and also in lizards (Cooper, 1989, 1991).

The basic pattern of a SICS is that when a viper strikes prey (generally stimulated by visual and/or thermal cues; Kardong, 1992) it releases the animal immediately and waits. After several minutes the snake begins to tongue flick at a high rate and then searches for the trail deposited by the departing prey. SICS is a highly reliable phenomenon that seems to occur only when the snake has actually struck (see reviews in Chiszar et al., 1983; Chiszar et al., 1992). It occurs in neonates and laboratory-reared snakes fed exclusively on dead prey, snakes that have never had to trail to find a meal. The strike-induced increase in tongue-flicking and searching can last for up to two hours.

Although SICS certainly reflects the general foraging behavior of vipers, SICS is not an obligatory response. Pit vipers occasionally scavenge on dead prey (Cowles and Phelan, 1958; Gillingham and Baker, 1981; Savitzky, 1992), and neonate *Thamnophis,* which normally hold on to their prey, will exhibit SICS when strike-eliciting chemosensory stimuli are removed (Burghardt and Chmura, 1993). Initial work with rattlesnakes indicated that visual and infrared stimuli were the sole cues needed from prey (Chiszar, 1977), but this was questioned by Graves and Duvall (1985), who showed that *C. viridis* with their vomeronasal ducts sutured would not feed. Gillingham and Clark (1981) also presented evidence that *C. atrox* use chemical cues to locate hidden prey. However, a recent field study (Hayes and Duvall, 1991) indicated that a model prey devoid of chemical cues not only led to predatory attacks but also SICS. Hayes and Duvall (1991) explain these disparate results by arguing that a functional vomeronasal system is necessary even if prey odors are not necessarily involved in prey strikes. Chemical cues may be used in locating and approaching prey, but if prey are brought close to the snake, or naturally pass by close to it, time might not allow for chemical sampling if visual and thermal cues indicate that the object is probably prey. This series of reports indicates the need for careful assessment of multiple sensory cues, the avoidance of reliance on one kind of experimental paradigm, and particularly the value of field settings for naturalistic experiments. Trailing of prey is probably common to many snakes and very important for some, i.e., Blind Snakes (Watkins et al., 1969; Webb and Shine, 1992), and the phenomenon deserves more serious study.

Prey responses to snake predators and exploitation by snakes of social responses of their prey have also led to a whole host of ecologically important questions about perception that have barely been addressed (Watkins et al., 1969; Coss and Owings, 1978; Dial, 1990; Mori, 1990; Towers and Coss, 1991).

Antipredator behavior

Snakes are themselves prone to being preyed upon by a wide variety of animals and Greene (1988) provides a survey of the myriad ways in which snakes respond to possible predators. All snakes have several responses, often organized sequentially, involving passive (freezing, tail-waving, even mimicking the head with the tail) and more active responses such as flight, striking, or voiding cloacal secretions (Weldon, 1990). Chemical, visual, and tactile (handling) cues have been examined to some degree as stimuli eliciting defensive responses in snakes (e.g., Scudder and Chiszar, 1977). Models and humans have been used in testing snakes primarily due to ethical problems with using live nonhuman predators (Huntingford, 1984; Donnelley and Nolan, 1990).

Tactile cues are readily shown to be important in the rapid change in behavior of many snakes once touched or picked up. When approached and touched, earlier phases of antipredator behavior such as freezing, fleeing, flattening, hissing, or striking may shift to writhing, cloacal gland secretion, and biting (Scudder and Burghardt, 1983). Nishimura (personal communication) has shown that *Trimeresurus flavoviridis* responds defensively to even the slightest touch, indicating that the tactile receptors of snakes may be extremely sensitive. Coiled brooding female Malayan pitvipers, *Calloselasma rhodostoma,* become particularly sensitive to touch and have a body-jerking behavior only seen at this time (York and Burghardt, 1988).

Pit vipers and colubrids have been shown to respond with increased tongue-flicking or other behaviors (i.e., body-bridging or rapid flight) when exposed to chemical cues of possible predators (Weldon, 1982, 1990; Weldon et al., 1990). The detection of odors of ophiophagous snakes by other snakes has been reviewed in Weldon (1990) and Weldon et al. (1992).

The use of models has allowed the characterization of some of the specific visual stimuli for defensive behaviors in a variety of snakes. Spitting Cobras are well known to direct spraying of venom towards the eyes of their molesters (Greene, 1988). Eyes also direct defensive attacks in young Common Garter Snakes, *T. sirtalis* (Herzog and Bern, 1992). In the sequentially increasing responses of the Eastern

Hognose Snake, *Heterodon platirhinos,* of bluffing, "dying," and final death feign, the performance is under visual control even though tactile stimulation is often needed for the death feign (Burghardt, 1991). Burghardt and Greene (1988) showed that recovery from the death feign took longer when Hognose Snakes were observed than when not observed, and longer when the observer was gazing directly at the snake than when his eyes were averted.

The role of movement of the predator has also been examined. In one-day-old *Thamnophis* a moving stimulus is generally more potent than a nonmoving one in eliciting defensive striking and flight (Herzog and Burghardt, 1986; Herzog et al., 1989b). Using mounted museum specimens of predators, Scudder and Chiszar (1977) found that two species of rattlesnakes habituated to models with repeated presentations when the stimuli were stationary, but not when the models were moving.

The defensive responses of pit vipers could rely to some degree on infrared reception, and Greene (1992) has argued that the pit may very well have originated for defense rather than for use in foraging.

A number of members of the genus *Thamnophis* have been tested for defensive behaviors and species differences are quite evident. For example, *T. melanogaster* responds with more strikes than does *T. sirtalis,* which, in turn, is more responsive than *T. butleri,* which rarely strikes or tries to bite even as an adult (Herzog and Burghardt, 1986). Individual and litter differences in defensive striking are also great, even among offspring from females caught within 20 m on the same lake, and in the absence of differential treatment such differences remain quite stable from day 1 to over a year of age (Herzog and Burghardt, 1988). Although the role of perceptual mechanisms in these differences has not been typically considered (but see Herzog et al., 1989a), such information could prove useful in models for examining the evolution of species differences in defensive displays (Cooper and Vitt, 1991).

Genetic Mechanisms

We have discussed to some degree the occurrence of individual, familial, and population differences in aspects of sensory perception. The implication of these differences is that the role of perception in controlling behavior in snakes is heritable. One way the basic intrinsic mechanisms of snake behavior have been examined has been to look at the responses of neonates to their first exposure to biologically relevant stimuli, be they prey, predators, or conspecifics, with the assumption that these are innate. To study these responses properly, however, it is necessary to have some understanding of development, learning, and possible roles of experience in these behaviors.

Although genetic factors are probably involved in behavioral diversity in all the response systems discussed in this chapter, formal genetic analyses are still uncommon. The best-known system in these terms is chemoreception in neonate *Thamnophis* (Burghardt, in press; Brodie and Garland, Chap. 8, this volume).

The initial indication that genetic factors were at work in the chemoreceptive prey responses of *Thamnophis* and other snakes was the demonstration that ingestively naive neonates showed increased tongue-flicking and attacks to cotton swabs dipped in prey extracts. Geographic and within-litter variation among snakes also pointed to the importance of genetics in individual differences (Burghardt, 1970b, 1975; Arnold, 1977), as did the fact that maternal diet did not influence offspring preference (Burghardt, 1971). However, the first strictly genetic analyses were performed by Arnold (1981a,b,c; Ayres and Arnold, 1983). Brodie and Garland (Chap. 8, this volume) provide a thorough review of behavioral genetic concepts and work to date as it relates to snakes.

It is clear that many aspects of snake biology and behavior are under genetic control, and heritabilities are being calculated for predatory responses, defensive behavior, stamina, and meristic characters among others (Arnold and Bennett, 1984; Garland, 1988; Schwartz, 1989; Brodie and Garland, Chap. 8, this volume). The major difficulty is that large numbers of litters from the same population are needed and the work involved is considerable. In addition, merely demonstrating heritability is only a first step in answering a biologically interesting question. Thus, we encourage small-scale studies to establish the reliability of measurements and experimental methods before embarking on large-scale breeding studies or those requiring capture of numerous gravid females. To avoid serious impact on natural populations such studies should be limited to species and populations known to be abundant. It would be worthwhile to study individual differences in responses of neonatal snakes to visual, chemical, thermal, and tactile cues in homing, orientation, habitat selection, antipredator, and other behaviors. All of these are prime areas for the interaction of genes, experience, and natural selection (see Brodie and Garland, Chap. 8, this volume).

Ontogenetic Processes

Ontogeny is an often neglected issue in behavioral and ecological work on perception in snakes. This omission derives from the challenge of even finding neonates and juveniles in the field, let alone observing them. That the young are living in different places than the adults is itself an indication that an important ontogenetic process is at work. Still, the view that snakes are basically instinctive machines with little cognitive capacity has fostered disregard. Compared to mammals and birds, neonatal snakes, with their lack of postnatal parental care, need to perform much important business of life on their own (Burghardt, 1978, 1988). All animals face as juveniles different problems than adults (Alberts and Cramer, 1988) and this may be even more evident for snakes. For example, a garter snake that grows from 2 to 200 g increases its mass 100-fold, and 1000-fold increases are not uncommon in larger species. This great size range has major consequences in terms of habitat selection, diet, and predator avoidance. Responses suitable when snakes are small are often less so when large. For example, juvenile Checkered Garter Snakes (*T. marcianus*) in Arizona are often preyed upon by Bullfrogs, but the adult snakes eat the large Bullfrog tadpoles (R. Seigel, personal communication). The perceptual mechanisms underlying these shifts are also likely to change. A major initial issue is to distinguish among the several processes underlying ontogenetic change.

Maturation

Maturation refers to behavioral change occurring throughout some period of ontogeny apparently as a consequence of normal physical development. For example, defensive behavior (strikes and flight) shows a developmental change in some natricines over the first 20 days of life (Burghardt, 1978; Herzog et al., 1992). *Thamnophis sirtalis* shows a dramatic increase in defensive striking but *Nerodia rhombifer* does not change. Striking occurs at a low level in *T. butleri* at all ages, and this species shows a small ontogenetic decline in escape behavior. These defensive responses are primarily elicited by visual cues, although tactile ones from handling also have an effect, especially in *Nerodia* (Scudder and Burghardt, 1983). Maturational changes in the perceptual mechanisms of snakes have not been carefully examined, and they are likely to be complex. For example, Burger (1989b, 1990a) has shown incubation temperatures have long-term effects on behavior of snakes (Fig. 4.6).

Neonatal Plainbelly Water Snakes, *Nerodia erythrogaster,* show a dietary switch to frogs from fish prey at a certain body size regardless of being raised on either a fish or frog diet, and this change is reflected in differences in preference for prey chemicals (Mushinsky and Lotz, 1980; Mushinsky et al., 1982). Whether this switch is based on age or body size is unknown. On the other hand, *T. sirtalis* continue to respond to frogs and chemical cues from frogs regardless of diet, indicating that prey preferences can be ontogenetically stable regardless of dietary experience (Burghardt and Hess, 1968; Arnold, 1978a).

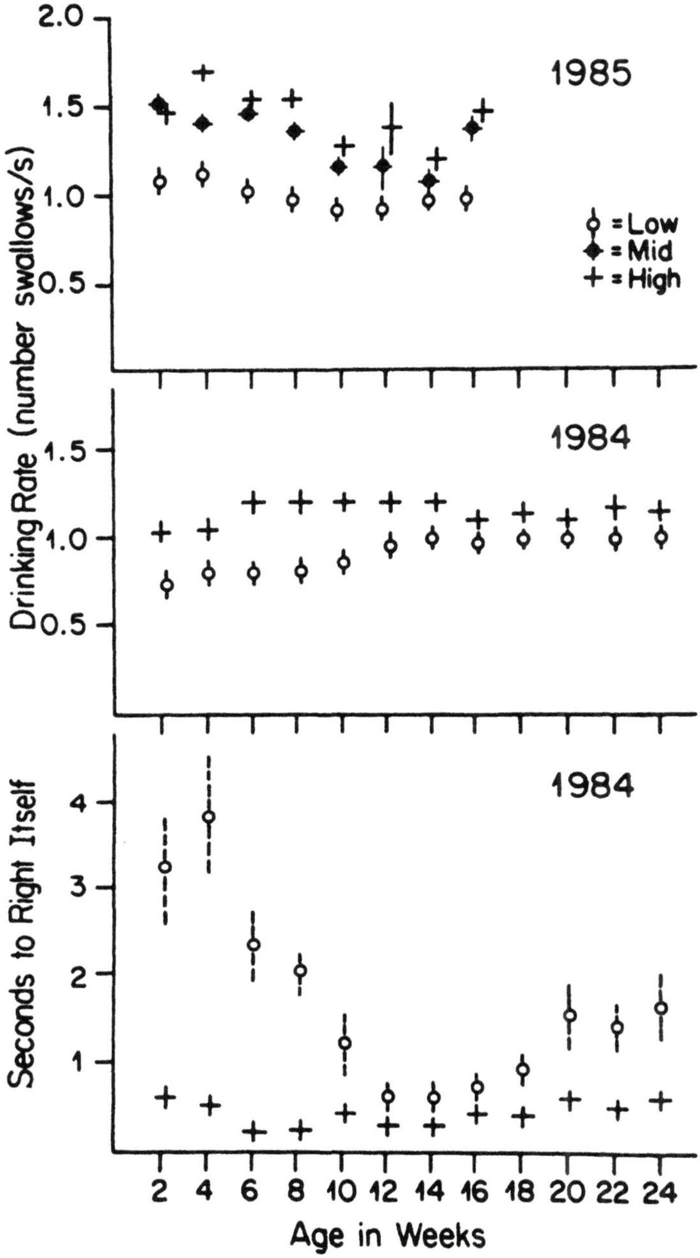

Figure 4.6 Righting response and drinking rate as a function of incubation temperature and age for 2–24-week-old Pine Snakes. (Burger, 1989; reprinted with permission.)

In stable or predictable environments, maturational changes in prey chemical preference may be the simplest way to ensure that snakes take appropriate prey. However, in more variable environments or with more generalized predators, predetermined changes might not be as useful. Arnold (1978b) discusses the theoretical issues of changing sensory cues from a herpetological perspective. To more fully understand the ontogeny of sensory systems, work is needed on not only more species but also on the nature of maturational changes in orientation, habitat selection, and other important behavioral systems.

Learning and experience

Snakes and other reptiles, although not considered highly intelligent animals, nevertheless show learning capabilities of some magnitude and rapidity (Burghardt, 1977). Both anecdotal isolated reports of learning in snakes (Bowers and Burghardt, 1992) and the limited information from experimental studies have been reviewed (Burghardt, 1977). Most tests of learning in poikilothermic animals have used traditional equipment developed for use with rats and other mammals. Generally snakes do poorly, being tested at inappropriate temperatures, with poorly selected reinforcers (i.e., electric shock). Furthermore, snakes have an *Umwelt* (sensory world) and response style so different from ours that anthropomorphic interpretation of their behavior is less common (Burghardt, 1991). It is thus useful to distinguish studies of ophidian learning using stimuli and responses that are not biologically relevant from learning phenomena that probably occur as a matter of course in the wild (i.e., learning landmarks for homing; Lawson, 1991).

The chemoreceptive prey preferences in snakes have been have been useful in studying learning. Basically, rearing snakes on selected diets can alter their preferences, but this is neither universal nor widely established (Fuchs and Burghardt, 1971; Gove and Burghardt, 1975; Burghardt, in press). As noted above, genetic and maturational factors play a role, and species (especially dietary specialists) may be well-buffered against experiential effects (Arnold, 1981c).

Dietary experience can increase snake's sensitivity to prey chemicals. *Thamnophis radix* raised for six months exclusively on earthworms were much more responsive to dilutions of an earthworm wash than were individuals reared on fish. Just the reverse was true for responses to fish extracts (Burghardt, 1990). However, Garter Snakes that specialize on fish and worms (*T. melanogaster* and *T. butleri*, respectively) did not increase their sensitivity to nonpreferred prey when fed those foods (Berghardt and Lyman-Herley, unpublished). Thus the ability to learn to discriminate and recognize traces of chemicals from prey may vary among species. Such data are germane to the issue of search image, a topic from ethology now being taken up with renewed interest in animal psychology (Shettleworth et al., in press).

The role of search-image formation also relates to work on SICS (discussed above). In one of the studies with Cottonmouths (*Agkistrodon piscivorus*) Chiszar et al. (1985) found that striking prey led to a prey-specific search image; in this case, the relative amount of time spent on the trail derived from envenomated prey. More recently, Melcer and Chiszar (1989) showed that in the brief contact of a strike, a rattlesnake learns distinctive chemical features of the prey (i.e., distinguishes mice reared on two different diets). A rattlesnake can even discriminate among littermates of wild mice (*Peromyscus*) reared together on the same diet (Furry et al., 1991). This is a rapid and highly precise type of learning that clearly has biological utility.

However, experiential effects can also be initially counterintuitive. Neonatal *Thamnophis sirtalis* raised for several days in tanks with prey (fish or worms) that were inaccessible, but whose volatile odors could emanate out, became less responsive to chemical cues from the exposed rather than the nonexperienced prey even many hours after the ambient prey odor had been removed (Burghardt, 1992) (Fig. 4.7). More specialist *Thamnophis* species did not show such shifts (Burghardt, unpublished). It might be adaptive for a generalist to

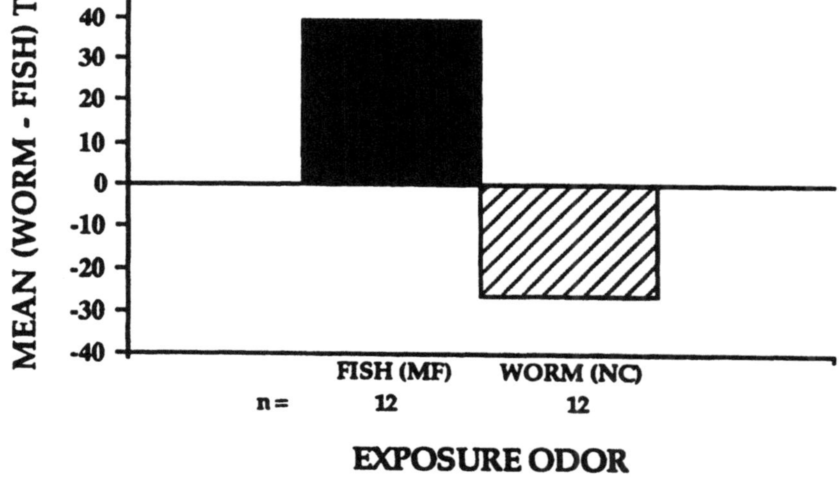

Figure 4.7 Differences in tongue-flick attack scores to earthworm and fish surface extracts presented on cotton swabs on the day after removal from cages that contained ambient odors of live fish or earthworms. Subjects were ingestively naive neonatal Common Garter Snakes from two litters ($n = 12$ each) balanced across exposure condition. (Based on data presented in Burghardt, 1992.)

decrease responses to prey types it has been prevented from obtaining. Whether adult snakes' feeding preferences are as easily modified as neonates' is unknown, as is the generality of the phenomenon outside of *Thamnophis*. In any event, these results suggest caution for studies of foraging in captive snakes where the prey is kept in the same room as the predator.

The effects of experience have also been studied in the antipredator responses of *Thamnophis* species as described above. Defensive striking in highly aggressive *T. melanogaster* declines with repeated stimulation both in the short term (within a day) and the long term (over days). However, a novel visual stimulus reinstates the response to some extent, showing that the habituation is stimulus specific (Herzog et al., 1989a). Snakes that Herzog et al. (1989b) used in a year-long feeding study showed more responses to a moving than a nonmoving predator (a human finger), while littermates not used in the study showed no discrimination (Herzog et al., 1989b). Herzog (1990) also showed that a slight harassing of a young Garter Snake can lead to greater flight tendencies weeks later, and that litter differences are apparent in this experiential effect. Bowers (1992) showed that species, age, experience, and stimulus differences occur in habituation responses in *Thamnophis*. While more studies could be discussed, the point should be amply clear that the role of postnatal experience in snake responses to stimuli is far from trivial and may mean major differences in survival.

Ecological Relevance

Snakes are more tied to certain aspects of their environment than are many endotherms and even other ectotherms. This may make links between ecological factors, sensory processes, and behavior more robust and the effects of experience more channeled and constrained. Certainly brain size is related in some way to learning capacity, and snakes have rather small brains.

Be that as it may, perhaps the best way to look at experience is to consider the problems a snake is likely to confront. For example, a generalist predator such as *T. sirtalis* might need to learn more things about different prey than a specialist would (i.e., where worms are available in the spring versus amphibians in the summer). Thus the former may show a great deal of switching in prey preference or foraging sites, while this is less likely with specialists within the same genus (Halloy and Burghardt, 1990).

As implied above, the generalist–specialist continua reflects more than just prey selection and can refer to habitat, microhabitat, or predator exposure. Thus it is important to know many aspects of a snake's life history. For example, Carpenter (1952) showed that *Thamnophis sirtalis* is more of a generalist in both diet and habitat than either sympatric *T. butleri* or *T. sauritus*. Detection and response to stimuli are important measures of the placement of animals ecologically, but ultimately field evaluation of the role of sensory cues is necessary.

Phylogeny and Ecology

How do we partition variation in perception and behavior of extant species into phylogenetic and ecological factors? Properly done, this could lead to very insightful studies of perception and behavioral functions, including the neural mechanisms involved at both the perceptual and effector levels (see Chiszar, 1986). Fortunately, new comparative tools for quantitatively partitioning this variation are becoming available (Gittleman, 1989). We encourage herpetologists and snake ethologists to apply them to both describing and understanding ophidian diversity in perceptual mechanisms at the phylogenetic, ecological, and individual levels.

Comparative studies fall into at least two categories (Burghardt and Gittleman, 1990). First, there are issues of phylogeny and phylogenetic reconstruction. These go back to the classic studies of Lorenz and other early ethologists and their interests in tracing behavioral homologies (Lauder, 1986). Modern cladistic methods with outgroup analysis will be powerful tools for continuing these studies (Brooks and McLennan, 1990). Although such methods have not commonly been applied in studies of responses to sensory cues, there is no reason not to do so (Burghardt, 1969).

The second category is to assess how much variation is due to ecological adaptation at the species level and individual adaptation (learning, experience) and how much by historical factors (phylogeny). This is a lively topic in comparative biology and the main new tools for contending with the problems include ANOVA and autocorrelation methods (see Gittleman and Kot, 1992). The critical concern using these statistical techniques is whether species are independent samples for testing hypotheses. If not, at what taxonomic level is independence a safe assumption?

Summary and Future Research

Summary statements concerning the role of perceptual mechanisms must remain general, descriptive, and even speculative at this time, primarily because of the lack of information on elapids, boids, and other families. It does appear that the chemosensory systems are of primary importance both in predation and for predator avoidance, particularly in thamnophiines. Vision and infrared reception are very important in boids and viperids; vision is also critical for many colubrids. It appears that the environment occupied (arboreal, aquatic, etc.) by specialists strongly influences the use of these cues. Tactile cues are likely important in all of these ecological activities. However, data on tangoreception are nearly nonexistent for snakes.

Clearly, we need careful behavioral studies investigating sensory mechanisms in a variety of species of snakes, particularly from families not traditionally kept in laboratories (i.e., boids, pythons, and elapids). It may also be useful to select species that are behavioral specialists, and this might be best achieved by looking to tropical snakes. We also need data on the interaction of the senses in specific actions. When basic information from a variety of animals and situations becomes available then we will be able to begin more integrated and theoretical assessment of the role of sensory mechanisms in the ontogeny, learning, genetics, and evolution of behavior in snakes.

Literature Cited

Akester, J., 1979, Male combat in captive gaboon vipers (Serpentes: Viperidae), *Herpetologica,* 35:124–128.

Alberts, A. C., 1992, Constraints on the design of vertebrate chemical communication systems, *Amer. Nat.,* 139:S62–S89.

Alberts, J. R., and C. P. Cramer, 1988, Ecology and experience: Sources of means and meaning of developmental change, in E. M. Blass, ed., *Developmental Psychobiology and Behavioral Ecology,* Vol. 9, Plenum, New York, pp. 1–62.

Allen, B. A., G. M. Burghardt, and D. S. York, 1984, Species and sex differences in substrate preference and tongue flick rate in three sympatric species of water snakes (*Nerodia*), *J. Comp. Psych.,* 98:358–367.

Andrén, C., 1976, Social behaviour during the reproductive period in *Vipera b. berus* (L.), *Norw. J. Zool.,* 24:234–235.

Andrén, C., 1982, The role of the vomeronasal organs in the reproductive behavior of the adder *Vipera berus, Copeia,* 1982:148–157.

Andry, M. L., and M. W. Luttges, 1972, Neural habituation in garter snakes, *Physiol. Behav.,* 9:107–111.

Arnold, S. J., 1977, Polymorphism and geographic variation in the garter snake *Thamnophis elegans, Science,* 197:676–678.

Arnold, S. J., 1978a, Some effects of early experience on feeding responses in the common garter snake, *Thamnophis sirtalis, Anim. Behav.,* 26:455–462.

Arnold, S. J., 1978b, The evolution of a special class of modifiable behaviors in relation to environmental pattern, *Amer. Nat.,* 112:415–427.

Arnold, S. J., 1981a, Behavioral variation in natural populations, I. Phenotypic, genetic, and environmental correlations between chemoreceptive responses to prey in the garter snake, *Thamnophis elegans, Evolution,* 35:489–509.

Arnold, S. J., 1981b, Behavioral variation in natural populations, II. The inheritance of a feeding response in crosses between geographic races of the garter snake, *Thamnophis elegans, Evolution,* 35:510–515.

Arnold, S. J., 1981c, The microevolution of feeding behavior, in A. Kamil and T. Sargent, eds., *Foraging Behavior: Ecological, Ethological, and Psychological Approaches,* Garland STPM, New York, pp. 409–453.

Arnold, S. J., and A. F. Bennett, 1984, Behavioral variation in natural populations: III. Antipredator displays in the garter snake *Thamnophis radix, Anim. Behav.,* 32:1108–1118.

Ayres, F. A., and S. J. Arnold, 1983, Behavioural variation in natural populations, IV. Mendelian models and heritability of a feeding response in the garter snake, *Thamnophis elegans, Heredity,* 51:405–413.

Baird, I. L., 1970, The anatomy of the reptilian ear, in D. Gans and T. S. Parsons, eds., *Biology of the Reptilia, Morphology B*, Vol. 2, Academic, New York, pp. 193–275.
Barker, D. G., J. B. Murphy, and K. W. Smith. 1979, Social behavior in a captive group of Indian pythons, *Python molurus* (Serpentes, Boidae) with formation of a linear social hierarchy, *Copeia*, 1979:466–471.
Barrett, R., 1970, The pit organ of snakes, in C. Gans, ed., *Biology of the Reptilia, Morphology B*, Vol. 2, Academic, New York, pp. 277–300.
Baumann, F., 1929, Experimente über den geruchssinn und der beuterwerb der viper (*Vipera aspis* L.), *Z. Vergleich. Physiol.*, 10:36–119.
Beer, T., 1898, Die akkommodation des auges b- den reptilien, *Arch. F. D. Ges. Physiol.*, 69:507–568.
Benson, D. M., and P. H. Hartline, 1988, A tectorotundo-telencephalic pathway in the rattlesnake: Evidence for a forebrain representation of the infrared sense, *J. Neurosci.*, 8:1074–1088.
Bowers, B. B., 1992, Habituation of antipredator behaviors and responses to chemical prey extracts in four species of garter snake, *Thamnophis* (Serpentes: Colubridae), Doctoral Thesis, Univ. Tennessee, Knoxville.
Bowers, B. B., and G. M. Burghardt, 1992, The scientist and the snake: Relationships with reptiles, in H. Davis and D. Balfour, eds., *The Inevitable Bond,* Cambridge Univ. Press, Cambridge, pp. 250–263.
Breidenbach, C. H., 1990, Thermal cues influence strikes in pitless vipers, *J. Herpetol.*, 24:448–450.
Brooks, D. R., and D. A. McLennan, 1990, *Phylogeny, Ecology, and Behavior,* Univ. Chicago Press, Chicago.
Brown, W. S., and F. M. MacLean, 1983, Conspecific scent-trailing by newborn timber rattlesnakes, *Crotalus horridus, Herpetologica,* 39:430–436.
Bullock, T. H., and R. Barrett, 1968, Radiant heat reception in snakes, *Comm. Behav. Biol.*, A, 1:19–29.
Bullock, T. H., and F. P. J. Diecke, 1956, Properties of an infrared receptor, *J. Physiol.*, 134:47–87.
Burchfield, P. M., 1982, Additions to the natural history of the crotaline snake *Agkistrodon bilineatus taylori, J. Herpetol.*, 16:376–382.
Burger, J., 1989a, Following of conspecific and avoidance of predator chemical cues by pine snakes (*Pituophis melanoleucus*), *J. Chem. Ecol.*, 15:799–806.
Burger, J., 1989b, Incubation temperature has long-term effects on behaviour of young pine snakes (*Pituophis melanoleucus*), *Behav. Ecol. Sociobiol.*, 24:201–297.
Burger, J., 1990a, Effects of incubation temperature on behavior of young black racers (*Coluber constrictor*) and kingsnakes (*Lampropeltis getulus*), *J. Herpetol.*, 24:158–163.
Burger, J., 1990b, Responses of hatchling pine snakes (*Pituophis melanoleucus*) to chemical cues of sympatric snakes, *Copeia,* 1990:1160–1163.
Burghardt, G. M., 1966, Stimulus control of the prey attack response in naive garter snakes, *Psychon. Sci.*, 4:37–38.
Burghardt, G. M., 1969, Comparative prey-attack studies in naive snakes of the genus *Thamnophis, Behaviour,* 33:77–114.
Burghardt, G. M., 1970a, Chemical perception of reptiles, in J. W. Johnston, Jr., D. G. Moulton and A. Turk, eds., *Communication by Chemical Signals,* Appleton-Century-Crofts, New York, pp. 241–308.
Burghardt, G. M., 1970b, Intraspecific geographical variation in chemical food cue preferences of newborn garter snakes (*Thamnophis sirtalis*), *Behaviour,* 36:246–257.
Burghardt, G. M., 1971, Chemical-cue preferences of newborn snakes: Influence of prenatal maternal experience, *Science,* 171:921–923.
Burghardt, G. M., 1975, Chemical prey preference polymorphism in newborn garter snakes, *Thamnophis sirtalis, Behaviour,* 52:202–225.
Burghardt, G. M., 1977, Learning processes in reptiles, in C. Gans and D. Tinkle, eds., *Biology of the Reptilia,* Vol. 7, Academic, New York, pp. 555–681.
Burghardt, G. M., 1978, Behavioral ontogeny in reptiles: Whence, whither, and why? in G. M. Burghardt and M. Bekoff, eds., *The Development of Behavior: Comparative and Evolutionary Aspects,* Garland STPM, New York, pp. 149–174.
Burghardt, G. M., 1980, Behavioral and stimulus correlates of vomeronasal functioning

in reptiles: Feeding, grouping, sex, and tongue use, in D. Müller-Schwarze and R. M. Silverstein, eds., *Chemical Signals in Vertebrates and Aquatic Invertebrates,* Plenum, New York, pp. 275–301.

Burghardt, G. M., 1983, Aggregation and species discrimination in newborn snakes, *Z. Tierpsychol.,* 61:89–101.

Burghardt, G. M., 1988, Precocity, play, and the ectotherm-endotherm transition: Profound reorganization or superficial adaptation? in E. M. Blass, ed., *Developmental Psychobiology and Behavioral Ecology, Handbook of Behavioral Neurobiology,* Vol. 9, Plenum, New York, pp. 107–148.

Burghardt, G. M., 1990, Chemically mediated predation in vertebrates: Diversity, ontogeny, and information, in D. McDonald, D. Müller-Schwarze, and S. Natynczuk, eds., *Chemical Signals in Vertebrates,* Vol. 5, Oxford Univ. Press, Oxford, pp. 475–499.

Burghardt, G. M., 1991, Cognitive ethology and critical anthropomorphism: A snake with two heads and hognose snakes that play dead, in C. A. Ristau, ed., *Cognitive Ethology: The Minds of Other Animals,* Lawrence Erlbaum, New Jersey, pp. 53–90.

Burghardt, G. M., 1992, Prior exposure to prey cues influences chemical prey preferences and prey choice in neonate garter snakes, *Anim. Behav.,* 44:787–789.

Burghardt, G. M., The comparative imperative: Genetics and ontogeny of chemoreceptive prey responses in natricine snakes, *Brain Behav. Evol.,* in press.

Burghardt, G. M., and P. J. Chmura, 1993, Strike-induced chemosensory searching by ingestively naive garter snakes (*Thamnophis sirtalis*), *J. Comp. Psych.,* 107:116–121.

Burghardt, G. M., and D. Denny, 1983, Effects of prey movement and prey odor on feeding in garter snakes, *Z. Tierpsychol.,* 62:329–347.

Burghardt, G. M., and J. L. Gittleman, 1990, Comparative behavior and phylogenetic analysis: New wine, old bottles, in M. Bekoff and D. Jamieson, eds., *Interpretation and Explanation in the Study of Animal Behavior,* Vol. 2, Westview Press, Boulder, Colorado, pp. 192–225.

Burghardt, G. M., and H. W. Greene, 1988, Predator simulation and duration of death feigning in neonate hognose snakes, *Anim. Behav.,* 36:1842–1843.

Burghardt, G. M., and E. H. Hess, 1968, Factors influencing the chemical release of prey attack in newborn snakes, *J. Comp. Physiol. Psych.,* 66:289–295.

Burghardt, G. M., S. E. Goss, and F. M. Schell, 1988, Comparison of earthworm- and fish-derived chemicals eliciting prey attack by garter snakes (*Thamnophis*), *J. Chem. Ecol.,* 14:855–881.

Burghardt, G. M., H. C. Wilcoxon, and J. A. Czaplicki, 1973, Conditioning in garter snakes: Aversion to palatable prey induced by delayed illness, *Anim. Learn. Behav.,* 1:317–320.

Burns, B., 1969, Oral sensory papillae in sea snakes, *Copeia,* 1969:617–619.

Carpenter, C. C., 1952, Comparative ecology of the common garter snake (*Thamnophis s. sirtalis*), the ribbon snake (*Thamnophis s. sauritus*) and Butler's garter snake (*Thamnophis butleri*) in mixed populations, *Ecol. Monogr.,* 22:235–258.

Carpenter, C. C., 1977, Communication and displays of snakes, *Amer. Zool.,* 17:217–223.

Carpenter, C. C., 1984, Dominance in snakes, in R. A. Seigel, L. E. Hunt, J. L. Knight, L. Malaret, and N. L. Zuschlag, eds., *Vertebrate Ecology and Systematics: A Tribute to Henry S. Fitch,* Univ. Kans. Mus. Nat. Hist. Spec. Publ., 10, pp. 195–202.

Carpenter, C. C., J. B. Murphy, and L. A. Mitchell, 1978, Combat bouts with spur use in the Madagascan boa (*Sanzinia madagascariensis*), *Herpetologica,* 34:207–212.

Charland, M. B., and P. T. Gregory, 1990, The influence of female reproductive status on thermoregulation in a viviparous snake, *Crotalus viridis, Copeia,* 1990:1089–1098.

Chiszar, D., 1977, Absence of prey-chemical preferences in newborn rattlesnakes *Crotalus cerastes, C. enyo* and *C. viridis, Behav. Biol.,* 21:146–150.

Chiszar, D., 1986, Motor patterns dedicated to sensory functions, in D. Duvall, D. Müller-Schwarze, and R. M. Silverstein, eds., *Chemical Signals in Vertebrates,* Vol. 4, Plenum, New York, pp. 37–44.

Chiszar, D., K. Stimac, and T. Boyer, 1983, Effect of mouse odors on visually-induced

and strike-induced searching in prairie rattlesnakes, *Crotalus viridis, Chem. Sen.*, 7:301–308.
Chiszar, D., C. W. Radcliffe, R. Overstreet, T. Poole, and T. Byers, 1985, Duration of strike-induced chemosensory searching in cottonmouths (*Agkistrodon piscivorus*) and a test of the hypothesis that striking prey creates a specific search image, *Can. J. Zool.*, 63:1057–1061.
Chiszar, D., C. W. Radcliffe, T. Boyer, and J. L. Behler, 1987, Cover-seeking behavior in red spitting cobras (*Naja mossambica pallida*): Effects of tactile cues and darkness, *Zoo Biol.*, 6:161–167.
Chiszar, D., K. Kandler, and H. M. Smith, 1988a, Stimulus control of predatory attack in the brown tree snake (*Boiga irregularis*), 1. Effects of visual cues arising from prey, *The Snake,* 20:151–155.
Chiszar, D., K. Kandler, R. Lee, and H. M. Smith, 1988b, Stimulus control of predatory attack in the brown tree snake (*Boiga irregularis*), 2. Use of chemical cues during foraging, *Amphibia-Reptilia,* 9:77–88.
Chiszar, D., R. Lee, H. Smith, and C. Radcliffe, 1992, Searching behaviors by rattlesnakes following predatory strikes, in J. A. Campbell and E. D. Brodie, Jr., eds., *Biology of the Pitvipers,* Selva, Tyler, Texas, pp. 369–382.
Cock Bunning, T. de., 1983, Thermal sensitivity as a specialization for prey capture and feeding in snakes, *Amer. Zool.,* 23:363–375.
Cohen, A. C., 1975, Some factors affecting water economy in snakes, *Comp. Biochem. Physiol.,* 51A:361–368.
Constanzo, J. P., 1989, Conspecific scent trailing by garter snakes (*Thamnophis sirtalis*) during autumn: Further evidence for use of pheromones in den location, *J. Chem. Ecol.,* 15:2531–2538.
Cooper, W. E., Jr., 1989, Strike-induced chemosensory searching occurs in lizards, *J. Chem. Ecol.,* 15:1311–1320.
Cooper, W. E., Jr., 1990, Prey odour discrimination by lizards and snakes, in D. W. MacDonald, D. Müller-Schwarze, and S. E. Natynczuk, eds., *Chemical Signals in Vertebrates,* Vol. 5, Oxford Univ. Press, Oxford, pp. 533–538.
Cooper, W. E., Jr., 1991, Responses to prey chemicals by lacertid lizard *Podarcis muralis:* Prey chemical discrimination and poststrike elevation in tongue-flick rate, *J. Chem. Ecol.,* 17:849–863.
Cooper, W. E., Jr., and G. M. Burghardt, 1990a, A comparative analysis of scoring methods for chemical discrimination of prey by squamate reptiles, *J. Chem. Ecol.,* 16:45–65.
Cooper, W. E., Jr., and G. M. Burghardt, 1990b, Vomerolfaction and vomodor, *J. Chem. Ecol.,* 16:103–105.
Cooper, W. E., Jr., and L. J. Vitt, 1986, Thermal dependence of tongue-flicking and comments on use of tongue-flicking as an index of squamate behavior, *Ethology,* 71:177–186.
Cooper, W. E., Jr., and L. J. Vitt, 1991, Influence of detectability and ability to escape on natural selection of conspicuous autonomous defenses, *Can. J. Zool.,* 69:757–764.
Cooper, W. E., Jr., S. G. McDowell, and J. Ruffer, 1989, Strike-induced chemosensory searching in the colubrid snakes *Elaphe g. guttata* and *Thamnophis sirtalis, Ethology,* 81:19–28.
Coss, R. G., and D. H. Owings, 1978, Snake-directed behavior by snake naive and experienced California ground squirrels in a simulated burrow, *Z. Tierpsychol.,* 48:421–435.
Cowles, R. B., and R. L. Phelan, 1958, Olfaction in rattlesnakes, *Copeia,* 1958:77–83.
Czaplicki, J. A., and R. H. Porter, 1974, Visual cues mediating the selection of goldfish *Carassius auratus* by two species of *Natrix, J. Herpetol.,* 8:129–134.
Dial, B., 1990, Predator-prey signals: Chemosensory identification of snake predators by eublepharid lizards and its ecological consequences, in D. W. MacDonald, D. Müller-Schwarze, and S. E. Natynczuk, eds., *Chemical Signals in Vertebrates,* Vol. 5, Oxford Univ. Press, Oxford, pp. 555–565.
Ditmars, R. L., 1936, *The Reptiles of North America,* Doubleday, Garden City.
Donnelley, S., and K. Nolan (eds.), 1990, Animals, science, and ethics, *Hastings Center Report Suppl.,* 20(3):1–32.
Drummond, H., 1979, Stimulus control of amphibious predation in the northern water snake (*Nerodia s. sipedon*), *Z. Tierpsychol.,* 50:18–44.

Drummond, H., 1983, Aquatic foraging in garter snakes: A comparison of specialists and generalists, *Behaviour,* 86:1–30.
Drummond, H., 1985, The role of vision in the predatory behaviour of natricine snakes, *Anim. Behav.,* 33:206–215.
Dundee, H. A., and M. C. Miller, III, 1968, Aggregative behavior and habitat conditioning by the prairie ringneck snake, *Diadophis punctatus arnyi, Tulane Stud. Zool. Bot.,* 15:41–58.
Duvall, D., and D. Chiszar, 1990, Behaviour and chemical ecology of venal migration and pre- and post-strike predatory activity in prairie rattlesnakes: Field and laboratory experiments, in D. W. MacDonald, D. Müller-Schwarze, and S. E. Natynczuk, eds., *Chemical Signals in Vertebrates,* Vol. 5, Oxford Univ. Press, Oxford, pp. 539–554.
Duvall, D., D. Chiszar, J. Trupiano, and C. W. Radcliffe, 1978, Preference for envenomated rodent prey by rattlesnakes, *Bull. Psychon. Soc.,* 11:7–8.
Duvall, D., M. B. King, and K. J. Gutzwiller, 1985, Behavioral ecology and ethology of the prairie rattlesnake, *Nat. Geogr. Res.,* 1:80–111.
Duvall, D., S. J. Arnold, and G. W. Schuett, 1992, Pitviper mating systems: Ecological potential, sexual selection, and microevolution, in J. A. Campbell and E. D. Brodie, Jr., eds., *Biology of the Pitvipers,* Selva, Tyler, Texas, pp. 321–336.
Ford, N. B., 1986, The role of pheromone trails in the sociobiology of snakes, in D. Duvall, D. Müller-Schwarze, and R. M. Silverstein, eds., *Chemical Signals in Vertebrates,* Vol. 4, Plenum, New York, pp. 261–278.
Ford, N. B., and V. Cobb, 1992, Timing of courtship in two colubrid snakes of the southern United States, *Copeia,* 1992:573–577.
Ford, N. B., and D. Holland, 1990, The role of pheromones in the spacing behaviour of snakes, in D. W. MacDonald, D. Müller-Schwarze, and S. E. Natynczuk, eds., *Chemical Signals in Vertebrates,* Vol. 5, Oxford Univ. Press, Oxford, pp. 465–472.
Ford, N. B., and J. R. Low, Jr., 1984, Sex pheromone source location by garter snakes: A mechanism for detection of direction in non-volatile trails, *J. Chem. Ecol.,* 10:1193–1199.
Ford, N. B., and M. L. O'Bleness, 1986, Species and sexual specificity of pheromone trails of the garter snake, *Thamnophis marcianus, J. Herpetol.,* 20:259–262.
Fuchs, J., and G. M. Burghardt, 1971, Effects of early feeding experience on the responses of garter snakes to food chemicals, *Learn. Motiv.,* 2:271–279.
Furry, K., T. Swain, and D. Chiszar, 1991, Strike-induced chemosensory searching and trail following by prairie rattlesnakes (*Crotalus viridis*) preying upon deer mice (*Peromyscus maniculatus*): Chemical discrimination among individual mice, *Herpetologica,* 47:69–78.
Garland, T., Jr., 1988, Genetic basis of activity metabolism, I. Inheritance of speed, stamina and antipredator displays in the garter snake, *Thamnophis sirtalis, Evolution,* 42:335–350.
Garton, J. D., and H. R. Mushinsky, 1979, Integumentary toxicity and unpalatability as an antipredator mechanism in the narrow mouthed toad, *Gastrophryne carolinensis, Can. J. Zool.,* 57:1965–1973.
Gartska, W. R., and D. Crews, 1981, Female sex pheromones in the skin and circulation of a garter snake, *Science,* 214:681–683.
Gibbons, J. W., and R. D. Semlitsch, 1987, Activity patterns, in R. A. Seigel, J. T. Collins, and S. S. Novak, eds., *Snakes: Ecology and Evolutionary Biology,* McGraw-Hill, New York, pp. 396–421.
Gibson, R. A., and J. Falls, 1979, Thermal biology of the common garter snake, *Thamnophis sirtalis* (L.), I. Temporal variations, environmental effects and sex differences, *Oecologia* (*Berlin*), 43:79–93.
Gillingham, J. C., 1979, Reproductive behavior of the rat snakes of eastern North America, genus *Elaphe, Copeia,* 1979:319–331.
Gillingham, J. C., 1987, Social behavior, in R. A. Seigel, J. T. Collins, and S. S. Novak, eds., *Snakes: Ecology and Evolutionary Biology,* McGraw-Hill, New York, pp. 184–209.
Gillingham, J. C., and R. R. Baker, 1981, Evidence of scavenging behavior in the western diamondback rattlesnake, *Crotalus atrox, Z. Tierpsychol.,* 55:217–227.
Gillingham, J. C., and J. A. Chambers, 1982, Courtship and pelvic spur use in the Burmese python, *Python molurus bivittatus, Copeia,* 1982:193–196.

Gillingham, J. C., and D. L. Clark, 1981, An analysis of prey-searching behavior in the western diamondback rattlesnake, *Crotalus atrox, Behav. Neural Biol.,* 32:235–240.

Gillingham, J. C., C. C. Carpenter, and J. B. Murphy, 1983, Courtship, male combat and dominance in the western diamondback rattlesnake, *Crotalus atrox, J. Herpetol.,* 17:265–270.

Gillingham, J. C., J. Rowe, and M. A. Weins, 1990, Chemosensory orientation and earthworm location by foraging eastern garter snakes, *Thamnophis s. sirtalis,* in D. McDonald, D. Müller-Schwarze, and S. Natynczuk, eds., *Chemical Signals in Vertebrates,* Vol. 5, Oxford Univ. Press, Oxford, pp. 522–532.

Gittleman, J. L., 1989, The comparative approach in ethology: Aims and limitations, in P. P. G. Bateson and P. H. Klopfer, eds., *Perspectives in Ethology,* Vol. 8, Plenum, New York, pp. 55–83.

Gittleman, J. L., and M. Kot, 1992, Adaptation: Statistics and a null model for estimating phylogenetic effects, *Syst. Zool.,* 39:227–241.

Gordon, D. M., 1980, An aggregation of gravid snakes in the Quebec Laurentians, *Can. Field Nat.,* 94:456–457.

Goris, R. C., and M. Nomoto, 1967, Infrared reception in oriental crotaline snakes, *Comp. Biochem. Physiol.,* 23:879–892.

Gove, D., and G. M. Burghardt, 1975, Responses of ecologically dissimilar populations of the water snake, *Natrix s. sipedon,* to chemical cues from prey, *J. Chem. Ecol.,* 1:25–40.

Graves, B. M., and D. Duvall, 1985, Avomic prairie rattlesnakes *Crotalus viridis* fail to attack rodent prey, *Z. Tierpsychol.,* 67:161–166.

Graves, B. M., and M. Halpern, 1988, Neonate plains garter snakes (*Thamnophis radix*) are attracted to conspecific skin extracts, *J. Comp. Psychol.,* 102:251–253.

Graves, B. M., and M. Halpern, 1990, Roles of vomeronasal organ chemoreception in tongue-flicking, exploratory and feeding behaviour of the lizard, *Chalcides ocellatus, Anim. Behav.,* 39:692–698.

Graves, B. M., D. Duvall, M. B. King, S. L. Lindstedt, and W. A. Gern, 1986, Initial den location by neonatal prairie rattlesnakes: Functions, causes, and natural history in chemical ecology, in D. Duvall, D. Müller-Schwarze, and R. M. Silverstein, eds., *Chemical Signals in Vertebrates,* Vol. 4, Plenum, New York, pp. 285–304.

Greene, H. W., 1988, Antipredator mechanisms in reptiles, in C. Gans and R. B. Huey, eds., *Biology of the Reptilia,* Vol. 16, Alan R. Liss, New York, pp. 1–152.

Greene, H. W., 1992, The ecological and behavioral context for pitviper evolution, in J. A. Campbell and E. D. Brodie, Jr., eds., *Biology of the Pitvipers,* Selva, Tyler, Texas, pp. 107–117.

Gregory, P. T., 1984, Habitat, diet, and composition of assemblages of garter snakes (*Thamnophis*) at eight sites on Vancouver Island, *Can. J. Zool.,* 62:2013–2022.

Gregory, P. T., J. M. McCartney, and K. W. Larsen, 1987, Spatial patterns and movements, in R. A. Seigel, J. T. Collins, and S. S. Novak, eds., *Snakes: Ecology and Evolutionary Biology,* McGraw-Hill, New York, pp. 366–395.

Halloy, M., and G. M. Burghardt, 1990, Ontogeny of fish capture and ingestion in four species of garter snakes (*Thamnophis*), *Behaviour,* 112:299–318.

Halpern, M., 1980, The telencephalon of snakes, in S. O. Ebbeson, ed., *Comparative Neurology of the Telencephalon,* Plenum, New York, pp. 257–295.

Halpern, M., 1983, Nasal chemical senses in snakes, in J. P. Ewert, R. R. Capranica, and D. J. Ingle, eds., *Advances in Vertebrate Neuroethology,* Plenum, New York, pp. 141–176.

Halpern, M., 1987, The organization and function of the vomeronasal system, *Ann. Rev. Neuroscience,* 10:325–362.

Halpern, M., 1992, Nasal chemical senses in reptiles: Structure and function, in C. Gans and D. Crews, eds., *Biology of the Reptilia,* Vol. 18, Univ. Chicago Press, Chicago, pp. 423–523.

Halpern, M., and J. L. Kubie, 1980, Chemical access to the vomeronasal organs of garter snakes, *Physiol. Behav.,* 24:367–371.

Halpern, M., and J. L. Kubie, 1983, Snake tongue flicking behavior: Clues to vomeronasal function, in D. Müller-Schwarze and R. M. Silverstein, eds., *Chemical Signals in Vertebrates,* Plenum, New York, pp. 45–72.

Halpern, M., and J. L. Kubie, 1984, The role of the ophidian vomeronasal system in species-typical behavior, *Trends Neurosci.,* 7:472–477.

Halpern, M., J. I. Morrell, and D. W. Pfaff, 1982, Cellular (3H)estradiol and (3H)testosterone localization in the brains of garter snakes: An autoradiographic study, *Gen. Comp. Endocrinol.*, 46:211–224.
Halpin, Z. T., 1990, Responses of juvenile eastern garter snakes (*Thamnophis sirtalis*) to own, conspecific, and clean odors, *Copeia*, 1990:1157–1160.
Hampton, R. E., and J. C. Gillingham, 1989, Habituation of the alarm reaction in neonatal eastern garter snakes, *Thamnophis sirtalis, J. Herpetol.*, 23:433–435.
Hartline, P. H., and H. W. Campbell, 1969, Auditory and vibratory responses in the midbrains of snakes, *Science*, 163:1221–1223.
Hartline, P. H., L. Kass, and M. S. Loop, 1978, Merging of modalities in the optic tectum: Infrared and visual integration in rattlesnakes, *Science*, 199:1225–1229.
Hayes, W. K., and D. Duvall, 1991, A field study of prairie rattlesnake predatory strikes, *Herpetologica*, 47:78–81.
Heller, S. B., and M. Halpern, 1982, Laboratory observations of aggregative behavior of garter snakes, *Thamnophis sirtalis:* Roles of the visual, olfactory, and vomeronasal senses, *J. Comp. Physiol. Psychol.*, 96:984–999.
Henderson, R. W., and M. H. Binder, 1980, The ecology and behavior of vine snakes (*Ahaetulla, Oxybelis, Thelotornis, Uromacer*): A review, *Milwaukee Publ. Mus. Contrib. Biol. Geol.*, 37:1–38.
Herzog, H. A., Jr., 1990, Experiential modification of defensive behaviors in garter snakes *Thamnophis sirtalis, J. Comp. Psychol.*, 104:334–339.
Herzog, H. A., Jr., and B. D. Bailey, 1987, Development of antipredator responses in snakes: II. Effects of recent feeding on defensive behaviors of juvenile garter snakes *Thamnophis sirtalis, J. Comp. Psychol.*, 101:387–389.
Herzog, H. A., Jr., and C. Bern, 1992, Do garter snakes strike at the eyes of predators?, *Anim. Behav.*, 44:771–773.
Herzog, H. A., Jr., and G. M. Burghardt, 1974, Prey movement and predatory behavior of juvenile western yellow-bellied racers, *Coluber constrictor mormon, Herpetologica*, 30:285–289.
Herzog, H. A., Jr., and G. M. Burghardt, 1986, Development of antipredator responses in snakes: I. Defensive and open-field behaviors in newborns and adults in three species of garter snakes *Thamnophis melanogaster, T. sirtalis, T. butleri, J. Comp. Psychol.*, 100:372–379.
Herzog, H. A., Jr., and G. M. Burghardt, 1988, Development of antipredator responses in snakes, III: Stability of individual and litter differences over the first year of life, *Ethology*, 77:250–258.
Herzog, H. A., Jr., B. B. Bowers, and G. M. Burghardt, 1989a, Development of antipredator responses in snakes: IV. Interspecific and intraspecific differences in habituation of defensive behavior in neonate garter snakes, *Dev. Psychobiol.*, 22:489–508.
Herzog, H. A., Jr., B. B. Bowers, and G. M. Burghardt, 1989b, Stimulus control of antipredator behavior in newborn and juvenile garter snakes *Thamnophis, J. Comp. Psychol.*, 100:372–379.
Herzog, H. A., Jr., B. B. Bowers, and G. M. Burghardt, 1992, Development of antipredator responses in snakes: V. Species differences in ontogenetic trajectories, *Dev. Psychobiol.*, 25:199–211.
Hess, E. H., 1973, Comparative sensory processes, in D. Dewsbury and D. A. Rethlingshafer, eds., *Comparative Psychology: A Modern Survey*, McGraw- Hill, New York, pp. 344–394.
Huey, R. B., C. R. Peterson, S. J. Arnold, and W. P. Porter, 1989, Hot rocks and not-so-hot rocks: Retreat site selection by garter snakes and its thermal consequences, *Ecology*, 70:931–944.
Huntingford, F., 1984, Some ethical issues raised by studies of predation and aggression, *Anim. Behav.*, 36:675–683.
Joy, J. E., and D. Crews, 1988, Male mating success in red-sided garter snakes: Size is not important, *Anim. Behav.*, 36:1839–1841.
Kardong, K. V., 1992, Proximate factors affecting guidance of the rattlesnake strike, *Zool. Jb. Anat.*, 122:233.
Kardong, K. V., and S. P. MacKessy, 1991, The strike behavior of a congenitally blind rattlesnake, *J. Herpetol.*, 25:208–211.

Kardong, K. V., and P. R. Smith, 1991, The role of sensory receptors in the predatory behavior of the brown tree snake, *Boiga irregularis* (Squamata: Colubridae), *J. Herpetol.*, 25:229–231.
King, M., D. McCarron, D. Duvall, G. Baxter, and W. Gern, 1983, Group avoidance of conspecific but not interspecific chemical cues by prairie rattlesnakes (*Crotalus viridis*), *J. Herpetol.*, 17:196–198.
Kroll, J. C., 1973, Taste buds in the oral epithelium of the blind snake, *Leptotyphlops dulcis* (Reptilia: Leptotyphlopidae), *Southwest. Nat.*, 17:365–370.
Kubie, J. L., and M. Halpern, 1979, The chemical senses used in garter snake prey trailing, *J. Comp. Physiol. Psychol.*, 93:648–667.
Kubie, J. L., A. Vagvolgyi, and M. Halpern, 1978, Roles of the vomeronasal and olfactory systems in courtship behavior of male garter snakes, *J. Comp. Physiol. Psych.*, 92:627–641.
Landreth, H., 1973, Orientation and behavior of the rattlesnake, *Crotalus atrox*, *Copeia*, 1973:26–31.
Larson, K. W., 1987, Movements and behavior of migratory garter snakes, *Thamnophis sirtalis*, *Can. J. Zool.*, 65:2241–2247.
Lauder, D., 1986, Homology, analogy, and the evolution of behavior, in M. H. Nitecki and J. A. Kitchell, eds., *Evolution of Animal Behavior: Paleontological and Field Approaches*, Oxford Univ. Press, New York, pp. 9–40.
Lawson, P. A., 1985, Preliminary investigations into the roles of visual and pheromonal stimuli on aspects of the behavior of the western plains garter snake, *Thamnophis radix haydeni*, Masters Thesis, Univ. Regina.
Lawson, P. A., 1989, Orientation abilities and mechanisms in a northern migratory population of the common garter snake (*Thamnophis sirtalis*), *Musk-Ox*, 37:110–115.
Lawson, P. A., 1991, Movement patterns and orientation mechanisms in garter snakes, Doctoral Thesis, Univ. Victoria.
Layne, J. R., and N. B. Ford, 1984, Flight distance of the queen snake, *Regina septemvittata*, *J. Herpetol.*, 18:496–498.
Lewke, R. E., 1979, Neck-biting and other aspects of reproductive biology of the Yuma kingsnake (*Lampropeltis getulus*), *Herpetologica*, 35:154–157.
Lillywhite, H. B., 1985, Trailing movements and sexual behavior in *Coluber constrictor*, *J. Herpetol.*, 19:306–308.
Lyman, L. P., 1990, The effects of dietary and social experience on the development of behavior in Butler's garter snake, *Thamnophis butleri*, Masters Thesis, Univ. Tennessee, Knoxville.
Marmie, W., S. Kuhn, and D. Chiszar, 1990, Behavior of captive-raised rattlesnakes (*Crotalus enyo*) as a function of rearing conditions, *Zoo Biol.* 9:241:246.
Mason, R. T., 1992, Reptilian pheromones, in C. Gans and D. Crews, eds., *Biology of the Reptilia*, Vol. 18, Univ. Chicago Press, Chicago, pp. 114–228.
Mason, R. T., H. M. Fales, T. H. Jones, J. W. Chinn, L. K. Pannell, and D. Crews, 1989, Sex pheromones in snakes, *Science*, 245:290–293.
McDowell, S. B., 1987, Systematics, in R. A. Seigel, J. T. Collins, and S. S. Novak, eds., *Snakes: Ecology and Evolutionary Biology*, McGraw-Hill, New York, pp. 3–50.
Melcer, T., and D. Chiszar, 1989, Striking prey creates a specific chemical search image in rattlesnakes, *Anim. Behav.*, 37:477–486.
Mell, R., 1929, *Grundzüge einer Ökologie der Chinesichen Reptilien*, Walter de Gruyter, Berlin.
Meredith, M., and G. M. Burghardt, 1978, Electrophysiological studies of the tongue and accessory olfactory bulb in garter snakes, *Physiol. Behav.*, 21:1001–1008.
Miller, D. E., and H. R. Mushinsky, 1990, Foraging ecology and prey size in the mangrove water snake, *Nerodia fasciata compressicauda*, *Copeia*, 1990:1099–1106.
Molenaar, G. J., 1992, Anatomy and physiology of infrared sensitivity of snakes, in C. Gans and P. S. Ulinski, eds., *Biology of the Reptilia*, Vol. 17, Chicago Univ. Press, Chicago, pp. 367–453.
Mori, A., 1990, Tail vibration of the Japanese grass lizard *Takydromus tachydromoides* as a tactic against a snake predator, *J. Ethol.*, 8:81–88.
Mushinsky, H. R., and K. H. Lotz, 1980, Responses of two sympatric water snakes to the extracts of commonly ingested prey species: Ontogenetic and ecological considerations, *J. Chem. Ecol.*, 6:523–535.

Mushinsky, H. R., J. J. Hebrard, and M. G. Walley, 1980, The role of temperature on the behavioral and ecological associations of sympatric water snakes, *Copeia,* 1980:744–754.

Mushinsky, H. R., J. J. Hebrard, and D. S. Vodopich, 1982, Ontogeny of water snake foraging ecology, *Ecology,* 63:1624–1629.

Newcomer, R., D. Taylor, and S. Guttman, 1974, Celestial orientation in two species of water snakes (*Natrix sipedon* and *Regina septemvittata*), *Herpetologica,* 30:194–204.

Newman, E. A., and P. H. Hartline, 1981, Integration of visual and infrared information in bimodal neurons of the rattlesnake optic tectum, *Science,* 213:789–791.

Newman, E. A., E. R. Gruberg, and P. H. Hartline, 1980, The infrared trigemino-tectal pathway in the rattlesnake and in the python, *J. Comp. Neurol.,* 191:465–477.

Noble, G. K., 1937, The sense organs involved in the courtship of *Storeria, Thamnophis,* and other snakes, *Bull. Amer. Mus. Nat. Hist.,* 73:673–725.

Noble, G. K., and H. J. Clausen, 1936, The aggregation behavior of *Storeria dekayi* and other snakes with especial reference to the sense organs involved, *Ecol. Monogr.,* 6:269–316.

O'Connell, B., R. Greenlee, J. Bacon, and D. Chiszar, 1982, Strike-induced chemosensory searching in Old World vipers and New World pit vipers at San Diego Zoo, *Zoo Biol.,* 1:287–294.

O'Connell, B., R. Greenlee, J. Bacon, H. M. Smith, and D. Chiszar, 1985, Strike-induced chemosensory searching in elapid snakes (cobras, taipans, tiger snakes, and death adders) at San Diego Zoo, *Psychol. Record,* 35:431–436.

Ota, H., 1986, Snake really an able hunter? Predatory behavior of Japanese striped snake, *Elaphe quadrivirgata,* in the field, *J. Ethol.,* 4:69–71.

Parker, W. S., and W. S. Brown, 1980, Comparative ecology of two colubrid snakes, *Masticophis t. taeniatus* and *Pituophis melanoleucus deserticola,* in northern Utah, *Milwaukee Publ. Mus. Publ. Biol. Geol.,* 7:1–104.

Perry-Richardson, J. J., 1987, Female selection of male fitness in checkered garter snakes, *Thamnophis marcianus,* Masters Thesis, Univ. Texas, Tyler.

Perry-Richardson, J. J., C. W. Schofield, and N. B. Ford, 1990, Courtship of the garter snake, *Thamnophis marcianus,* with a description of a female behavior for coitus interruption, *J. Herpetol.,* 24:76–78.

Peterson, E. H., 1992, Retinal structure, in C. Gans and P. S. Ulinski, eds., *Biology of the Reptilia,* Vol. 17, Chicago Univ. Press, Chicago, pp. 1–135.

Porter, K. R., 1972, *Herpetology,* Saunders, Philadelphia.

Porter, R. H., and J. A. Czaplicki, 1974, Responses of water snakes (*Natrix r. rhombifera*) and garter snakes (*Thamnophis s. sirtalis*) to chemical cues, *Anim. Learn. Behav.,* 2:129–132.

Quay, W. B., 1979, The parietal eye-pineal complex, in C. Gans, R. G. Northcutt, and P. Ulinski, eds., *Biology of the Reptilia,* Vol. 9, Academic, New York, pp. 245–406.

Reichenbach, N. G., 1983, An aggregation of female garter snakes under corrugated metal sheets, *J. Herpetol.,* 17:412–413.

Reinert, H. K., 1984a, Habitat separation between sympatric snake populations, *Ecology,* 65:478–486.

Reinert, H. K., 1984b, Habitat variation within sympatric snake populations, *Ecology,* 65:1673–1682.

Reinert, H. K., and L. M. Bushar, 1984, Foraging behavior of the timber rattlesnake, *Copeia,* 1984:976–981.

Reinert, H. K., and R. T. Zappalorti, 1988, Field observation of the association of adult and neonatal timber rattlesnakes, *Crotalus horridus,* with possible evidence for conspecific trailing, *Copeia,* 1988:1057–1059.

Repérant, J., J. Rio, R. Wards, S. Hergueta, D. Miceli, and M. Lemire, 1992, Comparative analysis of the primary visual system of reptiles, in C. Gans and P. S. Ulinski, eds., *Biology of the Reptilia,* Vol. 17, Chicago Univ. Press, Chicago, pp. 175–240.

Savitzky, B. A. C., 1992, Laboratory studies on piscivory in an opportunistic pitviper, the cottonmouth, *Agkistrodon piscivorus,* in J. A. Campbell and E. D. Brodie, Jr., eds., *Biology of the Pitvipers,* Selva, Tyler, Texas, pp. 347–368.

Schaeffel, F., and A. de Queiroz, 1990, Alternative mechanisms of enhanced underwater vision in the garter snakes *Thamnophis melanogaster* and *T. couchii, Copeia,* 1990:50–58.

Schieffelin, C. D., and A. de Queiroz, 1991, Temperature and defense in the common garter snake: Warm snakes are more aggressive than cold snakes, *Herpetologica,* 47:230–237.

Schroeder, D. M., 1981, Tectal projections of the infrared sensitive snake, *Crotalus viridis, J. Comp. Neurol.,* 195:477–500.

Schuett, G. W., and J. C. Gillingham, 1988, Courtship and mating of the copperhead, *Agkistrodon contortrix, Copeia,* 1988:374–381.

Schuett, G. W., and J. C. Gillingham, 1989, Male-male agonistic behaviour of the copperhead, *Agkistrodon contortrix, Amphibia-Reptilia,* 10:243–266.

Schwartz, J. M., 1989, Multiple paternity and offspring variability in wild populations of the garter snake *Thamnophis sirtalis,* Doctoral Thesis, Univ. Tennessee, Knoxville.

Schwenk, K., 1985, Occurrence, distribution and functional significance of taste buds in lizards, *Copeia,* 1985:91–101.

Scudder, K. M., and D. Chiszar, 1977, Effects of six visual stimulus conditions on defensive and exploratory behavior in two species of rattlesnakes, *Psychol. Rec.,* 27:519–526.

Scudder, K. M., N. J. Stewart, and H. M. Smith, 1980, Response of neonate water snakes (*Nerodia sipedon sipedon*) to conspecific chemical cues, *J. Herpetol.,* 14:196–198.

Scudder, R. M., and G. M. Burghardt, 1983, A comparative study of the defensive behavior in three sympatric species of water snakes *Nerodia, Z. Tierpsychol.,* 63:17–26.

Secor, S., 1987, Courtship and mating behavior of the speckled kingsnake, *Lampropeltis getulus holbrooki, Herpetologica,* 43:15–28.

Secor, S. M., 1990, Reproductive and combat behavior of the Mexican Kingsnake, *Lampropeltis mexicana, J. Herpetol.,* 24:217–221.

Shaw, C. E., 1951, Male combat in American colubrid snakes with remarks on combat in other colubrid and elapid snakes, *Herpetologica,* 7:149–168.

Shettleworth, S. J., P. J. Reid, and C. M. S. Plowright, The psychology of diet selection, in R. N. Hughes, ed., *Diet Selection,* Blackwell, London, in press.

Shine, R., 1991, Why do larger snakes eat larger prey?, *Funct. Ecol.,* 5:493–502.

Shine, R., G. C. Grigg, T. G. Shine, and P. Harlow, 1981, Mating and male combat in Australian blacksnakes, *Pseudechis porphyriacus, J. Herpetol.,* 15:101–107.

Stemmler-Morath, C., 1935, Beitrag zur fortflanzungungsbiologie europäisher Colubridae, *Zool. Gart.*, 8:38–41.

Stevenson, R. D., C. R. Peterson, and J. S. Tsuji, 1985, The thermal dependence of locomotion, tongue flicking, digestion, and oxygen consumption in the wandering garter snake, *Physiol. Zool.,* 58:46–57.

Teather, K. L., 1991, The relative importance of visual and chemical cues for foraging in newborn blue-striped garter snakes (*Thamnophis sirtalis simulus*), *Behaviour,* 117:255–261.

Thomas, E., 1972, *Vipera ammodytes montandoni* (Viperidae), kommentkampf der mannchen, *Publ. Wiss. Fil Sekt. Biol.,* 4:178–188.

Tinbergen, N., 1951, *The Study of Instinct,* Clarendon, Oxford.

Towers, S. R., and R. G. Coss, 1991, Antisnake behavior of Columbian ground squirrels (*Spermophilus columbianus*), *J. Mammal.,* 72:776–783.

Underwood, G., 1970, The eye, in C. Gans, ed., *Biology of the Reptilia, Morphology B,* Vol. 2, Academic, New York, pp. 1–98.

Underwood, G., 1992, Endogenous rhythms: in C. Gans and D. Crews, eds., *Biology of the Reptilia, Physiology E,* Vol. 18, Univ. Chicago Press, Chicago, pp. 229–297.

von Achen, P. H., and J. L. Rakestraw, 1984, The role of chemoreception in prey selection of neonate reptiles, in R. A. Seigel, L. E. Hunt, J. L. Knight, L. Malaret and N. L. Zuschlag, eds., *Vertebrate Ecology and Systematics—A Tribute to Henry S. Fitch,* Spec. Publ. Univ. Kansas Mus. Nat. Hist., 10, pp. 163–172.

Wall, F., 1921, *The Snakes of Ceylon,* H. R. Cottle, Colombo.

Walls, G. L., 1942, *The Vertebrate Eye and Its Adaptive Radiation,* Hafner, New York.

Wang, D., P. Chen, X. C. Jiang, and M. Halpern, 1988, Isolation from earthworms of a proteinaceous chemoattractant to garter snakes, *Arch. Biochem. Biophysics,* 267:459–466.

Warren, J. W., and U. Proske, 1968, Infrared receptors in the facial pits of the Australian python *Morelia spilotes, Science,* 159:439–441.

Watkins, H. J. F., F. R. Gehlbach, and J. C. Kroll, 1969, Attractant-repellent secretions in the intra- and interspecies relations of blind snakes *Leptotyphlops dulcis* and army ants *Neivamyrmex nigrescens, Ecology,* 50:1098–1104.

Weatherhead, P. J., and I. Robertson, 1990, Homing to food by black rat snakes (*Elaphe obsoleta*), *Copeia,* 1990:1164–1165.

Weaver, E. G., 1978, *The Reptile Ear: Its Structure and Function,* Princeton Univ. Press, New Jersey.

Weaver, E. G., and J. A. Vernon, 1960, The problem of hearing in snakes, *J. Auditory Res.,* 1:77–83.

Webb, J. K., and R. Shine, 1992, To find an ant: Trail-following behavior in Australian typhlopid snakes, *Anim. Behav.,* 43:941–948.

Weldon, P. J., 1982, Responses to ophiophagus snakes by snakes of the genus *Thamnophis, Copeia,* 1982:788–794

Weldon, P. J., 1990, Responses by vertebrates to chemicals from predators, in D. W. MacDonald, D. Müller-Schwarze, and S. E. Natynczuk, eds., *Chemical Senses in Vertebrates,* Vol. 5, Oxford Univ. Press, Oxford, pp. 500–521.

Weldon, P. J., N. B. Ford, and J. J. Perry-Richardson, 1990, Responses by corn snakes (*Elaphe guttata*) to chemicals from heterospecific snakes, *J. Chem. Ecol.,* 16:37–44.

Weldon, P. J., R. Ortiz, and T. S. Sharp, 1992, The chemical ecology of crotaline snakes, in J. A. Campbell and E. D. Brodie, Jr., eds., *Biology of the Pit Vipers,* Selva, Tyler, Texas, pp. 309–319.

Weldon, P. J., T. S. Walsh, and J. S. E. Kleister, Chemoreception in feeding behavior of reptiles: Relevance to maintenance and management, in J. B. Murphy, J. T. Collins, and K. Adler, eds., *Captive Management and Conservation of Amphibians and Reptiles,* Soc. Stud. Amphib. Rept., in press.

Wilde, W. S., 1938, The role of Jacobson's organ in the feeding reaction of the common garter snake, *Thamnophis sirtalis sirtalis* (Linn.), *J. Exp. Zool.,* 77:445–465.

Yeager, C., and G. M. Burghardt, 1991, The effect of food competition on aggregation: Evidence for social recognition in the plains garter snake (*Thamnophis radix*), *J. Comp. Psychol.,* 105:380–386.

York, D. S., and G. M. Burghardt, 1988, Brooding in the Malayan pit viper, *Calloselasma rhodostoma:* Temperature, relative humidity, and defensive behaviour, *Herpetol. J.,* 1:210–214.

Zimmerman, K., and H. Heatwole, 1990, Cutaneous photoreception: A new sensory mechanism for reptiles, *Copeia,* 1990:860–862.

Chapter

5

Ecology and Evolution of Snake Mating Systems

David Duvall

Gordon W. Schuett

Stevan J. Arnold

Introduction

A goal of evolutionary behavior analysis is to predict strategies and other actions of animals. Indeed, given an appropriate model organism, were an investigator to have a substantive understanding of its overall biology, one could, in theory, even predict accurately the behavioral strategy that would maximize reproductive success. Some of the elements that would comprise such a complete understanding for a population or taxon might include knowledge of (1) comparative phylogenetic relations and relevant trait derivation (Brooks and McLennan, 1991; Harvey and Pagel, 1991), (2) phylogenetic inertia manifested in proximate characteristics (including life-history patterns and more unitary adaptations; Seigel and Ford, 1987; Shine, 1988a; Harvey and Pagel, 1991; Duvall et al., 1992; Stearns, 1992), (3) the current panorama of variance in relevant trait values (Endler, 1986; Harvey and Pagel, 1991), (4) local ecological potential favoring some actions and strategies and not others (Orians, 1969; Emlen and Oring, 1977; Alcock, 1980; Thornhill and Alcock, 1983; Davies, 1991), (5) current natural selection forces (Endler, 1986), (6) current sexual selection forces and mating-system characteristics (Bateman, 1948; Wade, 1979; Wade and Arnold, 1980; Arnold, 1983; Arnold and Wade, 1984a,b; Bradbury and Andersson, 1987), and (7) relevant trait heritabilities (Lande, 1979; Arnold, 1983; Brodie and Garland, Chap. 8, this volume). Obviously, however, much more needs to be learned about the main effects and interactions of these many forces and variables in order to achieve a more precise predictability about reproductive strategy.

This is not to say that exciting progress on these many fronts has not been made, progress that allows good predictability even with much remaining to be learned. Particularly exciting progress has been made in the areas of sexual selection and mating-system analysis, headway germane to this chapter. And even though snakes only recently have become a focus of active research in this area, this dearth of knowledge about mating systems and sexual selection is bound not to last. Our own field (Graves and Duvall, 1990; Duvall et al., 1992), laboratory, and theoretical work (Duvall et al., 1992; Schuett, 1992; Arnold and Duvall, 1993) suggests potentially fruitful directions and possibilities for the analysis of snake mating systems and sexual selection patterns. Thomas Madsen, Richard Shine, and several other investigators (Gregory, 1974; Andrén, 1986; Slip and Shine, 1988a; Madsen et al., 1992; Shine, Chap. 2, this volume) now are making significant contributions in this area as well.

In this chapter we continue our dialogue on sexual selection, male and female strategy, and pattern and variation in mating systems, though in this chapter we broaden our focus to include snakes more generally. As in previous contributions, we focus primarily upon mating systems and sexual selection in the context of (1) local ecological potential and (2) phylogenetic inertia for certain mating-system patterns and not others (see Fig. 5.1). We also provide a treatment of some potential methodological approaches and kinds of data that

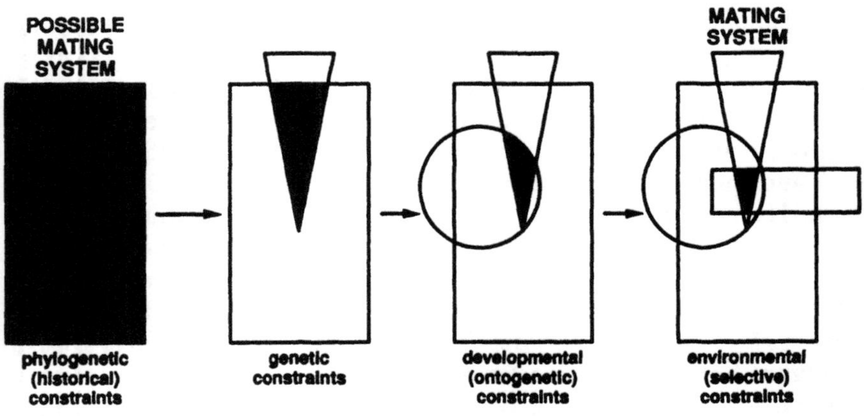

Figure 5.1 Several classes of constraints presumed to shape the ultimate form of an hypothesized, realized mating system. From Schuett and Duvall (unpub.). Modified from Brooks and Wiley (1988).

might be gathered were one to desire to work in this area. In the last major section of this chapter we survey some benchmark studies of snake reproductive biology that, in some instances, have a good deal to say about mating systems and/or sexual selection in snakes.

Theoretical and Conceptual Context

Though we do not wish simply to restate our theoretical propositions and reviews presented elsewhere (Duvall et al., 1992; Arnold and Duvall, 1993), nor review the voluminous, relevant general theoretical and empirical literature (Darwin, 1871; Bateman, 1948; Trivers, 1972; Emlen and Oring, 1977; Wells, 1977; Parker, 1978a,b; Shine, 1978; Howard, 1979; Payne, 1979; Wade, 1979; Alcock, 1980; Kluge, 1981; Andersson, 1982; Arnold, 1983; Thornhill and Alcock, 1983; Thornhill, 1986; Bradbury and Andersson, 1987; Shields, 1987; Clutton-Brock, 1988, 1989; Clutton-Brock and Vincent, 1991; Sullivan, 1989, 1991; Shine, Chap. 2, this volume) some treatment is necessary to provide context for one of our larger goals for this chapter, namely, reviewing and analyzing snake mating systems in abridged historical, theoretical, and empirical context. A second goal is to suggest directions for future study with descriptions of an approach and the kinds of data that might be utilized. The sample protocol we develop in next to the last section of this chapter would take advantage of potentially powerful DNA fingerprinting methods (Burke, 1989; Kirby, 1990). Refer to Table 5.1 for a list of variables mentioned throughout the text.

Sexual selection and mating systems

Though not always explicit in all treatments of mating systems, it is difficult not to consider simultaneously sexual selection forces in the context of reproductive strategy, whether they be intra- or intersexual, and whether these forces be greater on males or females. We (Duvall et al., 1992; Arnold and Duvall, 1993) have proposed that one good way to get a handle on symmetry or lack thereof in sexual selection forces acting on males and females in populations is to determine and compare sexual selection gradients for both males and females. We have defined these as the average slope of the partial regression of fecundity on mating success for members of each sex (see Fig. 5.2). The partial regression of fecundity on mating success must, by definition, simultaneously represent the final common pathway for all sexual selection forces. Thus mating success is more precisely definable as the *number of ego's mates that actually bear progeny* (see Arnold and Duvall, 1993). Accordingly, members of that sex in a population with the steepest sexual selection gradient will be those experiencing

TABLE 5.1. Variables Comprising Mating System and Related Models

Variable	Definition
$\overline{m}_f$	Average fitness (= fecundity or progeny count) of all females
$\overline{m}_f'$	Average fitness of females that mate once or more
$\overline{m}_m$	Average fitness of males
$\overline{X}_f$	Average female mating success (= number of mates)
$\overline{X}_m$	Average male mating success
H_x	Harmonic mean mating success of females with one or more mates
$\sigma^2_{x\text{-}f}$	Variance in mating success for all females (including the zero class with no mates)
$\sigma^2_{x\text{-}m}$	Variance in mating success for all males
q	Female mating failure (proportion of females capable of breeding that do not) (note that $p + q = 1$)
$\beta_{ss\text{-}f}$	Female sexual selection gradient
$\beta_{ss\text{-}m}$	Male sexual selection gradient
OSR	Operational sex ratio (the average over time of the number of sexually active males to the number of females capable of insemination)
BSR	Breeding sex ratio (= parental ratio; ratio of number of breeding males to number of breeding females)
L	Length of mating season
s	Mate searching time
h	Mate handling time
c	Mate cycling time (= $s + h$)
α_s	Average mate searching time
α_h	Average mate handling time
α	Average mate cycling time (= $\alpha_s + \alpha_h$)
α_p	Average mate persuasion efficacy
$\beta_{ss\text{-}\alpha_s}$	Sexual selection gradient for average searching time
$\beta_{ss\text{-}\alpha_h}$	Sexual selection gradient for average handling time
$\beta_{ss\text{-}\alpha}$	Sexual selection gradient for average mate cycling time
$\beta_{ss\text{-}\alpha p}$	Sexual selection gradient for average mate persuasion
σ^2	Variance in mate cycling time
$\sigma_p^{\ 2}$	Variance in mate persuasion efficacy
I_s	Potential for sexual selection (variance in mating success standardized to a mean of 1; = $\sigma^2/\overline{x}^2$)

the most intense sexual selection on traits associated with the mating-success–fitness-regression slope. Conversely, that sex with the relatively flatter sexual selection gradient will realize a smaller relative gain in fecundity as a function of continued matings, because selection on traits associated with mating success per se is less. In a very real sense, this explanation for inherent sex differences, based upon a determination and calculation of male and female popula-

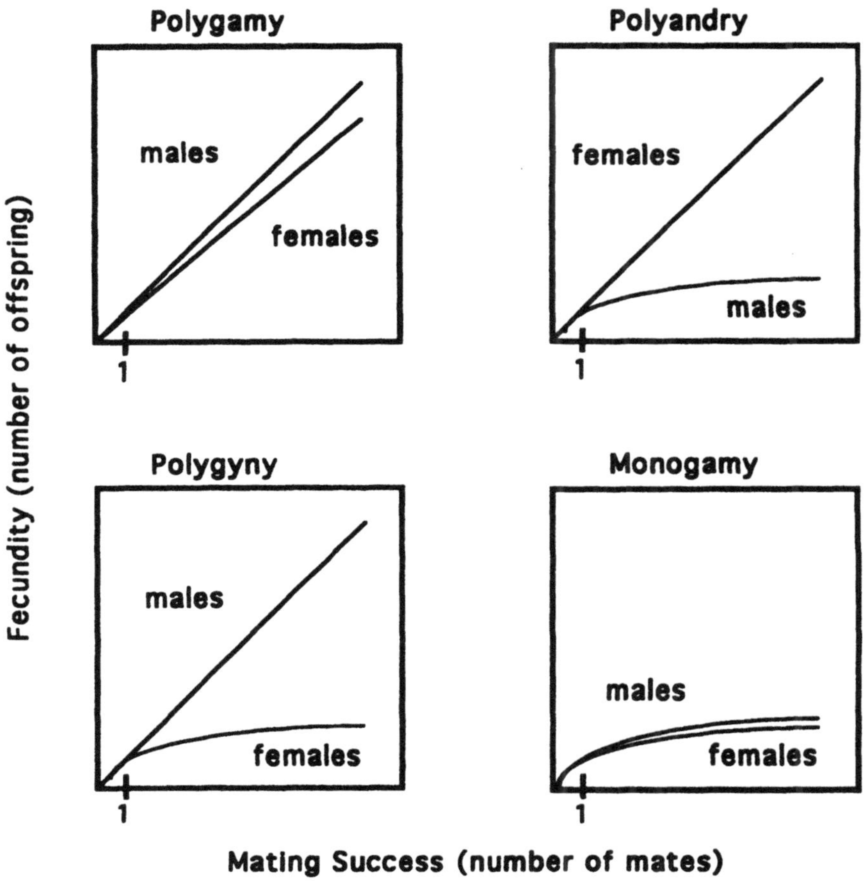

Figure 5.2 Predicted, broad sense mating-system types as a function of the relative association of fecundity and mating success for males and females. We term these associations sexual selection gradients. Modified from Arnold and Duvall (1993). See text for additional discussion.

tional sexual selection gradients, is a fresh explanation for proximate differences for male and female differences in sexual strategy. Most popular explanations focus on the ultimate origins of differences in sexual strategy, rather than on actions of sexual selection in progress on key proximate traits [compare Bateman (1948), Arnold and Wade (1984a,b), and Arnold and Duvall (1993), with Trivers (1972), Borgia (1979), Thornhill (1986), and Gwynne (1991)].

An advantage of our framework for mating systems and sexual selection analysis is that it provides an integrated, quantitative system that focuses upon empirically tractable parameters. Less specific and nonintegrative approaches, many of which include as the domain of mating-system analysis almost everything about the reproductive and mating biology of natural populations of animals, in fact may predict less.

We also advocate construction of a "parental table" (Arnold and Duvall, 1993), which comprises an empirical sampling of individual male and female parents of an acceptable sample of offspring in a target population. Such data represent the most direct method of determining precisely the relative male and female sexual selection gradients characteristic of a population. These data also would allow determination of the breeding sex ratio (BSR), or the ratio of actual male to female parents, which can provide an accurate index of sexual selection forces acting within populations (Duvall et al., 1992; Arnold and Duvall, 1993). Because each and every individual offspring in a population can have only two parents, the BSR reflects all such sets of parents and provides a quantitative index of the degree of constraint each sex places over the reproductive success of the other. Though we have also modeled conditions whereby behavioral data can be used to determine both the BSR and parental table (see below), protein electrophoresis and DNA fingerprinting data would be ideal (Gibbs et al., 1990). Indeed, the accuracy of the operational sex ratio [OSR, the ratio of potential mothers to fathers in a population, though we and most others cast this oppositely (see Table 5.1); Emlen and Oring, 1977] may lie in the extent to which it predicts the BSR. The OSR is often taken as a predictor of intrasexual competition and even sexual selection forces (Emlen and Oring, 1977), and relies heavily on the importance of female–mate monopolization potential. OSR theory predicts, for example, that male-biased OSR and increased male mating competition will be associated positively with increasing potential sexual selection forces acting on males. However, Ims (1988) has shown that female-biased OSRs, more so than male-biased OSRs, are more predictive of increased variance in male mating success. Ims' (1988) findings thus run counter to OSR theory, and appear to hold for mating systems akin to those seemingly more characteristic of snakes, and, interestingly, many mammals (Ims, 1988), namely those characterized by mate searching, guarding, and male fighting. OSR may, however, be more predictive of mating competition in territorial-type and perhaps hotspot mating systems (Arnqvist, 1992; see below). However, we know of no reports *confirming* territoriality in snakes (see below). See Duvall et al. (1992) and Arnold and Duvall (1993) for additional discussion.

Though perhaps only indirectly germane to analyses of snake mating systems, at least given our present knowledge of snake reproductive patterns and strategy in nature, nuptial gifts and parental care also can change the slope of sexual selection gradients, reflecting changing potential sexual selection forces. Arnold and Duvall (1993) take up these and related issues.

Kinds of mating systems: Broad and narrow senses

Arnold and Duvall (1993) present a quantitative scheme that holds some potential for analyzing and even classifying diverse animal mating systems (see Fig. 5.2). Two of the systems therein seem germane to snakes, especially *polygyny* and, to a lesser extent, effective *monogamy*. As we define it, in a polygynous snake mating system males realize increased fecundity with increased mating success. Females, conversely, experience an asymptotic fecundity curve because increased mating success will not increase fecundity. The road to female snake reproductive success instead will be paved with elements of efficacious foraging and feeding, and efficient energetic biology and pregnancy or gestation. This no doubt is due to the relatively extreme energetic costs per potential egg or offspring experienced by most female snakes, which greatly reduces their theoretical lifetime reproductive potential. Duvall et al. (1982, 1990, 1992) and Graves and Duvall (1990, 1993) take up these and related issues (see also Shine, 1988b). Accordingly, we should not be surprised as well to discover more robustness and variation in maternal investment adaptations among female snakes (Shine, 1988b; Graves and Duvall, 1993). Arnold and Duvall (1993), for example, find that maternal care of clutches or litters will decrease sexual selection forces acting on females.

When matings with multiple females are unlikely, for whatever reason(s), effective monogamy seems possible. This would be recognized in our system by roughly similar, asymptotic male and female partial regression slopes for fecundity and mating success. We imply nothing about pair bonding, or other proximate behavioral mechanisms that may mediate sustained gregariousness of male–female pairs. It is also relevant here that male snakes are largely unable to capture and supply food to pregnant or gravid females, by virtue of snakes' feeding biology and general morphology (see Duvall et al., 1992 for additional discussion).

Finally, because of the phylogenetic momentum for polygyny among the snakes, neither polyandry nor polygamy as defined in Fig. 5.2 are likely to occur.

It is appropriate, therefore, to consider the scheme diagrammed in Fig. 5.2 as providing a *broad sense* classification of potential snake or animal mating systems. More *narrow sense,* but less general, classifications would seem to fall naturally under the four broad sense rubrics portrayed in Fig. 5.3. Under polygyny and focusing on snakes, for example, and employing accepted narrow sense classifications such as the proven and popular scheme of Thornhill and Alcock (1983; see also Emlen and Oring, 1977; Bradbury et al., 1986), Duvall et al. (1992) describe and predict at least four forms of polygyny that,

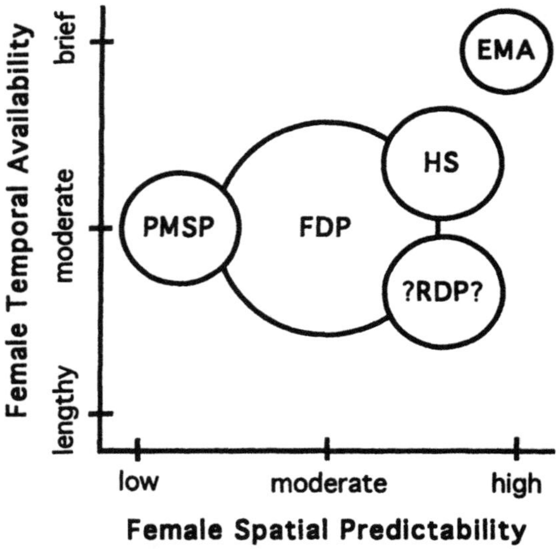

Figure 5.3 Some hypothesized, narrow-sense mating-system types, which, for most snakes, should fall under the rubric of polygyny. Circles and their areas are hypothesized to reflect the extent to which one or another form of polygyny should be expected, as a function of receptive female spatial and temporal predictability. FDP is female defense polygyny, or mate-guarding, and is probably very common among snakes. PMSP is prolonged mate-searching polygyny, EMA is explosive mating assemblage, and HS is hotspot polygyny. The entry for RDP, resource defense polygyny or territoriality, is qualified with question marks because it has yet to be demonstrated unequivocally in snakes. See text for additional discussion.

were they found to occur, may cover the gamut of types of snake mating systems (see Fig. 5.3). These subtypes of polygyny are predicted to be determined by distributions of receptive females both in time and space, and, thus, by similar distributions of resources important to females (Duvall et al., 1992). Again, a problem with narrow sense approaches is that they are somewhat idiosyncratic and not especially general.

Nevertheless, in the narrow sense, the four common snake mating systems likely are (1) female defense polygyny, or mate guarding, (2) hotspot polygyny (Bradbury et al., 1986), (3) prolonged mate searching polygyny, and (4) explosive mating assemblage. The latter two mating system types commonly are considered to be subcategories of scramble competition polygyny (Thornhill and Alcock, 1983). Territoriality, or resource defense polygyny, and Lek polygyny are, at present, unknown or unsubstantiated in snakes. Leks are male display areas where females may choose mates. The key test for those who are interested in the potential of territoriality in snakes will be to demonstrate that males compete for control of resources, habitats, and the like, that, in turn, are important to and/or selected by females. To show male–male fighting and competition in the presence of one or more receptive females, for example, is not sufficient. We expand upon this important issue below. Finally, we find little use for terminology and classification systems that are specific to one sex and that do not consider simultaneously symmetry and departures from symmetry that characterize strategy for both sexes (and on this one count we depart from Thornhill and Alcock, 1983).

As noted above, a variant of the hotspot polygynous mating system (Davies, 1991) may occur in some snakes (see below). Such a mating system would probably represent yet another subtype of scramble competition polygyny, where males move to and gather in locales where females are relatively abundant, the latter presumably occurring there in response to some limiting environmental feature, such as warmth or food, for example. Male–male fighting and even mate guarding of females may or may not occur in such locales (see below). However, we invoke nothing regarding potential female choice in hotspots, a component of initial formulations of this idea. We apply it here only to help explain relative spatial overabundance of receptive females in a locale or habitat unit.

A related and significant variable affecting any form of male–male competition that might occur in hotspots would be the phenomena of female multiple mating and mixed paternity, sperm competition and storage, and any mating-order effects on male fertilization success that may exist (Parker, 1970a, 1984; Dewsbury and Baumgardner, 1981; Devine, 1984; Schwagmeyer et al., 1987; Schwartz et al., 1989; Schwagmeyer, 1990; Schuett, 1992). For example, if females mate multiple times, if paternity is mixed, if no mating-order effects exist, if no physiological mechanisms (e.g., Garter Snake mating plugs; Devine, 1977) exist to enforce chastity or single male matings, and if the duration of female receptivity is even moderately lengthy, it is expected that males at a female hotspot will *not* fight. Rather, we might expect these hypothetical males simply to queue up and wait their turn. We expand upon this issue below.

Aside from effective female choice of males via potential sperm competition and mixed paternity of broods or litters, a suite of reproductive adaptations probably characteristic of many snakes (see Schuett, 1992), no reports currently exist indicating active, behavioral choice of mates by females. However, Schuett and Duvall (in review) discuss what may represent the first report of female mate choice among snakes (see below). Nevertheless, sexual selection among most snakes probably derives from various forms of male–male competition.

Sexual selection gradients, encounter rate theory, and mating systems

As noted above, the sexual selection gradient context for most snake mating systems is polygynous, as portrayed generally and graphically in Figure 5.2. This diagram tells us that, for many or most snake mating systems, males generally will realize increased fecundity with successively numerous matings but that females will not. Selection for traits boosting mating success therefore will have its greatest impact upon males rather than on comparable traits of females, and, thus, polygyny among snakes. Duvall et al. (1992) discuss in detail additional reasons for polygyny among snakes.

We can formalize this general situation in ways that allow us to study predictively polygynous snake mating systems and sexual selection. For purposes of theoretical development, we focus now only on selection operating within single seasons, with no formal accounting of population age-structuring and complex life histories. Arnold and Duvall (1993), however, do discuss mating systems in the context of age-structuring and mortality.

We begin by imagining a mating system in which males garner increased fecundity from multiple matings, but females do not (Fig. 5.4). Such a circumstance bears good resemblance to the reproductive functioning of most male and female snakes (Duvall et al., 1982, 1992; Schuett, 1992), and facilitates formal modeling that should further empirical analyses (Duvall and Schuett, in review). If a female mates once or more, her average fecundity equals some constant, say $\overline{m}_f'$ and is zero if she mates not at all. The average for all males in the population, both mated and unmated, is $\overline{m}_f$. Suppose that p is the proportion of females capable of breeding that mate once or more and bear progeny, and q is the proportion of females capable of breeding that do not mate ($p + q = 1$), then the average fecundity of females mating one or more times is

$$\overline{m}_f' = \overline{m}_f / p. \quad (1)$$

Females of many snake taxa long have been suspected of multiple mating (Madsen et al., 1992; Schuett, 1992), sperm storage, and probably competition, as well as mixed paternity of clutches or litters (Ludwig and Rahn, 1943; Gibson and Falls, 1975; Saint Girons, 1975; Devine, 1984; Schuett, 1982, 1992; Schwartz et al., 1989). It is unknown, however, if any mating order effects arise from multiple matings among snakes. Nevertheless, if the number of progeny an individual male can expect to sire is inversely proportional to the

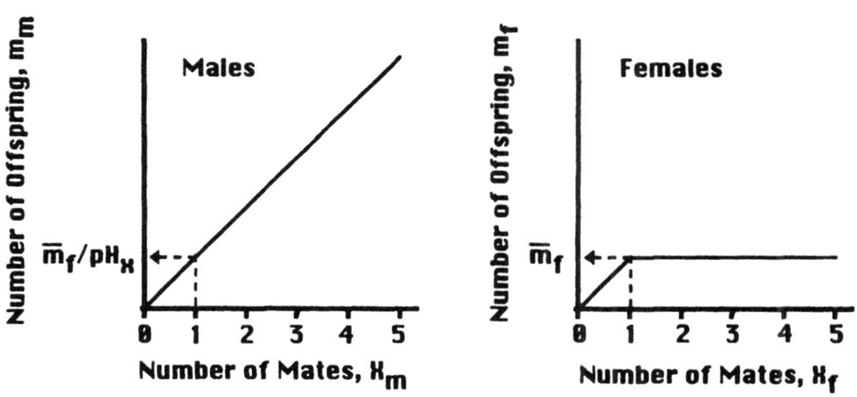

Figure 5.4 Model sexual selection gradients for a polygynous mating system, probably characteristic of most snakes, wherein average male fecundity ($\overline{m}_m$) does, but average female fecundity does not ($\overline{m}_f$), increase beyond that derived from the first mating. However, for taxa or populations with multiple sires, mixed paternity of clutches or litters, and no mating order effects, yet another system probably characteristic of many snakes, average male fitness is given by $\overline{m}_f/pH_x$, where p is the proportion of females that mate once or more and bear progeny and H_x is the harmonic mean number of male mates per female. See text for discussion. See also Arnold and Duvall (1993) and Duvall et al. (1992). Modified from the latter.

number of different males a target female actually mates with, then expected male fitness $\overline{m}_m$ is

$$\overline{m}_m = \overline{m}_f \overline{X}_m / pH_x, \qquad (2)$$

where H_x is the harmonic mean mating success of females with one or more mates (Fig. 5.4). Thus, $\overline{m}_m$ is inversely proportional to the harmonic average number of different males mated by an individual female.

As discussed by Duvall et al. (1992) and Arnold and Duvall (1993), by taking the first derivative of average male fecundity with respect to average male mating success, we find that the sexual selection gradient for males ($\beta_{ss\text{-}m}$) in the single male paternity system (i.e., where $H_x = 1$) is

$$\beta_{ss\text{-}m} = \overline{m}_f / p. \qquad (3)$$

In a system characterized by multiple female mating and mixed paternity this gradient becomes

$$\beta_{ss\text{-}m} = \overline{m}_f / pH_x. \qquad (4)$$

Thus, in the absence of other effects, multiple mating by females coupled with multiple paternity is expected to reduce the force of sexual selection on males.

For purposes of symmetry, and not because sexual selection gradients among females of any snake taxa are expected to be steeper than those of conspecific males (Duvall et al., 1992; Arnold and Duvall, 1993), it is useful to consider those parameters affecting sexual selection gradients for females ($ß_{ss\text{-}f}$). This relationship is more complex among females, and is given by

$$ß_{ss\text{-}f} = q\overline{m}_f \overline{X}_f / p\sigma^2_{x\text{-}f}, \tag{5}$$

where q is an index of female mating failure, or the proportion of females capable of breeding that do not, $\overline{X}_f$ is the average mating success of all females, and $\sigma^2_{x\text{-}f}$ is the variance among all females in mate number (including the zero class with no mates). Thus, so long as there is some variance in female mate numbers, mating failure is a prime factor promoting sexual selection in females. However, it is expected that among most female snakes mating failure should be minimal, which means that q will be small and $ß_{ss\text{-}f}$ will approach zero. This was, for example, in fact the case in our long-term field study of Prairie Rattlesnakes (Duvall et al., 1992; Duvall and Schuett, in review; see below).

Encounter rate phenomena and sexual selection gradients

Clearly, it would be useful to have available some means of extending formally sexual selection and mating systems theory in an empirically tractable fashion, whether a student of snakes were working in the laboratory or field. Next we describe some readily studied behavioral and phenological parameters, that may make possible a connection between relevant selection theory and behavioral empiricism. Again, consult Duvall et al. (1992) and Arnold and Duvall (1993) for additional information.

Like others (Holling, 1959; Parker, 1970a,b, 1974, 1978a,b; Baylis, 1981; Sutherland, 1985a,b, 1987; Real, 1990), we find it convenient and empirically tractable to cast key seasonal, behavioral, and reproductive phenomena in terms of accumulated time per activity and rates (Duvall et al., 1992; Arnold and Duvall, 1993). This is because encounter rate phenomena are easily observed and scored, a relevant body of renewal theory exists (Cox, 1962), some of which we employ below, and relevant data lend themselves nicely to statistical analysis (Duvall et al., 1992; Arnold and Duvall, 1993). The path diagram portrayed by Fig. 5.5 provides a fitness-related context in which such information can be considered.

We begin by defining arbitrarily distributed and statistically independent mate searching (s) and handling (h) time as key parameters. Furthermore, each round of s and h sum to accumulated mate cycling

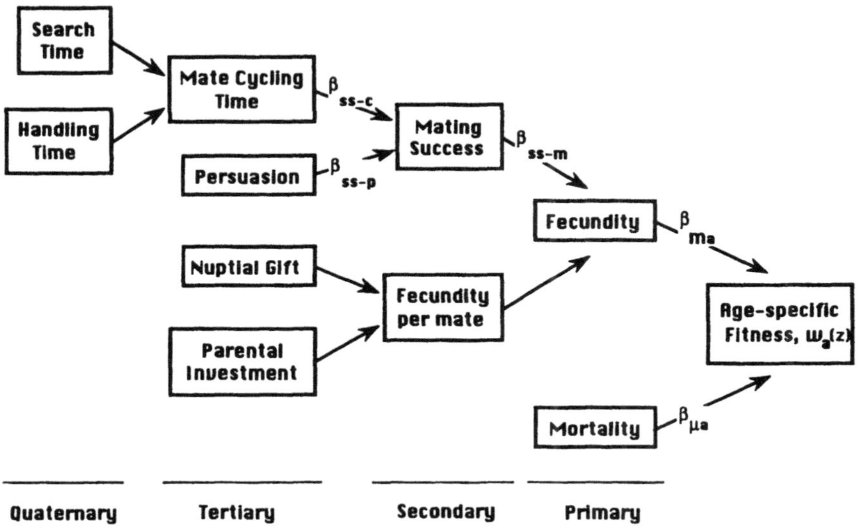

Figure 5.5 A hierarchy of traits (quaternary down through primary) affecting age-specific fitness [$w_a(z)$], with special focus on those descending to mating success. The symbols along respective paths denote selection coefficients known as selection gradients (ß values) and are discussed either in the text or in Arnold and Duvall (1993). Modified from Arnold and Duvall (1993). See text for additional discussion.

time (c) per potential mating ($c = s + h$). There may be few or many periods c per discrete reproductive period or season. Reproductive periods or seasons are of duration L, and there may be few or many of these as well. There may be one or several such periods L per year or season, depending on the taxon or population in question, and males and females may have different mate-cycling-time profiles. Handling includes activities such as mate accompaniment, courtship and copulation, mate guarding, male–male fighting, and the like. Period L is comprised entirely of searching and handling ($s + h$); in other words, once a unit of mate cycling time c is completed, whether or not a successful copulation results, another period c commences immediately. In our own studies of snake mating systems, where one focus has been variance in male fighting (= handling; Schuett and Duvall, in review) and male mate searching (Duvall and Schuett, in review), we "debit" and "credit" winners and losers of fights over potential mates, respectively, with appropriate negative increments of handling time h. Mate search failure likewise can be dealt with as a negative debit to accumulated search time. Such parameters also could be studied in the laboratory (see below). Additionally, since not all mates encountered will be mated, we incorporate what we assume to be a statistically independent measure of proportional mate persuasion efficacy p. Accordingly, we also can define means and variances for these parameters, and these are for c, s, h, and p, (1) α and σ^2, (2) α_s and σ_s^2, (3) α_h and σ_h^2, and (4) α_p and σ_p^2, respectively. Because of our assumption of independence, it follows that $\alpha = \alpha_s + \alpha_h$ and $\sigma^2 = \sigma_s^2 + \sigma_h^2$.

Focusing on males, as in Duvall et al. (1992) and Arnold and Duvall (1993), we find that average male mating success is

$$\overline{X}_m = \alpha_p L/\alpha. \tag{6}$$

Males thereby can increase mating success by (1) maximizing mate persuasion efficacy, (2) maximizing mating season duration, and (3) minimizing overall mate cycling time. Now, since variance in the number of mates cycled per season is $\sigma^2 L/\alpha^3$, as discussed by Cox (1962) and Arnold and Duvall (1993), variation in male mating success, $\sigma_{x\text{-}m}^2$, is

$$\sigma_{x\text{-}m}^2 = (\alpha_p^2 L^2/\alpha^2)\,\{(1/\alpha L)\sigma^2 + (1/\alpha_p)\sigma_p^2\}. \tag{7}$$

We assume that cycling time and mate persuasion are statistically independent. If we divide Eq. (7) by the square of mean mating success, $\alpha_p L/\alpha$, we obtain the standardized variance in relative mating success I, which is an index of the potential opportunity for sexual selection (Crow, 1958; Wade, 1979; Wade and Arnold, 1980; Arnold, 1986). We find that

$$I_{x\text{-}m} = (1/\alpha L)\sigma^2 + (1/\alpha_p^2)\sigma_p^2. \tag{8}$$

Accordingly, the opportunity for sexual selection is increased by (1) high variances in cycling time and persuasion, (2) brief average cycling time, (3) a short mating season, and (4) low average persuasive ability. Moreover, while the effects of short average cycling time and breeding season duration are to enhance the contribution of variance in cycling time to the opportunity for sexual selection, low average persuasion has the same effect on variance on this same parameter. And, as discussed by Arnold and Duvall (1993), since we can define the relative variance in number of mates cycled to be the ratio of absolute variance in numbers cycled to the squared mean number of cycled mates, $I_c = \sigma^2/L\alpha$, and the relative variance in persuasion to be $I_p = \sigma_p^2/\alpha_p^2$, then Eq. (8) becomes

$$I_{x\text{-}m} = I_c + I_p. \tag{9}$$

The opportunity for sexual selection, therefore, is simply the sum of selection opportunities arising from mate cycling and persuasion activities. Arnold and Duvall (1993) discuss a few approaches that can be employed when significant covariance arises between numbers of mates cycled per season and persuasion efficacy.

Sexual selection gradients and encounter rate parameters. There also is utility in casting encounter rate parameters in terms of sexual selection gradients. By calculating the partial derivative of $\overline{X}_m$ with respect to α, α_s, and α_h, we find that

$$\beta_{ss\text{-}c} = \beta_{ss\text{-}s} = \beta_{ss\text{-}h} = -\alpha_p L/\alpha^2, \tag{10}$$

and for α_p,

$$\beta_{ss\text{-}p} = L/\alpha. \tag{11}$$

Sexual selection gradients for L and L/α likewise can be determined by calculating first derivatives using Eq. (6).

Consequently, sexual selection on male rates is expected to increase as (1) average persuasion abilities increase, (2) mating season duration increases, and/or (3) average mate cycling times decrease. Sexual selection should act on males to (1) increase persuasion capacities, (2) extend duration of mating seasons, and (3) decrease the duration of average mate cycling times. Generally, sexual selection on encounter rates will be most intense when $L \gg \alpha$, and when average persuasion efficacy is high.

Perhaps most importantly for students of snakes, these formulations suggest some testable predictions for those interested in snake mating systems and sexual selection. In snake mating systems characterized by fighting or mate guarding, or even complex courtship activities, for example, the models we provide surrounding handling time h provide a means for integrating the analysis of such behavior, whether in the lab or field, into formal evolutionary theory. The same is true for searching time s, though this parameter may be studied more effectively in the field. The BSR also has practical and empirical utility because it may be more predictive of dynamic sexual selection forces in a variety of circumstances, than the OSR. DNA fingerprinting approaches may be especially useful in studying the latter, as we describe in some detail just below. And finally, the variance formulations defined above suggest clear-cut pathways for studying behavioral, reproductive, and genetic variables that increase dispersion from the mean, and, thus, potential sexual selection forces across a broad range of relevant phenomena.

Measuring Sexual Selection Gradients in Snake Populations

The data required to address many of the conceptual issues that we have discussed could be obtained in some snake populations. In particular, the denning habit of some species (Klauber, 1936; Gregory,

1984; Graves and Duvall, 1990) offers an exceptional opportunity to capture and monitor the reproductive success of an entire local population. We will focus on the situation in which all (or virtually all) of the mating partners in a local area can be captured at one or more dens. The research program is easiest to visualize if we imagine capturing and monitoring every breeding adult in the population. Such total sampling may not be practical in an actual situation, and so some statistical issues may arise that we will not address. Let us first consider the sequence of steps in obtaining data and then a series of steps in data analysis.

Obtaining the data

1. Enclose a hibernaculum and capture all emerging snakes. Enclosures can be built over hibernaculum openings that will fence in the emerging snakes so that they can be captured for processing. A number of authors describe apparatuses of this nature (Klauber, 1972; Fitch, 1987).

2. Equip all females capable of breeding with radiotransmitters. The aim here is to be able to find the females when they are gravid at the end of the season, so that they and their offspring may be captured. The details of fitting snakes with radiotransmitters are discussed by numerous authors (Reinert, 1992).

3. Take tissue and/or blood samples from all males capable of breeding. The tissue samples are then frozen for later use in paternity analysis. This data step relies on the fact that in humans and apparently many vertebrate populations almost every individual has a unique number of copies of certain highly repetitive genes (Jeffreys et al., 1985). That copy number can be determined and used as a "DNA fingerprint" in paternity analysis (Vassart et al., 1987; Wetton et al., 1987; Burke, 1989; Westneat, 1990). Fitting each male with a passive integrated transponder (PIT) tag is an option that could also be implemented at den capture time. PIT tags are small (ca. 1 × 4 mm) glass cylinders (enclosing an integrated circuit) that can be injected into a snake's body cavity. The snake's PIT tag responds with a unique identification number when activated by a detector that is held close to the snake's body. It is feasible to automate the detection process, so that snakes crawling past a detector, as they enter the enclosed den opening at the end of the season, are automatically recorded. Such automated detection would enable the investigator to score male survivorship, if it is safe to assume that a snake that fails to return is dead. Survivorship is an important issue in analyzing reproductive success in species that mate away from the den, because we will need to know which males were alive throughout the mating season.

4. Locate and capture all breeding females and their offspring. The period of egg-laying or birth lasts only a few weeks each year in many snake species. Consequently, the search for gravid females can be scheduled before the onset of the laying or birth season. Finding the females would be greatly facilitated by radiotransmitters. In some species (e.g., of *Thamnophis, Crotalus*) gravid females aggregate (Duvall et al., 1985; Huey et al., 1989; Graves and Duvall, 1993). In such species a good sample of gravid females can sometimes be obtained even without radiotransmitters. Even if only a few females are fitted with radiotransmitters, they may greatly help in locating female aggregations. The aim in capturing gravid females is to determine their brood sizes and obtain tissue from their progeny so that paternity analyses can be undertaken. One of us (S.J.A.) has had good success obtaining broods of natricine snakes (*Thamnophis* and *Nerodia*) by transporting gravid females to the laboratory and maintaining them individually on thermal gradients for several weeks until offspring are born. In this way about 1500 broods were obtained over a 15-year period. Production line methods for egg-laying species have been developed by commercial snake breeders.

5. Obtain tissue samples from all mothers and newborn offspring. Blood samples could be obtained without killing mothers or offspring and the snakes could be released afterward. The liver is easily biopsied in *Thamnophis* (and presumably in other snakes) without killing the snake. It would be important to obtain tissue samples from all available offspring in each brood because of the possibility of multiple paternity.

6. Determine the paternity of all offspring (e.g., using DNA fingerprinting). Molecular techniques are discussed in Vassart et al. (1987), Wetton et al. (1987), and Kirby (1990), and data analysis is reviewed by Kirby (1990).

7. Construct the parental table. This table lists individual adult females as labels for its columns and individual adult males as labels for its rows. Each cell in the table (intersection of a row and column) gives the number of offspring produced by a particular male and female pair. If all the offspring that a population produced are represented in the table, then the row totals give the fecundity of each male and the column totals give the fecundity of each female. The parental table is a useful summary of the paternity analysis even though many cells in the table will contain zeros. In particular, nonbreeding but adult animals of both sexes should be included in the table even though their rows or columns will be entirely composed of zeros. The inclusion of these animals is crucial because breeding failure is a potentially important component of some of the statistics of reproduction and sexual selection.

Analyzing the data

Although a formidable amount of effort would be required to execute the entire research program that we have outlined, a very large number of questions could be answered with such data. Several of these are listed next.

Measures of fecundity and mating success for each male and female. We have already discussed how we can obtain fecundity scores for individual males and females from the row and column totals of the parental table. Mating success scores also can be obtained by first making a revised version of the parental table. In this revised version we put a one in any nonzero cell and leave all the zero cells intact. To tally the mating success of any male (the number of mates that bore his progeny), we simply take the total for his row. Likewise, the mating success of any female is given by her column total.

Opportunities for selection in each sex mediated through fecundity and mating success. Opportunities for selection are simply standardized variances in fitness (or its components) that are useful because they place an upper bound on the magnitude of phenotypic selection (Crow, 1958; Arnold, 1986). For example, to compute the opportunity for fecundity selection in males, we calculate the variance in male fecundities (i.e., the variance of the row totals in the parental table) and divide this variance by the squared value of average male fecundity. The opportunity for sexual selection in males is computed by doing the analogous operations on the revised parental table. Selection opportunities for females are computed in the same way, using column totals. Arnold and Wade (1984a,b) discuss how the opportunity for fecundity selection in males can be partitioned into two parts: an opportunity for sexual selection (computed as discussed above) and a selection opportunity arising from variation in the average number of progeny per mate.

Sexual selection gradients for both sexes. To estimate the sexual selection gradient for males we compute an ordinary regression of fecundity (row totals from the parental table) on mating success (row totals from the revised parental table). In other words, treating fecundity as the Y variable and mating success as the X variable, we compute the regression slope for the least-squares line that predicts Y from X. This slope is our estimate of the male sexual selection gradient. The female sexual selection gradient is estimated in the same way, but using column totals instead of row totals from the parental tables. Arnold and Duvall (1993) give an example of such an analysis, using Bateman's (1948) published data.

Sexual selection gradients and fecundity selection gradients for various traits (determinants of fecundity and mating success). One can attempt to unravel the causes of variation in fecundity and mating success by measuring multiple traits and asking whether they are determinants of reproductive success. The exercise is statistical and statements about causation are merely inference. Lande and Arnold (1983) discuss the application of multiple regression to this problem and show how the resulting slope estimates correspond to coefficients of selection that appear in equations for evolutionary change. Imagine that we have measured a set of traits on each of the males and females whose reproductive success has been assessed. Body size, tail length, head width, and other sexually dimorphic attributes are good candidate determinants of fecundity and mating success. The sexual selection gradient for one of these traits (e.g., body size) in males could then be estimated as the partial regression of mating success on body size holding the other traits constant. Likewise, the fecundity selection gradient for males could be estimated as the partial regression of fecundity on body size holding the other traits constant. Similar estimates could be made for the females. The selection coefficients that we have estimated are for a single reproductive season and are based on the supposition that the same partial regression slopes prevail at all ages.

Tests for assortative mating. The trait measurements on males and females can also be used to test for assortative mating. One simple test is to compute the correlation between the values of the same attribute of mating partners (e.g., the correlation between the body size of a male and the body size of a female that bore his progeny). A more powerful approach is to use canonical correlation analysis to analyze all the traits at once (Harris, 1975). For some genetic interpretations it would be informative to compute weighted correlations in which the data points represented by each male and female pair that produced progeny are weighted by the number of offspring that they produced.

Tests for inbreeding are conceivable but probably impractical. The issue of inbreeding arises most urgently in very small populations such as the European Adder population studied by T. Madsen and colleagues in southern Sweden (Madsen et al., 1992). In such populations one might wish to know the pedigrees of mating partners so that the degree of inbreeding can be established. Unfortunately, although DNA fingerprints can often be used for paternity assignment (if fingerprints are available for all potential fathers) it is unlikely that the technique will yield unambiguous determinations of more distant relationships (Lynch, 1988).

Estimation of genetic parameters. It would be possible to estimate such genetic parameters as genetic variance and covariance, heritability and genetic correlation if traits are scored in offspring as well as in their parents (Falconer, 1989; Brodie and Garland, Chap. 8, this volume). Unless the age of parents is known and a very large sample is available, the best course is to focus on traits that do not change with age (e.g., meristic traits such as scale counts). Beatson (1976) and Arnold (1988), for example, provide some examples of data analysis.

While a number of potential research protocols come to mind, we favor an approach that takes advantage of the rise of genetic methods to assign paternity, an approach that is revolutionizing behavioral ecology and sociobiology (Burke, 1989). We do not dismiss continued focus upon behavioral, physiological, and other dependent measures as well, however. Indeed, as our encounter rate models noted above should suggest, we value and recommend such approaches as well. The ideal situation for future studies, to our minds, would be to attempt to integrate diverse approaches formally, as we have attempted to argue through the theoretical rationale we have presented in the previous sections. In the end, it is our hope that such integration will facilitate general comparison.

Benchmark Studies of Snake Mating Systems

Based upon the foregoing, the reader might incorrectly assume that little of worth has been done on mating systems and sexual selection in snakes. This is hardly the case. Our goals thus far in this chapter have been to outline some new approaches to the study of relevant phenomena, approaches that hold some potential for integrating the analysis of (1) dynamic sexual selection forces and (2) pattern and variation in snake mating systems.

Our goal in this section, however, is to touch upon some key studies of snake mating system-related phenomena that have been undertaken. Though work in this area really has taken off only in the past few years, some of it is substantive. Seigel et al. (1987) review most relevant studies through the mid-1980s.

European Adder

The European Adder (*Vipera berus*) is perhaps the best studied snake with respect to its mating systems in nature. Many of the traits (or variables) that we have already indicated as important to understanding mating systems have been described, including data on geographical variation in the timing of mating (Viitanen, 1967; Prestt, 1971), various aspects of the reproductive biology of both sexes

(Volsøs, 1944; Andrén and Nilson, 1981, 1983; Nilson, 1981; Andrén, 1982a,b; Saint Giron, 1982), multiple paternity (Stille et al., 1986), mate-searching and site(s) of mating (Viitanen, 1967; Prestt, 1971; Andrén, 1986; Saint Giron et al., 1989), mate competition among males (Kelleway, 1982; Andrén, 1986; Nilson and Andrén, 1982), life-history variables (Viitanen, 1967; Prestt, 1971; Andrén and Nilson, 1983), important aspects of natural history and the life cycle, such as seasonal movements and habitat selection (Viitanen, 1967; Prestt, 1971), predation (Andrén and Nilson, 1981), and foraging biology (Andrén, 1982b; Andrén and Nilson, 1983).

European Adders generally spend winters in communal dens and emerge in the spring, with adult males emerging prior to adult females (Viitanen, 1967; Prestt, 1971; Andrén, 1986; Madsen, 1988). Males disperse to areas termed "basking spots" (Viitanen, 1967), which are located only short distances from den sites. Males appear to select a particular individual basking spot where they will remain, on average, for several weeks (Viitanen, 1967). Foraging does not occur at this time. During the so-called basking period, males become reproductively competent (i.e., spermatozoa are present in the ductus deferens) and a complete skin shedding cycle occurs. After molting, males move a short distance to "mating areas," probably hotspots in our lexicon (see Bradbury et al., 1986), where receptive females reside and tend to cluster. Male fighting for priority access to females then occurs (Prestt, 1971; Madsen, 1988; Madsen et al., in press). The mating period lasts about a month, and both males (Viitanen, 1967; Prestt, 1971) and females (Stille et al., 1986) are known to engage in multiple matings with different individuals. Subsequent to the mating period, both sexes move to summer foraging grounds. At the end of the season they return to the wintering den sites.

Recent work on the European Adder by Madsen and colleagues (Madsen and Shine, 1992, 1993; in press; Madsen et al., 1992, in press; see Shine, this volume) has been quite successful in connecting key traits to mating success, a prime goal in integrating mating systems and selection theory (Duvall et al., 1992; Arnold and Duvall, 1993; Fig. 5.5). These studies have begun to provide the type of information we envision to be necessary to comprehend the accurate structure of mating systems. Importantly, their work also has provided insight into other general problems in animal reproductive biology, such as functions of multiple mating in females (Madsen et al., 1992).

Work on the European Adder by Madsen and colleagues also has revealed a number of insights into key determinants of mating success in males (Madsen and Shine, 1993, in press; Madsen et al., in press). Male–male combat is important in gaining access to receptive females, and larger males almost always win fights. Smaller (and likely younger) males generally do not engage in fights. These smaller males, however, have been observed to wait nearby (i.e., satellite tactics?) and return to females after rivals have departed. Of 148 observed matings, smaller males achieved 15 (10%) of the total.

North American Rattlesnakes

Rattlesnakes (*Crotalus* and *Sistrurus*) represent a group for which there exists an abundance of biological information on most aspects of their life cycles, as well as natural and life histories (Gloyd, 1940; Klauber, 1972; Campbell and Lamar, 1989; Campbell and Brodie, 1992; Ernst, 1992). Surprisingly, few studies provide information on the structure of mating systems (see Duvall et al., 1992; Schuett, 1992). There are, however, a number of studies on rattlesnakes that have been successful in collecting some data of the type important to describing and comprehending mating systems (Fitch, 1949, 1970; Landreth, 1973; Reinert, 1981; Diller and Wallace, 1984; Gannon and Secoy, 1984; Jacob et al., 1987; Macartney and Gregory, 1988; Reinert and Zappalorti, 1988; Brown, 1991, 1992; Brown and Lillywhite, 1992; Martin, 1992; Secor, 1992).

We will focus our discussion on a long-term and intensive study of free-ranging Prairie Rattlesnake (*Crotalus viridis viridis*) populations in the Red Desert (Haystack Mountains) of south-central Wyoming. Work by Duvall and colleagues on these populations has generated information on seasonal movements and space use patterns (Duvall et al., 1985; Graves and Duvall, 1990; King and Duvall, 1990), foraging ecology (Duvall et al., 1985, 1990; Brown, 1990), reproduction and social structure (Duvall et al., 1985, 1992; Graves et al., 1986; King and Duvall, 1990; Hayes et al., 1992; Graves and Duvall, 1993; Schuett et al., in press), and mating systems (Graves and Duvall, 1990; King and Duvall, 1990; Duvall et al., 1992; Duvall and Schuett, in review).

Shortly after adult Prairie Rattlesnakes emerge from dens in spring, they initiate long-distance (up to ca. 5 km from den sites) migrations away from dens in search of preferred prey in summer activity ranges (Duvall et al., 1985, 1990). Unlike the European Adder (see above), mating and shedding never have been observed to occur in spring near den sites when both sexes are thus clumped and in close physical proximity. Rather, in midsummer, when individuals of both sexes are widely dispersed, males begin prolonged mate searching activities. When receptive females are located, perhaps by pheromone trails produced by females (see Ford, 1986), males will accompany and attempt to court females for 2–4 days prior to actual copulation. Rarely will more than one male locate a female simultaneously; thus, male–male fights are uncommon. The period of mating is ca. 6–8 weeks in duration, usually ending by late August. During late summer and early fall, both sexes return to overwintering dens.

Although most of our preliminary results (Duvall and Schuett, in review) are presented in Duvall et al. (1992), a few additional points can be made here. First, the narrow sense mating system we have observed in the Red Desert is prolonged mate-searching polygyny, with most males failing to find even a single female to mate with in a season. Almost all females, conversely, are mated. Of interest (but not surprising), male body size does not predict mate location success. Rather, a number of movement- and orientation-related traits are associated with mate location success. Given the extremely male-biased OSR and/or the unpredictable, sparse, and clustered distribution of females at the onset of the mating season, it was expected that movement and not body-size factors per se would be more important (Duvall et al., 1989; Mintzer et al., 1993; Duvall et al., in review). In due course, we intend to apply DNA fingerprinting methodologies in future studies, such as those described above, so as to gain a better understanding of sexual selection and microevolutionary response to selection, factors that will be critical to an improved understanding of Prairie Rattlesnake mating systems.

North American Copperhead

Autecological study of the Copperhead, *Agkistrodon contortrix*, by Fitch (1960) has provided a wealth of data on various aspects of the natural and life history of this viperid snake, including much information relevant to mating systems (e.g., seasonal activity and movements, timing and location of mating, life-history traits).

Inspired by Fitch's (1960) field work and Carpenter's (1984) lab work, Schuett and his colleagues have investigated various aspects of this snake's reproductive biology in the laboratory, including analysis of sperm storage (Schuett, 1982, 1992), courtship (Schuett and Gillingham, 1988), male competitive tactics (Gillingham, 1987; Schuett and Gillingham, 1986, 1989; Schuett, in manuscript), mate choice (Schuett and Duvall, in review; see below), and hormone cycles (Schuett et al., in manuscript).

In both nature and captivity (Fitch, 1960; Schuett, 1982; Schuett and Gillingham, 1986, 1988), Copperheads exhibit two distinctive periods of mating: the first occurs in late summer or early fall, and the second in spring (Schuett, 1992). Captive individuals of both sexes will mate in either one or both periods, and females can exhibit multiple paternity within single litters (Schuett and Gillingham, 1986). Male fighting (Fig. 5.6) can occur for priority access to females

Figure 5.6 Sequence of fighting behavior in two captive male Copperheads (*Agkistrodon contortrix*): (*A*) approach and initial challenge displays; (*B*) both males in vertical posture (*sway*)—male in foreground is in the process of *hooking*; (*C*) initial phase of *entwinement*; (*D*) final phase of entwinement where separation will occur shortly thereafter (from Schuett, unpublished). See Schuett and Gillingham (1989) for a detailed description and analysis of the behavioral units and acts comprising fighting. See text for further details.

(Schuett and Gillingham, 1989; Schuett, in manuscript). Recently completed work on fighting dynamics (Schuett, in manuscript) has provided further insight into the determinants of agonism and mating success. Significantly, but not unexpectedly, body size (as measured by snout–vent length, SVL) was found to be an important determinant in winning staged fights. When SVL differences were 8–10%, larger males always won fights (Schuett and Gillingham, 1989; Madsen et al., in press; Schuett, in manuscript). Perhaps one of the most interesting findings was that losses in prior fights had a significant effect on subsequent fighting ability. In cases where contest losers were tested for fighting ability 24 h after a fight, and when paired with either the same male or a different male of the same SVL, prior losers never exhibited challenge displays and thus never engaged conspecifics in fighting. Even in cases where prior losers were tested at 24 h after a fight with males of smaller size (8–10% difference in SVL), again losers never challenged or fought. Thus it can be concluded that negative prior agonistic experience (i.e., losing fights) can override the normal positive influence of body size. In order to begin looking at the physiological mechanism(s) mediating winning and losing fights, Schuett et al. (in manuscript) have begun working on the effect of agonistic experience on circulating levels of testosterone and corticosterone.

Female mate choice. Schuett and Duvall (in review) have discovered an unusual type of female choice in the Copperhead, *Agkistrodon contortrix,* that has not been documented in any vertebrate. During a study of the determinants of fighting success in males, as described above, it was found that females may select for male quality by mimicking key behavioral acts characteristic of male combat, which may in turn facilitate female discrimination between prior winners and losers of fights. Specifically, it was found that, when males initiated courtship, most females lifted their heads from the substrate, sometimes quite dramatically (Fig. 5.7), in response to tactile stimuli (e.g., chin-rubbing) derived from the male. Head-lifting by females was very similar in form to head-lifting which comprised the initial stages of the "challenge display" typical of males (compare Figs. 5.6 and 5.7; cf., Schuett and Gillingham, 1988, 1989). Males often responded to head-lifting with a challenge display of their own. Even so, male–female fights never were observed, and all females that exhibited head-lifting resumed precourtship postures within several minutes.

Interestingly, when a receptive female is present in the test arena, the loser of a staged fight between two males attempts neither further fighting with the winning male nor courtship with the female. The latter also holds even when the winning male is removed from the arena, or when the loser is placed into a novel arena with a different receptive female (Schuett, in manuscript). It also was observed that almost any action on the part of females at these times caused the loser to retreat.

It thus appears that female Copperheads may mimic an important component of ritualized male combat, so as to facilitate the discrimination of male fighting ability and perhaps quality.

North American Garter Snakes

A multitude of field and laboratory studies have been done on North American Garter Snakes (*Thamnophis*), far more than can be described here. Several provide detailed information relevant to mating systems analyses (cf., Ford, 1986). The best studied taxon in this respect is the Red-Sided Garter Snake (*Thamnophis sirtalis parietalis*) of the Interlake region of Manitoba, Canada (see Crews, 1992; Halpern, 1992; Mason, 1992; Moore and Lindzey, 1992; Whittier and Tokarz, 1992, for the essential references).

Figure 5.7 Mate choice by male mimicry in a female Copperhead (*Agkistrodon contortrix*): (*A*) the male approaching phase of courtship and the female (see arrow) exhibiting the initial postures resembling the *challenge display* of fighting males; (*B*) the female continues to exhibit the challenge display during courtship, and the male has temporarily ceased courtship; (*C*) the female assumes an even more distinctive vertical posture, and the male continues courtship; (*D*) the male responds to the female's display by initiating the challenge display; (*E*) the female in process of terminating her display and the male resuming courtship; (*F*) the male is moving back from female and initiates a challenge display (from Schuett and Duvall, in review). See text for further details.

In populations of Red-Sided Garter Snakes from Manitoba, explosive mating assemblages (e.g., "mating balls") can occur on warm days in the spring when most males have already emerged and females are just beginning to do so (Gardner, 1955; Gregory, 1974; cf., Graves and Duvall, 1990). Dozens, sometimes hundreds, of males and just a handful of females may comprise an individual explosive mating ball, or scramble. Determinants of male mating success are not yet fully known, though male body size does not seem to be important (Joy and Crews, 1985, 1988; but cf., Madsen and Shine, in press). During the following coitus, which is fairly brief (5–15 min), copulatory materials are inserted and congeal in the female's reproductive tract, sometimes extruding from the cloaca (Devine, 1977, 1984). Copulatory plugs may enforce chastity and first-male advantage, and likely represent a form of mate-guarding (Ross and Crews, 1977). Contrary to some earlier views, multiple matings by females probably are common, and mating is not confined only to the spring (Whittier and Crews, 1986; Whittier and Tokarz, 1992). (Evidence for late-summer mating derives from females with copulatory plugs and/or sperm in the caudal regions of their reproductive tract later in the season.) In other populations of *T. sirtalis,* males have been observed to mate with multiple females, and mixed paternity has been verified (Gibson and Falls, 1975; Schwartz et al., 1989).

The seemingly obligate use of limited den sites in winter and the high densities of snakes that occupy them probably have been important selective factors in shaping the mating systems of these populations (Gregory, 1982, 1984). Although behavioral observations are limited, even late-summer mating activities might be influenced by migration episodes to the dens (P. Gregory, personal communication). Hopefully, future work on Interlake populations will include (1) investigations of the late-summer mating activities of larger samples of individuals and (2) analyses of DNA fingerprints of likewise large samples of individual neonates comprising litters. Such data could facilitate greatly analyses of the local mating system and dynamic sexual selection forces (Arnold and Duvall, 1993; see below).

Female mimicry recently has been described by Mason and Crews (1985) as an alternative mating tactic in the Red-Sided Garter Snake. In this case, a small fraction of males in the population, referred to as "she-males," mimic the courtship pheromones of females, attracting other males who thus waste their courtship efforts. Mason and Crews (1985) also determined that she-males not only are reproductively competent, but, in fact, mate more frequently with females than do normal males. In conclusion (p. 60), they argue that the "advantage to the she-male's attractiveness thus lies in an ability to gain a better position in the mating ball by confusing other males."

Male choice in snakes. Female garter snakes, like many other squamate reptiles, exhibit a well-known and strong positive association between increasing body size during growth and increased clutch or litter size (Duvall et al., 1982). Thus, to the extent that appropriate ecological and phylogenetic constraints apply, males should choose the largest, sexually mature female conspecifics with which to mate.

This has, in fact, been observed by several different investigators (Gregory, 1974; Hawley and Aleksiuk, 1976; Garstka and Crews, 1985). As yet, however, no one has considered how these patterns may have affected sexual selection forces acting on females. Though age–body-size covariance would have to be handled appropriately, the scheme we propose at the outset of this chapter (Arnold and Duvall, 1993) could be used to analyze this interesting problem quantitatively. Much more could be done with it.

Australian Diamond Python

The work of Slip and Shine (1988a,b,c,d,e; Shine, 1991, 1992; this volume) on the natural history of the Diamond Python, *Morelia spilota spilota,* is particularly important in that it represents the only published study of the mating system of a free-ranging boid (Slip and Shine, 1988a; Shine, 1992), and one of the best documents of habitat use, thermal ecology, and mating frequency in any wild snake. Using radiotelemetry, Slip and Shine (1988a), documented courtship and mating behaviors of *Morelia s. spilota* from 1982 to 1984. They found that during the mating season (spring; late September to early November) it was not uncommon for individual receptive females to be trailed, accompanied, and courted by multiple males (Shine, 1991). Females engaged in multiple matings with the same and/or different males, sometimes within very brief periods (e.g., 1–2 days). Interestingly, during periods when multiple males (e.g., 2–4 individuals) accompanied individual females, male combat (see Covacevich, 1975) was never observed. Slip and Shine (1988a) hypothesized that multiple-male accompaniment and the lack of male agonism resulted from (1) low numbers of receptive females, (2) a low likelihood of males locating additional receptive females, and (3) potential high costs of fighting. Although advantages of multiple mating by female snakes remains somewhat unresolved (Schuett, 1992; but see Madsen et al., 1992), Slip and Shine (1988a) suggest that potential advantages may include (1) ensuring sufficient quantities of viable sperm, and (2) reducing survival and/or energetic costs by not rejecting courting males when courtship and mating are low-risk activities (Gibson and Falls, 1975; but see Devine, 1984). Moreover, although they suggest that courtship and mating may attract predators, no such evidence is presented. Based upon correlated evolution alone, however, some degree of female multiple mating should not be unexpected (Lande, 1979; Lande and Arnold, 1983; Halliday and Arnold, 1987), particularly in light of the strong sexual selection for multiple mating likely acting on most male snakes (see polygyny rationale for most snakes in preceding sections).

Additional work on mating systems of Diamond Pythons and their close relatives, the Carpet Pythons, is in progress (Shine, 1992; R. Shine, personal communication). One goal is to identify proximate factors that result in the presence of fighting among Carpet Pythons and its absence among Diamond Pythons. The mating system of another species of python (*Liasis fuscus*) from tropical Australia (near Humpty Doo, Northern Territory) is currently being investigated by Shine and colleagues as well (Shine, 1991, 1992). Indeed, Shine and colleagues have examined mating systems and related phenomena in a number of other Australian snakes (Shine, 1977a,b, 1986, 1991; Shine et al., 1981).

Summary and Future Research

Clearly, work has only just begun in the area of snake mating systems and sexual selection. If the few studies that have been done so far are any indication of what is to come, however, the future looks bright indeed. In this spirit, herein we have attempted to present formal microevolutionary models that may facilitate future research, by showing interconnections between various bodies of relevant theory and at least a few empirical directions. We also have attempted to focus on some of the benchmark studies that have been done, that address key features of relevant microevolutionary processes. Another hope is that molecular genetic techniques such as DNA fingerprinting find a home in analyses of snake mating systems and sexual selection, especially in light of the power such approaches bring to the game of evolutionary behavioral research.

Acknowledgments

First we thank Rich Seigel and Joe Collins for patience and support well beyond what is normally reasonable. Next, of course, we thank our loved ones, who would have much preferred having us around a little (a lot!?) more. Thanks also to Tamara Wagar for editorial assistance. Finally, GWS thanks the Sigma Xi and the American Museum of Natural History for a Theodore Roosevelt Award, and DD and SJA the NSF (RII 86-10680 and BSR 89-18581, respectively).

Literature Cited

Alcock, J., 1980, Natural selection and the mating systems of solitary bees, *Amer. Sci.*, 68:146–153.

Andersson, M., 1982, Female choice selects for extreme tail length in a widowbird, *Nature*, 299:818–820.

Andrén, C., 1986, Courtship, mating and agonistic behaviour in a free-living population of adders, *Vipera berus* (L.), *Amphibia-Reptilia*, 7:353–383.

Andrén, C., 1982a, The role of the vomeronasal organs in the reproductive behavior of the adder *Vipera berus, Copeia,* 1982:148–157.

Andrén, C., 1982b, Effect of prey density on reproduction, foraging and other activities in the adder, *Vipera berus, Amphibia-Reptilia,* 3:81–96.

Andrén, C., and G. Nilson, 1981, Reproductive success and risk of predation in normal and melanistic colour morphs of the adder, *Vipera berus, Biol. J. Linn. Soc.,* 15:235–246.

Andrén, C., and G. Nilson, 1983, Reproductive tactics in an island population of adders, *Vipera berus* (L.), with a fluctuating food resource, *Amphibia-Reptilia,* 4:63–79.

Arnold, S. J., 1983, Sexual selection: The interface of theory and empiricism, in P. Bateson, ed., *Mate Choice,* Cambridge Univ. Press, Cambridge, pp. 67–107.

Arnold, S. J., 1986, Limits on stabilizing, disruptive, and correlation selection set by the opportunity for selection, *Amer. Nat.,* 128:143–146.

Arnold, S. J., 1988, Quantitative genetics and selection in natural populations: Microevolution of vertebral numbers in the garter snake *Thamnophis elegans,* in D. S. Weir, E. J. Eisen, M. M. Goodman and G. Namkoong, eds., *Proceedings of the Second International Conference on Quantitative Genetics,* Sinauer, Sunderland, Massachusetts, pp. 619–636.

Arnold, S. J., and D. Duvall, 1993, Animal mating systems: A synthesis based on selection theory, *Amer. Nat.,* in press.

Arnold, S. J., and M. J. Wade, 1984a, On the measurement of natural and sexual selection: Theory, *Evolution,* 38:709–719.

Arnold, S. J., and M. J. Wade, 1984b, On the measurement of natural and sexual selection: Applications, *Evolution,* 38:720–734.

Arnqvist, G., 1992, The effects of operational sex ratio on the relative mating success of extreme male phenotypes in the water strider *Gerris odontogaster* (Zett.) (Heteroptera; Gerridae), *Anim. Behav.,* 43:681–683.

Bateman, A. J., 1948, Intra-sexual selection in *Drosophila, Heredity,* 2:349–368.

Baylis, J. R., 1981, The evolution of parental care in fishes, with reference to Darwin's rule of sexual selection, *Env. Biol. Fish.,* 6:223–251.

Beatson, R. R., 1976, Environmental and genetic correlates of disruptive coloration in the water snake, *Natrix s. sipedon, Evolution,* 30:835–840.

Borgia, G., 1979, Sexual selection and the evolution of mating systems, in M. S. Blum and N. A. Blum, eds., *Sexual Selection and Reproductive Competition in Insects,* Academic, New York, pp. 19–80.

Bradbury, J. W., and M. B. Andersson (eds.), 1987, *Sexual Selection: Testing the Alternatives,* Wiley, New York.

Bradbury, J. W., R. M. Gibson, and I. M. Tsai, 1986, Hotspots and the evolution of leks, *Anim. Behav.,* 34:1694–1709.

Brooks, D. R., and D. A. McLennan, 1991, *Phylogeny, Ecology, and Behavior,* Univ. Chicago Press, Chicago, Illinois.

Brooks, D. R., and E. O. Wiley, 1988, *Evolution as Entropy: Toward a Unified Theory of Biology,* 2d ed., Univ. Chicago Press, Chicago, Illinois.

Brown, D. G., 1990, Observation of a prairie rattlesnake (*Crotalus viridis viridis*) consuming neonatal cottontail rabbits (*Sylvilagus nuttali*), with defense of the young cottontails by adult conspecifics, *Bull. Chicago Herp. Soc.,* 25:24–26.

Brown, T. W., and H. B. Lillywhite, 1992, Autecology of the Mojave desert sidewinder, *Crotalus cerastes cerastes,* at Kelso Dunes, Mojave Desert, California, USA, in J. A. Campbell and E. D. Brodie, Jr., eds., *Biology of the Pitvipers,* Selva, Tyler, Texas, pp. 279–308.

Brown, W. S., 1991, Female reproductive ecology in a northern population of timber rattlesnakes, *Crotalus horridus, Herpetologica,* 47:101–115.

Brown, W. S., 1992, Emergence, ingress, and seasonal captures at dens of northern timber rattlesnakes, *Crotalus horridus,* in J. A. Campbell and E. D. Brodie, Jr., eds., *Biology of the Pitvipers,* Selva, Tyler, Texas, pp. 251–258.

Burke, T., 1989, DNA fingerprinting and other methods for the study of mating success, *Trends Ecol. Evol.,* 4:139–144.

Campbell, J. A., and E. D. Brodie, Jr., 1992, *Biology of the Pitvipers,* Selva, Tyler, Texas.

Campbell, J. A., and W. W. Lamar, 1989, *The Venomous Reptiles of Latin America,* Cornell Univ. Press, Ithaca, New York.
Carpenter, C. C., 1984, Dominance in snakes, in R. A. Seigel, L. E. Hunt, J. L. Knight, L. Malaret, and N. L. Zuschlag, eds., *Vertebrate Ecology and Systematics—A Tribute to Henry S. Fitch,* Univ. Kansas Nat. Hist. Mus., Spec. Publ. 10, pp. 195–202.
Clutton-Brock, T. H., 1988, Reproductive success, in T. H. Clutton-Brock, ed., *Reproductive Success,* Univ. Chicago Press, Chicago, Illinois, pp. 472–485.
Clutton-Brock, T. H., 1989, Review lecture. Mammalian mating systems, *Proc. Royal Soc. London,* B236:339–372.
Clutton-Brock, T. H., and A. C. J. Vincent, 1991, Sexual selection and the potential reproductive rates of males and females, *Nature,* 351:58–60.
Covacevich, J., 1975, Snakes in combat, *Victorian Nat.,* 92:252–253.
Cox, D. R., 1962, *Renewal Theory,* Methuen, London.
Crews, D., 1992, Behavioural endocrinology and reproduction: An evolutionary perspective, in S. R. Milligan, ed., *Oxford Reviews of Reproductive Biology,* Vol. 14, Oxford Univ. Press, Oxford, pp. 303–370.
Crow, J. F., 1958, Some possibilities for measuring selection intensities in man, *Human Biol.,* 30:1–13.
Darwin, C., 1871, *The Descent of Man, and Selection in Relation to Sex,* Princeton Univ. Press, Princeton, New Jersey.
Davies, N. B., 1991, Mating systems, in J. R. Krebs and N. B. Davies, eds., *Behavioural Ecology,* 3d ed., Blackwell Scientific, Oxford, pp. 263–299.
Devine, M. C., 1977, Copulatory plugs, restricted mating opportunities, and reproductive competition among male garter snakes, *Nature,* 267:345–346.
Devine, M. C., 1984, Potential for sperm competition in reptiles: Behavioral and physiological consequences, in R. L. Smith, ed., *Sperm Competition and the Evolution of Animal Mating Systems,* Academic, Orlando, Florida, pp. 509–521.
Dewsbury, D. A., and D. J. Baumgardner, 1981, Studies of sperm competition in two species of muroid rodents, *Behav. Ecol. Sociobiol.,* 9:121–133.
Diller, L. V., and R. L. Wallace, 1984, Reproductive biology of the northern Pacific rattlesnake (*Crotalus viridis oreganus*) in northern Idaho, *Herpetologica,* 33:427–433.
Duvall, D., and G. W. Schuett, Competitive mate searching and orientation in a rattlesnake mating system, in review.
Duvall, D., S. J. Arnold, and G. W. Schuett, 1992, Pitviper mating systems: Ecological potential, sexual selection, and microevolution, in J. A. Campbell and E. D. Brodie, Jr., eds., *Biology of the Pitvipers,* Selva, Tyler, Texas, pp. 321–336.
Duvall, D., D. Chiszar, and R. A. Mintzer, Some plane geometry for naive searchers, in review.
Duvall, D., D. Chiszar, G. W. Schuett, and R. A. Mintzer, 1989, Migrations, movements, and activity of snakes: Results and predictions from the computer simulation RattleSnake©, First World Congress of Herpetology, Univ. Kent, Canterbury.
Duvall, D., M. J. Goode, W. K. Hayes, J. K. Leonhardt, and D. Brown, 1990, Prairie rattlesnake vernal migrations: Field experimental analyses and survival value, *Nat. Geog. Res.,* 6:457–469.
Duvall, D., L. J. Guillette, Jr., and R. E. Jones, 1982, Environmental control of reptilian reproductive cycles, in C. Gans and F. H. Pough, eds., *Biology of the Reptilia,* Vol. 13D, Academic, London, pp. 201–231.
Duvall, D., M. B. King, and K. J. Gutzwiller, 1985, Behavioral ecology and ethology of the prairie rattlesnake, *Nat. Geog. Res.,* 1:80–111.
Emlen, S. T., and L. W. Oring, 1977, Ecology, sexual selection, and the evolution of mating systems, *Science,* 197:215–223.
Endler, J. A., 1986, *Natural Selection in the Wild,* Princeton Univ. Press, Princeton, New Jersey.
Ernst, C. H., 1992, *Venomous Reptiles of North America,* Smithsonian Inst. Press, Washington, D.C.
Falconer, D. S., 1989, *Introduction to Quantitative Genetics,* 3d ed., Longman, Essex.
Fitch, H. S., 1949, Study of snake populations in central California, *Amer. Mid. Nat.,* 41:513–579.

Fitch, H. S., 1960, Autecology of the copperhead, *Univ. Kansas. Publ. Mus. Nat. Hist.*, 13:85–288.
Fitch, H. S., 1970, Reproductive cycles of lizards and snakes, *Univ. Kansas Mus. Nat. Hist. Misc. Publ.*, 52:1–247.
Fitch, H. S., 1987, Collecting and life-history techniques, in R. A. Seigel, J. T. Collins, and S. S. Novak, eds., *Snakes: Ecology and Evolutionary Biology,* McGraw-Hill, New York, pp. 143–164.
Ford, N. B., 1986, The role of pheromone trails in the sociobiology of snakes, in D. Duvall, D. Müller-Schwarze, and R. M. Silverstein, eds., *Chemical Signals in Vertebrates,* Vol. 4, Plenum, New York, pp. 261–278.
Gannon, V. P. J., and D. M. Secoy, 1984, Growth and reproductive rates of a northern population of the prairie rattlesnake, *Crotalus viridis viridis, J. Herpetol.*, 18:13–19.
Gardner, J. B., 1955, A ball of garter snakes, *Copeia,* 1955:310.
Garstka, W. R., and D. Crews, 1985, Mate preference in garter snakes, *Herpetologica,* 41:9–19.
Gibbs, H. L., P. J. Weatherhead, P. T. Boag, B. N. White, L. M. Tabak, and D. J. Hoysak, 1990, Realized reproductive success of polygynous red-winged blackbirds revealed by DNA markers, *Science,* 250:1394–1397.
Gibson, A. R., and J. B. Falls, 1975, Evidence for multiple insemination in the common garter snake, *Thamnophis sirtalis, Can. J. Zool.*, 53:1362–1368.
Gillingham, J. C., 1987, Social behavior, in R. A. Seigel, J. T. Collins, and S. S. Novak, eds., *Snakes: Ecology and Evolutionary Biology,* McGraw-Hill, New York, pp. 184–209.
Gloyd, H. K., 1940, The rattlesnakes, genera *Sistrurus* and *Crotalus, Chicago Acad. Sci. Spec. Publ.*, 4:1–266.
Graves, B. M., and D. Duvall, 1990, Spring emergence patterns of wandering garter snakes and prairie rattlesnakes in Wyoming, *J. Herpetol.*, 24:351–356.
Graves, B. M., and D. Duvall, 1993, Reproduction, rookery use, and thermoregulation in free-ranging, pregnant *Crotalus v. Viridis, J. Herpetol.*, 27:33–41.
Graves, B. M., D. Duvall, M. B. King, S. L. Lindstedt, and W. A. Gern, 1986, Initial den location by neonatal prairie rattlesnakes: Functions, causes, and natural history in chemical ecology, in D. Duvall, D. Müller-Schwarze, and R. M. Silverstein, eds., *Chemical Signals in Vertebrates,* Vol. 4, Plenum, New York, pp. 285–304.
Gregory, P. T., 1974, Patterns of emergence of the red-sided garter snake (*Thamnophis sirtalis parietalis*) in the Interlake region of Manitoba, *Can. J. Zool.*, 52:1063–1069.
Gregory, P. T., 1982, Reptilian hibernation, in C. Gans and F. H. Pough, eds., *Biology of the Reptilia,* Vol. 13D, Academic, London, pp. 53–154.
Gregory, P. T., 1984, Communal denning in snakes, in R. A. Seigel, L. E. Hunt, J. L. Knight, L. Malaret, and N. L. Zuschlag, eds., *Vertebrate Ecology and Systematics—A Tribute to Henry S. Fitch,* Univ. Kansas Nat. Hist. Mus. Spec. Publ. 10, pp. 57–75.
Gwynne, D. T., 1991, Sexual competition among females: What causes courtship-role reversal?, *Trends Ecol. Evol.*, 6:118–121.
Halliday, T., and S. J. Arnold, 1987, Multiple mating by females: A perspective from quantitative genetics, *Anim. Behav.*, 35:939–941.
Halpern, M., 1992, Nasal chemical senses in reptiles: Structure and function, in C. Gans and D. Crews, eds., *Biology of the Reptilia, Physiology E,* Vol. 18, Univ. Chicago Press, Chicago, Illinois, pp. 423–523.
Harris, R. J., 1975, *A Primer of Multivariate Statistics,* Academic, New York.
Harvey, P. H., and M. D. Pagel, 1991, *The Comparative Method in Evolutionary Biology,* Oxford Univ. Press, Oxford.
Hawley, A. W. L., and M. Aleksiuk, 1976, Sexual receptivity in the female red-sided garter snake (*Thamnophis sirtalis parietalis*), *Copeia,* 1976:401–404.
Hayes, W. K., D. Duvall, and G. W. Schuett, 1992, A preliminary report on the courtship behavior of free-ranging prairie rattlesnakes, *Crotalus viridis viridis* (Rafinesque), in south-central Wyoming, in P. D. Strimple and J. L. Strimple, eds., *Contributions in Herpetology,* Publ. Greater Cincinnati Herpetol. Soc., Cincinnati, Ohio, pp. 45–48.
Holling, C. S., 1959, Some characteristics of simple types of predation and parasitism,

Can Entomol., 91:385–398.
Howard, R. D., 1979, Estimating reproductive success in natural populations, *Amer. Nat.*, 114:221–231.
Huey, R. B., C. R. Peterson, S. J. Arnold, and W. P. Porter, 1989, Hot rocks and not-so-hot rocks: Retreat-site selection by garter snakes and its thermal consequences, *Ecology*, 70:931–944.
Ims, R. A., 1988, The potential for sexual selection: Effect of sex ratio and spatiotemporal distribution of receptive females, *Evol. Ecol.*, 2:338–352.
Jacob, J. S., S. R. Williams, and R. P. Reynolds, 1987, Reproductive activity of male *Crotalus atrox* and *C. scutulatus* (Reptilia: Viperidae) in northeastern Chihuahua, Mexico, *Southwest. Nat.*, 32:273–276.
Jeffreys, A. J., V. Wilson, and S. L. Thein, 1985, Hypervariable 'minisatellite' regions in human DNA, *Nature*, 314:67–73.
Joy, J. E., and D. Crews, 1985, Social dynamics of group courtship behavior in male red-sided garter snakes (*Thamnophis sirtalis parietalis*), *J. Comp. Psychol.*, 99:145–149.
Joy, J. E., and D. Crews, 1988, Male mating success in red-sided garter snakes: Size is not important, *Anim. Behav.*, 36:1839–1841.
Kelleway, L. G., 1982, Competition for mates and food items in *Vipera berus* (L.), *Brit. J. Herpetol.*, 16:225–230.
King, M. B., and D. Duvall, 1990, Prairie rattlesnake seasonal migrations: Episodes of movement, vernal foraging, and sex differences, *Anim. Behav.*, 39:924–935.
Kirby, L. T., 1990, *DNA Fingerprinting: An Introduction*, Stockton Press, New York.
Klauber, L. M., 1936, A statistical study of the rattlesnakes, I. Introduction, *Occas. Pap. San Diego Soc. Nat. Hist.*, 1:1–24.
Klauber, L. M., 1972, *Rattlesnakes, Their Habits, Life Histories, and Influence on Mankind*, 2nd ed., Univ. California Press, Berkeley.
Kluge, A. G., 1981, The life history, social organization, and parental behavior of *Hyla rosenbergi* Boulenger, a nest-building gladiator frog, *Univ. Michigan Mus. Zool. Misc. Publ.*, 160:1–170.
Lande, R., 1979, Quantitative genetic analysis of multivariate evolution, applied to brain:body size allometry, *Evolution*, 33:402–416.
Lande, R., and S. J. Arnold, 1983, The measurement of selection on correlated characters, *Evolution*, 37:1210–1226.
Landreth, H. F., 1973, Orientation and behavior of the rattlesnake, *Crotalus atrox*, *Copeia*, 1973:26–31.
Ludwig, M., and H. Rahn, 1943, Sperm storage and copulatory adjustment in the prairie rattlesnake, *Copeia*, 1943:15–18.
Lynch, M., 1988, Estimation of relatedness by DNA fingerprinting, *Mol. Biol. Evol.*, 5:584–599.
Macartney, J. M., and P. T. Gregory, 1988, Reproductive biology of female rattlesnakes (*Crotalus viridis*) in British Columbia, *Copeia*, 1988:58–64.
Madsen, T., 1988, Reproductive success, mortality and sexual size dimorphism in the adder, *Vipera berus*, *Holarctic Ecol.*, 11:77–80.
Madsen, T., and R. Shine, 1992, A rapid, sexually selected shift in mean body size in a population of snakes, *Evolution*, 46:1220–1224.
Madsen, T., and R. Shine, Male mating success and body size in European grass snakes, *Copeia*, in press.
Madsen, T., and R. Shine, 1993, Temporal variability in sexual selection on reproductive tactics and body size in male snakes, *Amer. Nat.*, 141:167–171.
Madsen, T., R. Shine, J. Loman, and T. Hakansson, 1992, Why do female adders copulate so frequently? *Nature*, 355:440–441.
Madsen, T., R. Shine, J. Loman, and T. Hakansson, Determinants of mating success in male adders, *Vipera berus*, *Anim. Behav.*, in press.
Martin, W. H., 1992, Phenology of the timber rattlesnake (*Crotalus horridus*) in an unglaciated section of the Appalachian Mountains, in J. A. Campbell and E. D. Brodie, Jr., eds., *Biology of the Pitvipers*, Selva, Tyler, Texas, pp. 259–277.
Mason, R. T., 1992, Reptilian pheromones, in C. Gans and D. Crews, eds., *Biology of the*

Reptilia, Physiology E, Vol. 18, Univ. Chicago Press, Chicago, Illinois, pp. 114–228.
Mason, R. T., and D. Crews, 1985, Female mimicry in garter snakes, *Nature,* 316:59–60.
Mintzer, R. A., D. Duvall, and D. Chiszar, 1993, RattleSnake©: Simulation software for Apple Macintosh.
Moore, M., and J. Lindzey. 1992, The physiological basis of sexual behavior in male reptiles, in C. Gans and D. Crews, eds., *Biology of the Reptilia, Physiology E,* Vol. 18, Univ. Chicago Press, Chicago, Illinois, pp. 70–113.
Nilson, G., 1981, Ovarian cycle and reproductive dynamics in the female adder, *Vipera berus* (Reptilia Vipendae), *Amphibia-Reptilia,* 2:63–820.
Nilson, G., and C. Andrén, 1982, Function of renal sex secretion and male hierarchy in the adder, *Vipera berus,* during reproduction, *Horm. Behav.,* 16:404–413.
Orians, G. H., 1969, On the evolution of mating systems in birds and mammals, *Amer. Nat.,* 103:589–603.
Parker, G. A., 1970a, Sperm competition and its evolutionary consequences in the insects, *Biol. Rev.,* 45:525–567.
Parker, G. A., 1970b, The reproductive behaviour and the nature of sexual selection in *Scatophaga stercoraria* L. (Diptera: Scatophagidae), *J. Anim. Ecol.,* 39:205–228.
Parker, G. A., 1974, Courtship persistence and female-guarding as male time investment strategies, *Behaviour,* 48:157–184.
Parker, G. A., 1978a, Searching for mates, in J. R. Krebs and N. B. Davies, eds., *Behavioural Ecology,* Sinauer, Sunderland, Massachusetts, pp. 214–244.
Parker, G. A., 1978b, Evolution of competitive mate searching, *Ann. Rev. Entomol.,* 23:173–196.
Parker, G. A., 1984, Sperm competition and the evolution of animal mating strategies, in R. L. Smith, ed., *Sperm Competition and the Evolution of Animal Mating Systems,* Academic, Orlando, Florida, pp. 1–60.
Payne, R. B., 1979, Sexual selection and intersexual differences in variance of breeding success, *Amer. Nat.,* 114:447–452.
Prestt, I., 1971, An ecological study of the viper, *Vipera berus,* in southern Britain, *J. Zool. London,* 164:373–418.
Real, L., 1990, Search theory and mate choice, I. Models of single-sex discrimination, *Amer. Nat.,* 136:376–404.
Reinert, H. K., 1981, Reproduction of the massasauga (*Sistrurus catenatus catenatus*), *Amer. Mid. Nat.,* 105:393–395.
Reinert, H. K., 1992, Radiotelemetric field studies of pitvipers: data acquisition and analysis, in J. A. Campbell and E. D. Brodie, Jr., eds., *Biology of the Pitvipers,* Selva, Tyler, Texas, pp. 185–197.
Reinert, H. K., and R. T. Zappalorti, 1988, Timber rattlesnakes (*Crotalus horridus*) of the Pine Barrens: Their movement patterns and habitat preference, *Copeia,* 1988:964–978.
Ross, P., and D. Crews, 1977, Influence of the seminal plug on mating behavior in the garter snake, *Nature,* 267:344–345.
Saint Girons, H., 1975, Sperm survival and transport in the female genital tract of reptiles, in E. S. E. Hafez and C. G. Thibault, eds., *The Biology of the Spermatozoa,* S. Karger, Basel, Switzerland, pp. 106–113.
Saint Girons, H., 1982, Reproductive cycles of male snakes and their relationships with climate and female reproductive cycles, *Herpetologica,* 38:5–16.
Saint Girons, H., R. Dughy, and G. Naullean, 1989, Spatio-temporal aspects of the annual cycle of temperate Viparinae, First World Congress of Herpetology, University of Kent at Canterbury, U.K.
Schuett, G. W., 1982, A copperhead (*Agkistrodon contortrix*) brood produced from autumn copulations, *Copeia,* 1982:700–702.
Schuett, G. W., 1992, Is long-term sperm storage an important component of the reproductive biology of temperate pitvipers?, in J. A. Campbell and E. D. Brodie, Jr., eds., *Biology of the Pitvipers,* Selva, Tyler, Texas, pp. 169–184.
Schuett, G. W., and D. Duvall, Male mimicry in a viperid snake: A potential mechanism for female choice, in review.
Schuett, G. W., and J. C. Gillingham, 1986, Sperm storage and multiple paternity in

the copperhead, *Agkistrodon contortrix*, *Copeia*, 1986:807–811.
Schuett, G. W., and J. C. Gillingham, 1988, Courtship and mating of the copperhead, *Agkistrodon contortrix*, *Copeia*, 1988:374–381.
Schuett, G. W., and J. C. Gillingham, 1989, Male-male agonistic behaviour of the copperhead, *Agkistrodon contortrix*, *Amphibia-Reptilia*, 10:243–266.
Schuett, G. W., P. A. Buttenhoff, D. Duvall, and A. T. Holycross, Geographical variation in timing of mating in free-ranging prairie rattlesnakes (*Crotalus viridis*), *Herp. Nat. Hist.*, in press.
Schwagmeyer, P. L., 1990, Ground squirrel reproductive behavior and mating competition: A comparative perspective, in D. A. Dewsbury, ed., *Contemporary Issues in Comparative Psychology*, Sinauer, Sunderland, Massachusetts, pp. 175–196.
Schwagmeyer, P. L., K. A. Coggins, and T. C. Lamey, 1987, The effects of sperm competition on variability in male reproductive success: some preliminary analyses, *Amer. Nat.*, 130:485–492.
Schwartz, J. M., G. F. McCracken, and G. M. Burghardt, 1989, Multiple paternity in wild populations of the garter snake, *Thamnophis sirtalis*, *Behav. Ecol. Sociobiol.*, 25:269–273.
Secor, S. M., 1992, A preliminary analysis of the movement and home range size of the sidewinder, *Crotalus cerastes*, in J. A. Campbell and E. D. Brodie, Jr., eds., *Biology of the Pitvipers*, Selva, Tyler, Texas, pp. 389–393.
Seigel, R. A., and N. B. Ford, 1987, Reproductive ecology, in R. A. Seigel, J. T. Collins and S. S. Novak, eds., *Snakes: Ecology and Evolutionary Biology*, McGraw-Hill, New York, pp. 210–252.
Seigel, R. A., J. T. Collins, and S. S. Novak, eds., 1987, *Snakes: Ecology and Evolutionary Biology*, McGraw-Hill, New York.
Shields, W. M., 1987, Dispersal and mating systems: Investigating their causal connections, in B. D. Chepko-Sade and Z. T. Halpin, eds., *Mammalian Dispersal Patterns*, Univ. Chicago Press, Chicago, Illinois, pp. 3–24.
Shine, R., 1977a, Reproduction in Australian elapid snakes I. Testicular cycles and mating seasons, *Aust. J. Zool.*, 25:647–653.
Shine, R., 1977b, Reproduction in Australian elapid snakes II. Female reproductive cycles, *Aust. J. Zool.*, 25:655–666.
Shine, R., 1978, Sexual size dimorphism and male combat in snakes, *Oecologia (Berlin)*, 33:269–277.
Shine, R., 1986, Ecology of a low-energy specialist: Food habits and reproductive biology of the Arafura filesnake (Acrochordidae), *Copeia*, 1986:424–437.
Shine, R., 1988a, Constraints on reproductive investment: A comparison between aquatic and terrestrial snakes, *Evolution*, 42:17–27.
Shine, R., 1988b, Parental care in reptiles, in C. Gans and R. B. Huey, eds., *Biology of the Reptilia*, Vol. 16 C, Alan R. Liss, New York, pp. 275–330.
Shine, R., 1991, *Australian Snakes: A Natural History*, Cornell Univ. Press, Ithaca, New York.
Shine, R., 1992, Ecological studies on Australian pythons, in M. J. Uricheck, ed., *15th International Herpetological Symposium on Captive Propagation and Husbandry*, Western Connecticut State Univ., Danbury, pp. 29–41.
Shine, R., G. C. Grigg, T. G. Shine, and P. Harlow, 1981, Mating and male combat in Australian blacksnakes, *Pseudechis porphyriacus*, *J. Herpetol.*, 15:101–107.
Slip, D. J., and R. Shine, 1988a, The reproductive biology and mating system of diamond pythons, *Morelia spilota* (Serpentes: Boidae), *Herpetologica*, 44:396–404.
Slip, D. J., and R. Shine, 1988b, Reptilian endothermy: A field study of thermoregulation by brooding diamond pythons, *J. Zool. London*, 216:367–378.
Slip, D. J., and R. Shine, 1988c, Habitat use, movements and activity patterns of free-ranging diamond pythons, *Morelia s. spilota* (Serpentes: Boidae): A radiotelemetric study, *Aust. Wildl. Res.*, 15:515–531.
Slip, D. J., and R. Shine, 1988d, Feeding habits of the diamond python, *Morelia s. spilota*: Ambush predation by a boid snake, *J. Herpetol.*, 22:323–330.
Slip, D. J., and R. Shine, 1988e, Thermoregulation of free-ranging diamond pythons, *Morelia spilota* (Serpentes, Boidae), *Copeia*, 1988:984–995.
Stearns, S. C., 1992, *The Evolution of Life Histories*, New York, Oxford Univ. Press.
Stille, B. T., T. Madsen, and M. Niklasson, 1986, Multiple paternity in the adder,

Vipera berus, Oikos, 47:173–175.

Sullivan, B. K., 1989, Desert environments and the structure of anuran mating systems, *J. Arid Env.,* 17:175–183.

Sullivan, B. K., 1991, Parasites and sexual selection: separating causes and effects, *Herpetologica,* 47:250–264.

Sutherland, W. J., 1985a, Chance can produce a sex difference in variance in mating success and explain Bateman's data, *Anim. Behav.,* 33:1349–1352.

Sutherland, W. J., 1985b, Measures of sexual selection, *Oxford Surv. Evol. Biol.,* 2:90–101.

Sutherland, W. J., 1987, Random and deterministic components of variance in mating success, in J. W. Bradbury and M. B. Andersson, eds., *Sexual Selection: Testing the Alternatives,* Wiley, New York, pp. 209–219.

Thornhill, R., 1986, Relative parental contribution of the sexes to their offspring and the operation of sexual selection, in M. H. Nitecki and J. A. Kitchell, eds., *Evolution of Animal Behavior: Paleontological and Field Approaches,* Oxford Univ. Press, New York, pp. 113–136.

Thornhill, R., and J. Alcock, 1983, *The Evolution of Insect Mating Systems,* Harvard Univ. Press, Cambridge, Massachusetts.

Trivers, R. L., 1972, Parental investment and sexual selection, in B. Campbell, ed., *Sexual Selection and the Descent of Man,* 1871–1971, Aldine, Chicago, Illinois, pp. 136–179.

Vassart, G., M. Georges, R. Monsieur, H. Brocas, A. S. Lequarre, and D. Christophe, 1987, A sequence in M13 phage detects hypervariable minisatellites in human and animal DNA, *Science,* 235:683–684.

Viitanen, P., 1967, Hibernation and seasonal movements of the viper, *Vipera berus* (L.), in southern Finland, *Ann. Zool. Fenn.,* 4:472–546.

Volsøe, H., 1944, Structure and seasonal variation of the male reproductive organs of *Vipera berus* (L.), *Spolia Zool. Mus. Hauniensis,* 5:1–157.

Wade, M. J., 1979, Sexual selection and variance in reproductive success, *Amer. Nat.,* 114:742–746.

Wade, M. J., and S. J. Arnold, 1980, The intensity of sexual selection in relation to male behaviour, female choice, and sperm precedence, *Anim. Behav.,* 28:446–461.

Wells, K. D., 1977, The social behaviour of anuran amphibians, *Anim. Behav.,* 25:666–693.

Westneat, D. F., 1990, Genetic parentage in the indigo bunting: A study using DNA fingerprinting, *Behav. Ecol. Sociobiol.,* 27:67–76.

Wetton, J. H., R. E. Carter, D. T. Parkin, and D. Walters, 1987, Demographic study of a wild house sparrow population by DNA fingerprinting, *Nature,* 327:147–149.

Whittier, J. M., and D. Crews, 1986, Ovarian development in red-sided garter snakes, *Thamnophis sirtalis parietalis:* relationship to mating, *Gen. Comp. Endocrinol.,* 61:5–12.

Whittier, J. M., and R. R. Tokarz, 1992, Physiological regulation of sexual behavior in female reptiles, in C. Gans and D. Crews, eds., *Biology of the Reptilia, Physiology E,*

Chapter

6

Habitat Selection in Snakes

Howard K. Reinert

Introduction

The surface of the earth is a complex structural and climatic mosaic landscape. Environmental heterogeneity occurs at practically every scale or level of habitable space. It is general knowledge that most animals demonstrate a specificity for certain characteristic portions of this mosaic, and students of nature quickly learn to recognize "suitable" and "unsuitable" areas for the organisms that they encounter most frequently. The suitable areas, or those places where one would go to find the species of interest, define the habitat or "address" of the species within the broader landscape (Odum, 1959). We can refine this simple idea into a working definition if we recognize that natural landscapes are comprised of numerous environmental gradients that can affect the distribution and abundance of a particular species. The habitat factors (or variables) responsible for such gradients are usually scenopoetic or stage-setting physical or chemical factors such as (but clearly not limited to) altitude, canopy density, plant communities, soil moisture, and pH. Consequently, the occurrence of a particular species with respect to such factors can be used to delineate the habitat of that species. More concisely, the habitat of a species can be described as that portion of a multidimensional hyperspace (defined by any number of habitat factors) that is occupied by a given species (Whittaker et al., 1973).

Although similar in terms of their multidimensionality, the habitat and niche of a species should not be confused. The niche is an abstract space defined by the sum total of a species' requirements and interactions with other species in a community (Hutchinson, 1978). Niches are characteristics of species, whereas habitats are actual, physically discernible places. Whittaker et al. (1973, 1975) and Hutchinson (1978) discuss the distinction between habitat and niche in greater detail.

Occasionally the term *microhabitat* is used to refer to the specific location of an organism within its habitat or to factors that define the internal structure or patterns of habitat variation within a community (Whittaker, 1975; Ehrlich and Roughgarden, 1987). For example, the height and diameter of branches selected as perching sites by Rough Green Snakes (*Opheodrys aestivus*) could conveniently be considered microhabitat factors (Plummer, 1981a). Because such factors vary on an intracommunity scale, they may also be considered components of the niche (Kulesza, 1975; Whittaker et al., 1975).

Habitat Selection and Community Structure

The concept of habitat specificity of organisms has played a major role in the formulation of theories regarding the evolution of species, the maintenance of species diversity, and the organization of community structure (Miller, 1942; Thorpe, 1945; MacArthur, 1958, Klopfer and MacArthur, 1961; Southwood, 1977; Rosenzweig, 1981, 1987). Habitat differences are often linked to the principle of competitive exclusion, which concerns the inability of ecologically similar species to coexist stably (Gause, 1934). Considerable emphasis has been placed upon documenting differential habitat use by animals (MacArthur, 1958; Schoener, 1974; Cody, 1978; Dueser and Shugart, 1979), and partitioning of habitats has been theoretically considered and empirically found to be the most common form of separation among sympatric species (Schoener, 1974, 1977).

Data pertaining to interspecific niche partitioning by snakes has lagged behind that of other vertebrate groups, notably lizards and birds (Schoener, 1977; Toft, 1985). This is largely due to the secretive nature of snakes and the difficulty of observing snakes under natural conditions. The work that has been performed does not dispute the potential role of resource partitioning in structuring snake communities (Henderson et al., 1979; Brown and Parker, 1982; Reynolds and Scott, 1982; Vitt, 1987). However, studies of sympatric snake populations have produced contrasting results regarding the importance of interspecific habitat differences. The habitat separation of sympatric Northern Copperheads (*Agkistrodon contortrix mokasen*) and Timber Rattlesnakes (*Crotalus horridus*) was found to be far in excess of that theoretically required for stable species coexistence in a stochastic environment (Reinert, 1984a). Likewise, Shine (1977a) reported significant interspecific habitat differences among six species of Australian elapids, and Gregory and McIntosh (1980) speculated that differential habitat selection was more important than thermal considerations in separating the niches of three Garter Snake species (*Thamnophis* sp.) on Vancouver Island (cf. Fleharty, 1967; Gregory, 1984a). Several studies of snake assemblages have reported habitat differences among some species but not among others (Carpenter, 1952; Pough, 1966; Platt, 1969; Hebrard and Mushinsky, 1978; Reynolds and Scott, 1982). In studies where there appeared to be extensive habitat overlap among species, resource partitioning has frequently been attributed to alternative niche axes such as time of activity, temperature, or, most commonly, food (Mushinsky and Hebrard, 1977a, 1977b; Moore, 1978; Magnuson et al., 1979; Mushinsky et al., 1980; Brown and Parker, 1982).

There are several possible reasons for discrepancies concerning the role of habitat partitioning among sympatric snake species. The first is that competitive conflicts among snake species in different communities may be solved by different mechanisms (Shine, 1977a). Although theoretically the range of habitats occupied should shrink with increased competition while the range of food items should not (i.e., the compression hypothesis of MacArthur and Wilson, 1967), this solution may be less common in snakes given their evolutionary propensity for specializations in feeding mechanisms (Greene, 1983; Pough, 1983; Savitzky, 1983).

Second, differing levels of precision in assessing resource use may produce erroneous results. For example, the determination of the preferred diet is usually performed quite precisely through identification of stomach or scat contents, whereas the evaluation of preferred habitat is often based upon imprecise, broad categories subjectively defined on the basis of associated vegetation. This could result in artificially broad overlap on the habitat dimension.

Third, it is highly unlikely that niche axes such as food and habitat are independent. As predators, snakes must utilize habitats having prey densities suitable to supply their energy requirements. Consequently, habitat selection by snakes may reflect the habitat selected by their prey and result in food and habitat axes being highly correlated (Carpenter, 1952; Gregory, 1984a; Shine and Lambeck, 1985).

Fourth, competition for resources may be temporally limited to periods of extreme environmental conditions such as drought (Dunham, 1980). Such a situation would make the detection of competitive interactions or their impacts possible only through long-term studies.

Finally, the concept of resource partitioning along niche dimensions is based upon the assumption of present or past competition for limited resources. This often unfalsifiable hypothesis has received much warranted criticism (Wiens, 1977; Schoener, 1982); nevertheless, a large number of field experiments suggest that competition is a pervasive factor in both plant and animal relationships (Schoener, 1983). However, the occurrence of competition has not been clearly documented between any snake species (Reichenbach and Dalrymple, 1980). Indeed, several careful studies of snake communities have failed to find clear evidence of interspecific competition for food resources (Kephart, 1982; Reichenbach and Dalrymple, 1986). Consequently, the apparent partitioning of resources may be an artifact of other evolutionary factors (Bogert, 1949; Hart, 1979; Reinert, 1984a, 1984b).

The role of competition in snake communities remains fertile ground for research in light of improved methods of studying snake behavior, and the issue is far from solved (Vitt, 1987). Studies of competition that concentrate on a specific taxonomic group instead of trophic guilds (Jaksic et al., 1981) may incur the danger of missing important interactions. However, research on feeding guilds has suggested that this may not be of great concern in certain communities in which snakes have been reported to interact mainly among themselves (Valverde, 1967; Jaksic and Delibes, 1987).

Despite potential methodological or theoretical shortcomings, existing studies of snake assemblages have produced a wealth of empirical evidence that substantiates the propensity of snakes within communities to have complementary habitats. Future studies will benefit from the recognition that differences in preferred habitat may play an important role in determining intracommunity relationships among snake species. However, more precise methods of defining preferred habitat will be necessary to assess clearly the role of habitat in such relationships.

Habitat Selection versus Habitat Correlation

Recognition of a relationship between the spatial distribution of an organism and specific environmental factors constitutes the first step in the study of habitat. Such information is generally termed *habitat correlation* (Klopfer, 1969) and comprises the bulk of the literature pertaining to snake habitat. These descriptions range from general, qualitative assessments of dominant environmental features in the vicinity of chance field encounters (Wright and Wright, 1957) to specific, quantitative evaluations of microhabitat obtained from intensive research efforts (Plummer, 1981a; Reinert, 1984a). In all cases, however, the purpose of the description is to distinguish the portion of the larger landscape that each species utilizes.

Unfortunately, information on habitat correlation suffers from interpretive problems similar to those of statistical correlation in which significant relationships do not necessarily indicate cause and effect (Sokal and Rohlf, 1981). For example, Striped Crayfish Snakes (*Regina alleni*) are found in high densities in water hyacinth mats. Godley (1980) suggested that this habitat preference may be related to receding water levels in the surrounding marshes, high prey densities in the mats, decreased competition with birds, increased protection from avian predators, or thermal properties of the mats. The habitat relationship suggests several hypotheses, and the preference of *R. alleni* for water hyacinth mats may be related to any, all, or none of these factors.

In extreme cases, habitat correlation need not even reflect an active preference or selection on the part of the organism being studied. For example, the apparent preference of a green insect for foliage as opposed to tree trunks, branches or flowers could result simply from predators quickly consuming members of the population that venture onto contrasting backgrounds (Goodhart, 1958; Kettlewell, 1965).

In contrast to studies of habitat *correlation*, studies of habitat *selection* seek to determine the proximal environmental cues used by animals in the process of occupying specific habitats, and the underlying ultimate factors responsible for such habitat propensities. Although habitat correlations may contain such information, evaluation of the *how* and *why* of habitat selection requires thorough knowledge of the ecology of the species, intensive behavioral observation, and, in most cases, experimentation. As a result, studies of actual habitat selection are relatively rare and often speculative. This is particularly true for snakes. For the purpose of clarity, only field or laboratory studies that seek to elucidate the causes or mechanisms responsible for observed patterns of habitat use should be termed *studies of habitat selection*. Field studies that define the spatial pattern of habitat occupancy (habitat correlation) for a species should be termed *studies of habitat utilization*.

The concept of habitat selection implies that an animal has a conscious perception of its surroundings (the *Umwelt* of von Uexkull, 1921) and seeks to limit its activity to specific subunits of the available environment (Wiens, 1976). One way of determining if snakes actively select habitats or simply distribute themselves at random within the environment is to compare the habitat in the vicinity of observed snakes with the available habitat within the prescribed study area. If the two sets of observations differ, the hypothesis that the animals are randomly distributing themselves among available habitat types is rejected.

Not surprisingly, studies of several snake species suggest that individual snakes do actively select preferred portions of their environment (Reinert, 1984b; Weatherhead and Charland, 1985; Burger and Zappalorti, 1988; Weatherhead and Prior, 1992). That is, their use of available habitat is nonrandom, and this nonrandom distribution among habitat types does not seem to be the immediate result of differential survival in adjacent habitats (e.g., removal by predators).

Intraspecific Variation in Habitat Preference

As a result of the influence of niche theory, there has been a strong emphasis placed upon interspecific comparisons in ecological studies. Consequently, it is not surprising that habitats are frequently defined as species-specific attributes. However, this implies that the mean habitat type is that most commonly used by typical specimens, and that the associated variance reflects the variation exhibited by a typical representative of the species (Reinert, 1984b). Such assumptions concerning habitat use patterns are frequently unrealistic and may result in a misinterpretation of the information obtained. Large habitat variances are frequently a result of differential habitat use by specific subunits of the population (i.e., different color morphs, sexes, or ages), which may actually demonstrate small variances about their respective means (Selander, 1966; Schoener, 1968; Preston, 1980; Reinert, 1984b). Consequently, studies designed to examine intraspecific differences in habitat use have the potential to produce valuable insight into the process of habitat selection and its biological significance (Ehrlich and Holm, 1962; Grant and Price, 1981).

Although intraspecific habitat variation in lizards has received a great deal of attention (Heatwole, 1977; Schoener, 1977; Huey et al., 1983), only a few recent studies have specifically addressed this topic in snakes (Reinert, 1984b; Shine,1986; Reinert and Zappalorti, 1988a; Burger and Zappalorti, 1989). However, many other studies have provided pertinent information on habitat variation, often within the context of broader ecological investigations. These works indicate that many snake species exhibit significant intrapopulation variation in habitat use. More importantly, observations of intraspecific variation have often produced insight into factors responsible for observed habitat utilization patterns.

Geographic variation

Geographic variation in habitat use is probably the most broadly documented form of intraspecific habitat variation in snakes. Different populations often appear, at least to human observers, to exhibit widely divergent habitat propensities. This is evident from the literature descriptions of the habitats of species having wide geographic ranges and many recognized subspecies, such as *Thamnophis sirtalis* (Wright and Wright, 1957) and *Crotalus viridis* (Klauber, 1972). It is usually assumed that such variation is the result of local adaptation to existing habitat conditions through natural selection. This suggests that any possible genetic component of habitat selection may be subject to varying degrees of plasticity. Such statements must be made cautiously, however, because our bias is to view habitats from a human perspective. To a snake, the habitats may be similar when perceived on the basis of important structural cues or through the use of sensory modalities different from our own (Ford and Burghardt, Chap. 4, this volume).

Sweet (1985) gives an excellent example of geographic variation in habitat use by Western Rattlesnakes (*C. viridis*) in California (cf. Fitch, 1949). He also examines the potential evolutionary implications of such variation for both the Western Rattlesnake and sympatric Gopher Snake (*Pituophis melanoleucus*). In coastal and montane portions of their broadly overlapping ranges, these species differ significantly in their habitats, with *P. melanoleucus* being found more frequently in grasslands and *C. viridis* in chaparral and woodland. However, in inland areas (e.g., Carrizo Plain) where chaparral and woodland are absent, both snakes utilize grassland habitats. In this habitat the species show a convergence in pattern that is often attributed to a selection for mimicry of *C. viridis* by *P. melanoleucus*. Sweet's investigation, however, emphasizes the role of habitat and makes a stronger case for a convergent selection for a cryptic pattern due to similar habitat propensities in inland areas (Sweet, 1985).

Seasonal variation

Seasonal variation in habitat use has been most commonly documented for temperate-zone snake species whose hibernating sites are located in habitats that differ considerably from the habitats used during the active season. Sidewinders (*Crotalus cerastes*) principally occupy areas of soft, shifting sand during the active season but overwinter in habitats having more solid substrates of rock and gravel. Burrows in such areas are less ephemeral, offer greater protection from potential predators, and possibly maintain higher winter temperatures than burrows in soft sand (Secor, 1992). The summer foraging habitat of the European Adder (*Vipera berus*) is often in the vicinity of boggy areas and low-lying meadows, whereas hibernating sites are located in higher, drier habitats (Viitanen, 1967; Prestt, 1971). Interestingly, the Massasauga (*Sistrurus catenatus*) has been observed to make a directly opposite habitat shift by hibernating in wetlands and summering in slightly higher elevations in fields and prairies where small mammal prey are more abundant (Reinert and Kodrich, 1982; Seigel, 1986). Viitanen (1967) felt that the absence of ground water at the hibernacula of *V. berus* was important to avoid possible drowning, whereas *S. catenatus* has been observed to hibernate at or under the level of the water table (Maple, 1968; Reinert, 1978).

Not only may the overwintering habitat have different features from the active season habitat, but it may also be distantly located, necessitating lengthy migrations in both the spring and fall (Gregory, 1982, 1984b). In addition, the accurate selection of suitable overwintering sites is more crucial for survival than the selection of foraging areas, because the latter may simply be changed if foraging success is low. The cost in that instance is a loss of time and energy that can possibly be reimbursed at the next site. The cost of selecting an inadequate hibernaculum is death. Most likely, the recognition of suitable hibernating habitat requires the evaluation of substantially different cues from those used for assessing the quality of active season habitats.

For snakes that aggregate at communal hibernacula, imprinting, scent-trailing, and habitat-conditioning (Dundee and Miller, 1968; Burghardt, 1983) may be important factors in the location of such sites (Graves et al., 1986; Reinert and Zappalorti, 1988b). Recent telemetric observations of Timber Rattlesnakes translocated into distant populations indicated that they were capable of locating local, communal hibernacula, apparently by trailing resident snakes (H. Reinert and R. Rupert, personal observation). Similarly, displaced *Vipera b. berus* were assumed to have trailed resident snakes to local communal aggregation sites (Viitanen, 1967). However, learning and/or communication are not adequate to explain how solitary hibernators locate suitable sites or how communal sites initially become established by a founder individual. That individual snakes apparently have the ability to assess the suitability of new hibernacula is suggested by the observation that artificial refugia were rapidly used by several species of native snakes in New Jersey (Zappalorti and Reinert, in press).

Shine and Lambeck (1985) observed extreme wet and dry season habitat shifts of telemetrically monitored Arafura File Snakes (*Acrochordus arafurae*) in Australia. Although aquatic during both seasons, these snakes responded rapidly to rising water levels during the wet season and utilized inundated grassland that did not exist during the dry season. They also reported daily shifts to shallower water in the grasslands at night and attributed both seasonal and daily alteration of habitat use to prey (fish) movements (Shine and Lambeck, 1985). Seasonal variation in the utilization of arboreal and aquatic habitats by Green Water Snakes (*Nerodia cyclopion*) was considered to be a mechanism for the maintenance of preferred body temperatures (Mushinsky et al., 1980).

Nesting sites may also differ significantly from surrounding habitat (Sexton and Claypool, 1978; Madsen, 1984; Burger and Zappalorti,

1986). The selection of oviposition sites that are safe from predators and provide optimal climatic conditions for embryonic development is necessary for reproductive success. However, the conditions of a nest site by the end of the incubation period may be difficult to predict at the time of egg-laying. Consequently, nest site selection must be based upon incomplete information (Orians and Wittenberger, 1991). It appears that some snakes may rely on past experience or site imprinting for this decision. For example, Northern Pine Snakes (*Pituophis m. melanoleucus*) frequently return in successive years to the same nest site (Burger and Zappalorti, 1991; R. T. Zappalorti, personal communication). Although supporting information is presently lacking, it is possible that these individuals may be returning to the site of their own hatching. Conspecific cues (e.g., pheromones, freshly laid eggs, old egg shells, shed skins of neonates) may also function in site selection, and communal nesting has been frequently reported for several snake species (Cook, 1964; Brodie et al., 1969; Covacevich and Limpus, 1972; Parker and Brown, 1972; Swain and Smith, 1978; Plummer, 1981b; Burger and Zappalorti, 1986, 1991).

An interesting and highly unusual case of seasonal habitat use was reported by Viitanen (1967) for the European Adder (*Vipera b. berus*) in Finland. He recognized a distinct mating area where both males and females aggregated over several consecutive years for the purpose of courtship and copulation. The site was described as a rather sparsely vegetated hilltop 50 to 100 m from hibernacula. The habitat in the mating area differed markedly from that used for spring basking and summer foraging. Reproductive females moved to the site shortly after emergence from hibernation. Males, on the other hand, initially moved to more distant basking sites after emergence and later made rapid, directed movements to the mating area (a distance of 150–200 m). Viitanen likened the area to the "display ground of the gallinaceous birds" (Viitanen, 1967).

Ontogenetic and morphological variation

Very few observations have been made concerning ontogenetic variation in the use of habitat by snakes. However, it is quite likely that it is a very common phenomenon that could have a considerable influence upon the density and spatial distribution of snake populations as well as the structure of snake communities. The paucity of information on this topic is undoubtedly a consequence of the tremendous difficulty of finding and observing young snakes under natural conditions and points to a major gap in our understanding of snake habitat selection. The difficulty in finding and observing the young of many snake species may not simply be a consequence of their small size, because several small species of snakes are rather abundant and easily found in certain habitats (e.g., *Storeria dekayi* and *Diadophis punctatus*).

The fact that many snakes show distinctive changes in color and pattern with age suggests concurrent changes in habitat propensities (Beatson, 1976). For example, it is quite likely that the red-colored young of Emerald Tree Boas (*Corallus caninus*) and Green Tree Pythons (*Chondropython viridis*) use different microhabitats from the bright green adults. Similar hypotheses could be postulated for species whose young are strongly blotched but whose adults are uniform or striped (e.g., *Coluber* and *Elaphe*).

Adult Timber Rattlesnakes (*C. horridus*) exhibit extreme color and pattern polymorphism that ranges from a pale yellow ground color with prominent crossbands to a nearly patternless, uniform black (for illustrations refer to Conant and Collins, 1991). However, all young *C. horridus* have a prominent pattern on a light ground color. Dark and light adult *C. horridus* in the same population were found to utilize different habitats with light specimens preferring sites dominated by leaf litter and dark specimens preferring the proximity of large fallen logs (Reinert, 1984b). Whether or not these differences are the result of active selection for background color matching or inherent habitat preferences remains to be examined. The habitat propensity of young *C. horridus* is currently unknown, but knowledge of their preference would provide useful information. If active background selection occurs, it could be hypothesized that the habitat preference of the young specimens should be more similar to that of light adults. If this were found to be correct, dark specimens would undergo an ontogenetic alteration in habitat preference.

Likewise, ontogenetic changes in mass and length can alter the ability of a snake to exploit certain habitats. This is particularly obvious for arboreal species (Lillywhite and Henderson, Chap. 1, this volume), but may also influence the habitat use of terrestrial species. For example, Clark (1970) found a positive relationship between the snout-to-vent length of Western Worm Snakes (*Carphophis vermis*) and the size of rocks used as shelter. He also reported consistent preferences of small specimens for soils with higher moisture content. The former relationship may have been associated with the size of the foraging areas, whereas the latter was suggested as a response to the potentially higher desiccation rates of smaller snakes (Elick and Sealander, 1972).

Changes in body size may also result in dietary changes that could serve as an impetus for habitat divergence. Sexton (1956–1957) reported that juvenile Fer-de-Lance (*Bothrops atrox*) in Venezuela used streamside habitats and foraged for small frogs while adults shifted to upland habitats to feed upon lizards and small mammals.

Although some of the small specimens observed in the study may actually have been *B. venezuelensis* instead of juvenile *B. atrox,* it was thought that the habitat shift was still a reality for the latter species (Sexton and Heatwole, 1965).

Factors Influencing Habitat Use

Sex and reproductive condition

Reproductive condition often has an obvious effect on habitat use, and gravid individuals of several species have been observed to frequent habitat that is divergent from that of males and nongravid females. Keenlyne (1972) reported that gravid Timber Rattlesnakes (*C. horridus*) in Wisconsin remained at large rocks near hibernacula during the summer while males and nongravid females dispersed. Duvall et al. (1985) observed similar behavior in Prairie Rattlesnakes (*Crotalus v. viridis*). Significant habitat divergence was subsequently documented for gravid *C. horridus* (as well as *Agkistrodon contortrix*) in Pennsylvania (Reinert, 1984b) and New Jersey (Reinert and Zappalorti, 1988a). The effect of reproductive condition was strongly demonstrated by the clear habitat shift of individual specimens between reproductive and nonreproductive years (see Fig. 2 in Reinert, 1984b). The shift from forested habitats to open habitats when gravid was thought to be a selection for sites with warmer or less variable thermal conditions to assist in embryonic development. Similarly, gravid elapid snakes of several species were encountered closer to water than nongravid individuals (Shine, 1979), and it was suggested that elapids required increased water intake during gestation (based on the results of Shine, 1977b).

Habitat divergence between male and nongravid female snakes does not appear as pronounced (Reinert, 1984b), but only limited data are currently available on this topic. Shine (1986) found that female Arafura File Snakes (*Acrochordus arafurae*) used deep-water habitats (>1 m) while males usually limited their activity to shallow water (<1 m). Males of this species are shorter with disproportionately smaller heads than females, and the differences in habitat were correlated to the habitat preferences of suitably sized piscine prey. Likewise, Fitch and Shirer (1971) determined that male Racers (*Coluber constrictor*) used arboreal habitats more frequently than females, possibly because males were shorter and lighter.

Foraging and digestive state

Obviously the selection of appropriate foraging habitat is necessary for maintaining adequate energy intake for survival. Because all snakes are predatory, the location and distribution of their prey has undoubtedly played an important role in the evolution of habitat propensities and selection in snakes. As mentioned previously, Shine and Lambeck (1985) stressed the role of prey location in the seasonal and daily habitat use patterns of Arafura File Snakes (*Acrochordus arafurae*), and Andrén (1982) associated habitat use with foraging success in the European Adder (*Vipera berus*).

Because of their well-developed chemosensory perception (Ford and Burghardt, Chap. 4, this volume), snakes probably assess the distribution and abundance of prey directly from proximal chemical cues (animal odors) in the environment and select foraging sites based upon this information (Madison, 1978). In an interesting field experiment, Duvall et al. (1990b) demonstrated that Western Rattlesnakes (*C. viridis*) terminated their outward spring migration from hibernacula upon locating chemical cues from their primary prey (Deer Mice). The snakes stopped their traveling and remained in areas (i.e., they effectively selected habitat) when they contacted cages containing either mice or cage bedding soiled by mice. Strong site preference responses to prey odors were also observed for *C. viridis* under controlled laboratory situations (Duvall et al., 1990a). Similarly, observations suggest that Timber Rattlesnakes (*C. horridus*) may actively select ambush sites on the basis of existing chemical cues left on the upper surface of logs used as runways by small mammals (Reinert et al., 1984).

A thermophilic response following feeding is commonly observed in snakes under laboratory conditions (Regal, 1966; Lysenko and Gillis, 1980; Slip and Shine, 1988; Gibson et al., 1989; Lutterschmidt and Reinert, 1990; Smucny and Gibson, in review; Peterson, Gibson, and Dorcas, Chap. 7, this volume). Such a response probably plays a role in habitat selection under natural conditions as well, although it has not been clearly documented.

Ecdysis

Under laboratory conditions, Common Garter Snakes (*Thamnophis sirtalis*) exhibit a strong thermophilic response during the five days immediately prior to molting (Gibson et al., 1989; Smucny and Gibson, in review; Peterson, Gibson, and Dorcas, Chap. 7, this volume). Such observations suggest that the onset of ecdysis could stimulate snakes to select habitats that are divergent from typical foraging habitats. This should be particularly obvious in temperate-zone woodland species whose foraging habitats may be generally cool. However, such habitat shifts are not well-documented for free-ranging snakes. This may be a result of inadequate sampling or it may indicate that, under natural conditions, the cost of such behavior (i.e., exposure to predators, energy loss in traveling to suitable sites) may outweigh the benefits (Huey and Slatkin, 1976).

Long-term telemetric observations of Timber Rattlesnakes (*Crotalus horridus*) indicated that specimens that were beginning the moulting cycle frequently ceased foraging in forested habitat and moved to open habitats (H. Reinert, unpublished results). These habitats provided both shelter (in the form of rocks or logs) and warmer environmental conditions than the forested sites (Reinert, 1984b). Shelters may provide protection from predators during the vulnerable "opaque" period when vision is impaired or help reduce cutaneous water loss (Lillywhite and Maderson, 1982). Higher temperatures may be important for the physiological processes involved in epidermal cell growth (Gibson et al., 1989). In *C. horridus,* the selection for thermal factors was assumed to be of paramount importance, because adequate shelters were also numerous in the forested habitat (H. Reinert, personal observation).

Disease and injury

The potential role of disease and injury in altering the habitat preference of snakes remains largely unexamined. However, several species of iguanid lizards inoculated with pathogenic bacteria demonstrated a preference for elevated temperatures (Bernheim and Kluger, 1976; Firth et al., 1980). The maintenance of a behavioral fever may function to depress bacterial growth and could also necessitate a selection of divergent habitat from that normally occupied.

Although very few species have been similarly studied, such a response may not be as widespread among reptiles as originally suspected (Laburn et al., 1981; Zurovsky et al., 1987a). Specifically, Olive Grass Snakes (*Psammophis phillipsii*) and Brown House Snakes (*Lamprophis fuliginosus*) both failed to demonstrate any clear thermophilic response following inoculation with bacterial pyrogens (Zurovsky et al., 1987b).

Kitchell (1969) reported that injured and heavily parasitized snakes retreated to cooler portions of a thermal gradient, but he presented no quantitative data indicating lowered temperature preferences for such specimens or hypotheses about the potential value of this behavior. However, if such thermophobic responses are present under natural conditions, they could also necessitate an alteration in preferred habitat.

Social relationships

Although snakes are often considered solitary and asocial (Brattstrom, 1974), a growing body of experimentation and observation suggests that this is probably inaccurate (reviewed in Gillingham, 1987). As already mentioned above (see the section

"Seasonal variation," p. 207), overwintering aggregations, aggregations of gravid viviparous snakes, and communal nesting have been reported for numerous snakes (reviewed in Gregory, 1984b; Gillingham, 1987; Gregory et al., 1987). Social interaction adds yet another, largely unexplored dimension to habitat selection in snakes.

In some regions, habitats that allow snakes to meet specific physiological requirements (e.g., during winter or while gravid) may be relatively scarce. Thus, the aggregation phenomenon results from the cumulative selection of a relatively rare but important habitat. Individual snakes may locate such sites fortuitously and aggregations could result from the independent selection of the same site by several individuals. However, aggregations often occur when suitable sites are apparently numerous (Dundee and Miller, 1968; Plummer, 1981b).

An increasing number of observations and experiments demonstrate the ability of snakes to actively follow conspecific scent trails (Ford and Burghardt, Chap. 4, this volume). Such behavior may not only be important in locating mates (Lillywhite, 1985) or selecting suitable hibernacula, but it may also function in the location of suitable habitat for other requirements such as food, shelter, or warmth. It is quite possible that juvenile trailing of adults functions as a habitat learning experience.

Learning

Tradition and site fidelity may also play an important but poorly understood role in habitat selection. Many snake species appear to develop well-defined activity ranges that may remain stable over many years and possibly over an individual snake's lifetime (Stickel et al., 1980; Reinert, personal observation). Individuals may continue to use the same sites within their range despite obvious structural changes in the habitat. Burger and Zappalorti (1991) reported the nesting of a Pine Snake (*Pituophis melanoleucus*) in the same location as previous years despite complete alteration of the area by construction equipment. Such site fidelity suggests a facet of habitat selection that is less plastic than the recognition of structural, climatic, or biotic aspects of the environment. However, the importance and frequency of such behavior cannot currently be assessed.

To the contrary, it is clear that some snakes are capable of exploiting new habitats quite rapidly. In 1986, 25 artificial, underground refugia were constructed at a study site in southern New Jersey in an attempt to improve the habitat for endangered and threatened snake species. During the first year following construction, snakes were observed at 5 of the refugia. By the end of 1989, 139 snakes of nine different species had been observed basking, shedding, foraging, mating, ovipositing, or hibernating at 17 of the 25 structures (Zappalorti and Reinert, in press). In May of 1985 a large load of grass clippings was illegally dumped at a study site in southern New Jersey. Within one month three different Timber Rattlesnakes (*C. horridus*) were observed at the pile of grass, and by September a total of five rattlesnakes (four gravid females and one male) had been observed at this novel environmental structure. Two of the gravid females remained in or around the grass pile for nearly three months until parturition (H. Reinert, personal observation). Its selection as a rookery and parturition site was probably associated with suitable climatic conditions generated by the decaying vegetation. It was later discovered that the pile also contained several clutches of Fence Lizard eggs (*Sceloporus undulatus*) and the shed skin of an Eastern Kingsnake (*Lampropeltis g. getula*). Madsen (1984) made similar observations concerning the use of manure piles by ovipositing Grass Snakes (*Natrix natrix*).

Relevant Cues in Habitat Selection

Temperature

Early studies of reptilian thermoregulation did much to dispel the idea that the body temperature of reptiles is solely a reflection of ambient environmental conditions (Cowles and Bogert, 1944; Bogert, 1949; Pearson, 1954). Active snake species in environments with high ambient temperatures (e.g., desert-dwelling *Masticophis* and *Coluber*) are heliothermic and maintain fairly stable body temperatures (Hammerson, 1979; Parker and Brown, 1980). Narrow body temperature ranges are maintained through a variety of behavioral mechanisms including differential habitat selection. Indeed, the role of thermoregulation for the maintenance of optimal physiological function may be inextricably linked to habitat selection in snakes (Huey, 1991; Peterson, Gibson, and Dorcas, Chap. 7, this volume). Because digestive efficiency, speed of locomotion, foraging efficiency, and reproductive success are all related to body temperature, thermal conditions may be one of the most important proximate factors in the habitat selection process of many snakes.

Although habitat use mediated through physiological requirements often appears to involve selection of habitat based mainly upon its thermal properties, several studies have demonstrated that some reptiles expend little effort in precise temperature regulation (Ruibal, 1961; Rand and Humphrey, 1968; Hertz, 1974; Lee, 1980). Some terrestrial snake species in cool, temperate-zone habitats (*Thamnophis* and *Crotalus*), some semiaquatic snakes (*Nerodia* and *Acrochordus*), and arboreal Rough Green Snakes (*Opheodrys aestivus*) appear to exhibit variable body temperatures and/or thermal passivity (Gibson and Falls, 1979; Mushinsky et al., 1980; Shine and Lambeck, 1985; Plummer, in press; H. Reinert, personal observation). Similarly, many fossorial snakes may live much of the time in suboptimal thermal environments (Elick et al., 1979). It is clear that the thermal biology of any particular species or individual specimen is a result of complex interactions among biophysical, biotic, "economic," and phylogenetic concerns (Huey, 1982; Peterson, Gibson, and Dorcas, Chap. 7, this volume), and may range from thermoconformity to precise thermoregulation (Huey and Slatkin, 1976). Consequently, the importance of thermoregulatory behavior and thermal preference in habitat selection should not simply be assumed without careful evaluation. Habitats that have optimal thermal regimes and suitable shelters are useless without adequate food supplies. This situation may be a major factor in the evolution of eurythermy and low temperature tolerance in some snake species.

Structural features

Habitat structure refers to the physical arrangement of objects in space (McCoy and Bell, 1991). Structural features of the environment appear to play a major role in habitat selection in many vertebrates, including reptiles (Sexton et al., 1964; Heatwole, 1977). For example, the influence of vegetation structure on site selection has been experimentally demonstrated for hatchling turtles (Sexton, 1958; Meseth and Sexton, 1963) and lizards (Heatwole, 1966; Kiester et al., 1975). Numerous studies have defined snake habitats both qualitatively and quantitatively in terms of their physical structure (Hebrard and Mushinsky, 1978; Plummer, 1981a; Reinert and Zappalorti, 1988a), and it is clear that structural environmental features can often be used successfully to describe snake habitat (Chandler and Tolson, 1990) and to distinguish the habitats of different species (Reinert, 1984a).

If visual or tactile assessment of structural habitat cues is important, then the key features involved in habitat selection should be relatively stable features (Heatwole, 1977). Such features may possibly include trees and other perennial vegetation, substrate surface features (rocks, logs, leaf litter, standing water), or, for fossorial species, soil structure. The preferred habitat of Timber Rattlesnakes (*C. horridus*) can be distinguished from that of Northern Copperheads (*Agkistrodon contortrix mokasen*) on the basis of structural components of the canopy and forest floor in Pennsylvania (Reinert, 1984a). However, the question remains as to whether or not snakes actually use such structural features as cues during their selection of habitat.

Identification of relevant structural features

If structural cues are important in the process of habitat selection by snakes, and if the recognition of such cues has an innate, species-specific component, then it should be possible to recognize stable structural attributes of the habitat selected by the same species of snake in different geographic regions. Studies of the same species of snake in different geographic areas provide an opportunity to investigate not only the potential plasticity of habitat selection, but also the factors that may function as relevant cues in the habitat selection process. For example, Timber Rattlesnakes (*C. horridus*) occur over a wide geographic range in which populations often occupy distinctly different habitats (Klauber, 1972), and the habitat use of this species has been studied in both the deciduous hardwood forests of the Appalachian Mountains of Pennsylvania (Reinert, 1984a,b) and in the coastal plain pine forests (Pine Barrens) of New Jersey (Reinert and Zappalorti, 1988a). Using statistical techniques of classification and ordination, it is possible to compare simultaneously the habitat use of both populations with respect to a common set of structural features. The following is an example of the application of such methods to examine whether or not consistent, stable structural features exist that could play a common role in the habitat selection process of *C. horridus* in both mountain and coastal plain environments.

Habitat comparisons indicated that the mountain and coastal plain study sites of *C. horridus* differed significantly in their structure. A discriminant function analysis (Pimentel, 1979) of ten, shared, structural variables indicated significant differences between the habitats and that the major factors distinguishing these two habitats were the composition of the forest floor (log cover, vegetation, and leaf litter) and the canopy structure. Generally, the deciduous forest in the mountainous region had a denser canopy, more fallen logs, and less surface woody vegetation than the pine-dominated forest of the coastal plain. Although the statistical confirmation is reassuring, these differences were evident even from casual observation (Fig. 6.1).

Intrapopulation habitat relationships among Timber Rattlesnakes within each geographic area were found to be remarkably similar at both locations. Figure 6.2 shows the relationship of males, nongravid females, and gravid females along the major habitat gradient derived by a discriminant function analysis of the same ten structural variables at each site. Nearly identical intrapopulation habitat relationships exist in the two different habitats and the separating habitat gradient is similar. In both locations the preference of males and nongravid females for heavily forested habitats contrasts with the preference of gravid females for open sites. This habitat difference has been

A

B

Figure 6.1 Typical habitat of the Timber Rattlesnake (*C. horridus*) in (*A*) mountainous regions of Pennsylvania and (*B*) the coastal plain of New Jersey, illustrating the difference in the canopy and forest floor structure.

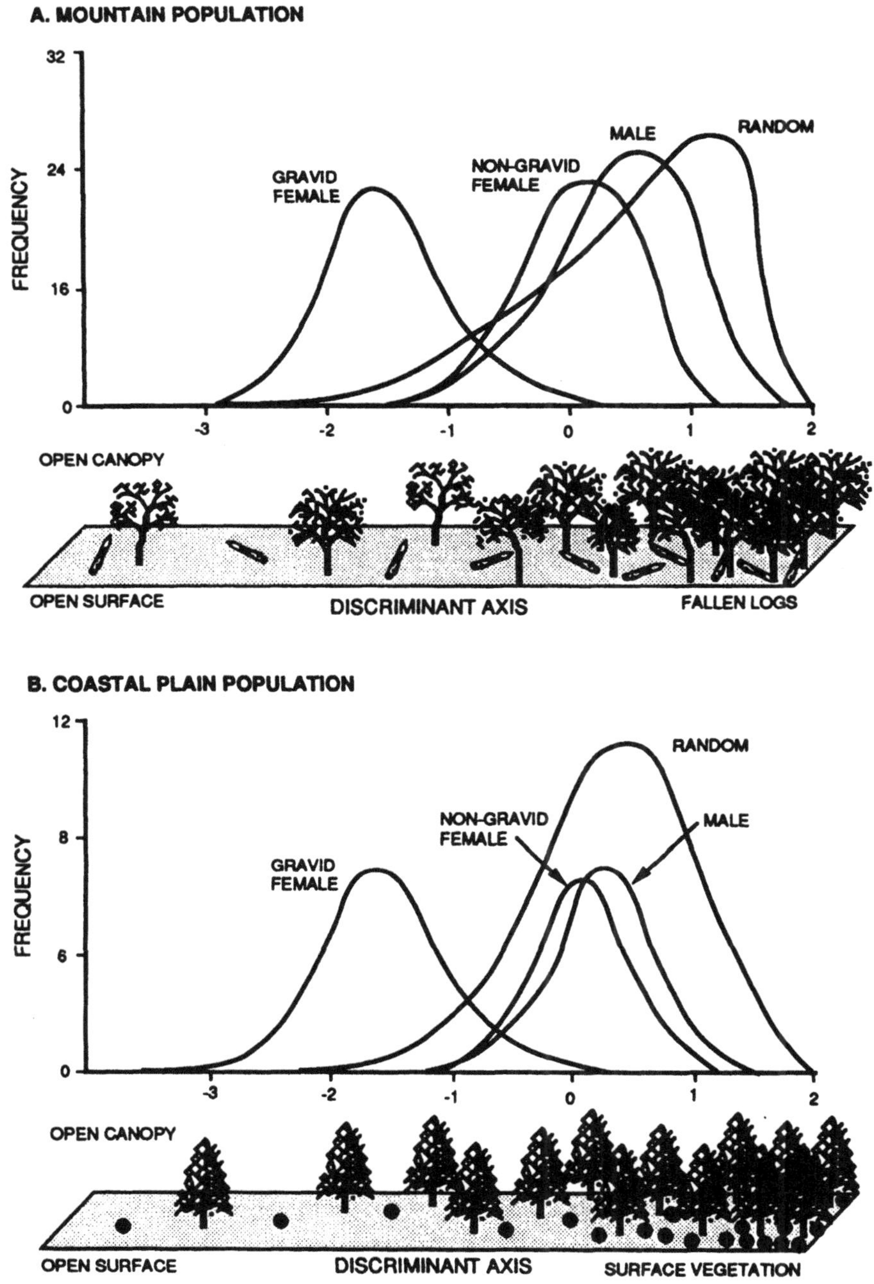

Figure 6.2 A comparison of intrapopulation habitat relationships of Timber Rattlesnakes (*C. horridus*) in (*A*) upland deciduous forests of Pennsylvania and (*B*) coastal plain pine forests of New Jersey. Illustrated are the frequency curves of discriminant scores and pictorial interpretations of the first discriminant axes based upon the same ten structural variables sampled in both populations.

suggested to result from the selection of warmer or less thermally variable habitats by gravid females for optimal embryonic development (Reinert, 1984b; Reinert and Zappalorti, 1988a). However, such a relationship might also result from the selection of optimal foraging sites by males and nongravid females. Overall, the similarity of intrapopulation habitat relationships in such different environments suggests some underlying stability in habitat preference by Timber Rattlesnakes. At the very least, it serves to demonstrate the pervasive effect that physiological condition has upon habitat selection.

This investigation can be taken a step further in an attempt to determine if Timber Rattlesnakes in both areas selected sites with similar structural attributes despite the general divergence between the two habitats. However, this is a more difficult and complex relationship to assess than the overall intrapopulation relationships in the previous analysis. For example, when the structural habitat data for male rattlesnakes from the coastal plain and mountain populations were subjected to a discriminant function analysis, there was significant separation of the two populations on the basis of several variables related to forest floor structure. From this analysis it would be correctly concluded that the habitats utilized by male rattlesnakes in the two areas were structurally different. This finding is not surprising in light of our knowledge that the general habitat structure of the two sites differed in several obvious and statistically significant ways. In fact, several of the variables that differed between mountain snake locations and coastal plain snake locations also differed between random samples from both study sites (percentage of surface vegetation cover, percentage of leaf litter, and fallen log diameters). This analysis merely demonstrated that structural features of the forest floor were major factors in distinguishing between mountain and coastal plain male snake locations. In short, coastal plain snakes occupied habitats with more vegetation cover, less leaf litter, and smaller logs. However, the coastal plain study area had generally more surface vegetation, less leaf litter, and smaller logs than the mountain site (Fig. 6.1). Consequently, if snakes were selecting habitat on the basis of other features without regard to these three variables, such a result might be expected simply by chance as long as the selected feature was not strongly correlated with these variables. This type of analysis does not result in the recognition of any variables that may actually be important to the snakes in their process of habitat selection.

Determining the potentially important structural cues involved in habitat selection requires a somewhat different approach than simply comparing samples. Important factors in habitat selection are more likely to be those that remain similar between sites selected by

Timber Rattlesnakes within both habitats despite the fact that they may differ in general between the two geographic locations.

Assessment of such factors in the current example began by determining the primary environmental gradients in each habitat through two separate principal components analyses (Pimentel, 1979) of the same ten variables collected at 150 random sites within both study areas. At the mountain study area the first component accounted for 32.7% of the variance in the data set and defined a gradient from sites with dense canopy to open sites. The second component (12.9% of the total variance) described a gradient from sites with numerous fallen logs to sites without logs. Once these gradients were defined, the position of each rattlesnake location was projected onto these general habitat gradients by generating their principal component scores using the factor score matrix from the principal component analysis of the random samples (Reinert, 1992). This method placed the sites selected by males from the mountain population into the available mountain habitat. More interestingly, it was used to "place" the sites selected by coastal plain males into the mountain habitat. When this was performed, it was discovered that coastal plain Timber Rattlesnakes statistically placed into the mountain habitat did not differ from the resident mountain Timber Rattlesnakes with respect to the two habitat gradients of canopy structure and fallen log cover on the forest floor (Fig. 6.3).

When the reciprocal exercise was performed, i.e., males from the mountain population were "placed" into the coastal plain habitat, different but enlightening results were obtained. First, a principal component analysis of 150 random coastal plain samples produced a structurally different set of components. This resulted from the previously mentioned structural distinctions between the two habitats. The first principal component (22.0% of the variance) was not a canopy structure component, but a forest floor structure gradient based upon ground surface woody vegetation cover. In other words, the surface vegetation structure was the most variable aspect of the pine barrens study site, whereas canopy structure was the most variable aspect of the mountain site. However, the second component (16.0% of the variance) was a canopy component similar to the first component generated from the mountain random sample, while the third component (14.7% of the variance) was a fallen log component similar to the second component at the mountain study area. Not surprisingly, mountain Timber Rattlesnakes "placed" onto the first, surface vegetation gradient differed significantly from native coastal plain Timber Rattlesnakes (Fig. 6.4). However, the sites selected by mountain males did not differ from coastal plain males on the second and third components (Fig. 6.5). This analysis again emphasizes the

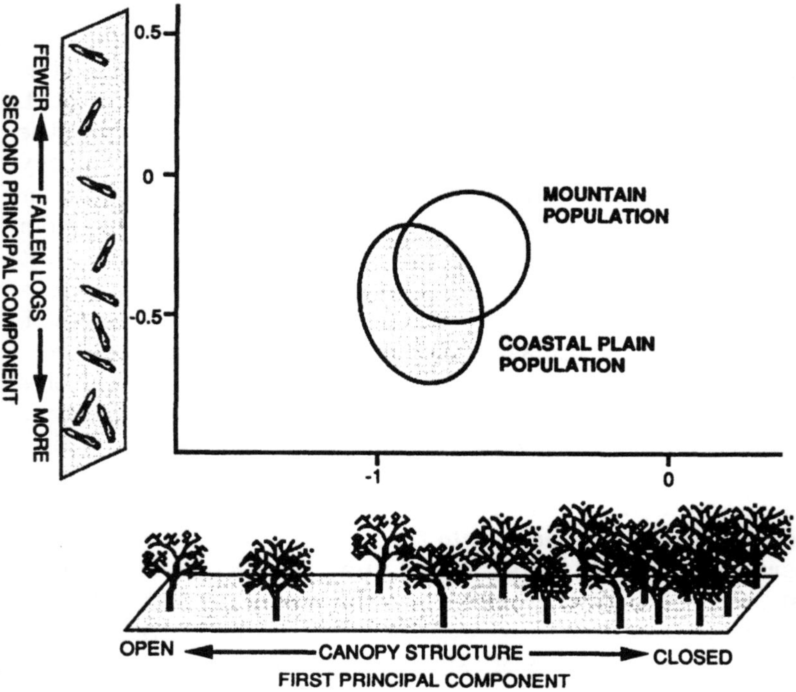

Figure 6.3 The estimated habitat position of coastal plain Timber Rattlesnakes (*C. horridus*) relative to the actual habitat position of native Timber Rattlesnakes in the mountain study site. Illustrated are 95% confidence ellipses of the mean principal component scores on the first and second principal components derived from an analysis of 150 random sampling sites and ten variables. Overlap of ellipses indicates that the mean habitat positions do not differ significantly ($P < 0.05$) on these component axes. Compare with Fig. 6.5 and refer to text for further discussion.

stability of canopy structure and fallen log density with respect to habitat use by *C. horridus*.

Despite the rather large differences between the canopy structure of the two forests, there was no difference in the percentage of canopy closure at sites selected by male Timber Rattlesnakes in both areas. Likewise, in both forests the snakes demonstrated a similar affinity for fallen logs despite the broad disparity of fallen log distribution between the two habitats. In contrast, there were significant differences between populations with regard to surface vegetation cover, suggesting that this structural feature may be unimportant in the evaluation of preferred habitat by these snakes.

The results of these analyses strongly implicate canopy structure and fallen logs as important cues in the habitat selection process of male Timber Rattlesnakes. The discovery of stable habitat features further suggests the possibility of an innate (i.e., genetic) component

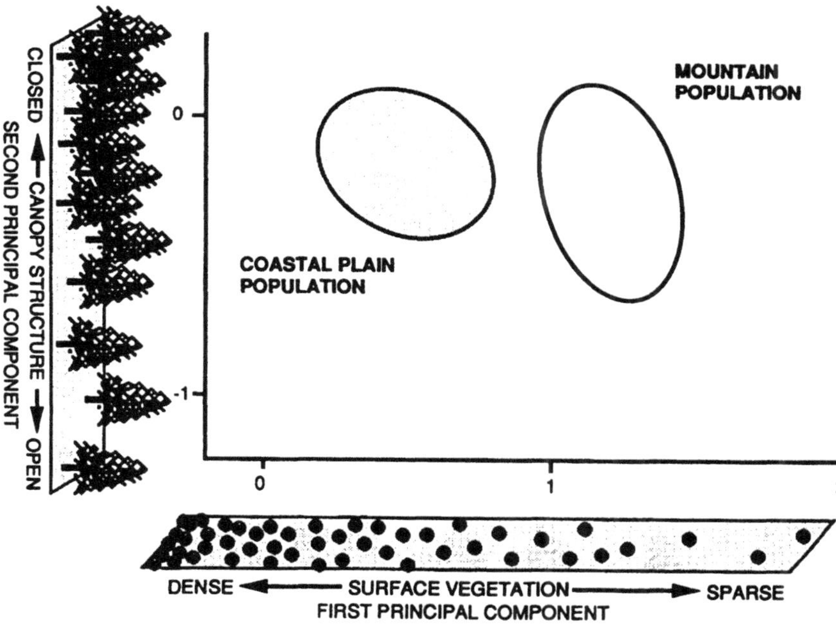

Figure 6.4 The estimated habitat position of mountain Timber Rattlesnakes (*C. horridus*) relative to the actual habitat position of native Timber Rattlesnakes in the coastal plain study site. Illustrated are 95% confidence ellipses of the mean principal component scores on the first and second principal components derived from an analysis of 150 random sampling sites and ten variables. Nonoverlap of ellipses indicates that the mean habitat positions differ significantly ($P < 0.05$) on these component axes. See text for further discussion.

to habitat selection in this snake species. Although to a human observer the mountain and coastal plain habitats appear to be greatly divergent both botanically and structurally, to a Timber Rattlesnake the two areas may offer nearly identical habitats because of the availability of similar key structural features. Analyses such as these reduce the complexity of habitat relationships to several, potentially critical features underlying the habitat selection process, and open the door to field and laboratory experimentation based upon a manageable number of habitat factors.

Model of Habitat Selection in Snakes

It is clear that the process by which an individual snake selects habitat may be influenced by a large number of complex biotic and abiotic factors. At this time, any generalities concerning the process of habitat selection in snakes must be based largely on conjecture. The model illustrated in Fig. 6.6 is proposed in an effort to stimulate discussion and further investigation. The basic attributes of this model

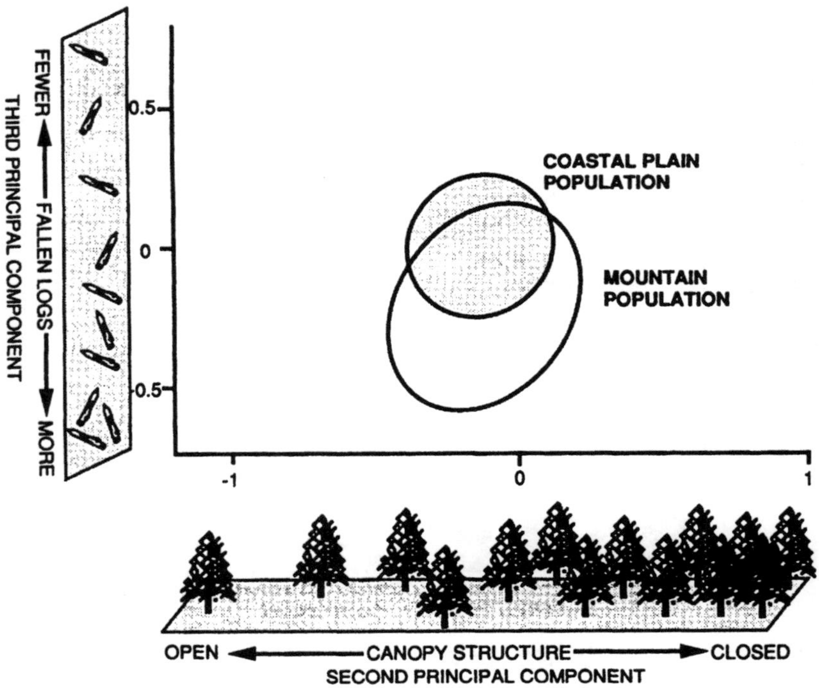

Figure 6.5 The estimated habitat position of mountain Timber Rattlesnakes (*Crotalus horridus*) relative to the actual habitat position of native Timber Rattlesnakes in the coastal plain study site. Illustrated are 95% confidence ellipses of the mean principal component scores on the second and third principal components derived from an analysis of 150 random sampling sites and ten variables. Overlap of ellipses indicates that the mean habitat positions do not differ significantly ($P<0.05$) on these component axes. Compare with Fig. 6.3 and refer to text for further discussion.

have been borrowed heavily from studies of habitat selection in birds and general models of optimal foraging (Charnov, 1976; Cody, 1985; Orians and Wittenberger, 1991).

This model recognizes two components of habitat selection: macrohabitat and microhabitat. It suggests that there is a hierarchical response of snakes to these two scales of spatial heterogeneity. As a result of the potential influence of habitat selection on the physiological performance of ectotherms (Huey, 1991) and the documented influence of physiological condition upon snake habitat use, physiological factors are seen as the major initiators in the habitat selection process. Under this model, physiological requirements prompt snakes to select among divergent habitat types. Such choices are potentially based upon easily perceived, stable environmental features, i.e., habitat structure. It should be understood that at this macrohabitat level such selection and the potential range in variation of the habitats utilized may be more

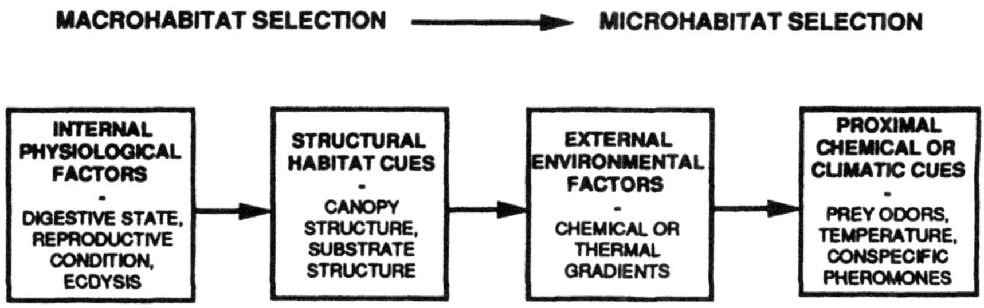

Figure 6.6 A proposed model of the habitat selection process in snakes.

strongly subject to genetic programming and imprinting than the term *choice* implies, and basic habitat propensities may be shaped by natural selection through competition, predation, and coevolution.

Once a snake is in the preferred macrohabitat, site selection proceeds through the evaluation of relevant environmental gradients. Final microhabitat site selection is based upon proximal chemical or climatic cues necessary to satisfy immediate physiological requirements. Selection of optimum habitats results in improved performance and individual fitness (Dunham et al., 1989; Huey, 1991).

Based upon the proposed hierarchical model of habitat selection and our present knowledge of behavior and habitat use patterns, we can conceive of simplified, stepwise procedures for habitat selection by snakes. Using the Timber Rattlesnake (*C. horridus*) as an example, the following scenarios may be predicted.

In the first example, a hungry snake must select suitable foraging habitat on the basis of its potential to contain prey. Suitable foraging sites are those that contain small, woodland mammals and adequate ambush stations where the snake is appropriately camouflaged (Reinert et al., 1984). Consequently, the snake is attracted to heavily forested areas by the perception of canopy closure (or light intensity) and/or forest floor structure (leaf litter, logs, etc.). In most cases prey are patchily distributed even in suitable habitat as a result of environmental heterogeneity. Therefore, once the snake is in suitable forested habitat, microhabitat selection proceeds through the evaluation of chemical gradients of prey odors. When an adequate foraging patch (i.e., one containing an appropriate quantity or quality of prey odors) is encountered, ambush site selection proceeds through the identification of specific, active, small mammal runways.

In the second example, the snake is gravid. Because it does not feed or forage in this condition, there is no necessity to select habitats containing potential prey (Keenlyne, 1972; Reinert et al., 1984). However, there is a preference for higher or less variable temperatures during gestation (Reinert, 1984b; Reinert and Zappalorti, 1988a). The snake is attracted to open areas by the perception of breaks in canopy structure (or increased light intensity). Once in open habitats, suitable microhabitat is evaluated on the basis of thermal gradients. Ultimate site selection is based upon optimal conditions for both basking and retreat sites (Huey et al., 1989), and possibly upon the presence of conspecific pheromones.

The idea that habitat selection may proceed in a stepwise fashion has also been proposed for birds (Burger and Gochfeld, 1982; Klopfer and Ganzhorn, 1985; Orians and Wittenberger, 1991). A process of sequential habitat evaluation allows for the occurrence of variation in habitat use frequently observed among and within individuals. Such variation is somewhat incompatible with the concepts of *Gestalt* recognition and the occurrence of specific releasers that have previously been suggested to explain the process of habitat selection (Klopfer and Ganzhorn, 1985). Experimental field and laboratory research focusing on this process holds great promise for contributing important information regarding habitat selection by snakes.

Habitat and Conservation

The protection and maintenance of suitable habitat are of paramount importance for the long-term survival of wild snake species (Dodd, 1987; Dodd, Chap. 9, this volume). To be effective, conservation and management programs must be based upon a thorough knowledge of the behavior and ecology of the species of concern. As already indicated, however, our knowledge of the habitat requirements of even the most common and numerous species is severely limited. Studies of endangered and threatened species are additionally constrained by factors such as small populations, limited distributions, and restrictive legislation. However, basic field studies should be the first step in any conservation program. Such studies need to be designed to produce unbiased descriptions of preferred habitat as well as seasonal and intrapopulation variation in habitat use patterns.

Quantitative habitat descriptions can be extremely useful in evaluating habitat suitability and designing habitat improvement programs. Once the required habitat of a species or population segment has been defined, it can be compared with existing habitat at that site or at other sites. For example, analysis of the habitat used by Copperheads and Timber Rattlesnakes (Reinert, 1984b) produced results that have potential application in habitat conservation and management. By comparing the habitat in which snakes were found with the mean available habitat determined from a large sample of random sites in the study area, it was possible to assess the suitability of the existing habitat. Discriminant function analysis allowed for the positioning of group centroids (or means) in discriminant space relative to a centroid (or mean) representing available resources. This was used as a measure of habitat (or niche, depending upon the nature of the variables) position (Shugart and Patten, 1972; Dueser and Shugart, 1979). The habitat position is a measure of the habitat used relative to the available habitat. The closer a group is to the mean of available resources, the less difficulty that group would experience in finding its preferred habitat. The results indicated that gravid Timber Rattlesnakes and gravid Copperheads both used a narrow range of habitat that was relatively scarce (far from random centroid), whereas the other population subunits, especially male Timber Rattlesnakes, utilized a broader range of habitats that more closely approximated the available conditions (see Table 6 in Reinert, 1984b). From a habitat management standpoint, these findings suggest that the careful creation of forest openings may increase the availability of the reproductive habitat used by both species. However, clear-cutting of large tracts could severely reduce the foraging habitat of rattlesnakes.

Martinka (1972) showed that discriminant function analysis could be used with a high degree of success to distinguish the territories of Blue Grouse (*Dendragapus obscurus pallidus*) from unused habitats. A similar approach could be used to evaluate the habitat suitability of pine barrens sites for Timber Rattlesnakes (*C. horridus*) in New Jersey (H. Reinert, personal observation). An area could be evaluated for its suitability as rattlesnake habitat by randomly sampling for a set of quantitative structural features identical to the ones used in prior studies of the species (i.e., Reinert and Zappalorti, 1988a). Discriminant scores could then be generated from the data obtained (Norusis, 1985), and those scores could be projected onto the known habitat axes of rattlesnakes to assess the habitat structure at the site in question. A similar method of habitat evaluation could be performed using the results of principal component analysis and principal component score coefficients (Reinert, 1992).

Summary and Future Research

For most snake species our knowledge of preferred habitat use and habitat variation is minimal and largely based upon qualitative, anecdotal observations. Over a quarter of a century ago, Elton (1966) lamented that the "definition of habitats, or lack of it, is one of the chief blind spots in zoology." This statement still applies to studies of snakes for which concise information is equally lacking for tropical, temperate, aquatic, terrestrial, arboreal, rare, and abundant species. Basic, detailed natural history studies (*sensu* Greene, 1986) that produce unbiased descriptions of preferred habitat and examine seasonal and intrapopulation variation in habitat use are sorely needed. Unfortunately, such studies require intensive field work to achieve adequate sample sizes over meaningful time periods, but there is no substitute for having an intimate knowledge of the organism that is being studied (Evans, 1976). Improved sampling methods (e.g., radiotelemetry) have begun to make such investigations feasible, but no less time-consuming.

A necessary requirement for any study of habitat utilization or habitat selection is to assess snake distributions in the environment accurately without observational bias in sampling. Chance observation and even systematic search and capture efforts in different habitats will often fail to produce unbiased information of the type required because snakes may simply appear more numerous in habitats where they are less well camouflaged and/or where they exhibit different behavior patterns.

This sampling problem exists for most snakes, even large, relatively numerous species. For example, Timber Rattlesnakes (*Crotalus horridus*) occasionally occupy rocky, open habitats where their bold pattern of crossbands makes them quite obvious against the solid rock background. In such habitats they often lie exposed in the sunlight and are quick to rattle and retreat, behaviors that easily attract the attention of human observers. However, these snakes spend the greatest proportion of their time in heavily forested habitat (Reinert, 1984a) where their color and pattern make them nearly indistinguishable from the forest floor (Fig. 6.7). In addition, in forested habitats they combine this cryptic coloration with a practically nonresponsive behavior (even upon close approach). Consequently, even systematic search efforts in forested habitats would not reveal the Timber Rattlesnake's habitat proclivities.

Over the past 20 years, radiotelemetry has offered an unparalleled opportunity to remove such observational biases. It has certainly become the method of choice for studying habitat use in larger snakes. Burger and Zappalorti (1988) demonstrated the value of radiotelemetry in adequately assessing the habitat use of Pine Snakes (*Pituophis melanoleucus*) by comparing their telemetric observations with data obtained through the intensive search efforts of experienced field observers. The latter method significantly underestimated this species' use of undisturbed forest habitats where monitored snakes were often found buried beneath leaves or hidden in logs.

However, radiotelemetry is not without its own difficulties and biases (summarized in Reinert, 1992). Probably the greatest of these is size bias because commercial transmitters are still too large for use in snakes less than 30 to 50 g in body mass. This limits their use to larger species or adult specimens. Radioactive tagging and drift fence trapping in different habitats are the current alternatives for study-

A

B

Figure 6.7 Timber Rattlesnakes (*C. horridus*) in (*A*) rocky, open habitat and (*B*) forested habitat, illustrating the potential for observational bias in determining habitat preference based upon visual search and capture sampling regimes.

ing habitat preference in small snakes (Barbour et al., 1969; Semlitsch et al., 1981). Technological advancements in transmitters and passive transponder systems (Van Dyke, 1974; Masconzoni and Wallin, 1986) may eventually aid in eliminating the current bias toward large snakes.

As outlined in an earlier section ("Structural features," p. 216) many animals, including reptiles, appear to respond to the geometric configuration of environmental features more strongly than to the biological composition of the habitat components. For the purpose of meaningful comparisons, predictions, and hypothesis testing, it is imperative that future descriptions of habitat include quantification of structural environmental features that may influence habitat selection. Some standardization of the structural habitat features evaluated would facilitate habitat comparisons and broader generalizations, particularly among investigators working on the same species in different geographic areas.

Unfortunately, most past studies have relied heavily upon habitat descriptions that seem to assume that snakes are good plant taxonomists. For example, habitat use by Eastern Massasaugas (*Sistrurus c. catenatus*) has been examined in Illinois (Wright, 1941), Ohio (Maple, 1968), and Pennsylvania (Reinert and Kodrich, 1982). In all three investigations, habitats were defined botanically on the basis of the dominant vegetation. In Illinois, Massasaugas were found most frequently in areas dominated by Dogwood (*Cornus*), Hawthorn (*Cratageus*), and Bluegrass (*Poa*). In Ohio, the preferred areas had Cinquefoil (*Potentilla*), Timothy (*Phleum*), and Sedge (*Carex*). In Pennsylvania, preferred habitat at one site was characterized by Goldenrod (*Solidago*), Aster (*Aster*), and Dogwood (*Cornus*), whereas at a second site the preferred habitat was composed of Poverty Grass (*Danthonia*), Goldenrod (*Solidago*), and Cinquefoil (*Potentilla*). Such descriptions can be useful for identifying Massasauga habitat at the local level, but it is exceedingly difficult to directly compare the preferred habitats of these Illinois, Ohio, and Pennsylvania rattlesnakes on the basis of this information.

Reinert and Kodrich (1982) suggested that there were obvious structural similarities between the preferred habitats of snakes at their two Pennsylvania sites. They also believed that the preferred habitats in Illinois and Ohio probably shared many structural similarities with Pennsylvania habitats. Unfortunately, these speculations cannot be substantiated or even investigated using the available data. The ability to test these hypotheses would lead to a better understanding of habitat selection by the Massasauga. If similar, quantitative structural variables had been measured in these three studies, a direct comparison of the preferred habitat of Massasaugas could be performed. Such comparisons of structure could provide interesting insight into the factors that may be responsible for habitat preferences and the process of habitat selection. Additionally, it could also contribute practical information that may prove useful in the conservation of this dwindling species (Ashton, 1976; Seigel, 1986).

The data presented previously for Timber Rattlesnakes provide another example of the value of collecting quantifiable structural data versus delineating vegetational zones. In Pennsylvania, male Timber Rattlesnakes preferred sites with 68.8% canopy closure (Reinert, 1984b) in a mountain forest dominated by Chestnut Oak (*Quercus prinus*), Red Oak (*Q. rubra*), Black Birch (*Betula lenta*), and Red Maple (*Acer rubrum*). Male Timber Rattlesnakes in New Jersey preferred sites with 68.4% canopy closure (Reinert and Zappalorti, 1988a) in a coastal plain forest that was dominated by Pitch Pine (*Pinus rigida*), Shortleaf Pine (*Pinus echinata*), Blackjack Oak (*Q. marilandica*), Scrub Oak (*Q. ilicifolia*), and Black Oak (*Q. velutina*). The two forests are not even remotely comparable on the basis of the dominant tree species because they share none. However, the quantitative measure of canopy structure in the two habitats is comparable and, clearly, very similar at snake-selected sites.

Investigators must also recognize the inherently multivariate nature of habitat and the value of multivariate statistical methods in analyzing habitat data (Carey, 1981). Collecting a multivariate data set and then analyzing it with univariate methods (Burger et al., 1988) ignores the very real occurrence and importance of factor correlation and interaction. Although it has been nearly 20 years since Reagan (1974) used multivariate analyses in his study of Three-Toed Box Turtle (*Terrapene carolina triunguis*) habitat, the application of these statistical methods in ecological studies of snakes has remained rather limited (Hart, 1979; Reinert, 1984a,b; Reinert and Zappalorti, 1988a; Chandler and Tolson, 1990). An increasing number of mainframe and microcomputer statistical packages (e.g., SYSTAT, SPSS, SAS, BMDP) that perform multivariate procedures such as principal component analysis and discriminant function analysis are making these methods easily accessible to all researchers. Examples of the use of multivariate statistics in the study of snake habitat have been given elsewhere (Reinert, 1992; and this chapter). General methodological descriptions are available in numerous statistical texts (Pimentel, 1979; Norusis, 1985; Kachigan, 1986; Johnson and Wichern, 1988). James and McCulloch (1990) provide a useful discussion of the basic assumptions and interpretations of multivariate analysis.

Studies investigating the process of habitat selection in snakes are almost totally lacking. As more information is gained about individual habitat preference and dominant sensory modalities (Ford and Burghardt, Chap. 4, this volume), the door will be opened to insightful experimental approaches (Duvall et al., 1990b). Experimental field and laboratory studies designed to examine the comparative role of innate preference, learning, social, and psychological factors in snake habitat selection (Heller and Halpern, 1982; Chiszar et al., 1987; Gillingham et al., 1990) would provide invaluable information. Many of the experimental designs used with other organisms (Sexton, 1958; Sluckin and Salzen, 1961; Meseth and Sexton, 1963; Wecker, 1963; Klopfer, 1965; Klopfer and Hailman, 1965) could be modified or directly applied to snakes to further our understanding of this most basic element of snake ecology.

Finally, one of the highest priority endeavors in habitat studies must be the accurate and unbiased description of the required habitat of the many rapidly declining snake species. Such information is urgently needed in order that meaningful conservation programs may be designed and initiated. The burgeoning human population with its associated consumption of natural resources and resultant environmental degradation is causing rapid destruction of wildlife habitat and the extinction of an ever-increasing number of wildlife species (Dodd, 1987; Dodd, Chap. 9, this volume). Studies of the *how* and *why* of the habitat selection process in snakes are clearly in their infancy and the future holds great promise for fascinating discoveries. However, the answers to these questions can never be obtained and will be of little value without the existence of wild snakes in natural habitats.

In summary, most snakes exhibit nonrandom use of the available landscape, and this usage forms the basis for the habitat correlations so familiar to field biologists. Specific habitat propensities can play a major role in the organization of communities and may be shaped by natural selection through competition, predation, and coevolution. Such active selection may be a consequence of innate or acquired responses, and the selection process may involve the evaluation of visual, chemical, or tactile stimuli. Numerous factors such as reproductive condition, nutritional state, seasonal requirements, and social interactions can strongly influence habitat use in snakes.

Despite the obvious importance of habitat selection in snakes, it has received limited scientific attention. Descriptions of the preferred habitat of most snake species have generally been derived from casual observations and often represent qualitative assessments of visually dominant physical or vegetational features. Accurate and concise habitat descriptions are currently available for very few species. This lack of information is directly attributable to the difficulty of consistently finding and observing snakes in the field.

The study of habitat use and selection by snakes is entering an exciting phase. The increasing availability of unbiased sampling methods (i.e., radiotelemetry, radioactive tagging) is providing a unparalleled opportunity to examine the daily lives and behavioral patterns of these secretive animals under natural conditions. In addition, computer-assisted analysis of multivariate data sets now allows for the evaluation of complex quantitative habitat information. Application of these techniques to studies of snake habitat are sorely needed in both temperate and tropical regions and for both common and rare species alike. Quantitative habitat descriptions are encouraged as a means of making meaningful inter- and intraspecific comparisons, to allow for evaluation of possible key features in the habitat selection process, and for use in conservation management. Experimental approaches, both in the field and laboratory, aimed at understanding the habitat selection process are practically nonexistent and offer many challenging opportunities for researchers.

T. R. E. Southwood's presidential address to the British Ecological Society in 1977 stressed the general paucity of knowledge concerning habitat use by animals, and his closing words are especially fitting at the terminus of this chapter: "I must detain you no longer, there is much to be done" (Southwood, 1977).

Literature Cited

Andrén, C., 1982, Effect of prey density on reproduction, foraging, and other activities of the adder, *Vipera berus*, *Amphibia-Reptilia,* 3:81–96.

Ashton, R. E., Jr., 1976, Endangered and threatened species of amphibians and reptiles in the United States, *SSAR Herpetol. Circ.*, 5:1–65.

Barbour, R. W., M. J. Harvey, and J. W. Hardin, 1969, Home range, movements, and activity of the eastern worm snake, *Carphophis amoenus amoenus, Ecology,* 50:470–476.

Beatson, R. R., 1976, Environmental and genetical correlates of disruptive coloration in the water snake, *Natrix s. sipedon, Evolution,* 30:241–252.

Bernheim, H. A., and M. J. Kluger, 1976, Fever and antipyresis in the lizard *Dipsosaurus dorsalis, Amer. J. Physiol.*, 231:198–203.

Bogert, C. M., 1949, Thermoregulation in reptiles, a factor in evolution, *Evolution,* 3:195–211.

Brattstrom, B. H., 1974, The evolution of reptilian social behavior, *Amer. Zool.*, 14:35–49.

Brodie, E. D., R. A. Nussbaum, and R. M. Storm, 1969, An egg-laying aggregation of five species of Oregon reptiles, *Herpetologica,* 25:223–227.

Brown, W. S., and W. S. Parker, 1982, Niche dimensions and resource partitioning in a Great Basin Desert snake community, in N. J. Scott, Jr., ed., *Herpetological Communities,* U. S. Fish Wildl. Serv. Wildl. Res. Rep., pp. 59–81.

Burger, J., and M. Gochfeld, 1982, Host selection as an adaptation to host-dependent foraging success in the cattle egret (*Bubulcus ibis*), *Behaviour,* 79:212–229.

Burger, J., and R. T. Zappalorti, 1986, Nest site selection by pine snakes, *Pituophis melanoleucus,* in the New Jersey pine barrens, *Copeia,* 1986:116–121.

Burger, J., and R. T. Zappalorti, 1988, Habitat use in free-ranging pine snakes, *Pituophis melanoleucus,* in New Jersey pine barrens, *Herpetologica,* 44:48–55.

Burger, J., and R. T. Zappalorti, 1989, Habitat use by pine snakes (*Pituophis melanoleucus*) in the New Jersey pine barrens: Individual and sexual variation, *J. Herpetol.*, 23:68–73.

Burger, J., and R. T. Zappalorti, 1991, Nesting behavior of pine snakes (*Pituophis m. melanoleucus*) in the New Jersey pine barrens, *J. Herpetol.*, 25:152–160.

Burger, J., R. T. Zappalorti, M. Gochfeld, W. I. Boarman, M. Caffrey, V. Doig, S. D.

Garber, B. Lauro, M. Mikovsky, C. Safina, and J. Saliva, 1988, Hibernacula and summer den sites of pine snakes (*Pituophis melanoleucus*) in the New Jersey pine barrens, *J. Herpetol.*, 22:425–433.
Burghardt, G. M., 1983, Aggregation and species discrimination in newborn snakes, *Z. Tierpsychol.*, 61:89–101.
Carey, A. B., 1981, Multivariate analysis of niche, habitat, and ecotope, in D. E. Capen, ed., *The Use of Multivariate Statistics in Studies of Wildlife Habitat,* USDA For. Serv. Gen. Tech. Rep., RM-87, pp. 104–113.
Carpenter, C. W., 1952, Comparative ecology of the common garter snake (*Thamnophis s. sirtalis*), the ribbon snake (*Thamnophis s. sauritus*) and Butler's garter snake (*Thamnophis butleri*) in mixed populations, *Ecol. Monogr.*, 22:235–258.
Chandler, C. R., and P. J. Tolson, 1990, Habitat use by a boid snake, *Epicrates monensis,* and its anoline prey, *Anolis cristatellus, J. Herpetol.*, 24:151–157.
Charnov, E. L., 1976, Optimal foraging: The marginal value theorem, *Theor. Popul. Biol.*, 9:129–136.
Chiszar, D., C. W. Radcliffe, T. Boyer, and J. L. Behler, 1987, Cover seeking behavior in red spitting cobras (*Naja mossambica pallida*): Effects of tactile cues and darkness, *Zoo Biol.*, 6:161–167.
Clark, D. R., 1970, Ecological study of the worm snake *Carphophis vermis* (Kennicott), *Univ. Kansas Publ. Mus. Nat. Hist.*, 19:85–194.
Cody, M. L., 1978, Habitat selection and interspecific territoriality among the sylviid warblers of England and Sweden, *Ecol. Monogr.*, 48:351–396.
Cody, M. L., 1985, *Habitat Selection in Birds,* Academic, New York.
Conant, R., and J. T. Collins, 1991, *A Field Guide to Reptiles and Amphibians of Eastern and Central North America,* 3rd ed., Houghton Mifflin, Boston, Massachusetts.
Cook, F. R., 1964, Communal egg laying in the smooth green snake, *Herpetologica,* 20:206.
Covacevich, J., and C. Limpus, 1972, Observations on community egg-laying by the yellow-faced whip snake, *Demansia psammophis* (Schlegel) 1837 (Squamata: Elapidae), *Herpetologica,* 28:208–210.
Cowles, R. B., and C. M. Bogert, 1944, A preliminary study of the thermal requirements of desert reptiles, *Bull. Amer. Mus. Nat. Hist.*, 82:265–296.
Dodd, C. K., Jr., 1987, Status, conservation, and management, in R. A. Seigel, J. T. Collins, and S. S. Novak, eds., *Snakes: Ecology and Evolutionary Biology,* McGraw-Hill, New York, pp. 478–513.
Dueser, R. D., and H. H. Shugart, Jr., 1979, Niche pattern in a forest-floor small mammal fauna, *Ecology,* 60:108–118.
Dundee, H. A., and M. C. Miller, III, 1968, Aggregative behavior and habitat conditioning by the prairie ringneck snake, *Diadophis punctatus arnyi, Tulane Stud. Zool. Bot.*, 15:41–58.
Dunham, A., 1980, An experimental study of interspecific competition between the iguanid lizards *Sceloporus merriami* and *Urosaurus ornatus, Ecol. Monogr.*, 50:309–330.
Dunham, A. E., B. W. Grant, and K. L. Overall, 1989, Interfaces between biophysical and physiological ecology and the population biology of terrestrial vertebrate ectotherms, *Physiol. Zool.*, 62:335–355.
Duvall, D., M. B. King, and K. J. Gutzwiller, 1985, Behavioral ecology and ethology of the prairie rattlesnake, *Nat. Geogr. Res.*, 1:80–111.
Duvall, D., D. Chiszar, W. K. Hayes, J. K. Leonhardt, and M. J. Goode, 1990a, Chemical and behavioral ecology of foraging in prairie rattlesnakes (*Crotalus viridis viridis*), *J. Chem. Ecol.*, 16:87–101.
Duvall, D., M. J. Goode, W. K. Hayes, J. K. Leonhardt, and D. G. Brown, 1990b, Prairie rattlesnake vernal migrations: Field experimental analysis and survival value, *Nat. Geogr. Res.*, 6:457–469.
Ehrlich, P. R., and R. W. Holm, 1962, Patterns and populations, *Science,* 137:652–657.
Ehrlich, P. R., and J. Roughgarden, 1987, *The Science of Ecology,* Macmillan, New York.
Elick, G. E., and J. A. Sealander, 1972, Comparative water loss in relation to habitat selection in small colubrid snakes, *Amer. Midl. Nat.*, 88:429–439.

Elick, G. E., J. A. Sealander, and R. J. Beumer, 1979, Temperature preferenda, body temperature tolerances and habitat selection of small colubrid snakes, *Trans. Mo. Acad. Sci.*, 13:21–32.
Elton, C., 1966, *The Pattern of Animal Communities*, Methuen, London.
Evans, G. C., 1976, A sack of uncut diamonds: The study of ecosystems and the future resources of mankind, *J. Appl. Ecol.*, 13:1–39.
Firth, B. T., C. L. Ralph, and T. J. Boardman, 1980, Independent effects of the pineal and a bacterial pyrogen in behavioural thermoregulation in lizards, *Nature*, 285:399–400.
Fitch, H. S., 1949, Study of snake populations in central California, *Amer. Midl. Nat.*, 41:513–579.
Fitch, H. S., and H. W. Shirer, 1971, A radiotelemetric study of spatial relationships in some common snakes, *Copeia*, 1971:118–128.
Fleharty, E. D., 1967, Comparative ecology of *Thamnophis elegans, T. cyrtopsis,* and *T. rufipunctatus* in New Mexico, *Southwest. Nat.*, 12:207–230.
Gause, G. F., 1934, *The Struggle for Existence,* Dover Publ., New York (1971 reprint).
Gibson, A. R., and J. B. Falls, 1979, Thermal biology of the common garter snake, *Thamnophis sirtalis* (L.), *Oecologica (Berlin)*, 43:79–97.
Gibson, A. R., D. A. Smucny, and J. Kollar, 1989, The effects of feeding and ecdysis on temperature selection by young garter snakes in a simple thermal mosaic, *Can. J. Zool.*, 67:19–23.
Gillingham, J. C., 1987, Social behavior, in R. A. Seigel, J. T. Collins, and S. S. Novak, eds., *Snakes: Ecology and Evolutionary Biology,* McGraw-Hill, New York, pp. 184–209.
Gillingham, J. C., J. Rowe, and M. A. Weins, 1990, Chemosensory orientation and earthworm location by foraging eastern garter snakes, *Thamnophis s. sirtalis,* in D. McDonald, D. Müller-Schwarze, and S. Natynczuk, eds., *Chemical Signals in Vertebrates,* Vol. 5, Oxford Univ. Press, Oxford, pp. 522–532.
Godley, S. J., 1980, Foraging ecology of the striped swamp snake, *Regina alleni,* in southern Florida, *Ecol. Monogr.*, 50:411–436.
Goodhart, C. B., 1958, Thrush predation on the snail *Cepaea hortensis, J. Anim. Ecol.*, 27:47–57.
Grant, P. R., and T. D. Price, 1981, Population variation in continuously varying traits as an ecological genetics problem, *Amer. Zool.*, 21:795–811.
Graves, B. M., D. Duvall, M. B. King, S. L. Lindstedt, and W. A. Gern, 1986, Initial den location by neonatal prairie rattlesnakes: Functions, causes, and natural history in chemical ecology, in D. Duvall, D. Müller-Schwarze, and R. M. Silverstein, eds., *Chemical Signals in Vertebrates,* Vol. 4, Plenum, New York, pp. 285–304.
Greene, H. W., 1983, Dietary correlates of the origin and radiation of snakes, *Amer. Zool.*, 23:431–441.
Greene, H. W., 1986, Natural history and evolutionary biology, in M. E. Feder and G. V. Lauder, eds., *Predator-Prey Relationships,* Univ. Chicago Press, Chicago, Illinois, pp. 99–108.
Gregory, P. T., 1982, Reptilian hibernation, in C. Gans and F. H. Pough, eds., *Biology of the Reptilia,* Vol. 13, Academic, New York, pp. 53–154.
Gregory, P. T., 1984a, Habitat, diet, and composition of assemblages of garter snakes (*Thamnophis*) at eight sites on Vancouver Island, *Can. J. Zool.*, 62:2013–2022.
Gregory, P. T., 1984b, Communal denning in snakes, in R. A. Seigel, L. E. Hunt, J. L. Knight, L. Malaret, and N. L. Zuschlag, eds., *Vertebrate Ecology and Systematics: A Tribute to Henry S. Fitch,* Univ. Kans. Mus. Nat. Hist. Spec. Publ. 10, pp. 57–75.
Gregory, P. T., and A. G. D. McIntosh, 1980, Thermal niche overlap in garter snakes (*Thamnophis*) on Vancouver Island, *Can. J. Zool.*, 58:351–355.
Gregory, P. T., J. M. Macartney, and K. W. Larsen, 1987, Spatial patterns and movements, in R. A. Seigel, J. T. Collins, and S. S. Novak, eds., *Snakes: Ecology and Evolutionary Biology,* McGraw-Hill, New York, pp. 366–395.
Hammerson, G. A., 1979, Thermal ecology of the striped racer, *Masticophis lateralis, Herpetologica,* 35:267–273.
Hart, D. R., 1979, Niche relationships of *Thamnophis radix haydeni* and *Thamnophis sirtalis parietalis* in the interlake district of Manitoba, *Tulane Stud. Zool. Bot.*, 21:125–140.

Heatwole, H., 1966, Factors affecting orientation and habitat selection in some geckos, *Z. Tierpsychol.*, 23:303–314.
Heatwole, H., 1977, Habitat selection in reptiles, in C. Gans and F. H. Pough, eds., *Biology of the Reptilia,* Vol. 7, Academic, New York, pp. 137–155.
Hebrard, J. J., and H. R. Mushinsky, 1978, Habitat use by five sympatric water snakes in a Louisiana swamp, *Herpetologica,* 34:306–311.
Heller, S. B., and M. Halpern, 1982, Laboratory observations of aggregative behavior of garter snakes, *Thamnophis sirtalis:* roles of the visual, olfactory, and vomeronasal senses, *J. Comp. Physiol. Psychol.*, 96:984–999.
Henderson, R. W., J. R. Dixon, and P. Soini, 1979, Resource partitioning in Amazonian snake communities, *Milw. Pub. Mus. Contrib. Biol. Geol.*, (22):1–11.
Hertz, P. E., 1974, Thermal passivity of a tropical lizard, *Anolis polylepis, J. Herpetol.*, 8:323–327.
Huey, R. B., 1982, Temperature, physiology, and the ecology of reptiles, in C. Gans and F. H. Pough, eds., *Biology of the Reptilia,* Vol. 12, Academic, New York, pp. 25–91.
Huey, R. B., 1991, Physiological consequences of habitat selection, *Amer. Nat.*, 137:S91–S115.
Huey, R. B., and M. Slatkin, 1976, Cost and benefits of lizard thermoregulation, *Q. Rev. Biol.*, 51:363–384.
Huey, R. B., E. R. Pianka, and T. W. Schoener, 1983, *Lizard Ecology: Studies of a Model Organism,* Harvard Univ. Press, Cambridge, Massachusetts.
Huey, R. B., C. R. Peterson, S. J. Arnold, and W. P. Porter, 1989, Hot rocks and not-so-hot rocks: Retreat site selection by garter snakes and its thermal consequences, *Ecology,* 70:931–944.
Hutchinson, G. E., 1978, *An Introduction to Population Ecology,* Yale Univ. Press, New Haven, Connecticut.
Jaksic, F. M., and M. Delibes, 1987, A comparative analysis of food-niche relationships and trophic guild structure in two assemblages of vertebrate predators differing in species richness: Causes, correlations, and consequences, *Oecologia (Berlin)*, 71:461–472.
Jaksic, F. M., H. W. Greene, and J. L. Yanez, 1981, The guild structure of a community of predatory vertebrates in central Chile, *Oecologia (Berlin)*, 49:21–28.
James, F. C., and C. E. McCulloch, 1990, Multivariate analysis in ecology and systematics: Panacea or Pandora's Box?, *Annu. Rev. Ecol. Syst.*, 21:129–166.
Johnson, R. A., and D. W. Wichern, 1988, *Applied Multivariate Statistical Analysis,* Prentice-Hall, Englewood Cliffs, New Jersey.
Kachigan, S. K., 1986, *Statistical Analysis: An Interdisciplinary Introduction to Univariate and Multivariate Methods,* Radius, New York.
Keenlyne, K., 1972, Sexual differences in the feeding habits of *Crotalus horridus horridus, J. Herpetol.*, 6:234–237.
Kephart, D. G., 1982, Microgeographic variation in the diets of garter snakes, *Oecologia (Berlin)*, 52:287–291.
Kettlewell, H. B. D., 1965, Insect survival and selection for pattern, *Science,* 148:1290–1296.
Kiester, H. R., G. C. Gorman, and D. C. Arroya, 1975, Habitat selection behavior of three species of *Anolis* lizards, *Ecology,* 56:220–225.
Kitchell, J. F., 1969, Thermophilic and thermophobic responses of snakes in a thermal gradient, *Copeia,* 1969:189–191.
Klauber, L. M., 1972, *Rattlesnakes: Their Habits, Life Histories, and Influence on Mankind,* 2 Vols., Univ. Calif. Press, Berkeley.
Klopfer, P. H., 1965, Behavioral aspects of habitat selection, I, *Wilson Bull.*, 77:376–381.
Klopfer, P. H., 1969, *Habitats and Territories: A Study of the Use of Space by Animals,* Basic Books, New York.
Klopfer, P. H., and J. U. Ganzhorn, 1985, Habitat selection: Behavioral aspects, in M. L. Cody, ed., *Habitat Selection in Birds,* Academic, New York, pp. 435–453.
Klopfer, P. H., and J. P. Hailman, 1965, Habitat selection in birds, in D. S. Lehrman, R. A. Hinde and E. Shaw, eds., *Advances in the Study of Behavior,* Vol. 1, Academic, New York, pp. 279–303.
Klopfer, P. H., and R. H. MacArthur, 1961, On the causes of tropical species diversity: Niche overlap, *Amer. Nat.*, 95:223–226.

Kulesza, G., 1975, Comment on "niche, habitat, and ecotope," *Amer. Nat.*, 109:476–479.
Laburn, H., D. Mitchell, E. Kenedi, and G. N. Louw, 1981, Pyrogens fail to produce fever in a cordylid lizard, *Amer. J. Physiol.*, 241:R198–R202.
Lee, J. C., 1980, Comparative thermal ecology of two lizards, *Oecologica (Berlin)* 44:171–176.
Lillywhite, H. B., 1985, Trailing movements and sexual behavior in *Coluber constrictor*, *J. Herpetol.*, 19:306–308.
Lillywhite, H. B., and P. F. A. Maderson, 1982, Skin structure and permeability, in C. Gans and F. H. Pough, eds., *Biology of the Reptilia*, Vol. 12, Academic, New York, pp. 397–442.
Lutterschmidt, W. I., and H. K. Reinert, 1990, The effect of ingested transmitters upon the temperature preference of the northern water snake, *Nerodia s. sipedon*, *Herpetologica*, 46:39–42.
Lysenko, S., and F. E. Gillis, 1980, The effect of ingestive status on thermoregulatory behavior of *Thamnophis sirtalis sirtalis* and *Thamnophis sirtalis parietalis*, *J. Herpetol.*, 14:155–159.
MacArthur, R. H., 1958, Population ecology of some warblers of northeastern coniferous forests, *Ecology*, 39:599–619.
MacArthur, R. H., and E. O. Wilson, 1967, *The Theory of Island Biogeography*, Princeton Univ. Press, Princeton, New Jersey.
Madison, D. M., 1978, Behavioral and sociochemical susceptibility of meadow voles (*Microtus pennsylvanicus*) to snake predators, *Amer. Midl. Nat.*, 100:23–28.
Madsen, T., 1984, Movements, home range size and habitat use of radio-tracked grass snakes (*Natrix natrix*) in southern Sweden, *Copeia*, 1984:707–713.
Magnuson, J. J., L. B. Crowder, and P. A. Medvick, 1979, Temperature as an ecological resource, *Amer. Zool.*, 19:331–343.
Maple, W. T., 1968, The overwintering adaptations of *Sistrurus c. catenatus* in northeastern Ohio, Masters Thesis, Kent State Univ., Kent, Ohio.
Martinka, R. R., 1972, Structural characteristics of blue grouse territories in southwestern Montana, *J. Wildl. Manage.*, 36:498–510.
Masconzoni, D., and H. Wallin, 1986, The harmonic radar: A new method of tracing insects in the field, *Ecol. Entomol.*, 11:387–390.
McCoy, E. D., and S. S. Bell, 1991, Habitat structure: The evolution and diversification of a complex topic, in S. S. Bell, E. D. McCoy, and H. R. Mushinsky, eds., *Habitat Structure: The Physical Arrangement of Objects in Space*, Chapman and Hall, New York, pp. 3–27.
Meseth, E. H., and O. J. Sexton, 1963, Response of painted turtles, *Chrysemys picta*, to removal of surface vegetation, *Herpetologica*, 19:52–56.
Miller, A. H., 1942, Habitat selection among higher vertebrates and its relation to intraspecific variation, *Amer. Nat.*, 76:25–35.
Moore, R. G., 1978, Seasonal and daily activity patterns and thermoregulation in the southwestern speckled rattlesnake (*Crotalus mitchelli pyrrhus*) and the Colorado desert sidewinder (*Crotalus cerastes laterorepens*), *Copeia*, 1978:439–442.
Mushinsky, H. R., and J. J. Hebrard, 1977a, Food partitioning by five species of water snakes, *Herpetologica*, 33:162–166.
Mushinsky, H. R., and J. J. Hebrard, 1977b, The use of time by sympatric water snakes, *Can. J. Zool.*, 55:1545–1550.
Mushinsky, H. R., J. J. Hebrard, and M. G. Walley, 1980, The role of temperature on the behavioral and ecological associations of sympatric water snakes, *Copeia*, 1980:744–754.
Norusis, M. J., 1985, *SPSS^x Advanced Statistics Guide*, McGraw-Hill, New York.
Odum, E. P., 1959, *Fundamentals of Ecology*, Saunders, Philadelphia.
Orians, G. H., and J. F. Wittenberger, 1991, Spatial and temporal scales in habitat selection, *Amer. Nat.*, 137:S29–S49.
Parker, W. S., and W. S. Brown, 1972, Telemetric study of movements and oviposition of two female *Masticophis t. taeniatus*, *Copeia*, 1972:892–895.
Parker, W. S., and W. S. Brown, 1980, Comparative ecology of two colubrid snakes, *Masticophis t. taeniatus* and *Pituophis melanoleucus deserticola*, in northern Utah, *Milw. Public Mus. Publ. Biol. Geol.*, (7):1–104.
Pearson, O. P., 1954, Habits of the lizard *Liolaemus m. multiformis* at high altitudes in southern Peru, *Copeia*, 1954:111–116.

Pimentel, R. A., 1979, *Morphometrics: The Multivariate Analysis of Biological Data,* Kendall/Hunt, Dubuque, Iowa.
Platt, D. R., 1969, Natural history of the hog-nose snakes, *Heterodon platyrhinos* and *Heterodon nasicus, Univ. Kans. Publ. Mus. Nat. Hist.,* 18:253–420.
Plummer, M. V., 1981a, Habitat utilization, diet and movements of a temperate arboreal snake (*Opheodrys aestivus*), *J. Herpetol.,* 15:425–432.
Plummer, M. V., 1981b, Communal nesting of *Opheodrys aestivus* in the laboratory, *Copeia,* 1981:243–246.
Plummer, M. V., 1993, Thermal ecology of arboreal green snakes (*Opheodrys aestivus*), *J. Herpetol.,* in press.
Pough, F. H., 1983, Feeding mechanisms, body size, and the ecology and evolution of snakes: Introduction to the symposium, *Amer. Zool.,* 23:339–342.
Pough, H., 1966, Ecological relationships of rattlesnakes in southeastern Arizona with notes on other species, *Copeia,* 1966:676–683.
Preston, C., 1980, Differential perch site selection by color morphs of the red-tailed hawk (*Buteo jamaicensis*), *Auk* 97:782–789.
Prestt, I., 1971, An ecological study of the viper *Vipera berus* in southern Britain, *J. Zool. (London),* 164:373–418.
Rand, A. S., and S. S. Humphrey, 1968, Interspecific competition in the tropical rain forest: Ecological distribution among lizards at Belem, Para, *Proc. U.S. Nat. Mus.,* 125:1–17.
Reagan, D. P., 1974, Habitat selection in the three-toed box turtle, *Terrepene carolina triunguis, Copeia,* 1974:512–527.
Regal, P. J., 1966, Thermophilic responses following feeding in certain reptiles, *Copeia,* 1966:588–590.
Reichenbach, N. G., and G. H. Dalrymple, 1980, On the criteria and evidence for interspecific competition in snakes, *J. Herpetol.,* 14:409–412.
Reichenbach, N. G., and G. H. Dalrymple, 1986, Energy use, life histories, and the evaluation of potential competition in two species of garter snake, *J. Herpetol.,* 20:133–153.
Reinert, H. K., 1978, The ecology and morphological variation of the massasauga rattlesnake, *Sistrurus catenatus,* Masters Thesis, Clarion State College, Clarion, Pennsylvania.
Reinert, H. K., 1984a, Habitat separation between sympatric snake populations, *Ecology,* 65:478–486.
Reinert, H. K., 1984b, Habitat variation within sympatric snake populations, *Ecology,* 65:1673–1682.
Reinert, H. K., 1992, Radiotelemetric field studies of pitvipers: Data acquisition and analysis, in J. A. Campbell and E. D. Brodie, eds., *The Biology of Pitvipers,* Selva, Tyler, Texas, pp. 185–197.
Reinert, H. K., and W. R. Kodrich, 1982, Movements and habitat utilization by the massasauga, *Sistrurus catenatus catenatus, J. Herpetol.,* 16:162–171.
Reinert, H. K., and R. T. Zappalorti, 1988a, Timber rattlesnakes (*Crotalus horridus*) of the pine barrens: Their movement patterns and habitat preference, *Copeia,* 1988:964–978.
Reinert, H. K., and R. T. Zappalorti, 1988b, Field observation of the association of adult and neonatal timber rattlesnakes, *Crotalus horridus,* with possible evidence for conspecific trailing, *Copeia,* 1988:1057–1059.
Reinert, H. K., D. Cundall, and L. M. Bushar, 1984, Foraging behavior of the timber rattlesnake, *Crotalus horridus, Copeia,* 1984:976–981.
Reynolds, R. P., and N. J. Scott, Jr., 1982, Use of a mammalian resource by a Chihuahuan snake community, in N. J. Scott, Jr., ed., *Herpetological Communities,* U.S. Fish Wildl. Serv. Wildl. Res. Rep., 13, pp. 99–118.
Rosenzweig, M. L., 1981, A theory of habitat selection, *Ecology,* 62:327–335.
Rosenzweig, M. L., 1987, Editor's coda: Central themes of the symposium, *Evol. Ecol.,* 1:401–407.
Ruibal, R., 1961, Thermal relations of five species of tropical lizards, *Evolution,* 15:98–111.
Savitzky, A. H., 1983, Coadapted character complexes among snakes: Fossoriality, piscivory, and durophagy, *Amer. Zool.,* 23:397–409.

Schoener, T. W., 1968, The *Anolis* lizards of Bimini: Resource partitioning in a complex fauna, *Ecology,* 49:704–726.
Schoener, T. W., 1974, Resource partitioning in ecological communities, *Science,* 174:27–37.
Schoener, T. W., 1977, Competition and niche, in C. Gans and D. W. Tinkle, eds., *Biology of the Reptilia,* Vol. 7, Academic, New York, pp. 35–136.
Schoener, T. W., 1982, The controversy over interspecific competition, *Amer. Sci.,* 70:586–595.
Schoener, T. W., 1983, Field experiments on interspecific competition, *Amer. Nat.,* 122:240–285.
Secor, S. M., 1992, Activities and energetics of a sit-and-wait foraging snake, *Crotalus cerastes,* Doctoral Dissertation, Univ. Calif., Los Angeles.
Selander, R. K., 1966, Sexual dimorphism and niche utilization, *Condor,* 68:113–151.
Seigel, R. A., 1986, Ecology and conservation of an endangered rattlesnake (*Sistrurus catenatus*), in Missouri, USA, *Biol. Conserv.,* 35:333–346.
Semlitsch, R. D., K. L. Brown, and J. P. Caldwell, 1981, Habitat utilization, seasonal activity, and population size structure of the southeastern crowned snake *Tantilla coronata, Herpetologica,* 37:40–46.
Sexton, O. J., 1956–1957, The distribution of *Bothrops atrox* in relation to food supply, *Bol. Mus. Sci. Nat. (Venezuela)* 2–3:47–54.
Sexton, O. J., 1958, The relationship between the habitat preferences of hatchling *Chelydra serpentina* and the physical structure of the vegetation, *Ecology,* 39:751–754.
Sexton, O. J., and L. Claypool, 1978, Nest sites of a northern population of an oviparous snake, *Opheodrys vernalis* (Serpentes, Colubridae), *J. Nat. Hist.,* 12:365–370.
Sexton, O. J., and H. Heatwole, 1965, Life history notes on some Panamanian snakes, *Carib. J. Sci.,* 5:39–43.
Sexton, O. J., H. Heatwole, and D. Knight, 1964, Correlation of microdistribution of some Panamanian reptiles and amphibians with structural organization of the habitat, *Carib. J. Sci.,* 4:261–295.
Shine, R., 1977a, Habitats, diets, and sympatry in snakes: A study from Australia, *Can. J. Zool.,* 55:1118–1128.
Shine, R., 1977b, Reproduction in Australian elapid snakes, II. Female reproductive cycles, *Aust. J. Zool.,* 25:655–666.
Shine, R., 1979, Activity patterns in Australian elapid snakes (Squamata: Serpentes: Elapidae), *Herpetologica,* 35:1–11.
Shine, R., 1986, Sexual differences in morphology and niche utilization in an aquatic snake, *Acrochordus arafurae, Oecologia (Berlin),* 69:260–267.
Shine, R., and R. Lambeck, 1985, A radiotelemetric study of movements, thermoregulation and habitat utilization of Arafura filesnakes (Serpentes, Acrochordidae), *Herpetologica,* 41:351–361.
Shugart, H. H., Jr., and B. C. Patten, 1972, Niche quantification and the concept of niche pattern, in B. C. Patten, ed., *Systems Analysis and Simulation in Ecology,* Vol. 2, Academic, New York, pp. 238–327.
Slip, D. J., and R. Shine, 1988, Thermophilic response to feeding of the diamond python, *Morelia s. spilota* (Serpentes: Boidae), *Comp. Biochem. Physiol.,* 89A:645–650.
Sluckin, W., and E. H. Salzen, 1961, Imprinting and perceptual learning, *Q. J. Exp. Psychol.,* 13:65–77.
Smucny, D. A., and A. R. Gibson, Patterns of heat use by female garter snakes in a laboratory thermal mosaic, *J. Herpetol.,* in review.
Sokal, R. R., and F. J. Rohlf, 1981, *Biometry,* Freeman, San Francisco.
Southwood, T. R. E., 1977, Habitat, the templet for ecological strategies?, *J. Anim. Ecol.,* 46:337–365.
Stickel, L. F., W. H. Stickel, and F. C. Schmid, 1980, Ecology of a Maryland population of black rat snakes (*Elaphe o. obsoleta*), *Amer. Midl. Nat.,* 103:1–14.
Swain, T. A., and H. M. Smith, 1978, Communal nesting in *Coluber constrictor* in Colorado (Reptilia: Serpentes), *J. Herpetol.,* 34:175–177.
Sweet, S., 1985, Geographic variation in *Pituophis* and *Crotalus, J. Herpetol.,* 19:55–67.
Thorpe, W. H., 1945, The evolutionary significance of habitat selection, *J. Anim. Ecol.,* 14:67–70.

Toft, C. A., 1985, Resource partitioning in amphibians and reptiles, *Copeia,* 1985:1–21.
Valverde, J. A., 1967, *Estructura de una Comunidad Mediterranea de Vertebrados Terrestres,* Consejo Superior de Investigaciones Cientificas, Madrid.
Van Dyke, D. L., 1974, A low cost very high frequency animal transponder tracking system, *Biotelemetry,* 1:110.
Viitanen, P., 1967, Hibernation and seasonal movements of the viper, *Vipera berus berus* (L.), in southern Finland, *Ann. Zool. Fenn.,* 4:472–546.
Vitt, L. J., 1987, Communities, in R. A. Seigel, J. T. Collins, and S. S. Novak, eds., *Snakes: Ecology and Evolutionary Biology,* McGraw-Hill, New York, pp. 335–365.
von Uexkull, J., 1921, *Umwelt und Innenwelt der Tiere,* Springer, Berlin.
Weatherhead, P. J., and M. B. Charland, 1985, Habitat selection in an Ontario population of the snake, *Elaphe obsoleta, J. Herpetol.,* 19:12–19.
Weatherhead, P. J., and K. A. Prior, 1992, Preliminary observations of habitat use and movements of the eastern massasauga rattlesnake (*Sistrurus c. catenatus*), *J. Herpetol.,* 26:447–452.
Wecker, S. C., 1963, The role of early experience in habitat selection by the prairie deer mouse, *Peromyscus maniculatus bairdi, Ecol. Monogr.,* 33:307–325.
Whittaker, R. H., 1975, *Communities and Ecosystems,* Macmillan, New York.
Whittaker, R. H., S. A. Levin, and R. B. Root, 1973, Niche, habitat, and ecotope, *Amer. Nat.,* 107:321–338.
Whittaker, R. H., S. A. Levin, and R. B. Root, 1975, On the reasons for distinguishing "niche, habitat, and ecotope," *Amer. Nat.,* 109:479–482.
Wiens, J. A., 1976, Population responses to patchy environments, *Annu. Rev. Ecol. Syst.,* 7:81–120.
Wiens, J. A., 1977, On competition and variable environments, *Amer. Sci.,* 65:590–597.
Wright, A. H., and A. A. Wright, 1957, *Handbook of Snakes,* 2 Vols., Comstock, Ithaca, New York.
Wright, B. A., 1941, Habit and habitat studies of the massasauga in northeastern Illinois, *Amer. Midl. Nat.,* 25:659–672.
Zappalorti, R. T., and H. K. Reinert, Artificial refugia as a habitat improvement strategy for snake conservation, in J. B. Murphy, J. T. Collins, and K. Adler, eds., *Captive Management and Conservation of Amphibians and Reptiles,* SSAR Contrib. Herpetol., in press.
Zurovsky, Y., D. Mitchell, and H. Laburn, 1987a, Pyrogens fail to produce fever in the leopard tortoise *Geochelone pardalis, Comp. Biochem. Physiol.,* 87A:467–469.
Zurovsky, Y., T. Brain, H. Laburn, and D. Mitchell, 1987b, Pyrogens fail to produce fever in the snakes *Psammophis phillipsii* and *Lamprophis fuliginosus, Comp. Biochem. Physiol.,* 87A:911–914.

Chapter

7

Snake Thermal Ecology: The Causes and Consequences of Body-Temperature Variation

Charles R. Peterson

A. Ralph Gibson

Michael E. Dorcas

Introduction

Our purpose in writing this chapter is to show how studies of the temperature relationships of snakes contribute to an understanding of their ecology and evolution. These studies are not only critical to understanding snake ecology but they also provide excellent opportunities for studying general questions in thermal biology. We begin with a discussion of why body-temperature (T_b) variation is important. We then identify what we believe are the key questions regarding the thermal ecology of snakes and provide a conceptual framework for addressing them. There are several recent reviews that address snake thermal ecology (Avery, 1982; Huey, 1982; Lillywhite, 1987). We regard these reviews as essential companions to the present chapter and have not sought to duplicate them here. Instead, we illustrate our conceptual approach with selected examples from studies of snakes and other ectothermic vertebrates and evaluate the methodologies that have been applied. Throughout the chapter, we identify what we consider to be important areas for future research and approaches likely to be useful in answering these questions.

Importance of thermal ecology

A combination of several snake characteristics makes studies of temperature relationships critical to understanding many aspects of snake ecology. First, snake body temperatures are determined primarily by heat obtained directly from the physical environment (i.e., snakes are ectothermic). With few exceptions, the body temperatures of snakes are constrained by the range of thermal conditions present in the environment. Second, because most snakes live in thermally variable environments, their body temperatures also are subject to variation, even though they can behaviorally control their body temperatures within limits set by that environment. Third, the resulting variation in body temperature affects most developmental, physiological, and behavioral processes (e.g., dates of birth, feeding requirements, and strike speed). This combination of ectothermy, thermally variable environments, and the thermal dependency of biological processes thus affects the ecology of snakes either directly (by changing the rate of biochemical reactions) or indirectly (because of costs associated with thermal compensation or thermoregulation). For example, in a cool environment, snakes may be less successful at capturing prey (a direct effect of T_b on strike speed) and have less time available for foraging (an indirect effect resulting from constraints on when and where the snakes can effectively thermoregulate). Their feeding requirements may be lower, however, because of the direct effect of T_b on metabolic rate. Low temperatures might also result in snakes growing more slowly and reproducing less frequently but living longer. Thus, nearly all aspects of snake ecology (from development to population dynamics) are potentially affected by T_b either directly or indirectly.

Key ecological questions

Because of the complexity of the causes and the importance of the consequences of body-temperature variation, there exists an extensive literature on the thermal biology of snakes. For reviews of this literature, see Fitch, 1956; Brattstrom, 1965; Avery, 1982; Bartholomew, 1982; Huey, 1982; and Lillywhite, 1987. From the perspective of an ecologist, we consider five questions to be particularly relevant:

1. What is the range of possible body temperatures under natural conditions? Answering this question not only requires environmental measurements but also requires an understanding of the behavioral and physiological mechanisms snakes use to control their body temperatures (e.g., microhabitat selection and cardiovascular adjustments).

2. What are the proximate factors that determine which body temperatures a snake selects from that range of possibilities?
3. How do the body temperatures of individual snakes actually vary under natural conditions over extended periods of time?
4. What are the functional effects (i.e., developmental, physiological, and behavioral) of body-temperature variation?
5. What are the ecological consequences of body-temperature variation (or of the physiological and behavioral adjustments employed to control or compensate for T_b variation)? In other words, how do T_b variation and thermoregulatory adjustments affect the distribution and abundance of snakes?

Studies of the ecological consequences of body-temperature variation and thermoregulatory adjustments naturally lead to questions concerning the evolution of the thermal characteristics of snakes. Because many thermal characteristics affect other aspects of a snake's biology, an understanding of their evolution has implications far beyond thermal biology alone (e.g., the evolution of body size, coloration, and reproductive mode).

Conceptual framework

Developing answers to these questions requires a variety of concepts, methods, data, and analyses. Our conceptual approach to the thermal ecology of snakes is illustrated in Figure 7.1, the causes and consequences of body-temperature variation. This diagram illustrates how studies from a variety of fields (natural history, environmental physics, embryology, physiology, behavior, genetics, and evolution) interrelate and contribute to our understanding of snake thermal ecology. The integration of information from such diverse disciplines is one of the strengths of this approach. Key concepts include environmental constraints and thermal budget analysis (Porter et al., 1973; Tracy and Christian, 1986), behavioral thermoregulation (Cowles and Bogert, 1944; Heath, 1965), thermal preferences (Dawson, 1975; Kluger, 1979; Gibson et al., 1989), thermoregulatory precision (DeWitt, 1967; Bowker and Johnson, 1980), regional heterothermy (Regal, 1966; Johnson, 1973), thermal dependency of performance (Brett, 1971; Huey and Stevenson, 1979; Pough, 1980), thermal acclimation (Prosser and Heath, 1991), cost–benefit analysis (Huey and Slatkin, 1976), interindividual variation, heritability, and fitness (Arnold, 1983; Bennett, 1987), coadaptation (Huey and Bennett, 1987), and evolutionary constraints (Huey, 1987).

In this scheme, body temperature (center of Fig. 7.1) is viewed as both a dependent and an independent variable. Body temperatures

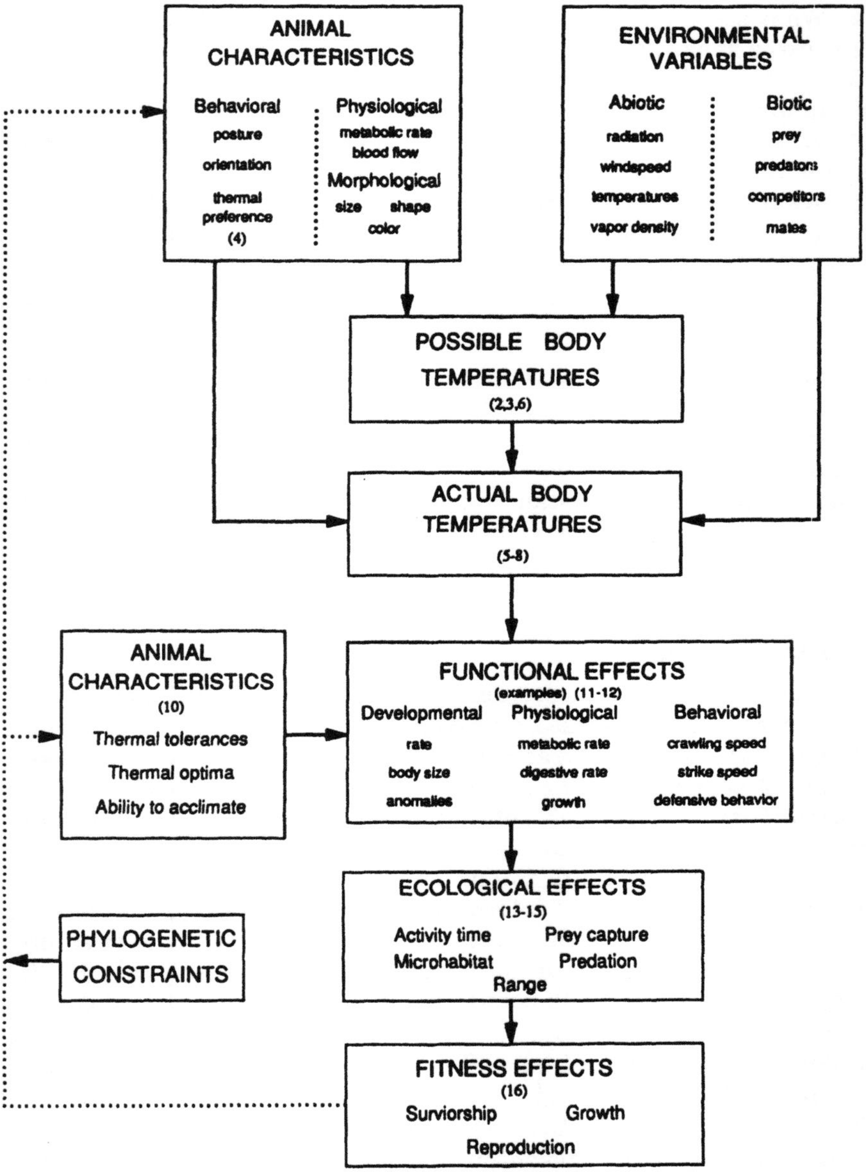

Figure 7.1 The causes and consequences of body-temperature variation. See the Introduction for a detailed explanation of this diagram.

can be measured by capturing the snake or through radiotelemetry. The interpretation and prediction of body-temperature variation is complex and requires data on both the causes and consequences of T_b variation.

The upper portion of Fig. 7.1 identifies the proximate causes of body-temperature variation. Abiotic environmental variables (such as radiation and air temperature) and various snake characteristics (such as size and color) interact to determine the range of possible body temperatures. For example, under high solar radiation levels, a dark snake will heat faster and have higher steady-state T_b's than will an otherwise identical light snake. The range and distribution of possible temperatures can be measured through the use of physical models and/or thermal budget analysis. Biotic environmental variables (such as the presence of predators and prey) and the condition of the snakes (such as nutritional and reproductive status) determine what temperatures they will accept or select from the range of possible body temperatures, thus partially determining actual patterns of body-temperature variation. For example, a snake might accept relatively low body temperatures when foraging but then select higher temperatures when digesting a prey item. Field observations, field experiments, and laboratory studies of temperature-selection behavior are required to determine the effects of these variables.

The lower portion of Fig. 7.1 indicates the functional (developmental, physiological, and behavioral) and ecological (distribution and abundance) consequences of body-temperature variation. Functional effects are generally determined using controlled laboratory experiments. Field and laboratory data can be combined to make ecological predictions that can be tested using computer modeling, laboratory simulations, field observations, field comparisons, and field experiments.

Finally, the broken arrows in Fig. 7.1 indicate that fitness effects are related to snake characteristics (such as thermal preferences and tolerances) via natural selection. These effects can be studied through computer modeling, selection experiments, and comparative studies. In general, more information exists concerning the causes of body-temperature variation than about its consequences; especially lacking are studies linking functional effects with ecology and fitness.

For each of the areas illustrated in the diagram, we will provide examples from studies of snakes or other ectothermic vertebrates. We will especially focus on Garter Snakes (genus *Thamnophis*) because, with respect to temperature relationships, they are the most well-studied group of snakes and can provide the most integrated picture of snake thermal biology. We also discuss what areas need further investigation, how to obtain that information, and some of the problems involved.

Why snakes are good animals for studying general questions in thermal ecology

Several characteristics of snakes make them attractive animals for addressing questions of general interest to physiological and evolutionary ecologists. Because many snakes tolerate radiotransmitters well, it is easier to obtain continuous T_b data. These data allow identification of individual patterns of daily and seasonal variation in body temperature, studies of thermoregulatory precision, and analyses of interindividual variation. It also is relatively easy to make physical models of snakes so that their thermal environments can be better described. Because of the elongate, limbless body form of snakes, studies of their thermal ecology may complement rather than simply replicate those of most other ectothermic vertebrates. For example, the elongate bodies of snakes allow them to adjust easily the amount of their bodies exposed to solar radiation, establish large temperature differences within their bodies, and, through coiling, adjust their surface area to volume ratios. The absence of limbs, however, constrains their ability to modify conductive heat transfer (e.g., snakes cannot raise themselves off of the substrate as many lizards do or adjust heating and cooling rates by varying blood flow to limbs). In comparison to other ectothermic vertebrates, the effects of physiological condition on thermoregulatory behavior in snakes are easier to detect because of characteristics such as infrequent feeding, large meal size, the synchrony of skin shedding, and the occurrence of viviparity in many common species. Some species of snakes are well suited for selection studies because they have large, relatively confined populations that can be effectively sampled. Finally, snakes can be maintained and bred in the laboratory so that controlled physiological, behavioral, and genetic studies can be conducted.

Possible Body Temperatures

Understanding the causes of T_b variation requires knowledge of what body temperatures are possible and what factors determine the temperatures a snake will select or accept from that range of possibilities. Describing the thermal environment of a snake consists of determining what a snake's body temperature could be at different locations in its habitat at any given time period. If the distribution of possible T_b's is known, then the relative importance of snake behavior and environmental constraints on T_b variation can often be distinguished (Peterson, 1987). One of the major problems in evaluating body-temperature data in past studies has been the lack of information about what T_b's were possible. For example, without environmental data, you would not know if a snake with a T_b of 25°C was voluntarily selecting that temperature or if it was constrained to that temperature by the environment. Fortunately, the concepts and techniques needed to determine the distribution of possible body temperatures are now available. These techniques still need to be refined, validated for specific situations, and applied to a variety of species and environments.

Determinants of snake body temperature

The distribution of body temperatures that is possible in a given ecological situation is determined by the interactions of snake characteristics and abiotic environmental variables. Important environmental variables include air temperature, substrate or water temperature, long- and short-wavelength radiation, wind velocity, and humidity. Reptile properties that influence body temperature include size, shape, reflectivity, metabolic rate, and skin permeability to water (Porter and Tracy, 1983). Routes of heat exchange include absorption of long- and short-wavelength radiation, emission of long-wavelength radiation, conduction, convection, and evaporation (Fig. 7.2). For a

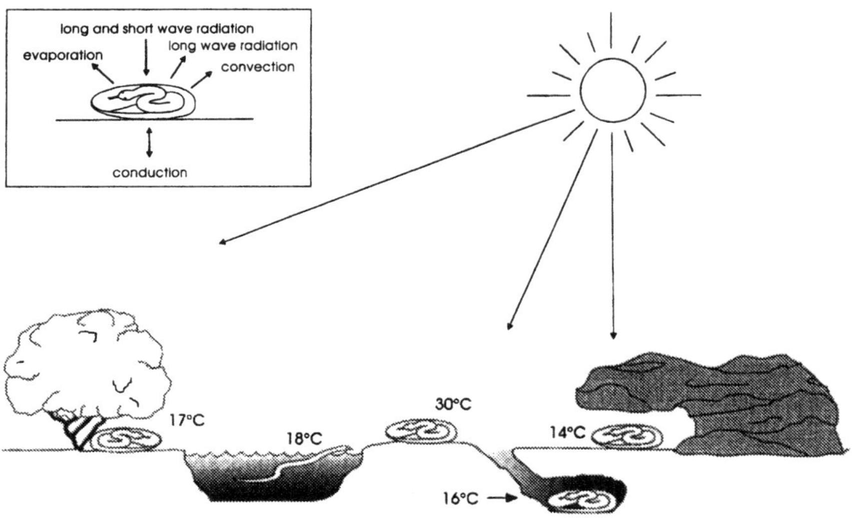

Figure 7.2 Spatial variation in the operative temperatures of snakes at a meadow in northern California at 0930 on 6 August 1986. Microhabitat types include a shaded area under vegetation, a stream pool, an exposed surface of meadow, a burrow (soil), and underneath a large rock. Figure 7.3 indicates daily variation in operative temperatures in these microhabitats (C. R. Peterson, S. Arnold, and W. Porter, unpublished data). The inset diagram indicates the routes and typical directions of heat exchange for a snake.

detailed description of heat exchange between animals and their environments, see Campbell (1977) and the references therein.

Snakes can control their body temperatures both behaviorally and physiologically. For a discussion of the relative importance of behavioral and physiological adjustments and the effects of size on body-temperature control in terrestrial ectotherms, see Stevenson (1985a,b). In snakes, behavioral thermoregulation is usually most important and includes selecting appropriate microhabitats, adjusting posture and orientation, and aggregating with other snakes (see Lillywhite, 1987 for references; Graves and Duvall, 1987). Some snakes can exert significant physiological control over rates of heat gain and loss by altering their heart rates and blood-flow patterns (Bartholomew, 1982). Some large species, such as Indian Pythons (*Python molurus*), can regulate body and egg temperatures by increasing their metabolic rates through muscular contractions (Hutchison et al., 1966a; Bartholomew, 1982). Larger snakes generally are more able to control T_b physiologically than smaller snakes.

Regional body-temperature variation

Although the body temperatures of reptiles are usually measured in only one location (e.g., the cloaca or body cavity), considerable variation may exist in temperatures from different regions of the body. Consequently, caution must be exercised when generalizing about body temperatures measured from only one location within an individual (Dill, 1972). This is especially true for snakes because their attenuated body form allows relatively large regional differences in body temperature to be maintained. Because of their size, large snakes are generally able to maintain greater regional temperature differences than are small snakes. However, even some relatively small snakes, such as Rubber Boas (*Charina bottae*), may exhibit substantial differences (up to 3°C) between their cloaca and anterior esophagus much of the time, even at night (M. Dorcas, unpublished data). Regional heterothermy is generally attributed to behavioral mechanisms such as differential basking (Bartholomew, 1982; Gregory, 1990), but it may also be due to physical differences between regions of the body (Pough, 1974) and physiological mechanisms, such as blood shunts and countercurrent heat exchangers (Webb and Heatwole, 1971).

Methods for measuring snake thermal environments

A variety of measurements have been used to characterize the thermal environments of snakes. The most common approach has been to measure air and substrate temperatures at the site where a snake has been captured. Although this is easy to do, it is usually inadequate for two reasons. First, single environmental temperature measurements fail to indicate the range of possible T_b's and tend to bias analyses of the relationships between environmental and body temperatures (Heath, 1964). Second, because many factors interact to determine T_b and considerable temperature differences may exist between the snake and its surroundings (Fig. 7.2), air or ground temperatures may be poor predictors of snake temperatures, even if they are correlated with them. This is especially true in sunlit, terrestrial habitats where the differences between air temperature and an ectotherm's T_b may exceed 10°C (Stevenson, 1985a). Temperature differences between a snake and the environment will usually be much less under low radiation levels (at night, in the shade, on overcast days, and underground) or in the water.

Operative temperatures. The operative temperature (T_e) is a better index for characterizing an animal's thermal environment than is any single environmental temperature (e.g., air or substrate temperature). The operative temperature concept was derived from heat-transfer theory and incorporates both environmental variables and animal characteristics. The operative temperature can be defined as the temperature of an inanimate object of zero heat capacity with the same size, shape, and radiative properties as the animal exposed to the same microclimate and thus represents the true environmental temperature experienced by the animal (Bakken and Gates, 1975). For most snakes, the heat lost via evaporation usually is small and is generally offset by heat generated via metabolism. Consequently, for a dry-skinned ectotherm like a snake, T_e equals the steady-state T_b of a snake under the same environmental conditions. The steady-state body temperature of a snake for a given environmental situation can be determined through the use of physical snake models or with mathematical models of heat transfer. It is important to realize that these models are used to predict the range of body temperatures possible under given environmental conditions, not the temperatures that the snakes actually select or accept.

Physical models. Operative temperature can be measured using a suitable physical model of the animal as a "T_e thermometer" (Bakken and Gates, 1975; Bakken, 1976, 1992; Bakken et al., 1985). Several studies have used physical models of snakes to describe the thermal environment (Peterson, 1982, 1987; Stevenson, 1983; Huey et al., 1989; Charland, 1991). To encourage the use of this important technique, we briefly describe how to construct snake models. Some of our suggestions are based on a preliminary sensitivity analysis conducted on several clear summer days in Pocatello, Idaho, using models with various lengths, diameters, paint types, postures, and orientations to the sun.

Although casting and electroforming procedures can be used to make detailed models of reptiles (Bakken and Gates, 1975), we have found that appropriately painted, hollow, copper pipes can be used with good results. Copper is the preferred material because it is highly conductive and minimizes the occurrence of temperature gradients within the model. The temperature of the model can be measured by suspending a temperature sensor such as a thermocouple in its center. The ends of the model should be sealed (e.g., with rubber stoppers) to minimize convective heat transfer within the model.

Although the diameter of the pipe should be similar to the snake's diameter, the length of the model can be considerably shorter than the snake's length. In our field tests, we found that smaller-diameter models had lower temperatures, but this effect decreased at larger diameters. Above 10 cm, the length of the pipe had little effect on model temperature. Therefore, to make placement of the model in the field easier, we suggest a model length approximately the width of a coiled snake rather than the total length of the snake.

Models should be painted to match the average reflectance of the dorsal surface of a snake. The average reflectances for a variety of paint samples, measured with a Beckman DK-2a spectroreflectometer using the procedures of Porter (1967), ranged from 1.5% to 68% (Table 7.1). Our field tests indicated that temperatures of white (68% reflectance) models were as much as 15°C lower than those of models painted black (2% reflectance). Gray spray primers appear to be suitable for many snake species. The average reflectances (290–2600 nm) for eight species of snakes ranged from 6.0% to 24.9% and averaged 18.2% (Table 7.2). Models with paint reflectances similar to those of light (24.9%) and dark (6.0%) snakes showed temperature differences of about 5°C during the day.

To measure maximal T_e's at the surface, we usually use straight models oriented as close to perpendicular to the path of the sun as possible. Coiled models are difficult to shape and result in temperatures intermediate to those of straight models placed perpendicular and parallel to the sun. Although we have not looked specifically at the effect of contact area between the model and the substrate, it is quite important and can affect model temperature by several degrees. The bottom surface of the model can be flattened to improve contact between the model and substrate.

It is important to realize that under transient (nonsteady-state) conditions, T_e may not equal a snake's T_b because a hollow model will respond faster than an actual snake. For small snakes (e.g., less than

TABLE 7.1. Paint Reflectances (290–2600 nm). Determined in Warren Porter's laboratory for P. Hertz, C. Peterson, and R. Stevenson

Total reflectance (%)	Total absorptivity (%)	Paint type
68.6	31.4	Krylon No. 500 White Primer
68.0	32.0	Krylon No. 1502 Flat White
61.1	38.9	Krylon No. 1503 Antique White
53.4	46.6	Krylon No. 1403 Dull Aluminum
49.8	50.2	Krylon No. 1403 Dull Aluminum
47.8	52.2	Krylon No. 1506 Almond
45.0	55.0	Krylon No. 2003 Jade Green
35.1	64.9	Krylon No. 2503 Taupe
27.3	72.7	Zynolyte No. 0741 Desert Gold Flat Enamel
23.5	76.5	Rustoleum No. 7773 Clean Metal Primer
18.1	81.9	Western Auto Gray Auto Primer No. 73-1819-9
17.1	82.9	Krylon No. 1314 All Purpose Platinum Spray Primer
16.7	83.3	Rustoleum No. 7769 Rusty Metal Primer
14.1	85.9	Krylon No. 1317 Ruddy Brown Primer
11.8	88.2	Rustoleum Navy Gray No. 975
10.6	89.4	Krylon No. 1318 All Purpose Gray Spray Primer
8.6	91.4	Spar Var All Purpose Gray No. S-203
8.3	91.7	Krylon No. 1318 All Purpose Gray Spray Primer
7.3	95.8	Krylon No. 1318 Gray Primer
3.6	96.4	Krylon No. 902 Black Primer
1.5	98.5	Krylon No. 1602 Ultra Flat Black

TABLE 7.2. Snake reflectances (290–2600 nm)

Reflectance (%)	Species	Source
24.9	*Crotalus viridis lutosus*—ground color	Peterson and Cobb (unpublished data)
24.0	*Trimorphodon vandenburghi*	Porter (1967)
23.6	*Arizona elegans*	Porter (1967)
20.7	*Charina bottae*	Peterson and Dorcas (unpublished data)
19.7	*Salvadora hexalepis*	Porter (1967)
18.7	*Hypsiglena torquata*	Porter (1967)
18.4	*Thamnophis elegans*—light	Peterson (1987)
15.7	*Thamnophis elegans*—dark	Peterson (1987)
15.1	*Crotalus viridis lutosus*—body blotch	Peterson and Cobb (unpublished data)
13.9	*Pituophis melanoleucus*	Porter (1967)
6.0	*Thamnophis elegans*—melanistic	Peterson and Fabian (1984)

about 50 g and 50 cm snout–vent length), models made with hollow pipes resemble snake temperatures closely (Peterson, 1987). The larger the snake, the greater the lag between the hollow model and the snake. The model will better reflect current environmental conditions (i.e., T_e) but will usually overestimate the possible T_b of a snake that is heating and underestimate the possible T_b of a snake that is cooling. Filling a model with water basically reverses this problem, and such models may greatly underestimate possible snake T_b's under some conditions (e.g., when a snake emerges from a relatively warm burrow onto the surface following a cold night during which the water freezes in the model). A possible solution to this problem involves mathematically combining data on the cooling and heating rates of live snakes (i.e., their time constants) with hollow-model temperatures to generate predicted snake temperatures. This approach allows accurate measurement of current conditions, calculation of possible temperatures (corrected for the mass of the snake), and it incorporates physiological aspects of T_b control (i.e., cardiovascular adjustments).

Regardless of their characteristics, all models should be validated carefully. This can be done by comparing the temperatures of models and dead or live snakes (either restrained or anesthetized) placed under similar conditions in the field or laboratory. If possible, the comparisons should include conditions with high radiation levels and relatively low air temperatures and low wind speeds because the greatest differences (i.e., errors) should be evident then. Using this procedure, Peterson (1982, 1987) and Stevenson (1983) found that model and anesthetized or restrained *Thamnophis elegans* temperatures agreed well (e.g., $n = 31$, $r^2 = 0.96$, slope not significantly different from 1, and intercept not significantly different than 0).

The number and placement of models into the environment will depend on the resources available and the particular question being asked. For most studies, the objective will be to bound the T_e range (i.e., measure the minimal and maximal operative temperatures). Although data from only a single model are useful, two or more models are required to determine the range of T_b's possible in the environment. Thermocouples alone (i.e., not in models) can usually be used to indicate operative temperatures in water or underground. A data logger can be used to measure and record the temperatures from a number of models (Peterson and Dorcas, in press). The use of multiple models allows comparisons of thermal characteristics among different microhabitats (Fig. 7.3).

Even when many models are used, however, snakes may find sites where they can attain body temperatures outside of the range indicated by the models (Fig. 7.6). It is particularly difficult to locate and measure the warmest temperatures available to snakes when they are in retreat (e.g., at night or during the winter; note Fig. 7.6*A*).

Although knowledge of the actual distribution of possible T_b's is desirable, it is more difficult to obtain, requiring the use of many models and considerable care in how they are placed in the environment. Examples of this approach include studies of *T. elegans* by Stevenson (1983) and of lizards by Grant (1990) and Hertz (1992a,b). It is unclear how to best measure the distribution of possible T_b's for subterranean environments. Another problem in describing the distribution of T_e's for snakes involves characterizing the spatial relationships of the thermal environment. Snakes have the ability to position their elongate bodies in more than one microhabitat (e.g., in a sun–shade mosaic, into a burrow, or along the shoreline of a stream). Consequently, it is necessary to think in terms of the thermal gradients available rather than simply the temperatures occurring at a single location (Huey et al., 1989). Infrared thermography could prove very useful in visualizing and quantifying spatial variation in surface environmental temperatures.

Mathematical models. Mathematical models of heat transfer also can be used to calculate possible body temperatures, either for steady-state or varying conditions (Porter et al., 1973; Tracy, 1982). Important animal characteristics such as size, shape, and reflectivity need to be measured or estimated. Values for environmental variables such as long-wavelength and short-wavelength radiation, air temperature, substrate temperature, and wind speed are measured in the field or calculated. These environmental data are then combined with the information on animal characteristics into a heat-transfer equation that is solved for body temperature using a computer.

Mathematical modeling can be very effective at quantitatively evaluating the factors influencing body temperature (Muth, 1977). It also has great potential for modeling large-scale phenomena (e.g., variation in activity times on a continental scale [Porter and Tracy, 1983; Adolph and Porter, 1993] or the effects of global warming) and for estimating thermal conditions from weather station records (Porter and Tracy, 1983). There are, however, several disadvantages to this approach (Bakken, 1992). It requires a considerable amount of equipment and technical expertise to make the necessary measurements and to use the computer models properly. The size of the instruments may prevent measurements of some microenvironments. Restrictions on the number of environmental sensors available limit how many microenvironments can be characterized and the spatial scale of resolution.

Heat-transfer analysis has successfully been used to characterize the thermal environments of a number of ectothermic vertebrates, including amphibians (Tracy, 1976) and turtles, crocodilians, and lizards (Tracy, 1982). This approach has rarely been used to study snake T_b relationships (Scott et al., 1982), probably because of the difficulty in modeling conductive heat transfer. Nevertheless, we believe

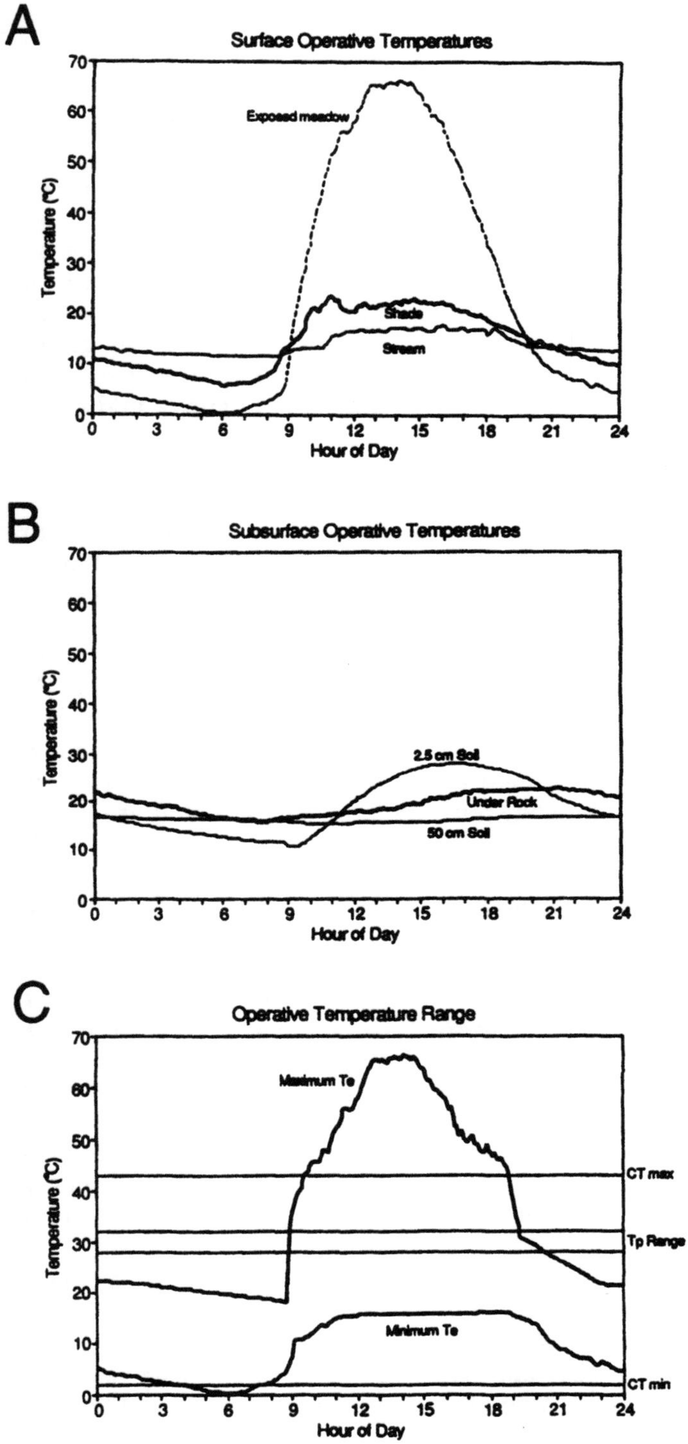
A
Surface Operative Temperatures
Exposed meadow
Shade
Stream
Temperature (°C)
Hour of Day
B
Subsurface Operative Temperatures
2.5 cm Soil
Under Rock
50 cm Soil
Temperature (°C)
Hour of Day
C
Operative Temperature Range
Maximum Te
Minimum Te
CT max
Tp Range
CT min
Temperature (°C)
Hour of Day

that the development and application of heat-transfer models to snakes has great potential for improving our understanding of the factors affecting snake temperatures and for characterizing thermal environments on a large scale or for times when more direct measurements are unavailable.

Spatial and temporal variation in the thermal environment

To illustrate the use of physical models to describe spatial and temporal variation in the thermal environments of snakes, we next examine a series of plots of model temperature versus time of day from *Thamnophis elegans* studies conducted in western North America.

Daily variation in T_e's from a variety of microhabitats in a mountain meadow in northeastern California on a clear day is shown in Fig. 7.3. Each line indicates the operative temperature for a different microhabitat (exposed surface, shaded surface, water, underneath a rock, in a burrow). All of the sites showed sinusoidal variation in T_e over the course of 24 h. The greatest variation occurred at the exposed surface where thermal conditions were surprisingly harsh, varying from below freezing to over 60°C within 24 h. Less variation would occur on a cloudy day. Intermediate variation occurred at the shaded surface and in the water. The least variation in T_e occurred underground. The warmest temperatures during most of the day occurred at the surface, but the situation reversed during the night when the highest T_e's were found below the surface. The actual times at which minimal and maximal T_e's occur depend on factors such as degree of exposure, rock thickness, and depth underground (Huey et al., 1989). The distance between the lines indicates the approximate range of possible T_b's. This difference was greatest (approximately 50°C) in the early afternoon and then decreased to the smallest difference (approximately 5°C) just before sunrise (approximately 0900 because this site is located at the base of a steep, west-facing, talus slope). Because of the great spatial and temporal variation in T_e's, snakes can exert considerable control over T_b by selecting the appropriate microhabitat and times of activity. This may require only short movements (<1 m) if an exposed area is adjacent to a retreat under a rock or in a burrow.

Figure 7.3 Among microhabitat variation in operative environmental temperatures for *Thamnophis elegans* at a meadow in northern California on 6 August 1986. (*A*) Surface operative temperatures. (*B*) Subsurface operative temperatures. (*C*) The maximal and minimal operative temperatures from 25 locations, including both surface and subsurface microhabitats. The critical thermal minimum, critical thermal maximum, and the preferred temperature range are indicated by horozontal lines (C. R. Peterson, S. Arnold, and W. Porter, unpublished data).

The range of T_e will also vary with weather conditions and season of the year. Overcast conditions result in a decreased T_e range (e.g., Fig. 7.6*C*). In hot environments, however, cloudy days may reduce maximal T_e's and thereby increase the time and the amount of the environment in which snakes can be active. During the winter, surface conditions in temperate regions may be too cool for activity or even survival (e.g., Fig. 7.6*A*).

The characterization of thermal environments with physical models has not been done at enough sites to make large-scale geographic comparisons (latitudinal and altitudinal). Comparisons of a meadow in northeastern California with a nearby site, 600 m higher in elevation, indicated that snake temperatures at both sites would be similar during summer days but that snakes at the lower altitude could stay several degrees warmer during the night because of warmer subsurface temperatures. As the thermal environments for more species, geographic locations, and habitats are measured, we will be better able to detect patterns of variation with respect to time of year, latitude, altitude, and habitat type.

Factors Determining Selected Body Temperatures

The body temperatures displayed by free-ranging snakes are a subset of the possible temperatures defined by a snake's thermal environment (Fig. 7.1). This subset of actual body temperatures is restricted by those physiological characteristics of the animal that influence the temperatures a snake will select (behavioral temperature regulation), and by those elements of the snake's biotic environment that affect the temperatures that a snake will accept. For ecologists interested in predicting snake body temperatures, these physiological factors and elements of the biotic environment represent important independent variables in a predictive equation. Similarly, the variation in T_b induced by some factors can confound investigations of others: a study of acclimation and preferred T_b could have real effects masked by uncontrolled variation in the reproductive status or condition of the study animals. Finally, a snake's thermoregulatory responses to the physiological and biotic factors, an expression of intrinsic behavioral programs, presumably reflect the action of natural selection. As such, they can guide evolutionary ecologists in forming hypotheses which address issues such as the adaptive significance of ectothermy (Pough, 1980).

There are several recent reviews of thermoregulation in ectotherms (Avery, 1982; Huey, 1982; Lillywhite, 1987; Hutchison and Dupré,

1992), and these have identified a large number of factors that are known or suspected of influencing body-temperature selection. These include feeding, condition, reproductive status, ecdysis, disease, water balance, hypoxia, hypercapnia, acclimation, and diurnal and seasonal rhythms. As noted in the Introduction, we have not endeavored to replicate these reviews here. Instead, in this section we have attempted to identify those studies that, in our opinion, provide the strongest evidence for the operation of particular factors in snakes. We also have attempted to determine why some studies may have been especially successful or convincing in documenting those factors. As is the case for so many aspects of snake ecology, we cannot determine now if these selected studies of thermoregulatory behavior are generally characteristic of snakes.

Methods

Diverse approaches have been used effectively to investigate these physiological and biotic factors in snakes. Not surprisingly, laboratory studies are more common than are studies of free-ranging animals. In our opinion, the most convincing demonstrations have tended to share four methodological attributes. First, successful studies have taken steps to accommodate their particular study animals, and thereby to elicit snake behavior that is as near to normal as possible. Second, the assessment of temperature selection generally has been nondisruptive or even indirect. Third, convincing demonstrations typically have benefitted from thoughtful experimental design. Finally, the resolving power of these studies often has been enhanced by strong, sometimes long-term, sampling efforts.

Three studies, or groups of studies, can be used to illustrate both the range of approaches and the desired methodological attributes. Slip and Shine (1988) used surgically implanted radiotransmitters to examine the thermoregulatory effects of feeding in captive Diamond Pythons (*Morelia spilota*). The study animals were given not less than several weeks in which to accommodate to captivity and to recover from surgical implantation of the transmitters. A regular feeding schedule helped reduce any differences in general condition, thereby better isolating the effect of feeding. The relatively long, spacious gradient was placed in a quiet room. The pythons were placed in the gradient for two days prior to data acquisition and were tested individually. The trials themselves typically lasted several days, with individuals serving as their own controls. Snakes were not tested if they were nearing ecdysis.

Lutterschmidt and Reinert (1990) also used a laboratory thermal gradient to describe temperature selection in Northern Water Snakes (*Nerodia sipedon*). The authors measured the substrate temperatures selected by snakes, rather than T_b itself. The careful yet simple experimental design applied by Lutterschmidt and Reinert avoided errors (Mathur and Silver, 1980; Sievert, 1989) that have reduced the discriminating power and interpretability of many previous studies. Their's is an important illustration of the gains that can be had at little or no cost in experimental effort. Each individual snake had three replicate trials in the gradient under each of the three treatment conditions (unfed, natural food, dummy transmitter). The sequence of nine trial and treatment combinations was randomized for each individual, thereby reducing possible confounding effects from the order of the treatments, and from any learning by the snakes. Similarly, the interval between each trial was varied randomly to control for possible carryover effects. Sequential measurements of the substrate temperature being selected by the snake were averaged over the three replicate trials, giving a single value for each snake under each treatment condition. This conservative step removes concerns about lack of independence among the sequential measurements. The individual snakes were included as a factor ("repeated measures") in the analysis of variation (ANOVA) design, thereby controlling for individual differences and achieving the appropriate improvement in resolving power for the comparison of the three treatment conditions.

Gibson and co-workers have applied an atypical, thermal mosaic methodology to their laboratory studies of thermoregulatory behavior in Garter Snakes (Kollar, 1988; Gibson et al., 1989; Gibson et al., in review; Smucny and Gibson, in review). Study animals are housed in large arenas where they can choose from among shelter boxes presenting a discrete, usually small array of temperatures (e.g., 20 or 30°C). The relative frequency with which the snakes occupy the warmer (or warmest) shelters is used as an index of thermoregulation. The thermal mosaic methodology has the disadvantage that T_b is not known precisely because it is not measured directly. However, as used in these studies, the method has four attributes that may have contributed to its effectiveness. First, snakes in the thermal mosaic were not forced to remain exposed; they could retreat to cover at any time and could select either warm or cool shelters when they did so. Similarly, each of the studies has lasted for about 100 days, allowing the snakes near maximal accommodation to the test apparatus, and making it possible to increase resolving power by collecting large sample sizes. Third, thermoregulatory behavior can be assessed without disturbances that may result from the surgical implantation of transmitters in the coelom, the chronic placement of thermocouples through the cloaca, or the repeated capture and handling required for capture measurements of body temperature. Finally, the mosaic design allows the snakes to move without encountering temperature extremes; exploratory movement, for example, can be distinguished from body-temperature selection (Slip and Shine, 1988).

Physiological effects

Feeding. Following the early report by Regal (1966) that feeding can induce thermophily in snakes, research from three separate laboratories has provided particularly strong confirming evidence. Slip and Shine (1988) used coelomically implanted radiotransmitters to assess thermoregulatory behavior by Diamond Pythons (*Morelia spilota*) in a laboratory thermal gradient. They compared the body temperatures selected immediately before and after feeding in 15 individuals and found an average increase of about 2.5°C.

Lutterschmidt and Reinert (1990) compared substrate temperatures selected by Northern Water Snakes (*Nerodia sipedon*) under each of three treatments: (1) unfed for ten days, (2) ingested natural food item, and (3) dummy transmitter palpated into the snake's stomach. They confirmed a strong postprandial (after feeding) thermophily (Sievert, 1989), finding that fed snakes selected substrates 8.3°C warmer than did unfed snakes. It is worth noting that snakes with dummy transmitters in their stomachs selected substrate temperatures that were indistinguishable from those selected by the snakes when they had ingested natural food items.

Studies applying the thermal mosaic methodology to four separate groups of animals (*Thamnophis sirtalis* and *T. elegans*) have consistently identified a strong thermophily that persists for about 36 to 72 h following feeding (Kollar, 1988; Gibson et al., 1989; Gibson et al., in review; Smucny and Gibson, in review). For example, Gibson et al. (in review), using time-lapse video recording to monitor shelter selection continuously, described postprandial thermophily in young *T. elegans* (Figure 7.4*A*). In this study, the pattern of near-constant use of the heated (30°C) shelter for the first 36 h, followed by about 36 h of declining use, characterized both the light and dark phases of the 12L:12D light cycle.

The three approaches discussed here (Slip and Shine, 1988; Lutterschmidt and Reinert, 1990; Gibson et al., in review) have identified postprandial thermophily in two natricines and a python, and there are at least indications of its occurrence in other major taxa of snakes (Saint-Girons, 1975; Huey, 1982; Lillywhite, 1987). At present,

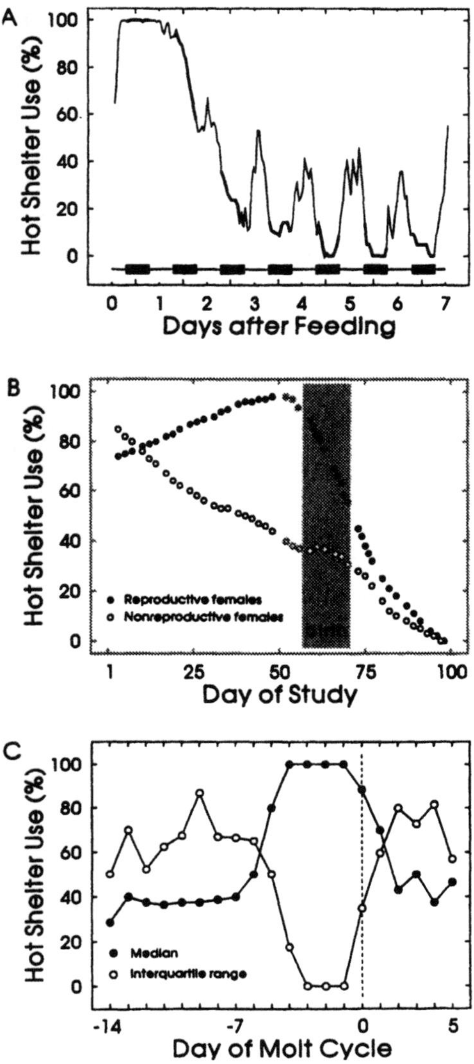

Figure 7.4 Patterns of heat use by captive Garter Snakes (*Thamnophis* spp.) housed in large laboratory arenas. The animals could choose from among shelter boxes presenting a discrete, usually small, array of temperatures (e.g., 20 or 30°C) maintained 24 h a day. Hot shelter use (HSU), an index of thermoregulatory behavior, is the relative frequency with which the study animals occupied the warmer or warmest shelters. (*A*) Average hourly HSU by seven young *T. elegans,* plotted as a function of days following weekly feedings. The dark phases of the 12L:12D photoperiod are marked by the bold portion of the data line and by the solid bars. Taken from Gibson et al. (in review). (*B*) Median HSU during the photophase by 22 adult female *T. sirtalis,* plotted across 100 days of study. The snakes were removed from artificial hibernation one week before the study began. Reproductive females (solid circles, n = 11) are those that mated and subsequently gave birth (shaded bar); nonreproductive females (open circles, n = 11) are those that did not become gravid. The values plotted are the trend lines fitted using LOWESS smoothing. Adapted from Smucny and Gibson (in review). (*C*) HSU during the photophase by 39 young *T. sirtalis,* plotted with respect to the day on which ecdysis occurred (day 0). The days preceding molt are represented by negative values; days −5 to −1 encompassed the "dull skin, cloudy eyes" states. Solid circles: daily medians; open circles: daily interquartile ranges. Adapted from Kollar (1988) with permission.

however, we cannot say if the phenomenon is rare or nearly universal in snakes. If strong methodologies were applied to additional species, the results would allow estimation of its systematic and ecological distribution.

Extension of laboratory results to the field will be difficult but postprandial thermophily could be an important influence on free-ranging snakes. For example, seasonal changes in feeding frequency could result in seasonal changes in the frequency of feeding thermophily. The resulting pattern of T_b variation might be mistaken for circannual changes in preferred body temperature. Feeding differences among years, locations, or even species could produce similar effects. Future laboratory work might explore whether or not snakes maintained on an *ad libitum* feeding regime exhibit a constant thermophily. Is there a lower threshold of meal size needed to elicit a thermophilic response? Clearly there is added value in careful studies of feeding frequencies in natural populations of snakes.

Condition. The cost–benefit analysis of ectotherm thermoregulation provided by Huey and Slatkin (1976) and Huey (1982) can predict a decrease in body-temperature selection in response to reductions in overall feeding success: the costs associated with thermoregulatory behavior and the metabolic costs resulting from elevated body temperature may exceed the energy obtained from feeding. Despite this conceptual framework and the widespread characterization of reptiles as high-efficiency, low-energy organisms (Pough, 1980), little attention has been paid to assessing the impact of feeding frequency (here equated with condition) on temperature selection in snakes.

Gibson et al. (1989), using the thermal mosaic methodology described above, compared the heat use pattern of young Common Garter Snakes (*Thamnophis s. sirtalis*) fed once a week with that obtained when the animals were fed once every two weeks (biweekly). Both groups showed a strong postprandial thermophily of similar duration. Over the remainder of the feeding schedule, however, snakes on the weekly regime used the 30°C shelters about twice as frequently as they did when fed biweekly. Kollar (1988) replicated and extended this effect of condition, in a study comparing twice-weekly, weekly, and biweekly feeding regimes. In one comparison, 10% use of the 30°C shelters on the biweekly regime contrasted with 60% use on the twice-weekly regime. Shine and Lambeck (1990) used radiotelemetry to compare thermoregulatory behavior of free-ranging Australian Black Snakes (*Pseudechis porphyriacus*) between disjunct spring and summer seasons. Despite the higher environmental temperatures of the summer period, the Black Snakes selected body temperatures that were on average 5°C lower than those during an earlier spring period. The authors attributed this difference and a reduction in activity levels by the snakes to a sharp decline in the availability of the Black Snake's prey, itself a consequence of a drought that characterized the summer of the study.

As with postprandial thermophily, the effects of condition on thermoregulation in snakes could be a general and important phenomenon in their ecology. Variation in condition might produce variation in the temperatures selected by free-ranging snakes that appeared to be attributable to different seasons, sites, and systematic groupings. The thermoregulatory behavior of animals captured in the field and tested in the laboratory could be influenced for some time by past foraging success. For several reasons, then, it would be valuable to determine the distribution of a condition effect on temperature selection over an ecological and systematic sampling of snakes, and the frequency and magnitude in natural populations of variation in condition. It is not difficult to restrict ration in laboratory studies, but field experiments here present challenges. Perhaps it would be possible instead to supplement the diet of marked individuals in populations undergoing irregular food shortages, such as those caused by drought.

Reproductive status. Gravid snakes of several species have been shown to maintain body temperatures that on average are higher and less variable than those displayed by nongravid females, postpartum females, or males. Strong demonstrations of these differences now are available from laboratory and field studies. In captivity, female pythons of several species use shivering thermogenesis to maintain their body temperature and that of their eggs at a relatively high, constant level (e.g., Hutchison et al., 1966a; reviewed by Shine, 1988). Female Diamond Pythons (*Morelia spilota*) also have been shown to display this capacity in the wild (Slip and Shine, 1988). Gravid Timber Rattlesnakes (*Crotalus horridus*) differ from males and nongravid females by selecting warmer, more open habitats, basking frequently, and by maintaining higher body temperatures (Reinert, 1984; Reinert and Zappalorti, 1988; Reinert, Chap. 6, this volume; Reinert, personal communication). Free-ranging gravid Great Basin Rattlesnakes (*Crotalus viridis lutosus*) maintain body temperatures that are higher and more precise than those of nongravid females (Fig. 7.7*A*); this careful, intense thermoregulation appears to change abruptly following parturition (Fig. 7.7*B*). Other recent telemetric studies also have found similar clear effects of pregnancy (Charland, 1991; Graves and Duvall, 1993). In a laboratory thermal mosaic, adult female *Thamnophis sirtalis* displayed a high level of heat use following emergence from artificial hibernation. Hot shelter use by reproductive females (those that mated and became gravid) slowly increased until it reached nearly 100% at parturition and then decreased abruptly. Nonreproductive females (those that did not become gravid) instead displayed a continuous decline in use of the warmest shelters. Following parturition by the reproductive females, the heat-use patterns of postpartum and nonreproductive females could not be distinguished (Fig. 7.4*B*).

The recent studies discussed above suggest two reasons why the thermophilic effects of pregnancy might have been difficult to detect using capture measurements of the body temperatures of free-ranging snakes. First, the laboratory study (Smucny and Gibson, in review) compared reproductive and nonreproductive females that had been removed from hibernation on the same day. In field populations, we can expect much greater variation in emergence times (Gregory, 1982); on a given date, then, the later emerging nonreproductive females could be selecting higher temperatures than their earlier emerged counterparts. Relative to the synchronized laboratory study, the effect could be to raise the mean and the variance of the body temperatures of the nonreproductive females, thus reducing the difference with the gravid females. Capture measurements from field studies face a second bias that could have reduced their ability to detect thermoregulatory differences between gravid and nongravid females. Behavioral thermoregulation often requires that the animal be exposed on the surface (but not always—see Huey et al., 1989), and it is these exposed individuals that are most likely to be sampled in nontelemetric field studies. Gravid, postpartum, and nonreproductive females might share a common preferred body temperature, yet diverge substantially in the time spent thermoregulating. In this case, capture measurements would sample only actively thermoregulating individuals and detect no differences among the reproductive classes. Telemetric and laboratory studies, which are able to sample animals more representatively, thereby might detect real differences if they exist (Smucny and Gibson, in review).

As with all the physiological factors that appear to affect body-temperature selection in snakes, there is a great need for a broader ecological and systematic representation. And in those species for which pregnancy has been shown to enhance thermoregulatory behavior, more specific "follow-up" questions are possible. For example, does the thermoregulatory behavior of gravid females intensify along an environmental gradient of increasing thermal constraint, such as that often imposed by elevation or latitude? Similarly, the effects of reproductive status on thermoregulation by oviparous females and on males of all species warrants greater attention. Many oviparous snakes are large enough to carry modern transmitters easily, and many also breed well in the laboratory. Both field and laboratory approaches should be applicable. Prior to egg-laying, do gravid oviparous females maintain higher, less variable body temperatures than nongravid females? Presented with a choice of potential nest sites differing only in temperature, do gravid females select a particular thermal environment for their eggs? Male *Thamnophis elegans* apparently select higher body temperatures earlier in the season than they do later (Scott, 1978; Scott and Pettus, 1979), displaying a pattern similar to that of nonreproductive female *T. sirtalis* (Fig. 7.4*B*). Is their thermoregulation therefore at its most intense during courtship, immediately after emergence in spring? Assessing any thermoregulatory effect of the spermatogenic cycle will require monitoring that cycle, which usually will be a more difficult undertaking than determining whether a given female is gravid or not. Is the basking period that follows emergence in male European Adders (*Vipera berus*) an expression of their spermatogenic cycle or of the shed that typically ends the basking period (Viitanen, 1967)?

Ecdysis. In light of Kitchell's (1969) long-standing report that ecdysis induced a thermophobic response in four species of colubrid snakes, it was unexpected that recent field and laboratory studies would identify molt as a particularly strong thermophilic effect. Long-term telemetric field studies now demonstrate that male and nongravid female Timber Rattlesnakes (*Crotalus horridus*) alter their habitat selection prior to molt. The two classes shift from forested foraging habitats to the more open habitats suitable for basking that also are characteristic of gravid females (Reinert, 1984; Reinert and Zappalorti, 1988; Reinert, Chap. 6, this volume; H. Reinert, W. Brown, W. Martin, personal communication). Young *Thamnophis sirtalis* in a laboratory thermal mosaic display a pronounced, clearly defined thermophily beginning about six days before molt and ending the day of molt or the first day thereafter. The four days before ecdysis, which encompass the stages of dull skin and cloudy eyes, show extremely low variation in hot shelter use (Fig. 7.4*C*). During the thermophilic period before shedding, snakes frequently were observed resting their heads directly on the heated substrate of their shelters, an unusual behavior not observed, for example, in gravid or recently fed animals (Kollar, 1988). Ecdysis has exerted an equally strong effect on thermoregulatory behavior with every group of snakes examined using the mosaic methodology (Kollar, 1988; Gibson et al., 1989; Gibson et al., in review; Smucny and Gibson, in review).

It is worth considering why ecdysis could emerge as a major thermophilic influence in field telemetric work of Timber Rattlesnakes and laboratory investigations of Garter Snakes, and yet has gone essentially unnoticed in so many previous laboratory and field studies. For example, thermophilic effects have long been attributed to feeding and pregnancy, even if the strongest demonstrations are much more recent. One possible explanation for the discrepancy, suggested by Kitchell himself (1969), is that molt induces a tendency to seek cover, so as to reduce exposure to predators at a time when the snake's vision is impaired. If cover-seeking is a stronger tendency than heat-seeking in molting snakes, then these animals would be undersampled in nontelemetric field studies. And if cover objects do not offer the range of thermoregulatory opportunities available to snakes exposed on the surface, the thermophilic effects of molt could be masked or reversed. Field telemetric studies could address the question if animals can be visually checked often enough to detect molt events. Laboratory studies could evaluate the relative priorities assigned to thermophily and retreat by molting snakes.

Other factors. Other physiological influences on thermoregulatory behavior (e.g., disease, parasitism, dehydration, hypoxia, hypercapnia, and acclimation), as well as ontogenetic effects, have been identified in a variety of ectotherms but have received little or no attention in snakes (Huey, 1982; Lillywhite, 1987; Hutchison and Dupré, 1992). Using the thermal mosaic methodology, Kollar (1988) described a slight decline in heat use by young female Garter Snakes as they grew to a size where they attained sexual maturity, although the result has other ready explanations.

Behavioral fever (Kluger et al., 1975) has been described for a wide variety of ectotherms (Hutchison and Dupré, 1992), but Zurovsky et al. (1987) found no such effect in their study of the Olive Grass Snake (*Psammophis phillipsii*) and the Brown House Snake (*Lamprophis fuliginosis*). Behavioral fever should be sought in additional species of snakes, perhaps with a combination of the careful microbiological techniques of Zurovsky et al. (1987) and thermoregulatory testing procedures that are more accommodating to normal snake behavior. It would be particularly helpful if such studies, when carried out, included a description of the biochemical and physiological responses to bacterial infection. If behavioral fever is shown to occur in snakes, the question will shift to estimating its importance in natural populations. Doing so likely will depend on data from controlled laboratory studies because of the difficulty in determining in the field if snakes have bacterial infections.

Two considerations suggest that students of the behavior and ecology of snakes should remain alert to the possibility of other, as yet unidentified, influences on thermoregulation. First, field and laboratory studies often identify sufficient unexplained variation in thermoregulatory behavior to accommodate unknown factors with effects of considerable magnitude (Shine, 1987; Slip and Shine, 1988; Gibson et al., 1989; Smucny and Gibson, in review). Second, new candidates do arise. Ecdysis remained essentially unsuspected for some time, and Geiser et al. (1992) recently have demonstrated that the levels of polyunsaturated fatty acids in the diet of a lizard (*Tiliqua rugosa*) can influence its selection of body temperatures. The normal gastrointestinal flora of mice influences temperature regulation in these endotherms (Conn et al., 1991) and could exert a similar effect in snakes.

Influences of the biotic environment

As depicted in Fig. 7.1, the prey, predators, competitors, and mates of a snake represent important potential influences on the body temperatures that it will accept. Unfortunately, these biotic variables have received much less attention than the physiological variables that affect behavioral temperature selection. A particularly welcome exception is provided by long-term telemetric studies of *Crotalus horridus,* which demonstrate the effect of prey distribution (Reinert, 1984; Reinert and Zappalorti, 1988; Reinert, Chap. 6, this volume). The foraging habitat and foraging mode of this snake largely preclude active thermoregulation, and as a result the animals apparently spend long periods of the active season as thermoconformers. As discussed elsewhere, pregnancy and molt in this species are associated with a shift to more open habitats and the adoption of behavioral thermoregulation. Ongoing work (Lutterschmidt, 1991; H. Reinert, personal communication) indicates that this habitat shift reflects primarily a change in the priority attached to thermoregulation, rather than a change in preferred body temperature. Similar constraints are imposed on Sidewinders (*Crotalus cerastes*) by the nocturnal activity patterns of their prey (Secor, 1992). On a much shorter time span, Garter Snakes foraging in cold water can be forced to accept body temperatures much below the preferred value of about 30°C (Gregory and Nelson, 1991; C. R. Peterson and S. Arnold, unpublished data); indeed, the range and frequency of these body-temperature changes can be dramatic and give the appearance of poor regulation (Fig. 7.6*D*).

There is very little information on the possible influence of predators on a snake's temperature selection. Peterson (1982, 1987) speculated that snakes may delay emergence until later in the morning when T_e's are higher and allow for quicker increases in body temperature. Elsewhere we discussed Kitchell's (1969) hypothesis that ecdysis indirectly may affect temperature selection if molting snakes seek cover to reduce their vulnerability to predation. Additional studies in this general area would be particularly instructive.

Temporal Influences

Many ectotherms display intrinsic daily or seasonal rhythms of temperature selection (Huey, 1982; Lillywhite, 1987; Hutchison and Dupré, 1992), as distinct from daily or seasonal patterns of constraint imposed by the animal's thermal environment (Peterson, 1987; following section on field body temperatures). Again, the situation for snakes is much less well established. Several laboratory studies reported changes in temperature selection across the photophase (Scott, 1978; Justy and Mallory, 1985; Kollar, 1988; Smucny and Gibson, in review), but the clearest demonstration of daily patterns was that of Gibson et al. (in review). Young *Thamnophis elegans* housed in a large thermal mosaic displayed a pronounced nocturnal hypothermia (Regal, 1967) with a peak in heat use about midway through the photophase (Fig. 7.4*A*). The unimodal pattern across the photophase appeared not to be simply a reflection of a diurnal activity pattern in this snake. Laboratory studies also identified longer-term or "seasonal" patterns of change in temperature selection. Nonreproductive female *Thamnophis sirtalis* in a thermal mosaic displayed a gradual, apparently regular decline in heat use over a 100-day study, a pattern quite different from that displayed by reproductive females (Fig. 7.4*B*). A remarkably similar pattern of seasonal decline characterized male *T. elegans* when tested in a laboratory thermal gradient (Scott, 1978; Scott and Pettus, 1979) and a mixed-sex sample of free-ranging *T. elegans* studied using radiotransmitters (Fig. 7.8*B*).

Body-Temperature Variation in the Field

In this section, we address the question of how the body temperatures of snakes actually vary under natural conditions. This question is of central importance because it describes the dependent variable of the causation studies and the independent variable of the consequences studies. In this section, the two principal methodologies for measuring field T_b's (capture measurements and radiotelemetry) are briefly described and evaluated. We then provide examples of daily and seasonal patterns of T_b variation and ways of categorizing that variation. We finish with a discussion about evaluating body-temperature variation of free-ranging snakes.

Methods

Capture measurements. Capture measurements are the most commonly used method for obtaining body-temperature measurements in the field. The goal is to capture a snake before or as quickly as possible after it has been disturbed. Cloacal or esophageal temperatures are then taken with a quick-reading thermometer (e.g., a mercury thermometer with a thin glass bulb, a thermistor, or a small gauge thermocouple). See Avery (1982) and the references therein for further details concerning making capture measurements.

Capture measurements have several advantages and disadvantages. Compared to radiotelemetry, the equipment is inexpensive and easy to use. Measurements can be made on small snakes and data can be collected from many individuals. Perhaps the greatest disadvantage of capture measurements is the difficulty in measuring the body temperatures of snakes while they are undercover or underground, which may be the majority of the time (Huey, 1982; Peterson, 1987; Huey et al., 1989). Capture measurements are usually hard to replicate, making it difficult to obtain data on intraindividual variation or to analyze interindividual variation in T_b statistically. Finally, capture measurements require the disturbance of the snake being measured and possibly other snakes in the population as well.

Capture measurements can provide population data on activity temperatures of the full size range of snakes, especially those that are surface-active. These data are often summarized as frequency distributions, which indicate the range and modal temperatures of active snakes. These distributions typically are negatively skewed with a modal temperature in the 28–32°C range. If the sample sizes are large enough, these data are also useful for evaluating the effects of time, sex, and location on the activity temperatures of populations (Gibson and Falls, 1979; Rosen, 1991; Plummer, 1993). Capture measurements are also useful for determining appropriate conditions for maintaining captive snakes and for thermal treatments in laboratory studies.

Summaries and reviews of capture measurement studies are found in Brattstrom (1965), Avery (1982), and Lillywhite (1987). Body-temperature measurements for less than 5% of all snake species have been reported. Consequently, it has proven difficult to make generalizations about activity body temperatures from these data, primarily because of differences in the way the data are obtained and a lack of accompanying data concerning environmental conditions and the physiological condition of the snakes.

Radiotelemetry. Although most telemetry studies of snakes have focused on movement patterns and habitat selection, radiotelemetry is very useful in studying snake temperature relationships as well (Reinert, 1992). Transmitters can be designed with a thermistor in the circuit that causes the pulse rate to increase with increasing temperature. Careful calibration of transmitters in the laboratory (using a temperature-controlled water bath) permits remote measurements of body temperatures accurate to within 0.2°C.

In most studies, transmitters are surgically implanted into the snake's peritoneal cavity in the posterior portion of the body. The use of a whip antenna placed under the skin (Reinert and Cundall, 1982) results in a greatly increased range over a contained, coiled antenna. Some biologists force-feed transmitters to snakes. This technique has the advantage of not requiring surgery but the disadvantage of transmitters being regurgitated or passed. Care must be taken to be sure that the transmitter does not alter the snake's normal thermoregulatory behavior. Relatively large, force-fed transmitters may cause snakes to behave as if they are digesting a meal (Lutterschmidt and Reinert, 1990). Reinert (1992) provides a detailed description of the use of radiotransmitters with snakes.

Radiotelemetry has several advantages over capture measurements. Telemetry makes it possible to measure the body temperatures of snakes even when they are in retreat. This is critical for obtaining a representative picture of field T_b variation. Because telemetry does not require recapturing the snake to measure its T_b, its use decreases disturbance of the telemetered snake and other snakes that may be in the area. It is also possible to measure repeatedly the body temperature of the same individual. Although radiotelemetry is usually employed to make measurements of snakes infrequently (e.g., daily), it is most informative when temperature measurements can be made at regular, short intervals (e.g., 15 min). Such continuous T_b data can reveal temporal patterns of T_b variation, correlations with weather conditions, the effects of physiological condition (e.g., reproductive status), and intra- and interindividual differences in T_b variation (Peterson and Arnold, 1986; Peterson, 1987; Shine and Lambeck, 1990; Charland, 1991).

To obtain continuous T_b data, a variety of systems can be used to automate the measurement of telemetered T_b's. The techniques include using timer-controlled tape recorders or data loggers to record the signals from radioreceivers (Peterson, 1987; Peterson and Dorcas, 1992, in press). Automated telemetry is especially useful when multiple snakes are within range and can be telemetered (e.g., during hibernation or with species that do not disperse long distances). Automated telemetry not only makes gathering T_b data easier, but it also may improve the quality of data for those species that are affected by the presence of a human observer. *Thamnophis elegans,* for example, will often retreat under cover in response to humans (even if the observer is 5 m or more away from the snake) and consequently experience lower T_b's (Peterson, 1982). Automated telemetry also makes it possible to study the effects of human disturbance on free-ranging snakes (i.e., do the T_b and activity patterns of snakes vary when humans are present?).

Radiotelemetry entails several disadvantages. A serious limitation is the size of snake that can be studied. The adults of small species and the juveniles of most species cannot be studied, and data sets are consequently biased toward the adults of large species. This situation has recently improved somewhat with the availability of temperature-sensitive transmitters under 2 g that permit the study of snakes weighing as little as 40 g. Transmitters may affect a snake's behavior (e.g., feeding rate, body-temperature selection, activity level, and locomotor performance), but few studies have addressed this problem and more controlled experiments are needed. A preliminary study of the effect of surgical implantation of radiotransmitters on the activity levels of two captive *Crotalus viridis* indicated that the snakes' activity levels decreased by about 70% following surgery and did not return to presurgery levels until at least 7 days later (L. Burns, unpublished data). Sample sizes in telemetry studies are often small because of equipment expenses and the amount of time required to keep track of free-ranging snakes. Sample sizes are often further reduced because of equipment failures (bad batteries, penetration of coelomic fluid into the electronics, poor electrical connections, etc.).

Daily patterns of body-temperature variation

Daily patterns of body-temperature variation measured via radiotelemetry can be divided into several categories based on the shape of the plots of T_b versus time of day (Peterson, 1987). This simple classification scheme is used to facilitate description, analysis, and discussion of T_b patterns. Comparisons of these patterns with environmental and behavioral data are required to evaluate the causes of any particular pattern and similar patterns may result from different causes. Intermediate patterns exist and changes in this system will undoubtedly be required as the T_b patterns of more snake species are described.

The first type, termed the plateau pattern, consists of three phases: a short, rapid heating phase in the morning, an extended, plateau phase that is relatively stable during the day, and a long, slow cooling phase from the late afternoon or evening until the following morning. (Figs. 7.5*B*, 7.6*E*, 7.7*A*, and 7.7*B*). This pattern usually occurs when the environment allows snakes to attain T_b's within their preferred

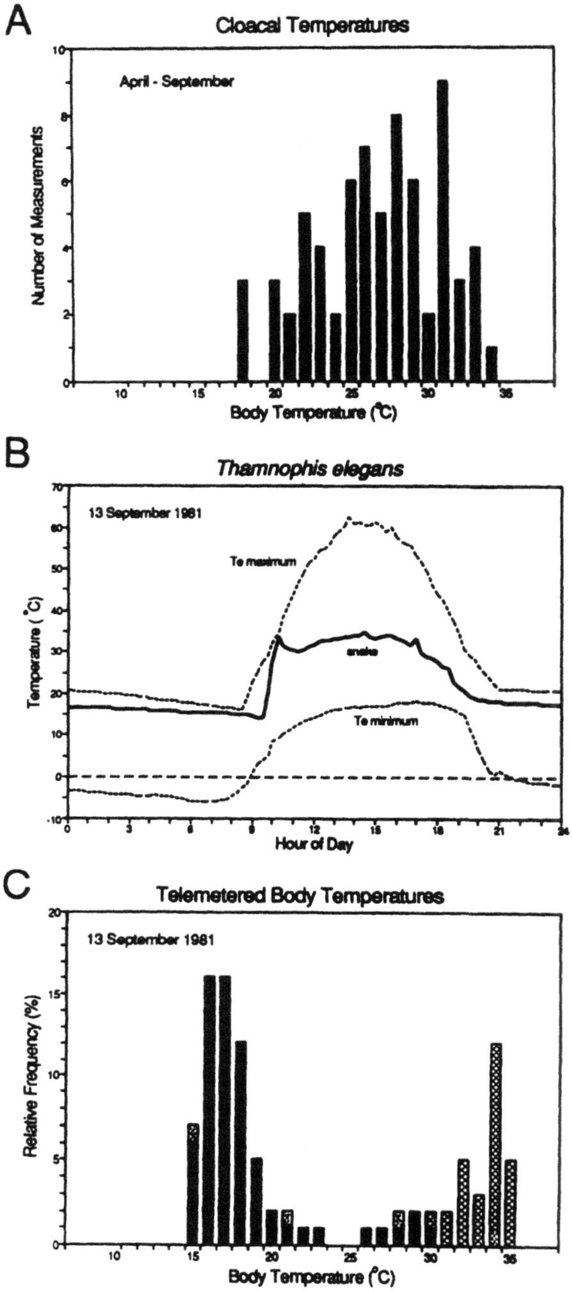

Figure 7.5 Comparison of capture (cloacal) and telemetered field T_b's of *Thamnophis elegans*. (*A*) Frequency distribution of cloacal temperatures for active snakes, April through September ($n = 70$). (*B*) Plateau T_b pattern for a nongravid female in eastern Washington. Minimal and maximal operative environmental temperatures (from 47 models) are indicated by the stippled lines. (*C*) Relative frequency distribution of telemetered body temperatures for the T_b data shown in *B*. Temperature readings from the cooling (solid), heating (stippled), and plateau (crosshatched) phases are distinguished (R. Stevenson and C. R. Peterson, unpublished data).

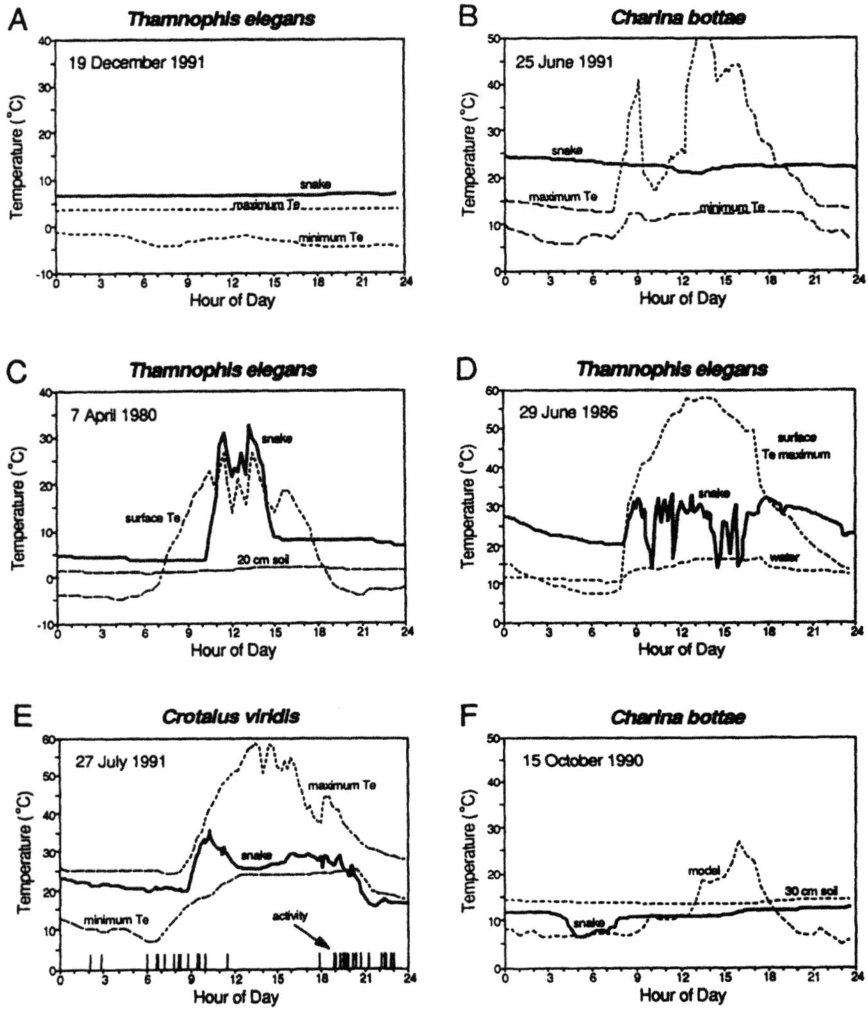

Figure 7.6 Examples of different types of daily patterns of snake T_b variation determined by radiotelemetry. Operative temperatures are indicated by broken lines. (*A*) Low, smooth T_b pattern of a hibernating *Thamnophis elegans* in southeastern Idaho. (*B*) High, smooth T_b pattern of a *Charina bottae* in southeastern Idaho; the snake remained in a burrow for the entire day. (*C*) An oscillating T_b pattern for a *Thamnophis elegans* in eastern Washington that resulted from varying cloud cover. Adapted with permission from Peterson (1987) (Ecological Society of America). (*D*) An oscillating T_b pattern for a female *Thamnophis elegans* that apparently resulted from foraging in a stream in northeastern California (C. R. Peterson and S. Arnold, unpublished data). (*E*) A plateau pattern for a female *Crotalus viridis* in southeastern Idaho. Low T_b's in the evening were associated with surface activity, indicated by the lines at the bottom of the figure that represent toggles of a mercury switch in the snake's transmitter (V. Cobb, unpublished data). (*F*) An oscillating T_b pattern for a nongravid female *Charina bottae* in southeastern Idaho that resulted from surface activity in the early morning.

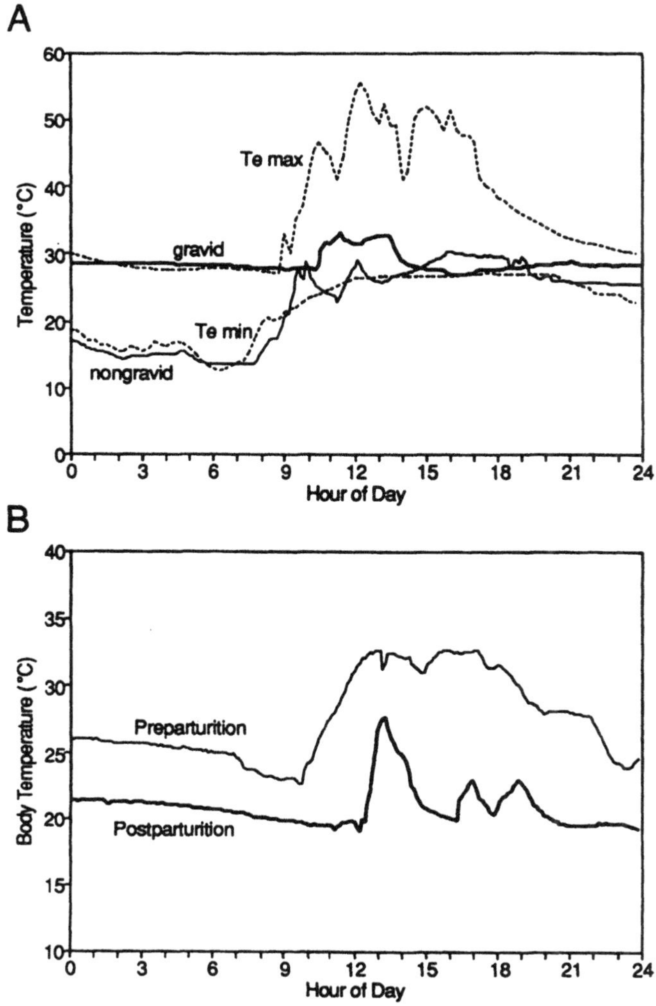

Figure 7.7 Daily body-temperature patterns for *Crotalus viridis* in southeastern Idaho. (*A*) Body temperatures of a gravid and a non-gravid *C. viridis* on the same day. Minimal and maximal operative environmental temperatures are also indicated. (*B*) Body temperatures of a female *C. viridis* just before and soon after birth (V. Cobb, unpublished data).

(selected) T_b range. This is the most common pattern observed in active *Thamnophis elegans* (Peterson, 1987; Charland, 1991) and has been observed in a variety of other species, including *Charina bottae* (Peterson and Dorcas, 1992), *Morelia spilota* (Slip and Shine, 1988), *Masticophis lateralis* (Hammerson, 1979), *Nerodia fasciata* (Osgood,

1970), *Thamnophis sirtalis* (Charland, 1991), *Pseudechis porphyriacus* (Shine and Lambeck, 1990), *Vipera aspis, V. berus* (Saint-Girons, 1975), *Agkistrodon contortrix* (Sanders and Jacob, 1981), and *Crotalus viridis* (Cobb and Peterson, 1991).

During the plateau phase, *Thamnophis elegans* usually select a relatively narrow range of temperatures. For example, the average plateau phase T_b range for *T. elegans* was 28.0 to 32.5°C (Peterson, 1987), which is similar to T_b selection in laboratory thermal gradients. The mean values of capture measurements of activity T_b's sometimes are lower than telemetered, plateau phase mean T_b's because of the inclusion of measurements taken when T_e's were lower than preferred temperatures or taken from snakes still in their heating phases. Body-temperature variation during individual plateau phases was relatively low (mean range = 4.5°C, standard deviation = 1.3°C). Limited data from snakes implanted with temperature and activity radiotransmitters suggest that the variation in T_b increases somewhat when the snakes are actively moving at the surface (C. R. Peterson and K. Brown, unpublished data).

Nighttime cooling phase T_b's may be relatively high or low, depending on the location of the snake (e.g., the T_b of a snake in a burrow would be warmer than a snake at the surface; Figs. 7.5*B* and 7.6*E*). The most pronounced differences among individuals in their field T_b's usually occur at night because of differences in nocturnal microhabitat selection (e.g., Fig. 7.7*A*).

Body-temperature frequency distributions for plateau phase patterns are generally bimodal; the low and high portions of the distributions correspond to the cooling and heating phases, respectively (Fig. 7.5*C*). Frequency distributions of activity T_b's from cloacal measurements are usually unimodal (Fig. 7.5*A*) because cooling phases T_b's are not usually measured. An awareness of the differences between these distributions is important when designing laboratory studies and when interpreting the thermal dependency of biological processes.

Oscillating patterns are characterized by one or more periods of marked variation in T_b, usually during the day. These oscillations may result from different causes. Variation is often imposed on snakes by the thermal environment. For example, variable cloudiness resulted in the T_b oscillations seen in Fig. 7.6*C*. Alternatively, some snakes may accept variations in T_b in order to be active. The oscillations in Fig. 7.6*D* presumably resulted when the Garter Snake was foraging for fish in the relatively cool water of a mountain stream. The early morning oscillation in the T_b of the *Charina bottae* in Fig. 7.6*F* resulted from the snake emerging from its relatively warm burrow and being active at the cooler surface for several hours. A third type of oscillating pattern occurs even when the thermal environment does not appear to be constraining snakes. The T_b curve for the postpartum Great Basin Rattlesnake in Fig. 7.7*B* indicates that the snake emerged late in the morning, heated up to about 30°C, and then retreated from the exposed surface and cooled down, even though it could have maintained its T_b at a higher level (V. Cobb, unpublished data). This type of pattern corresponds with some of the results observed in laboratory studies of hot shelter use by postpartum Garter Snakes (Fig. 7.4*B*).

Smooth body-temperature patterns are characterized by gradual or no change in T_b throughout an entire 24-h period. Smooth patterns may result when environmental conditions are too cool for snakes to be active at the surface (Fig. 7.6*A*). This pattern has been described in hibernating *Charina bottae* (M. Dorcas, unpublished data), *Thamnophis elegans* (M. Dorcas, unpublished data), and *Crotalus viridis* (Cobb and Peterson, 1991), and probably is typical of most species when they are hibernating or aestivating. Smooth patterns also occur in snakes that inhabit thermally stable environments such as Arafura File Snakes (*Acrochordus arafurae*) in billabongs (Shine and Lambeck, 1985). Warm, smooth patterns have been observed in *C. bottae* (Fig. 7.6*B*) and in gravid *T. elegans* (Huey et al., 1989), which remain in burrows or under rocks for the entire day. These snakes were able to maintain T_b's within the preferred range throughout most of a day without emerging from their retreat sites.

Seasonal patterns of body-temperature variation

Extensive seasonal as well as daily variation occurs in snake body temperatures. The percent of the different types of T_b patterns of *Thamnophis elegans* varied seasonally (Peterson, 1982), generally corresponding to thermal conditions. Plateau patterns predominated in the summer (87%); oscillating patterns were most common in spring and fall (45% and 46%, respectively); and smooth patterns predominated in the late fall and presumably persisted through hibernation into early spring.

The daily minimal, maximal, and mean body temperatures of *Thamnophis elegans* showed an increasing–decreasing pattern of seasonal variation from the spring through the fall (Fig. 7.8*A*). Minimal daily body temperatures showed the greatest seasonal effect because of the annual cycle in ground temperatures; the daily minimal tem-

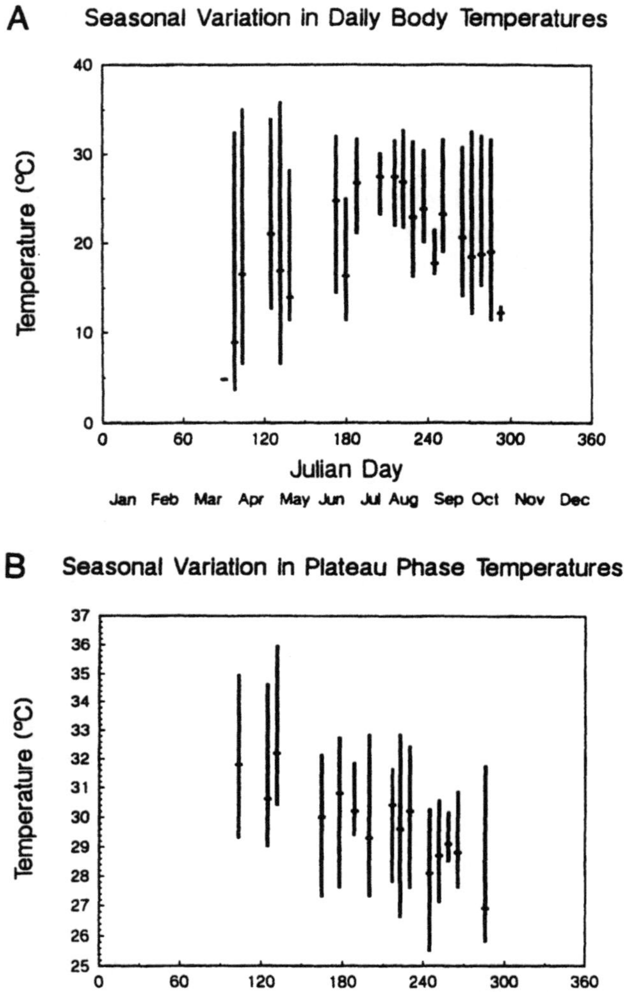

Figure 7.8 Seasonal change in body temperatures for male and female *Thamnophis elegans* in eastern Washington from June through October of 1979 and March through May of 1980. Each bar represents data from one snake for one day. (*A*) Minimal, maximal, and mean daily body temperatures. (*B*) Minimal, maximal, and mean T_b's recorded during the plateau phases; all measurements were taken when operative environmental temperatures were greater than 30°C (Peterson, 1982).

perature almost always occurred when the snakes were in retreat just before emergence (e.g., Fig. 7.5*B*). Maximal daily T_b's showed the least effect of season, presumably because they were determined more by the thermoregulatory behavior of the snakes than by environmental constraint. Seasonal variation in mean daily body temperatures was intermediate, but more closely resembled the pattern for the daily minimum because most of the snakes' time was spent in the cooling phases.

The duration of the different phases of plateau-pattern T_b records also showed marked seasonal variation. The duration of the plateau phases increased from spring to summer and then decreased into fall. Cooling phase durations showed the opposite pattern, decreasing from spring to summer and then increasing into the fall. Interestingly, heating phase durations did not vary seasonally because snakes emerged later in the day during the spring and fall.

Minimal, maximal, and mean snake temperatures during the plateau phase also varied seasonally, decreasing from spring to summer to fall (Fig. 7.8*B*). The range and standard deviation of plateau phase T_b's showed no seasonal change, however. This indicates that the seasonal changes in T_b's were due to changes in the snakes' behavior rather than the direct result of changing environmental temperatures. Note the correspondence between this pattern and the seasonal change in hot shelter use by nonreproductive Garter Snakes (Fig. 7.4*B*).

Although seasonal variation in snake T_b's is probably most extensive in high-latitude snake populations, it also occurs in less thermally variable environments. Snakes may utilize their thermal environment in different ways because of ecological factors other than the range of temperatures available. For example, *Pseudechis porphyriacus* select lower body temperatures in the summer, when it is actually warmer, than in the spring, presumably because snakes are feeding less in the summer when prey are less available (Shine and Lambeck, 1990). In lower latitudes, it may be especially important to determine the low as well as the high end of the available temperature range. Documentation of natural T_e ranges for snakes is one of the most important areas of snake thermal biology that needs to be addressed.

Although many species of snakes spend a considerable amount of time hibernating, it has been difficult to obtain T_b's from hibernating snakes because they are usually inaccessible underground. The use of radiotelemetry during the past 20 years has greatly improved our understanding of body-temperature variation in hibernating snakes. The general pattern of T_b variation for hibernating snakes studied so far appears to be a gradual decline in T_b from the fall into late winter and then a gradual increase in T_b until spring emergence (Jacob and Painter, 1980; Gregory, 1982; Macartney et al., 1989; Weatherhead, 1989). For example, *Crotalus viridis lutosus* in the desert in southeastern Idaho entered hibernation during October and emerged in May (Fig. 7.9). Mean body temperatures decreased from approximately 12°C in mid-November to a minimum of 6°C in

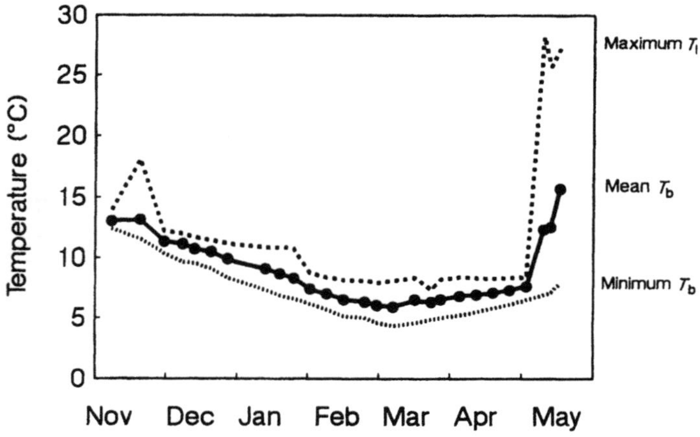

Figure 7.9 Minimal, maxiumal, and mean body temperatures of seven hibernating *Crotalus viridis lutosus* at a den site in southeastern Idaho during the fall, winter, and spring of 1989–1990. Measurements were taken about once per week (C. R. Peterson and V. Cobb, unpublished data).

late February and then increased to approximately 7.5°C by late April. The lowest T_b experienced by any of the snakes was 4.2°C. During the winter, differences in T_b's among individuals measured at the same time were as great as 4°C. These results indicate considerable interindividual variation in body temperatures of hibernating snakes. These data, in combination with observations of snake movements within dens (Sexton and Hunt, 1980; Sexton and Marion, 1981; C. R. Peterson and V. Cobb, unpublished data) raise the intriguing possibility of active T_b selection during hibernation. The extent to which snakes select T_b's from available gradients during hibernation, the consequences of this variation, and its relationship with the physiological condition of individual snakes deserves further study in both the field and laboratory. An alternative hypothesis to explain interindividual T_b variation during hibernation is that some snakes may aggregate and elevate their T_b's above den temperatures by trapping heat produced through metabolism (White and Lasiewski, 1971).

Evaluating field body-temperature data

Although temperature is one of the easiest variables to measure and many environmental and body temperature measurements have been made, the interpretation of those data has been surprisingly difficult. The proximate and ultimate causes of body-temperature variation are complex, and a variety of data are required to evaluate them. A general goal is to quantify the extent of thermoregulation. In most cases, categorizing a species simply as a thermoconformer or a thermoregulator is probably misleading because individuals may carefully control their body temperatures under some circumstances (e.g., gestating) and allow considerable body-temperature variation at other times (e.g., when foraging).

Many studies have attempted to evaluate temperature variation without having adequate data to do so. Ideally, four types of data are required: (1) the distribution of possible temperatures for the snakes' environment; (2) information on snake temperature-selection behavior (i.e., their thermoregulatory set points); (3) if physiological and behavioral factors affect T_b selection, then information on those relationships and the condition of the snakes in the field will also be required; and (4) measurements of T_b variation for animals under field conditions. Repeated measurements on the same individuals are especially useful.

Although no study has yet measured all of these variables, indices quantifying the extent of thermoregulation can be calculated by comparing actual T_b's and the distribution of possible T_b's with the temperatures snakes would select in a thermal gradient. Hertz et al. (in press) discuss this issue in detail, present a research protocol, and apply that protocol to lizards of the genus *Anolis*. Refinement and application of their approach to snakes should significantly improve our evaluation of the thermoregulatory behavior of snakes.

Functional Effects of Body-Temperature Variation

Importance

To understand the consequences of body-temperature variation in snakes, it is necessary to determine the extent to which temperature affects their ability to carry out various biological processes. Knowing the proximate consequences of body-temperature variation in snakes is the central link between body-temperature variation and ecological and evolutionary consequences (Fig. 7.1). Many snakes experience large fluctuations in body temperature over the course of a season (Fig. 7.8) and even a single day (Fig. 7.6; Peterson, 1987). These fluctuations can have serious consequences for snakes because many developmental, physiological, and behavioral processes are greatly affected by temperature, and thus are important to their survival, growth, and reproduction (Huey, 1982; Stevenson et al., 1985). For example, crawling speed is an important component of many snakes' ability to escape predators and capture prey. If the crawling speed of a snake is substantially reduced at low temperatures, then its ability to escape predators and capture prey is likewise reduced if the snake is active at low temperatures. Therefore, to understand the importance of body-temperature variation to the survival, growth, and reproduction of snakes, it is vital that we determine the thermal dependencies of their biological processes.

Methods

Typically, the thermal dependencies of biological processes such as digestive rate, crawling speed, and strike speed, are measured in the laboratory under controlled conditions. Studying the thermal dependencies of biological processes in snakes requires many considerations such as: (1) which processes should be tested; (2) the acclimation state of the snakes; (3) the range of test temperatures; (4) the temperature intervals; (5) the sequence of test temperatures; (6) sample sizes; and (7) how to elicit maximal responses. From an ecological standpoint, it is more informative to focus on whole-animal functions rather than on those at the cellular or tissue level (Bartholomew, 1958; Huey and Stevenson, 1979; Huey, 1982). Additionally, it is important to test functions that are ecologically relevant to the species being examined (Arnold, 1986). For example, determining the thermal dependency of swimming for a desert snake would be of limited value. Knowledge of the natural history of the species being studied (activity times, foraging behavior, etc.) allows one to choose ecologically relevant functions (Greene, 1986; Pough, 1989). Snakes should be tested at the time of day and during the season (usually the active season) appropriate to the function being studied. To increase the ecological relevance of laboratory studies and to reduce the effects of acclimation, snakes should be maintained on a photoperiod and temperature regime that matches what they would experience in the field (Stevenson et al., 1985). Snakes then are moved to an environmental chamber set at the appropriate temperature and allowed enough time to equilibrate with that temperature before testing. The time required for measurements may range from milliseconds (e.g., strike speed) to months (e.g., growth and development). It is generally best to measure each individual at each temperature and, whenever possible, to conduct several trials with each individual. It is also important, if appropriate, to allow enough rest time between separate trials (e.g., one hour for crawling speed) and temperature treatments (e.g., one test temperature per day). Additionally, the sequence of temperatures (e.g., random versus high to low) over which the snakes are tested might have important consequences for the results of the study (Stevenson et al., 1985).

To quantify the results of thermal dependency tests, curve-fitting techniques can be used to generate equations that accurately summarize the data (e.g., polynomials, skewed normals, exponentials, and cubic splines; Huey and Stevenson, 1979; Huey, 1982; R. Huey, personal communication). Frequently, curves are chosen that minimize the sums of squares (Huey, 1982; Stevenson et al., 1985), but other approaches such as the minimum convex polygon technique also are used (van Berkum, 1986). These thermal dependency curves then can be used to predict how temperature affects the functions being examined.

A variety of curve shapes may be generated from thermal dependency data. Generally, there is a positive relationship between the function and temperature (e.g., metabolism [Fig. 7.12*A*]). Some functions may exhibit a linear relationship over much of the range (e.g., crawling speed) and some may peak at one temperature and then decrease at higher temperatures (e.g., swimming speed; Fig. 7.12*C*). Other functions may show an exponential increase with temperature or may exhibit a plateau over which there is little change in performance over a range of temperatures (e.g., metabolism [Fig. 7.12*A*]). The thermal dependency curves can also be used to generate several indices that are useful in interpreting the data, especially when making comparisons.

Indices that can be generated from the thermal dependency curves include: (1) the tolerance range, (2) Q_{10} values, (3) the thermal performance breadth, and (4) the optimal temperature of performance (Huey and Stevenson, 1979; Huey, 1982; Fig. 7.10). The tolerance range is the range of temperatures outside of which the snake loses its ability to carry out the process being examined (Huey and Stevenson, 1979; Huey, 1982). The tolerance range is bounded by the upper and lower critical temperatures (Huey, 1982). A Q_{10} value is

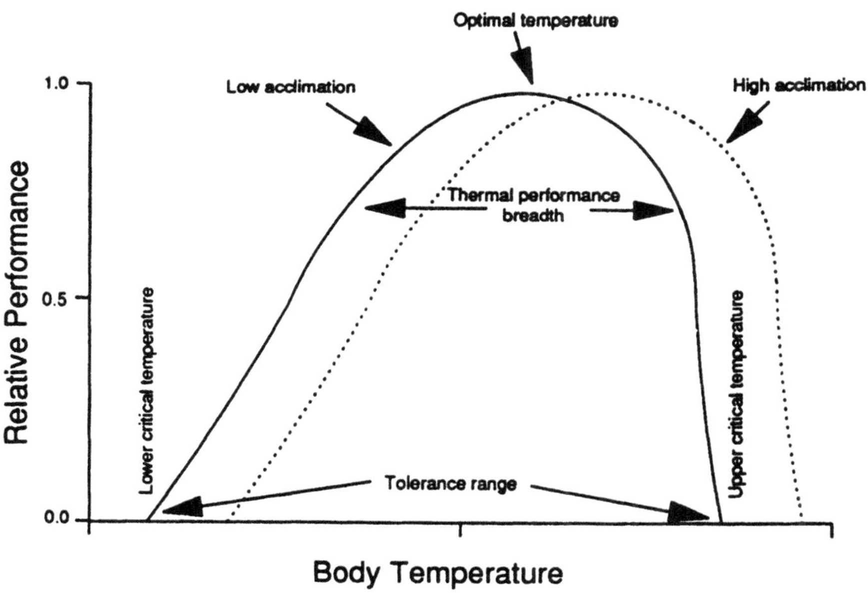

Figure 7.10 Hypothetical performance curves of a snake as a function of body temperature. Adapted with permission from Huey and Stevenson, (1979) (Allen Press). The hypothetical effect of warm acclimation is also indicated (dotted curve).

the amount of change in the rate of a process over a 10°C change in temperature (e.g., a snake whose metabolic rate doubled between 10 and 20°C would have a Q_{10} of 2 from 10 to 20°C; Schmidt-Nielsen, 1983). The thermal performance breadth is the range of temperatures over which an animal performs well (often arbitrarily set at ≥80% of maximal performance [Huey, 1982]). This index is often useful when compared to frequency distributions of body temperatures. For example, it may be useful to compare a frequency distribution of body temperatures for active snakes with the thermal performance breadth of crawling speed. The optimal temperature of performance is the temperature or temperature range at which maximal performance is achieved (Huey, 1982). It is important to remember that the maximal temperature may not necessarily represent the temperature that is optimal for the survival, growth, and reproduction of the animal (Huey, 1982). For example, the temperature at which the digestive rate is maximal may not be the same as the temperatures of maximal digestion efficiency or maximal energy gain (Stevenson et al., 1985).

The characteristics of thermal dependency curves are subject to the effects of acclimation and acclimatization (Prosser and Heath, 1991). Animals may physiologically adjust to changing environmental conditions at different times of the year through acclimatization (Schmidt-Nielsen, 1983). Consequently, these physiological adjustments should be reflected in the positions and shapes of the thermal dependency curves. For example, a snake might have a wider thermal performance breadth for digestion in the spring when it is cool than in the summer when it is warm. The conditions in which the snakes are maintained in captivity (acclimation) can also affect the shapes and positions of performance curves (Fig. 7.10). For example, the thermal dependency curve for the crawling speed of a snake maintained in captivity at low temperatures for an extended period of time might be shifted toward lower temperatures when compared to a snake maintained at higher temperatures. Inverse acclimation, in which metabolic rate decreases at low temperature (below that predicted from thermal dependency curves for snakes during their active season), occurs in several species of snakes (Gregory, 1982). Because of these confounding effects on the properties of thermal dependency curves, it is important to consider the effects of acclimation and acclimatization when conducting experiments on the thermal dependence of performance.

Development

Temperature has numerous effects on snake development. The tolerance range over which gravid *Thamnophis elegans* can produce living young when held at a constant temperature is relatively narrow, 23–32°C (S. Arnold and C. R. Peterson, unpublished data). Rates of development also are greatly affected by temperature. Blanchard and Blanchard (1940) reported that the gestation period of captive *T. sirtalis* was lengthened by approximately 8 days/°C difference in environmental temperature. When gravid snakes were held at constant temperatures, an increase of 1°C shortened the gestation period by about 5 days in *T. elegans* and by about 7 days in *T. sirtalis* (S. Arnold and C. R. Peterson, unpublished data; Fig. 7.11*A*).

Meristic characters also can be affected by gestation temperature (Fox, 1948; Fox et al., 1961; Osgood, 1978). Similar to the results obtained by Osgood (1978) for *Nerodia fasciata,* S. Arnold and C. R. Peterson (unpublished data) found that developmental temperature did not affect some scale counts (e.g., middorsals) but had a small, statistically significant effect on other scale counts. For example, the maximal number of subcaudals and the minimal number of ventral scales developed in the young of *T. elegans* held at a constant 27°C (Fig. 7.11*B*). The number of vertebral anomalies and scale asymmetries in *T. elegans* was minimized at a gestation temperature of 27°C (S. Arnold and C. R. Peterson, unpublished data). It is interesting to note that the temperature at the minimum or maximum for these norms of reactions more closely corresponds to the mean field body temperature of gravid snakes (i.e., 27°C) than the mean plateau temperature of gravid snakes under field conditions (i.e., about 30°C; Arnold, 1988; C. R. Peterson and S. Arnold, unpublished data).

Physiology

Survival tolerances. In addition to determining thermal dependencies in snakes, it is important to know the tolerance range within which snakes can survive because environmental temperatures frequently fall outside of that range (Lillywhite, 1987; Fig. 7.3). The thermal tolerance range (bounded by the lethal minimum and maximum) is relatively wide for many species of snakes (often greater than 35°C). Additionally, the lethal maximum of snakes is generally lower than that for lizards (Huey, 1982). Differences in tolerance ranges among species are often correlated with geographical range, habitat, and habits of the snake (Lillywhite, 1987). For example, a tropical snake might be expected to have a higher lethal minimum than that of a snake from a cooler environment. Snakes such as some *Thamnophis sirtalis* have a lethal minimum below 0°C. These snakes can survive freezing of up to 36% of their body water and can remain frozen for at least 48 h (Costanzo and Claussen, 1988).

Metabolism. Determining the effect of temperature on metabolic rate is a vital component in understanding the energetic relationships of snakes. Resting metabolic rate generally increases exponentially with

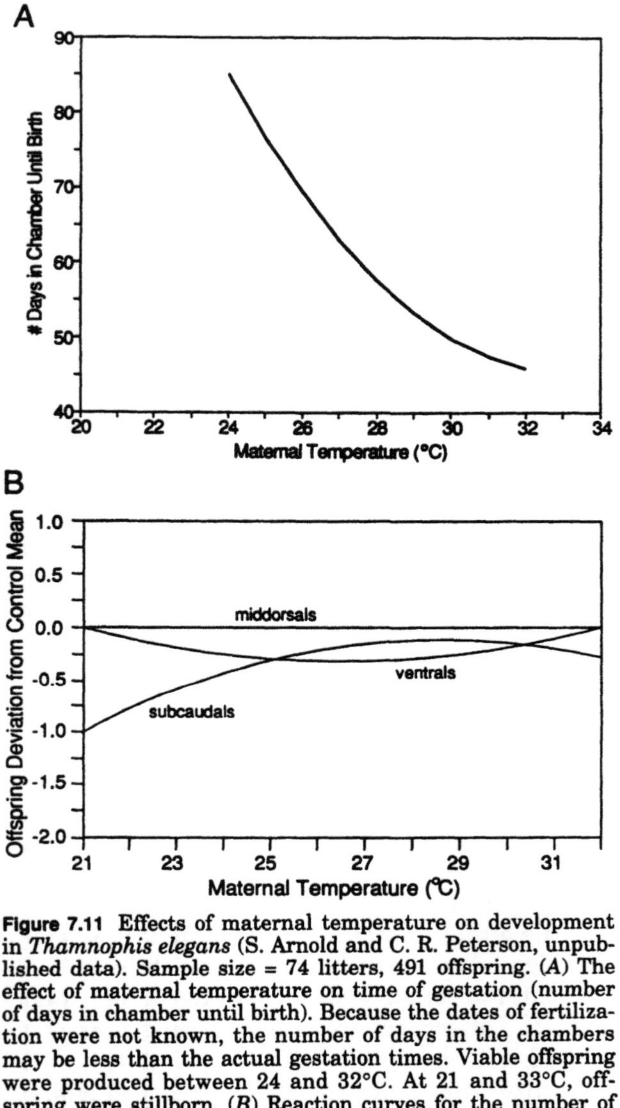

Figure 7.11 Effects of maternal temperature on development in *Thamnophis elegans* (S. Arnold and C. R. Peterson, unpublished data). Sample size = 74 litters, 491 offspring. (*A*) The effect of maternal temperature on time of gestation (number of days in chamber until birth). Because the dates of fertilization were not known, the number of days in the chambers may be less than the actual gestation times. Viable offspring were produced between 24 and 32°C. At 21 and 33°C, offspring were stillborn. (*B*) Reaction curves for the number of middorsals, subcaudals, and ventrals for neonatal Garter Snakes born to females maintained at the temperature indicated on the *x* axis. The units on the *y* axis are phenotypic standard deviations for the trait in question.

temperature (Bennett and Dawson, 1976). However, Aleksiuk (1971) reported that metabolic rate in *Thamnophis sirtalis* increased from 5 to 10°C, then decreased from 10 to 15°C, and then increased again as body temperature rose. Aleksiuk interpreted this unusual curve as instantaneous temperature compensation. Some snake species exhibit a plateau at intermediate temperatures over which resting metabolic rate is relatively unchanged (i.e., Q_{10} values near 1; Fig. 7.12*A*; Stevenson et al., 1985; M. Dorcas and M. Walton, unpublished data). Investigation into the adaptive significance of these "metabolic plateaus" may provide more detailed understanding of snake energet-

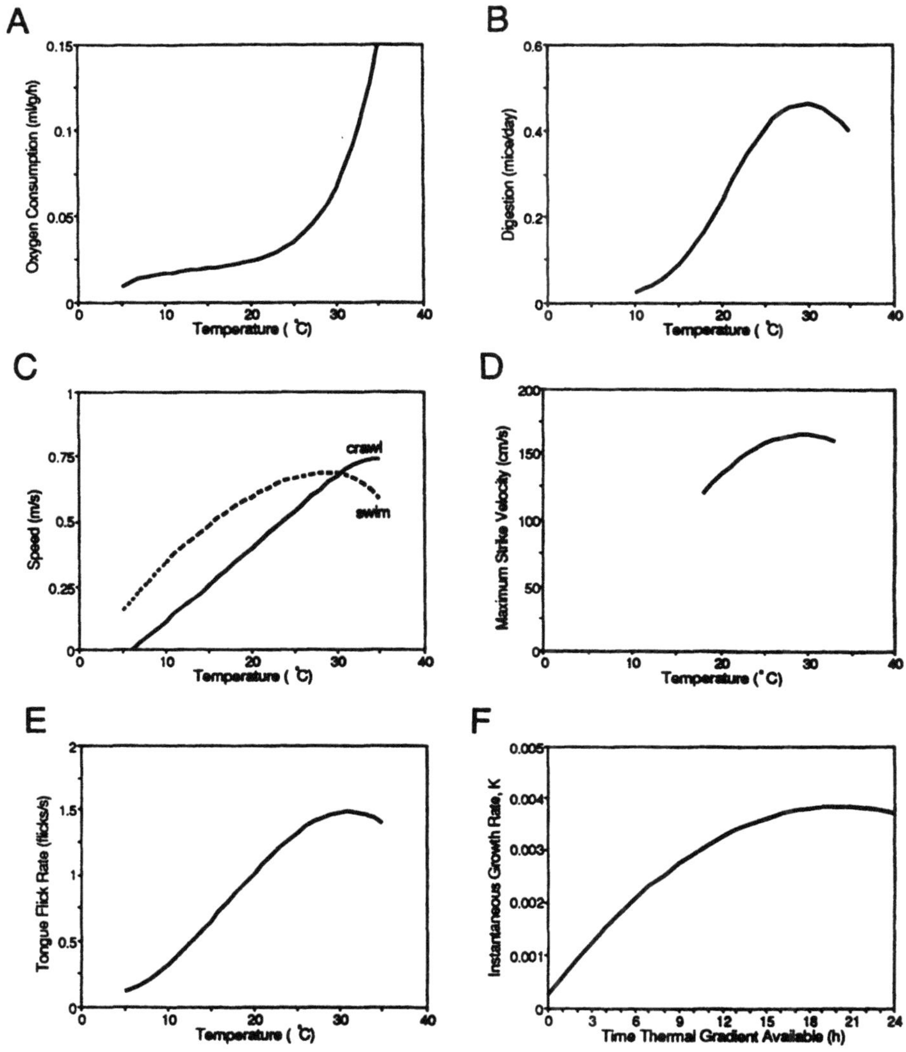

Figure 7.12 Thermal dependency of (*A*) oxygen consumption, (*B*) digestive rate, (*C*) crawling and swimming speed, and (*E*) tongue-flick rate in *Thamnophis elegans*. Adapted with permission from Stevenson et al. (1985), (University of Chicago Press). Thermal dependency of (*D*) strike speed in *Pituophis catenifer*. Adapted with permission from Greenwald (1974) (Allen Press). Effect of differential access to heat on growth (*F*) in *Thamnophis sirtalis* (C. R. Peterson and S. Arnold, in preparation).

ics (Lillywhite, 1987). Low Q_{10} values would allow body-temperature fluctuations without much change in metabolic cost. Therefore, if metabolic rates are adapted to activity T_b ranges (Greenwald, 1971), snakes that are active over a wide range of body temperatures would be expected to have a corresponding wide metabolic plateau.

Digestion. The effect of temperature on digestion has been examined both directly (i.e., using x-ray techniques [Skoczylas, 1970; Stevenson et al., 1985]) and indirectly (i.e., using passage time [Henderson, 1970; Greenwald and Kanter, 1979; Naulleau, 1983]). Stevenson et al. (1985) found that in *Thamnophis elegans,* digestive rate increased slowly between 10 and 20°C, dramatically between 20 and 25°C, leveled off between 25 and 30°C, and decreased slightly at 35°C (Fig. 7.12*B*). All snakes regurgitated at 10°C. Greenwald and Kanter (1979) found similar results for digestive rate in Corn Snakes (*Elaphe guttata*) but digestive efficiency was only slightly affected by temperature. Because snakes may hold feces in their large intestine for considerable periods of time, the accuracy of passage rate as an indicator of digestive rate is not clear. Thus, further studies comparing the effectiveness of different techniques (e.g., x-ray and passage time) are needed. The digestive abilities of snakes from high latitudes may be limited by environmental temperatures. It has been suggested that success of venomous snakes at high latitudes is due, in part, to their ability to increase digestion rates at low temperatures through envenomation (Thomas and Pough, 1979).

Growth. Surprisingly few studies have examined the effect of temperature on the growth of snakes. Presumably, growth rate should be maximal at the temperature where the difference between metabolic rate and digestive rate is greatest (e.g., 29°C for *Thamnophis elegans*; Porter, 1989). Peterson and Arnold (in review) varied the amount of time that neonatal *T. sirtalis* had access to heat. Although there was much variation in growth rates, snakes with longer access to heat grew faster (Fig. 7.12*F*). However, differential access to heat did not affect body shape (Arnold and Peterson, 1989)

Ecdysis. Studies of the effects of temperature on ecdysis in snakes are limited. Cliburn (1976) found that Pine Snakes (*Pituophis melanoleucus*) shed more frequently during the summer than in the fall or spring but found it difficult to separate this from temperature effects. Semlitsch (1979) found that two species of Water Snakes (*Nerodia*) also shed more frequently when maintained in captivity at higher temperatures. Maderson (1984) reported that moving *Thamnophis sirtalis* from a constant temperature of 21 to 28°C induced molting. It is possible that other environmental factors (especially humidity) affect shedding frequency more than does temperature (Maderson, 1984).

Immune system. To our knowledge, the effect of temperature on the ability of snakes to fight infection has not been determined. Kluger (1979) has shown that Desert Iguanas (*Dipsosaurus dorsalis*) infected with the bacteria *Aeromonas hydrophila* have a higher survival rate if maintained at 42°C than at lower temperatures. Certainly the investigation of temperature effects on the immune system of snakes warrants further investigation.

Behavior

Numerous studies have examined the effects of temperature on various behaviors of snakes (Ford and Burghardt, Chap. 4, this volume). These include studies of locomotion (Heckrotte, 1967), tongue-flicking (Stevenson et al., 1985), prey capture (Greenwald, 1974), and defensive behavior (Goode and Duvall, 1989). Studies of the thermal dependencies of behaviors often integrate several systems and so may have more ecological relevance than physiological studies (e.g., crawling speed integrates the thermal dependence of both nervous and muscular systems into one ecologically relevant and measurable behavior).

Locomotion. Determining the effect of temperature on the locomotor ability of snakes is important because many snakes rely on locomotion to escape predators and capture prey. Stevenson et al. (1985) found that the maximal crawling speed of *Thamnophis elegans* was strongly temperature dependent, and they crawled fastest at 34.5°C (Fig. 7.12*C*). Heckrotte (1967) recorded similar results for the maximal crawling speed of *Thamnophis sirtalis,* but showed that cruising speed was nearly independent of temperature from 15 to 33°C. The optimal temperature for swimming speed was 28.5°C in *T. elegans* (Fig. 7.12*C*), significantly less than the temperature for maximal crawling speed (Stevenson et al., 1985). The thermal dependency curve for swimming speed shows a relatively wide thermal performance breadth. The functional basis and ecological significance of the lower optimum and wider thermal performance breadth for swimming speed in *T. elegans* are not known.

Tongue-flicking. Determining the effects of temperature on the rate of tongue-flicking is important because tongue-flicking is necessary for snakes to fully utilize the chemoreceptive abilities of their vomeronasal organs. These organs are vitally important to many aspects of snakes' lives, especially foraging and prey capture

(Mushinsky, 1987). The tongue-flicking rate of *Thamnophis elegans* increased greatly between 5 and 20°C and was greatest at 30°C (Stevenson et al., 1985; Fig. 7.12*E*). Future studies of the thermal dependence of tongue-flicking are needed and, if possible, should include the ability of the snakes to detect prey.

Prey capture. The thermal dependency of prey capture is especially important because it not only integrates several physiological systems but also incorporates the thermal dependency of the prey's escape ability. However, surprisingly few studies have examined the effect of temperature on prey-capturing ability. Greenwald (1974) showed that the strike speed and prey-capture success of *Pituophis catenifer* increased linearly from 18 to 27°C but did not increase significantly above 27°C (Fig. 7.12*D*). Because of the ecological significance of this behavior to many species of snakes, more studies of this type are needed.

Defensive behavior. Like prey capture, the effect of temperature on the defensive behavior of snakes is ecologically important. Several studies of lizard defensive behavior have documented temperature effects (Christian and Tracy, 1981; Hertz et al., 1982; Mautz et al., 1992). For example, Christian and Tracy (1981) found that juvenile Galapagos Land Iguanas (*Conolophus pallidus*) were preyed upon more heavily when their body temperature was low (i.e., less than 32°C) than when their body temperature was higher. Schieffelin and de Queiroz (1991) found that *Thamnophis sirtalis* were less likely to demonstrate active antipredator behaviors (e.g., biting) at low temperatures. Goode and Duvall (1989) found that gravid Western Rattlesnakes (*Crotalus viridis*) were most aggressive at low temperatures but males and nongravid females were not. Burger (1991) found that hatchling Racers (*Coluber constrictor*) incubated at 28°C held their heads higher prior to striking, had longer strikes, and reached higher when striking than hatchlings incubated at 22°C. Burger also found that hatchling Common Kingsnakes (*Lampropeltis getula*) incubated at 28°C performed better in maneuverability, strike, and escape behavior tests than hatchlings incubated at 32°C.

Learning ability. There are several studies of the effect of temperature on the learning ability of lizards (Krekorian et al., 1968; Burghardt, 1977; Brattstrom, 1978). However, the thermal dependency of learning ability in any species of snake has not yet been determined.

Functional aspects of regional heterothermy

Regional body-temperature variation is an important factor to consider when conducting tests of the thermal dependency of performance. Because of their attenuated form, snakes are often able to maintain substantial differences between their head and body temperatures. These differences may have functional significance. For example, the thermal performance breadth for the functioning of the central nervous system may be more narrow than the thermal performance breadth for muscle contraction. Therefore, to maximize crawling speed, a snake might regulate its head temperature more precisely than its body temperature. Additionally, by examining the thermal dependencies of snakes with regional differences in body temperature, we might be able to determine what factors limit optimal functioning of the system being studied (J. Kauffman and A. Bennett, personal communication). An interesting way to examine this concept would be to test the locomotor performance of snakes with warm heads and cold bodies and contrast that with the performance of snakes with cold heads and warm bodies.

Functional Integration

It is noteworthy that, with some exceptions, many of the whole-animal functions of Garter Snakes are maximal or near maximal at similar temperatures. This parallels what has been found in the few species of lizards in which this has been examined (Huey, 1982). In *Thamnophis elegans* the optimal temperatures for these functions also generally correlate well with the thermal preference and activity temperatures (Stevenson et al., 1985). Detailed studies of snakes that differ both phylogenetically (e.g., elapids, typhlopids) and ecologically (e.g., tropical, nocturnal) are needed to determine if this is a general principle of snake thermal ecology.

Ecological Effects of Body-Temperature Variation

The primary goal of snake thermal ecology studies is to understand how variation in the thermal environment influences the distribution and abundance of snakes. Specifically, how does environmental temperature variation affect dependent ecological variables such as activity times, microhabitat selection, prey-capture rates, reproductive output, predator avoidance, population dynamics, and geographic distribution? Answers to these questions are not only of academic interest but also are becoming increasingly important because of the need to evaluate the effects of local and global environmental changes (such as thermal pollution and global warming) on natural populations of organisms.

Although clearly important, surprisingly few studies have actually quantified the ecological consequences of environmental temperature variation. Several reasons account for this lack of studies. First, it is difficult to measure many of the dependent ecological variables, such as activity patterns or predation rates, that we are seeking to explain. Second, the thermal dependency of physiological processes may not simply map onto ecology (Huey and Stevenson, 1979) because many factors besides temperature influence these ecological variables. For example, accurate predictions of growth rates require information not only about the thermal dependency of growth but also about prey types, prey availability, competitive interactions, and predation risk. Because so many factors may be involved, it is often difficult or impossible for an individual scientist to acquire the needed data. Even if the data are available, it is challenging to integrate them so that specific, testable predictions can be made.

Evaluating the ecological effects of environmental temperature variation must also take into account the indirect effects of environmental temperature variation. For example, even though a snake may be able to maintain stable body temperature in the face of fluctuating environmental temperatures, there may be costs involved such as decreased foraging time or increased exposure to predation. Time and energy costs are also associated with compensatory changes such as the synthesis of new enzymes with different thermal optima or the replacement of one form of lipid for another to maintain similar viscosity of cell membranes at different temperatures. Quantifying these indirect effects is even more difficult than estimating the direct effects of body-temperature variation.

Methods

The effects of environmental temperature variation on snake ecology can be studied in a variety of ways, including field observations, field comparisons, field experiments, laboratory simulations, and computer modeling. In many cases, questions can be answered by correlating environmental temperatures with field measurements of the ecological variable of interest (such as activity patterns, distance moved, prey capture rates, or dates of birth). For example, the hourly activity of individual *Pituophis catenifer,* measured with an automated telemetry system and motion sensitive radiotransmitters, was positively correlated with increasing operative temperatures (Fig. 7.13; Grothe, 1992). The correlation approach can be extended to multiple-site comparisons (e.g., measuring growth rates in populations at different latitudes or altitudes). Care must be taken, however, to consider intersite differences other than temperature (such as prey availability or predatory risk). For example, the growth rates of *Thamnophis elegans* at a small mountain lake in northern California are less than 70% of the growth

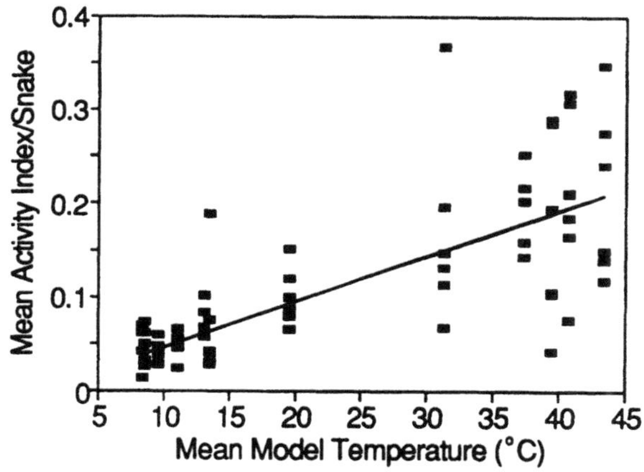

Figure 7.13 Regression of mean activity index of active telemetered Gopher Snakes (*Pituophis catenifer*) versus mean model temperatures over 2-h blocks of time from 11 May to 20 June 1991, on the Snake River Birds of Prey Area in southwestern Idaho. Six snakes were sampled every 5 min. Only time blocks with activity are included ($n = 68$). The activity index was calculated by dividing the number of toggles of a mercury switch in the radiotransmitter during a 2-h period by the number of toggles that could have occurred within that same time period. Adapted with permission from Grothe (1992).

rates in a population at a larger, lower elevation lake (S. Arnold, unpublished data). Adult snakes from the high elevation site also are about 25% shorter than adults from the low elevation site. Rough calculations, based on field T_e measurements at these sites and laboratory studies of the thermal dependency of growth, suggest that less than half of this difference is attributable to the direct effect of T_b on growth. Presumably, other factors such as prey type, prey availability, and parasitism account for the remaining differences.

Field experiments represent a potentially powerful but underutilized method for evaluating the ecological effects of temperature variation. For example, the effect of changing the range of available temperatures could be studied by supplying a supplemental heat source or by adding cover objects (Fitch, 1987; Grant et al., 1992) and measuring whether or not such manipulations attracted snakes and increased their growth rates. Supplemental feeding experiments could be performed to test the effect of feeding on T_b variation in free-ranging snakes.

Correlational and experimental studies could also be undertaken in the laboratory using simulated field conditions. For example, the effect of water temperature on fish-capture rates by Garter Snakes could be studied in the laboratory. Such a study would be more ecologically relevant than simply studying the thermal dependency of swimming or striking speed because it would also incorporate the effects of temperature on the prey's ability to escape.

Modeling represents another important approach to studying the effects of temperature on the ecology of reptiles (Porter et al., 1973; Tracy, 1982; Porter and Tracy, 1983; Porter, 1989; Dunham et al., 1989; Huey et al., 1989; Dunham 1993; Adolph and Porter, 1993). Mathematical modeling possesses several important features: (1) it helps identify which variables are important; (2) it allows complex problems to be studied because numerous variables can be incorporated into the model; (3) it permits sensitivity analyses to evaluate the relative importance of each independent variable; and (4) it facilitates making quantitative predictions.

Mathematical models based on environmental and field T_b data, and laboratory data on the thermal dependencies of development, physiology, and behavior can be used to make quantitative ecological predictions. The concepts of constraint (e.g., Porter et al., 1973) and optimality (e.g., Huey and Slatkin, 1976) are key elements in these models. The possible (constrained) and optimal solutions to the models are determined by both environmental conditions and the thermal dependency curves. For example, the locations in the environment where a snake could survive will be determined by the operative temperatures at those locations and by the snake's thermal tolerances. Similarly, the best place(s) for a gestating snake could be predicted by combining operative temperature data with the thermal dependency curves for development. These predictions can be tested using the methods described previously. Rigorous testing of a model is crucial to establishing its validity. Ideally, the approach is iterative with the results of the comparisons of predictions and observations being used to refine the model.

This modeling approach was used by Huey et al. (1989) to generate and test predictions of retreat site selection by *Thamnophis elegans* in a rocky area by a lake in northern California. Predictions were based on field measurements of the temperatures of rocks of different shapes and sizes, the thermal dependency of several physiological processes, and the thermoregulatory behavior of gravid Garter Snakes. Huey et al. tested their predictions of which rocks snakes would select for retreat sites by comparing the distribution of available rocks with the distribution of rocks selected by telemetered snakes. As predicted, most gravid snakes selected intermediate thickness rocks that allowed them to maintain their body temperatures within or near the preferred range for long periods of time throughout the day and night.

Examples

Activity patterns. Although the relationship between temperature and activity is obviously important, it has not been well-quantified. The activity patterns of individual snakes are an example of a dependent ecological variable that is poorly described (Gibbons and Semlitsch, 1987). The primary contribution that thermal ecology studies can make to the studies of activity patterns is to indicate thermal constraints on activity. Potential daily and seasonal activity patterns can be predicted by combining data on surface operative temperatures with information about snake thermoregulatory behavior (i.e., minimal and maximal voluntary temperatures; Porter et al., 1973; Porter and Tracy, 1983). More detailed predictions can be made if certain behaviors are linked with particular microhabitats that can be characterized with T_e measurements. For example, if snakes did not forage in water colder than 10°C, then potential aquatic foraging times could also be predicted. These calculations can be tested by regularly sampling for surface-active snakes or with automated activity telemetry (Peterson et al., 1989; Grothe, 1992; V. A. Cobb, unpublished data; Fig. 7.6*E*).

Habitat selection. The primary contributions that thermal ecology studies can make to studies of habitat selection are to indicate which microhabitats will allow snakes to maintain T_b's within acceptable limits and to indicate the potential functional consequences of selecting different microhabitats. For example, during most of a clear, summer day, Garter Snakes could not remain at the exposed surface in a meadow in northern California without exceeding their lethal maximum T_b (Fig. 7.3). Similarly, snakes would not be predicted to be active at night on the exposed surface because they would experience T_b's below their critical thermal minimum if they remained there for more than a short time.

This approach can be extended to considerations of where snakes could obtain preferred T_b levels (Huey et al., 1989). However, such predictions often may not be very useful for two reasons. First, there may be so many combinations of places and ways for snakes to obtain preferred T_b's that the number of possible microhabitats that snakes could use would still be large. Second, laboratory thermal preference studies may not indicate what snakes in the field will do because other activities such as foraging may take precedence over careful thermoregulation (Reinert, Chap. 6, this volume). Thermal considerations also may prove useful in broader-scale analyses of habitat selection, such as the location of suitable hibernacula (Gregory, 1982).

Feeding ecology. Thermal ecology studies are useful in understanding several aspects of snake-feeding ecology (Arnold, Chap. 3, this volume), including feeding requirements, frequency of feeding, foraging mode, foraging times, foraging success, and prey processing times. Energetic calculations, which incorporate the effects of body temperature on metabolic rates, can be used to estimate the feeding requirements of snakes (Porter and Tracy, 1974). One of the more interesting results from these models is the relatively small feeding requirements of snakes for maintenance. For example, we calculated that a 100-g, actively foraging *Thamnophis elegans* in eastern Washington in the summer would only have to capture a 15-g salamander every other week to stay in energy balance. If the snake were ambush-feeding, it would only need to capture a salamander once per month. In these calculations, we assumed that the snake's plateau-phase T_b was 30°C. A lower T_b would allow a snake to go longer between meals but might decrease its ability to capture prey items when they were encountered. Modeling may also be used to predict when snakes may be able to forage and when ectothermic prey are available (Porter et al., 1975). Information on environmental temperatures and the thermal dependency of prey capture could be used to predict foraging success. Finally, modeling can be used to calculate how seasonal variation in environmental temperatures would affect sna ke time budgets by changing the amount of time required to digest a prey item. For example, calculations indicate that in July, a *Thamnophis elegans* in eastern Washington can digest a rodent in less than half of the time it takes to do so in April (C. R. Peterson, unpublished data).

Another very important aspect of thermal ecology is the effect that T_b variation may have on energy balance. The relatively low energy requirements of snakes can be further reduced by voluntary hypothermia. Voluntary hypothermia might be limited to part of the day (e.g., at night for diurnal species) or might extend for months (e.g., during a dry season). Laboratory studies that indicate an inverse relationship between feeding frequency and heat use (Gibson et al., 1989), in conjunction with field observations such as those concerning the lower temperatures of snakes when prey are less available (Shine and Lambeck, 1990), suggest that snakes are capable of using T_b selection to change energetic demands. Such behavior could extend the fasting endurance of a snake over twice that of a snake with high T_b's and may be an important adaptation for surviving periods of low food and water availability such as droughts. For example, we calculated that a 100-g *Thamnophis elegans* with a 10-g fat reserve could increase its fasting endurance from approximately 160 to 430 days by selecting a T_b of 10°C rather than 30°C.

Reproduction. Temperature may influence snake reproductive ecology in a variety of ways. It is probable that the reproduction of many, perhaps most, snake species is limited by the availability of nutrients and energy. Environmental conditions may prevent snakes from obtaining sufficient energy so that females reproduce on a multiyear rather than an annual cycle. This is known to occur in a number of high-latitude species (e.g., *Charina bottae* [M. Dorcas, unpublished data], *Thamnophis elegans* [C. R. Peterson, unpublished data], *Crotalus horridus* [Brown, 1991], and *Crotalus viridis* [Klauber, 1972]). Temperature may also affect the time of egg-hatching or birth (Blanchard and Blanchard, 1940). A prominent feature of pregnancy in snakes is strong thermophily and reduced activity (e.g., Charland, 1991). An interesting experiment would be to maintain gravid females in environmental chambers programmed to mimic the T_b's of free-ranging, gravid, and nongravid snakes (determined by radiotelemetry). Such an experiment would directly test the effect of T_b variation on development rates, size of young at birth, meristic characters, and the occurrence of abnormalities.

Predator avoidance. The avoidance of predators by snakes is one of the most important but least understood areas of snake ecology. Consideration of the effects of temperature on snake locomotor and striking performance and on microhabitat selection can contribute to studies of predation on snakes. Snakes hibernating in dens appear to be particularly susceptible to predation by endothermic predators such as birds and mammals (Gregory, 1977; Burger et al., 1992; V. Cobb and C. R. Peterson, unpublished data). Snakes may also be especially vulnerable to predation when they emerge in the morning. At this time, they are exposed at the surface, their T_b's are low, and their ability to escape or defend themselves is reduced (Fig. 7.14). Peterson (1982) reported an apparent case of Magpie predation on an adult female *Thamnophis elegans* as she was heating in the morning. Peterson (1982) hypothesized that the delayed emergence typically seen in *T. elegans* in the spring and fall may be an adaptation to reducing the time when snakes with low T_b's are exposed to predation.

Grothe (1992) found that *Pituophis catenifer* in the Snake River Birds of Prey Area in southwestern Idaho were most likely to be preyed upon by Red-Tailed Hawks in the late morning and late afternoon when environmental temperatures were moderately high (30°C $<$ exposed $T_e < 40$°C) and when the snakes were active (Fig. 7.15). In contrast, small mammals were consistently captured over the full range of environmental temperatures. Grothe speculated that, in the summer when conditions were warmer, an increase in nocturnal snake activity might reduce the risk of hawk predation.

Population structure and dynamics. Environmental temperature variation may influence snake population dynamics by affecting growth, reproductive output, and survivorship. Snakes living in cooler environments may grow more slowly during the course of a year, repro-

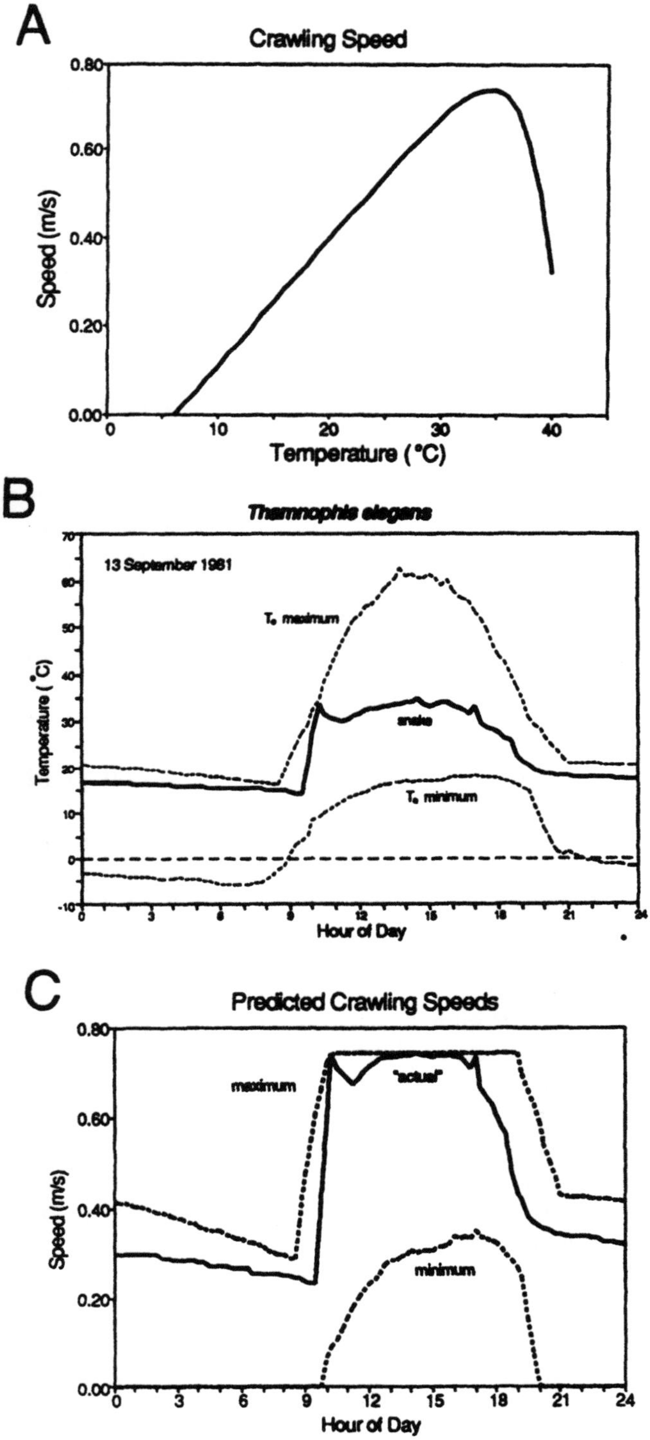
A
Crawling Speed
Speed (m/s)
Temperature (°C)
B
Thamnophis elegans
13 September 1981
T_e maximum
snake
T_e minimum
Temperature (°C)
Hour of Day
C
Predicted Crawling Speeds
maximum
"actual"
minimum
Speed (m/s)
Hour of Day

duce less frequently, and live longer (e.g., Brown, 1991). Future studies should look for correlations between environmental temperatures, growth rates, reproductive output, and age. Little is known about natural snake mortality resulting from exposure to lethal temperatures. It seems that mortality due to failure to avoid low lethal temperatures is more likely than deaths due to overheating because snakes will almost always be able to find retreats below their lethal thermal maxima.

Biophysical and physiological models of energy exchange between animals and their environments provide an important tool for studying the effects of environmental temperature variation on individuals (Porter, 1989). Population models can then be used to link individual models to predict emergent population dynamics (Dunham et al., 1989; Dunham, 1993). Dunham, Porter, and their colleagues are applying this approach to lizard populations, and we believe this approach would be useful in understanding snake population dynamics as well.

Geographic distribution. Thermal ecology studies can contribute to understanding the geographic distribution of snakes by clarifying the basis of correlations between species ranges and environmental temperatures, especially at the latitudinal and altitudinal limits for the species. Several hypotheses exist for why temperature may constrain the ranges of reptiles (R. Brooks, personal communication). The first hypothesis is that geographical limits occur where snake thermal tolerances are periodically exceeded and thus snakes cannot survive. A second hypothesis is that snakes would not be able to obtain sufficient amounts of energy within a year to reproduce. The last, and probably most likely hypothesis in many cases, is that thermal conditions do not allow adequate time for development to take place. These hypotheses can be tested by (1) measuring environmental temperatures to see if lethal temperatures could be avoided; (2) comparing the success of reproduction during cold and warm years in populations on the periphery of snake distributions; and (3) by combining laboratory data on the thermal dependency of development and environmental temperature data to calculate if sufficient time is available for complete development to occur in the field.

Figure 7.14 The potential effect of daily body-temperature variation on the crawling speed of *Thamnophis elegans*. (*A*) Thermal dependency of maximal crawling speed. Adapted with permission from Stevenson et al. (1985), (University of Chicago Press).(*B*) Daily T_b pattern of a nongravid female snake. Minimal and maximal operative environmental temperatures are indicated with broken lines. (*C*) Predicted crawling speeds for hypothetical snakes calculated by combining the thermal dependency of crawling speed with observed T_b's and minimal and maximal T_e's (R. Stevenson and C. R. Peterson, unpublished data).

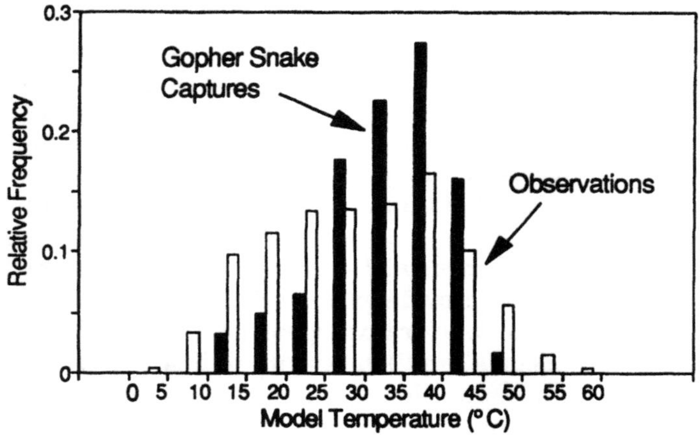

Figure 7.15 Effect of operative temperature on the frequency of predation on *Pituophis catenifer* by Red-Tailed Hawks. The relative frequencies of exposed model temperatures, when Red-Tailed Hawk nests were observed, are compared with the relative frequencies of model temperatures when snakes were captured. The study was conducted during 1990 and 1991 on the Snake River Birds of Prey Area in southwestern Idaho (n = 62 snake captures, 537 h of observation). This figure indicates that snakes were captured at a higher frequency when operative temperatures were relatively high (between 30 and 40°C; G =8.96, $p < 0.05$). If operative temperature did not have an effect on predation frequency, then the distributions would be expected to be the same. Adapted with permission from Grothe (1992).

Evolution of Snake Thermal Characteristics

The complexity and intensity of behavioral thermoregulation, the profound effects of temperature on biological function, and the ecological consequences of those effects suggest that adaptation to thermal variation has been an important theme in the evolution of snakes, especially the evolution of their thermal characteristics. Thermal characteristics include those traits that influence T_b variation (such as size, coloration, and thermal preference) and those traits influencing the consequences of T_b variation (such as thermal tolerances, thermal optima, and the ability to acclimate). See Fig. 7.1.

We identify four general evolutionary questions: (1) How do thermal characteristics vary among snake taxa? (2) What was the pattern of evolutionary change in these characteristics? (3) What mechanisms (e.g., natural selection, genetic drift, and phylogenetic or developmental constraints) have shaped the thermal characteristics of snakes? (4) How has the evolution of snake thermal characteristics affected the evolution of other traits?

Interesting examples of specific questions concerning the evolution of snake thermal characteristics include: (1) To what extent are the thermal characteristics of snakes correlated with their environments? (2) Why are the preferred temperatures of most snakes lower than those of most lizards? (3) Why are the preferred body temperatures of most snakes so similar? (i.e., about 30°C); (4) Are the thermal physiology and thermoregulatory behavior of snakes coadapted? (5) Has minimization of energy expenditure been a dominant theme in the evolution of snake thermoregulatory behavior?

These questions are difficult to address for a number of reasons and there are relatively few studies of the evolution of thermal characteristics in snakes. Chief among these difficulties is the absence of sufficient data to determine taxonomic and ecological patterns of variation in any thermal characteristic of snakes. For example, are there systematic differences among snakes in the T_b's selected while digesting a meal, or differences in the priority attached to thermoregulation by gravid snakes along latitudinal or altitudinal gradients? Data for snakes from the tropics and for snakes with limited access to heat (e.g., many nocturnal, fossorial, and aquatic species) are especially needed. Due to the lack of adequate information, the focus of this discussion is on approaches we feel show promise for answering evolutionary questions concerning the thermal biology of snakes.

Methods

Comparison among species is a common, but problematic, approach for examining the adaptive significance of organismal traits (Endler, 1986). Adaptation through natural selection is a frequent explanation for correlation between organismal features and environmental variables. For example, the thermal tolerances of snakes may be correlated with their geographic ranges (Lillywhite, 1987). Comparative studies, however, have been criticized for not removing the statistical bias that results from nonindependence of taxa sharing common ancestry (Felsenstein, 1985), and for failure to consider alternative explanations to adaptation, such as genetic drift, genetic correlation among traits, and phylogenetic or developmental constraints (Gould and Lewontin, 1979). Interspecific comparisons indicate the results of evolution, but may reveal little concerning the evolutionary processes and causal factors leading to that endpoint.

Fortunately, recent conceptual advances provide several investigative approaches that may be applied to the study of the evolution of thermal characteristics in snakes. We consider the most promising among these to include: (1) phylogenetically based comparative studies; (2) quantification of interindividual variation and natural selection within populations; and (3) modeling the fitness consequences of variation in thermal characteristics.

Phylogenetically based comparative studies. A variety of analytical methods has been proposed recently that seeks to incorporate phylogenetic or taxonomic information into comparative studies (reviewed by Brooks and McLennan, 1991, and Harvey and Pagel, 1991). Although the methods differ in detail, there are at least two objectives: (1) the reduction of statistical bias due to the nonindependence of taxa compared; and (2) the partitioning of variation into fractions attributable to evolutionary history and fractions that can be interpreted as the result of evolution in current environmental conditions. The correlation of this latter fraction with environmental variation can be examined for a more rigorous test of adaptive significance than is achieved through traditional, nonphylogenetic correlations (Martins and Garland, 1991).

Some methods also provide estimates of ancestral states of the trait in question, so that hypotheses concerning the rate and direction of change can be developed (Huey, 1987). Application of these techniques to two traits allows a calculation of correlation between traits that purportedly is free of phylogenetic bias. The only application of these methods to thermal characteristics is that reported by Huey and Bennett (1987) and Garland et al. (1991) in their analysis of the evolution of thermal preferences among Australian skinks.

Some of the drawbacks of these methods are that they may require information concerning phylogenetic relationships among the taxa compared, branch lengths of the phylogenetic tree, and the tempo of evolution (e.g., gradual or punctuational). Moreover, these methods have been applied to few data sets, so that all their properties and potential shortfalls are not well understood.

Studies of intrapopulational variation and selection. Studies of intrapopulational variation allow the direct quantification of natural selection on thermal characteristics. The method, as proposed by Arnold (1983), is conceptually simple. The thermal trait in question is scored in a large number of individual snakes, each of which is uniquely marked and released into the environment. The cohort may then be monitored by repeated recaptures to determine the repeatability of the thermal characteristic, growth, variation in reproductive output among individuals, and survival. Fitness consequences of the trait are assessed by comparing variation in the trait to variation in survival and reproduction. Thus, this technique provides a potentially powerful tool for testing adaptive hypotheses (Arnold, 1986).

Demonstration of interindividual variability and correlation of that variability with differential fitness are only the first steps in evaluating hypotheses of adaptation by natural selection. Even though selection may favor enhanced survival or reproduction of a particular variant, adaptive evolution toward increased representation of that variant in populations is possible only if some fraction of that variability has a heritable, genetic basis. Heritabilities of thermal characteristics can be estimated through comparison of traits among broods or by comparing the trait values of offspring to those of parents, using standard techniques of quantitative genetics (Arnold, 1987; Brodie and Garland, Chap. 8, this volume). No studies as yet have estimated the heritability of thermal characteristics in snakes, although Arnold and Peterson (unpublished data) have found significant among-litter differences in T_b selection in neonatal *Thamnophis elegans*. Heritable variation has been found in the thermal characteristics of other organisms (Huey and Bennett, 1990).

Despite their clear potential, few intrapopulational studies of selection have been attempted (Koenig and Albano, 1987; Arnold, 1988; Jayne and Bennett, 1990), probably because of logistical impediments. Foremost among these problems are the large sample sizes needed to detect selection on minor, but functionally important, variation in traits. The measurement of hundreds, if not thousands, of individuals is required for such studies. Consequently, studies may not be feasible if measurements take a long time to complete or cannot be mass-produced. A related problem is the need to recapture a large fraction of the marked cohort after its release into the environment.

Species that are best suited for these studies are those that occur in large numbers and concentrate their activity within a confined area that can be effectively sampled. Fortunately, some snake species may be among the best vertebrate candidates for such studies, as illustrated by the recent work of S. J. Arnold, A. F. Bennett, and their colleagues. Jayne and Bennett (1990), for example, detected significant selection on body size and locomotor performance variables in *Thamnophis sirtalis* based on a three-year study that amassed over 1500 snake captures and recaptures. In an ongoing study of selection on growth and metabolism in the same population of Garter Snakes, recapture success within a single season exceeded 80% (M. Walton, personal communication). Clearly, similar studies could examine variation in selected and field body temperatures, microhabitat selection, or the thermal dependence of function.

Even if the restrictive requirements mentioned above are met, caution is necessary in the interpretation of these studies. Selection may be episodic, confined to periods of environmental stress or to short intervals of an animal's lifetime. For example, Jayne and Bennett (1990) found that locomotor characteristics undergoing selection in yearling snakes may be less subject to selection at later ages and larger body sizes. Moreover, thermal characteristics certainly are complex, polygenic traits, and pleiotropy may cause correlations among traits that confound direct and correlated responses to selection (Arnold, 1983, 1987).

The choice of the thermal traits for analysis entails several considerations (Arnold, 1986; Pough, 1989). Detailed information on the natural history of the species and models of the fitness effects of thermal characteristics (described below) can aid in the choice of the appropriate thermal traits. Multivariate studies can be used to identify correlations among thermal characteristics (Arnold, 1987).

Modeling. Another approach to evaluating the effect of thermal characteristics on fitness involves the use of individual-based models that incorporate behavioral, physiological, and ecological data. For example, we modeled the effect of thermal preference on fitness in *Thamnophis elegans*. Our simulation began with an observed T_b pattern that had a plateau-phase mean T_b of approximately 30°C. Alternate T_b patterns were generated by adding 5°C and by subtracting 5 or 10°C from the observed T_b pattern (Fig. 7.16*A*) and then were combined with equations for the thermal dependency of metabolism and digestion (Stevenson et al., 1985). From this, we calculated how many offspring could be produced from the energy available to snakes with different thermal preferences. We assumed that snakes had continuous access to prey for four weeks, an ecologically realistic assumption similar to snakes preying on breeding or metamorphosing amphibians. The results (Fig. 7.16*B*) indicate that reproductive output would be strongly affected by thermal preference. For example, snakes with a thermal preference of 30°C would produce over 30% more young than snakes with a thermal preference of 20°C. To the extent that the model assumptions reflect real conditions, these results suggest that thermal preference should be under strong selection pressure. This type of modeling may be useful for predicting which thermal characteristics are most strongly influenced by natural selection, thereby directing the focus of subsequent intrapopulational studies.

A modeling approach also may yield insights into how the evolution of snake thermal characteristics may have influenced the evolution of other traits, such as color or foraging behavior. For example, cold environments may favor the evolution of viviparity (Shine, 1985) but limit snake size because of constraints on heating rates.

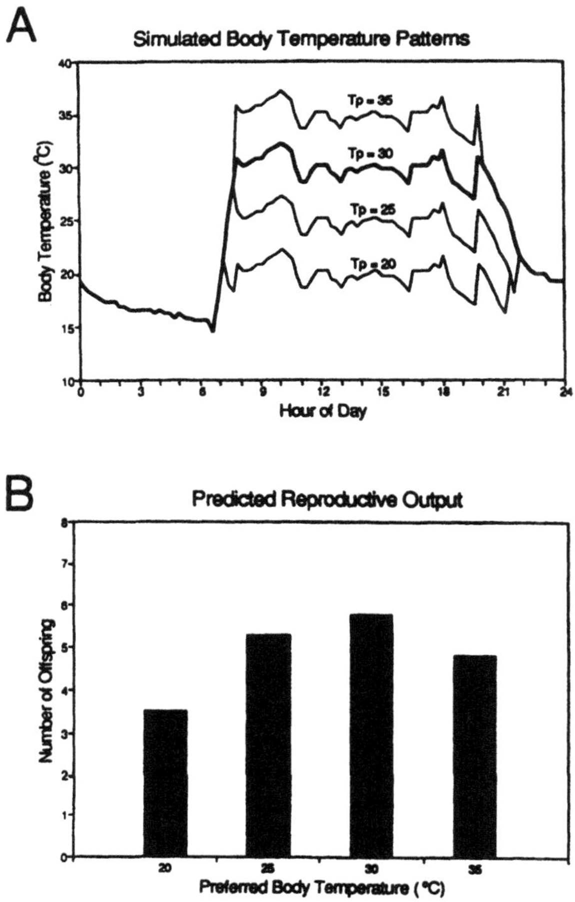

Figure 7.16 Predicted effect of variation in thermal preference on fitness in *Thamnophis elegans.* (*A*) Simulated body-temperature patterns (light lines) for *Thamnophis elegans* with different preferred temperatures (35, 25, and 20°C) generated by adding or subtracting multiples of 5°C to or from an observed plateau-phase T_b pattern with a mean of 30°C (heavy line). (*B*) Predicted reproductive output for *Thamnophis elegans* with different thermal preferences. See text for details.

Summary and Future Research

Because snakes are ectothermic and their biological processes are temperature dependent, nearly all aspects of their ecology are affected either directly or indirectly by variation in environmental and body temperatures. Thermal ecology studies of snakes are not only critical to understanding snake ecology but they also provide excellent opportunities for studying general questions in thermal biology.

Five questions concerning the thermal relationships of snakes are particularly relevant to understanding their ecology: (1) What is the range of possible body temperatures under natural conditions? (2) What are the proximate factors that determine which body temperatures a snake selects from that range of possibilities? (3) How do the body temperatures of individual snakes actually vary under natural conditions over extended periods of time? (4) What are the functional effects of body-temperature variation? (5) What are the ecological consequences of body-temperature variation?

We present an integrated approach to answering these questions. The key components of this approach include: (1) the use of biophysical models to determine the range of body temperatures possible under field conditions; (2) laboratory and field studies of the factors influencing the temperatures that snakes select or accept from within that range; (3) field studies of body-temperature variation in individual, free-ranging snakes; (4) laboratory studies of the developmental, physiological, and behavioral effects of body-temperature variation; and (5) the combination of field and laboratory data to generate specific hypotheses concerning the ecological consequences of T_b variation, and the testing of those hypotheses through computer modeling, laboratory simulations, field observations, field comparisons, and field experiments. Because the components of this approach have been applied to relatively few snake species, we need data from a wider taxonomic and ecological range of taxa.

Understanding the causes of T_b variation requires information about what body temperatures are possible so the relative importance of snake behavior and environmental constraints can be distinguished. One of the major problems in evaluating body-temperature data in past studies has been the lack of data on possible T_b's. The range of possible T_b's in the field can be measured with physical and mathematical models. The thermal environments of many snakes are highly variable spatially and temporally, some of the time constraining their body temperatures and at other times offering a wide range of possible T_b's. The elongate body form of snakes allows them to use this variation in a variety of ways to control their T_b's. Considerable opportunities exist for the application of biophysical modeling to evaluate the effects of snake characteristics such as size, color, and posture on T_b variation. Characterizations of a wide variety of snake thermal environments are needed, especially for tropical species.

Interpreting and predicting body-temperature variation also require data on those factors that determine the temperatures a snake will select or accept from the range of possibilities. The primary way of identifying these factors is through studies of the thermoregulatory behavior of captive snakes, supplemented by field observations. A variety of physiological and behavioral factors influences the temperatures that snakes actually select or accept. Based on the relatively few species that have been examined, it appears that snakes select a narrow range of T_b's under some conditions and accept a much wider range of T_b's under other conditions. Several studies indicate that digesting, shedding, and gravid snakes select a narrow range of relatively high T_b's. Snakes that are foraging or avoiding predators appear willing to accept a wider range of T_b's with a lower mean. Snakes that are fasting, aestivating, or entering hibernation generally appear to be selecting relatively lower temperatures. Temperature selection may also vary with time of day and season of year. The effects of various factors on T_b selection indicate that it may be important to provide thermal gradients for snakes kept in captivity. Future studies need to include more factors (such as condition, disease, and season) and a wider variety of species.

Determining how the body temperatures of free-ranging snakes vary is of central importance in studying snake thermal ecology. Capture measurements and radiotelemetry are the two principal techniques for measuring field T_b's. Although radiotelemetry is more expensive and cannot be used on small snakes, it allows repeated measurements on both active and inactive snakes so that temporal patterns of T_b variation and intra- and interindividual differences can be studied. Information on field T_b variation in snakes is very incomplete. Field measurements exist for less than 5% of all snake species, and the T_b patterns of individual snakes are known for less than 1% of all species. Many of the snakes that have been studied show clear daily and seasonal patterns of T_b variation that indicate both the constraints placed on them by the thermal environment and their ability to select a narrow subrange of temperatures from within that range. Activity temperatures for most of the species studied so far are largely restricted to a range of 25–35°C and are similar to preferred temperatures measured in the laboratory. Studies of a wide variety of species (e.g., fossorial, nocturnal, and aquatic forms) are needed to determine if this is a general pattern. The evaluation of T_b variation data is complex and requires a variety of data, including environmental temperature variation, thermoregulatory set-points, field T_b variation, and the physiological condition of the snakes. The development of quantitative indices for integrating these data promises to improve interpretations of T_b variation in free-ranging snakes.

Determining the functional consequences of T_b variation in snakes is the central link needed to evaluate the ecological and evolutionary importance of body-temperature variation. The functional consequences of T_b variation can be determined by measuring the thermal dependencies of whole-animal performances in the laboratory (e.g., developmental rate, digestion, and crawling speed). Resulting data can then be fitted to curves from which various useful indices can be calculated (e.g., thermal tolerances, performance breadths, and optima). Most performances studied so far are highly temperature dependent. Although in Garter Snakes the thermal optima for several functions occur within the range of activity and selected body temperatures, it is unknown whether this is a general phenomenon. Future research should characterize multiple, whole-animal performances for snakes from a range of taxa and a variety of ecological conditions. The shapes of the thermal dependency curves and their relationships to each other and to field and selected temperatures deserve particular attention.

The primary goal of snake thermal ecology studies is to understand how variation in the thermal environment influences the distribution and abundance of snakes. Few studies have quantified the ecological effects of temperature variation because many of the dependent ecological variables (e.g., activity time and habitat selection) are poorly described and influenced by many factors besides temperature. The ecological effects of T_b variation can be studied in a variety of ways, including field observations, field comparisons, field experiments, laboratory simulations, and computer modeling. The primary contribution of thermal studies is to indicate the times and locations where snakes can maintain T_b's within acceptable limits and the possible consequences of temperature variation on feeding, growth, reproduction, and predation.

Studies of the ecological consequences of body-temperature variation and thermoregulatory behavior naturally lead to evolutionary questions. These questions concern the thermal characteristics of snakes, i.e., those traits that influence T_b variation (such as size, coloration, and thermal preference) and those characteristics influencing the consequences of T_b variation (such as thermal tolerances, thermal optima, and the ability to acclimate). General evolutionary questions include: (1) How do thermal characteristics vary among snake taxa? (2) What was the pattern of evolutionary change in these characteristics? (3) What evolutionary mechanisms have shaped the thermal characteristics of snakes? (4) How has the evolution of snake thermal characteristics affected the evolution of other snake traits? These questions can be addressed using comparative studies, selection experiments, and modeling. The profound effects of temperature on biological function and the ecological consequences of those effects suggest that adaptation to the thermal environment has been an important theme in the evolution of snakes. At present, however, the data do not allow making general statements about the evolution of snake thermal characteristics. In addition to needing data on the thermal characteristics from a wide range of taxa, we also need data on independent variables such as snake phylogenies, thermal environments, and natural history (e.g., habitat selection, activity patterns, foraging mode, and reproductive mode).

Acknowledgments

Over the years, we have benefited greatly from discussing the ideas contained in this chapter with many of our colleagues. We would especially like to acknowledge the influence of Stevan Arnold, Theodore Garland, Raymond Huey, Warren Porter, Richard Shine, and Robert Stevenson. Michael Walton generously provided extensive help with writing the evolution section. Stevan Arnold, Vincent Cobb, Scott Grothe, Jay Kollar, Warren Porter, Howard Reinert, and Robert Stevenson graciously allowed us to use data that have not yet been published elsewhere. Paul Bartelt, Raymond Clark, Harvey Lillywhite, Debra Patla, Robert Stevenson, and Michael Walton reviewed the manuscript and provided helpful criticisms. We are responsible for any remaining errors. Finally, we would like to thank Cleveland State University and Idaho State University for supporting our research programs.

Literature Cited

Adolph, S. C., and W. P. Porter, 1993, Temperature, activity and lizard life histories, *Amer. Nat.*,142:273–295.

Aleksiuk, M., 1971, Temperature dependent shifts in the metabolism of a cool temperate reptile, *Thamnophis sirtalis parietalis, Comp. Biochem. Physiol.*, 50A:627–631.

Arnold, S. J., 1983, Morphology, performance and fitness, *Amer. Zool.*, 23:347–361.

Arnold, S. J., 1986, Laboratory and field approaches to the study of adaptation, in M. E. Feder and G. V. Lauder, eds., *Predator-Prey Relationships, Perspectives and Approaches from the Study of Lower Vertebrates*, Univ. Chicago Press, Chicago, pp. 157–179.

Arnold, S. J., 1987, Genetic correlation and the evolution of physiology, in M. E. Feder, A. F. Bennett, W. W. Burggren, and R. B. Huey eds., *New Directions in Physiological Ecology*, Cambridge Univ. Press, New York, pp. 189–211.

Arnold, S. J., 1988, Quantitative genetics and selection in natural populations: Microevolution of vertebral numbers in the garter snake *Thamnophis elegans, Proc. 2d. Int. Conf. Quant. Genetics*, pp. 619–638.

Arnold, S. J., and C. R. Peterson, 1989, A test for temperature effects on the ontogeny of shape in the garter snake, *Thamnophis sirtalis, Physiol. Zool.*, 62:1316–1333.

Avery, R. A., 1982, Field studies of body temperatures, in C. Gans and F. H. Pough, eds., *Biology of the Reptilia*, Vol. 12, Academic, New York, pp. 25–91.

Bakken, G. S., 1976, A heat transfer analysis of animals: Unifying concepts and the application of metabolism chamber data to field ecology, *J. Theoret. Biol.*, 60:337–384.

Bakken, G. S., 1992, Measurement and application of operative and standard operative temperatures in ecology, *Amer. Zool.*, 32:194–216.

Bakken, G. S., and D. M. Gates, 1975, Heat transfer analysis of animals: Some implications for field ecology, physiology, and evolution, in D. M. Gates and R. B. Schmerl, eds., *Perspectives of Biophysical Ecology*, Springer, New York, pp. 255–290.

Bakken, G. S., W. R. Santee, and D. J. Erskine, 1985, Operative and standard operative temperatures: Tools for thermal energetics studies, *Amer. Zool.,* 25:933–943.
Bartholomew, G. A., 1958, The role of physiology in the distribution of terrestrial vertebrates, in C. L. Hubbs, ed., *Zoogeography,* Amer. Assoc. Adv. Sci., Washington, D.C., pp. 81–95.
Bartholomew, G. A., 1982, Physiological control of body temperature, in C. Gans and F. H. Pough, eds., *Biology of the Reptilia,* Vol. 12, Academic, New York, pp. 167–211.
Bennett, A. F., 1987, The accomplishments of ecological physiology, in M. E. Feder, A. F. Bennett, W. W. Burggren, and R. B. Huey eds., *New Directions in Physiological Ecology,* Cambridge Univ. Press, New York, pp. 1–8.
Bennett, A. F., and W. R. Dawson, 1976, Metabolism, in C. Gans and W. R. Dawson, eds., *Biology of the Reptilia,* Vol. 5, Academic, New York, pp. 127–223.
Blanchard, F. N., and F. C. Blanchard, 1940, Factors determining the time of birth in the gartersnake *Thamnophis sirtalis sirtalis* (Linnaeus), *Pap. Mich. Acad. Sci. Arts Lett.,* 26:161–176.
Bowker, R. G., and O. W. Johnson, 1980, Thermoregulatory precision in three species of whiptail lizards (Lacertilia: Teiidae), *Physiol. Zool.,* 53:176–185.
Brattstrom, B. H., 1965, Body temperatures of reptiles, *Amer. Midl. Nat.,* 73:376–422.
Brattstrom, B. H., 1978, Learning studies in lizards, in N. Greenberg and P. D. MacLean, eds., *Behavior and Neurology of Lizards: An Interdisciplinary Colloquium,* Dept. Health, Educ. Welfare Publ. No. (ADM), pp. 173–181.
Brett, J. R., 1971, Energetic responses of salmon to temperature, A study of some thermal relations in the physiology and freshwater ecology of sockeye salmon (*Oncorhynchus nerka*), *Amer. Zool.,* 11:99–113.
Brooks, D. R., and D. A. McLennan, 1991, *Phylogeny, Ecology, and Behavior: A Research Program in Comparative Biology,* Univ. Chicago Press, Chicago.
Brown, W. S., 1991, Female reproductive ecology in a northern population of the timber rattlesnake, *Crotalus horridus, Herpetologica,* 47:101–115.
Burger, J., 1991, Effects of incubation temperature on behavior of young black racers (*Coluber constrictor*) and kingsnakes (*Lampropeltis getulus*), *J. Herpetol.,* 24:158–163.
Burger, J., R. T. Zappalorti, J. Dowdell, T. Georgiadis, J. Hill, and M. Gochfeld, 1992, Subterranean predation on pine snakes (*Pituophis melanoleucus*), *J. Herpetol.,* 26:259–263.
Burghardt, G. M., 1977, Learning processes in reptiles, in C. Gans and F. H. Pough, eds., *Biology of the Reptilia,* Vol. 7, Academic, New York, pp. 555–681.
Campbell, G. S., 1977, *An Introduction to Environmental Biophysics,* Springer, New York.
Charland, M. B., 1991, Reproductive ecology of female garter snakes (*Thamnophis*) in southeastern British Columbia, Doctoral Thesis, Univ. Victoria, British Columbia.
Christian, K. A., and C. R. Tracy, 1981, The effect of the thermal environment on the ability of hatchling Galapagos land iguanas to avoid predation during dispersal, *Oecologia (Berlin),* 49:218–223.
Cliburn, J. W., 1976, Observations of ecdysis in the black pine snakes, *Pituophis melanoleucus lodingi* (Reptilia, Serpentes, Colubridae), *J. Herpetol.,* 10:299–301.
Cobb, V. A., and C. R. Peterson, 1991, The effects of pregnancy on body-temperature variation in free-ranging western rattlesnakes, *Amer. Zool.,* 31:78A
Conn, C. A., R. Franklin, R. Freter, and M. J. Kluger, 1991, Role of gram-negative and gram-positive gastrointestinal flora in temperature regulation of mice, *Amer. J. Physiol. Reg. Integ. Comp. Physiol.,* 30:R1358–R1363.
Costanzo, J. P., and D. L. Claussen, 1988, Natural freeze tolerance in a reptile, *Cryo-Letters,* 9:380–385.
Cowles, R. B., and C. M. Bogert, 1944, A preliminary study of the thermal requirements of desert reptiles, *Bull. Amer. Mus. Nat. Hist.,* 83:265–296.
Dawson, W. R., 1975, On the physiological significance of the preferred body temperatures of reptiles, in D. M. Gates and R. Schmerl, eds., *Perspectives of Biophysical Ecology,* Springer, New York, pp. 443–473.
DeWitt, C. B., 1967, Precision of thermoregulation and its relation to environmental factors in the desert iguana, *Dipsosaurus dorsalis, Physiol. Zool.,* 40:49–66.

Dill, C. D., 1972, Reptilian core temperatures: variation within individuals, *Copeia,* 1972:577–579.

Dunham, A. E., 1993, Population responses to environmental change: Physiologically structured models, operative environments, and population dynamics, in P. M. Karieva, J. G. Kingsolver, and R. B. Huey, eds., *Biotic Interactions and Global Change,* Sinauer, Sunderland, Massachusetts, pp. 95–119.

Dunham, A. E., B. W. Grant, and K. L. Overall, 1989, Interfaces between biophysical and physiological ecology and the population ecology of terrestrial vertebrate ectotherms, *Physiol. Zool.,* 62:335–355.

Endler, J. A., 1986, *Natural Selection in the Wild,* Princeton Univ. Press, Princeton, New Jersey.

Felsenstein, J., 1985, Phylogenies and the Comparative Method, *Amer. Nat.,* 125:1–15.

Fitch, H. S., 1956, Temperature responses in free-living amphibians and reptiles of northeastern Kansas, *Univ. Kans. Publ. Mus. Nat. Hist.,* 8:417–476.

Fitch, H. S., 1987, Collecting and life history techniques, in R. A. Seigel, J. T. Collins, and S. S. Novak, eds., *Snakes: Ecology and Evolutionary Biology,* McGraw-Hill, New York, pp. 143–164.

Fox, W. W., 1948, Effect of temperature on development of scutellation in the garter snake, *Thamnophis elegans atratus, Copeia,* 1948:252–262.

Fox, W. W., C. Gordon, and M. H. Fox, 1961, Morphological effects of low temperatures during the embryonic development of the garter snake, *Thamnophis elegans, Zoologica,* 46:57–71.

Garland, T., Jr., R. B. Huey, and A. F. Bennett, 1991, Phylogeny and coadaptation of thermal physiology in lizards: A reanalysis, *Evolution,* 45:1969–1975.

Geiser, F., B. T. Firth, and R. T. Seymour, 1992, Polyunsaturated dietary lipids lower the selected body temperature of a lizard, *J. Comp. Physiol. B. Syst. Environ. Physiol.,* 162:1–4.

Gibbons, J. W., and R. D. Semlitsch, 1987, Activity Patterns, in R. A. Seigel, J. T. Collins, and S. S. Novak, eds., *Snakes: Ecology and Evolutionary Biology,* McGraw-Hill, New York, pp. 396–421.

Gibson, A. R., and J. B. Falls, 1979, Thermal biology of the common garter snake *Thamnophis sirtalis,* I. Temporal variation, environmental effects and sex differences, *Oecologia (Berlin),* 43:79–97.

Gibson, A. R., D. A. Smucny, and J. Kollar, 1989, The effects of feeding and ecdysis on temperature selection by young garter snakes in a simple thermal mosaic, *Can. J. Zool.,* 67:19–23.

Gibson, A. R., R. A. Bear, S. M. Mavroidis, and M. A. Gates, Daily cycles of thermoregulation and activity in captive garter snakes, *Thamnophis elegans, Herpetologica,* in review.

Goode, M. J., and D. Duvall, 1989, Body temperature and defensive behaviour of free-ranging prairie rattlesnakes (*Crotalus viridis viridis*), *Anim. Behav.,* 38:360–362.

Gould, S. J., and R. C. Lewontin, 1979, The spandrels of San Marco and the Panglossian paradigm: A critique of the adaptationist programme, *Proc. Roy. Soc. London B,* 205:581–598.

Grant, B. W., 1990, Trade-offs in activity time and physiological performance for thermoregulating desert lizards, *Sceloporus merriami, Ecology,* 71:2323–2333.

Grant, B. W., A. D. Tucker, J. E. Lovich, A. M. Mills, P. M. Dixon, and J. W. Gibbons, 1992, The use of coverboards in estimating patterns of reptile and amphibian biodiversity, in D. R. McCullough and R. H. Barrett, eds., *Wildlife 2001: Populations,* Elsevier Applied Science, London, pp. 379–403.

Graves, B. M., and D. Duvall, 1987, An experimental study of aggregation and thermoregulation in prairie rattlesnakes (*Crotalus viridis viridis*), *Herpetologica,* 43:253–258.

Graves, B. M., and Duvall, D., 1993, Reproduction, rookery use, and thermoregulation in free-ranging, pregnant *Crotalus viridis viridis, J. Herpetol.,* 27:33–41.

Greene, H. W., 1986, Natural history and evolutionary biology, in M. E. Feder and G. V. Lauder, eds., *Predator-Prey Relationships, Perspectives and Approaches from the Study of Lower Vertebrates,* Univ. Chicago Press, Chicago, pp. 99–108.

Greenwald, O. E., 1971, The effect of body temperature on oxygen consumption and heart rate in the Sonora gopher snake, *Pituophis catenifer affinis* Hallowell, *Copeia,* 1971:98–106.

Greenwald, O. E., 1974, The thermal dependence of striking and prey capture by gopher snakes, *Copeia,* 1974:141–148.

Greenwald, O. E., and M. E. Kanter, 1979, The effects of temperature and behavioral thermoregulation on digestive efficiency and rate in corn snakes (*Elaphe guttata guttata*), *Physiol. Zool.,* 52:398–408.

Gregory, P. T., 1977, Life-history parameters of the red-sided garter snake (*Thamnophis sirtalis parietalis*) in an extreme environment, the Interlake region of Manitoba, *Nat. Mus. Can. Publ. Zool.,* 13:1–44.

Gregory, P. T., 1982, Reptilian hibernation, in C. Gans and F. H. Pough, eds., *Biology of the Reptilia,* Vol. 12, Academic, New York, pp. 53–154.

Gregory, P. T., 1990, Temperature differences between head and body in garter snakes (*Thamnophis*) at a den in central British Columbia, *J. Herpetol.,* 24:241–245.

Gregory, P. T., and K. J. Nelson, 1991, Predation on fish and intersite variation in the diet of common garter snakes, *Thamnophis sirtalis,* on Vancouver Island, *Can. J. Zool.,* 69:988–994.

Grothe, S., 1992, Red-tailed hawk predation on snakes: The effects of weather and snake activity, Masters Thesis, Idaho State University, Pocatello.

Hammerson, G. A., 1979, Thermal ecology of the striped racer, *Masticophis lateralis, Herpetologica,* 35:267–273.

Harvey, P. H., and M. D. Pagel, 1991, *The Comparative Method in Evolutionary Biology,* Oxford Univ. Press, Oxford.

Heath, J. E., 1964, Reptilian thermoregulation: Evaluation of field studies, *Science,* 145:784–785.

Heath, J. E., 1965, Temperature regulation and diurnal activity in horned lizards, *Univ. Calif. Publ. Zool.,* 64:97–136.

Heckrotte, C., 1967, Relations of body temperature, size, and crawling speed of the common garter snake, *Thamnophis s. sirtalis, Copeia,* 1967:520–526.

Henderson, R. W., 1970, Feeding behavior, digestion, and water requirements of *Diadophis punctatus arnyi* Kennicott, *Herpetologica,* 26:520–526.

Hertz, P. E., 1992a, Evaluating thermal resource partitioning by sympatric lizards *Anolis cooki* and *A. cristatellus:* A field test using null hypotheses, *Oecologia (Berlin),* 90:127–136.

Hertz, P. E., 1992b, Temperature regulation in Puerto Rican *Anolis* lizards: A field test using null hypotheses, *Ecology,* 73:1405–1417.

Hertz, P. E., R. B. Huey, and E. Nevo, 1982, Fight versus flight: Thermal dependence of defensive behavior in a lizard, *Anim. Behav.,* 30:676–679.

Hertz, P. E., R. B. Huey, and R. D. Stevenson, Evaluating temperature regulation in field active ectotherms: The fallacy of the inappropriate question, *Amer. Nat.,* in press.

Huey, R. B., 1982, Temperature, physiology, and the ecology of reptiles in C. Gans and F. H. Pough, eds., *Biology of the Reptilia,* Vol. 12, Academic, New York, pp. 25–67.

Huey, R. B., 1987, Phylogeny, history, and the comparative method, in M. E. Feder, A. F. Bennett, W. W. Burggren, and R. B. Huey, eds., *New Directions in Physiological, Ecology,* Cambridge Univ. Press, New York, pp. 76–98.

Huey, R. B., and A. F. Bennett, 1987, Phylogenetic studies of coadaptation: Preferred temperatures versus optimal performance temperatures of lizards, *Evolution,* 41:1098–1115.

Huey, R. B., and A. F. Bennett, 1990, Physiological adjustments to fluctuating thermal environments: An ecological and evolutionary perspective, in R. I. Morimoto, A. Tissieres, and C. Georogopoulos, eds., *Stress Proteins in Biology and Medicine,* Cold Spring Harbor Lab. Press, New York, pp. 37-59.

Huey, R. B., and M. Slatkin, 1976, Costs and benefits of lizard thermoregulation, *Q. Rev. Biol.,* 51:363–384.

Huey, R. B., and R. D. Stevenson, 1979, Integrating thermal physiology and ecology of ectotherms: A discussion of approaches, *Amer. Zool,* 19:357–366.

Huey, R. B., C. R. Peterson, S. J. Arnold, and W. P. Porter, 1989, Hot rocks and not-so-hot rocks: Retreat-site selection by garter snakes and its thermal consequences, *Ecology,* 70:931–944.

Hutchison, V. H., and R. K. Dupré, 1992, Thermoregulation, in M. E. Feder and W. W. Burggren, eds., *Environmental Physiology of the Amphibians,* Univ. Chicago Press, Chicago, pp. 206-249.

Hutchison, V. H., H. G. Dowling, and A. Vinegar, 1966a, Thermoregulation in a brooding female Indian python *Python molurus bivitattus, Science,* 151:694–696.

Hutchison, V. H., A. Vinegar, and R. J. Kosh, 1966b, Critical thermal maxima in turtles, *Herpetologica,* 22:32–41.

Jacob, J. S., and C. W. Painter, 1980, Overwinter thermal ecology of *Crotalus viridis* in the north-central plains of New Mexico, *Copeia,* 1980:799–805.

Jayne, B. C., and A. F. Bennett, 1990, Selection on locomotor performance capacity in a natural population of garter snakes, *Evolution,* 44:1204–1229.

Johnson, C. R., 1973, Thermoregulation in pythons, II. Head-body temperature differences and thermal preferenda in Australian pythons, *Comp. Biochem. Physiol.,* 45A:1065–1087.

Justy, G. M., and F. F. Mallory, 1985, Thermoregulatory behavior in the northern water snake, *Nerodia s. sipedon,* and the eastern garter snake, *Thamnophis s. sirtalis, Can. Field-Nat.,* 99:246–249.

Kitchell, J. F., 1969, Thermophilic and thermophobic responses of snakes in a thermal gradient, *Copeia,* 1969:189–191.

Klauber, L. M., 1972, *Rattlesnakes: Their Habits, Life Histories, and Influence on Mankind,* 2 Vols., Univ. Calif. Press, Berkeley.

Kluger, M. J., 1979, Fever in ectotherms: Evolutionary implications, *Amer. Zool.,* 19:295–304.

Kluger, M. J., D. H. Ringler, and M. R. Anver, 1975, Fever and survival, *Science,* 188(4184):166–168.

Koenig, W. D., and S. S. Albano, 1987, Lifetime reproductive success, selection, and the opportunity for selection in the white-tailed skimmer *Plathemis lydia* (Odonata: Libellulidae), *Evolution,* 41:22–36.

Kollar, J., 1988, Influences on temperature selection in young common garter snakes (*Thamnophis sirtalis*), Masters Thesis, Cleveland State Univ., Cleveland, Ohio.

Krekorian, C. O., V. J. Vance, and A. M. Richardson, 1968, Temperature-dependent maze learning in the desert iguana, *Dipsosaurus dorsalis, Anim. Behav.,* 16:429–436.

Lillywhite, H. B., 1987, Temperature, energetics, and physiological ecology, in R. A. Seigel, J. T. Collins, and S. S. Novak, eds., *Snakes: Ecology and Evolutionary Biology,* McGraw-Hill, New York, pp. 442–477.

Lutterschmidt, W. I., 1991, The thermal preferenda of the timber rattlesnake, *Crotalus horridus,* with a discussion of thermoregulatory constraints within preferred microhabitats, Masters Thesis, Southeastern Louisiana Univ., Hammond, Louisiana.

Lutterschmidt, W. I., and H. K. Reinert, 1990, The effect of ingested transmitters upon the temperature preference of the northern water snake, *Nerodia s. sipedon, Herpetologica,* 46:39–42.

Macartney, J. M., K. W. Larsen, and P. T. Gregory, 1989, Body temperatures and movements of hibernating snakes (*Crotalus* and *Thamnophis*) and thermal gradients of natural hibernacula, *Can. J. Zool.,* 67:108–111.

Maderson, P. F. A., 1984, The squamate epidermis: New light has been shed, *Symp. Zool. Soc. Lond.,* 52:111–126.

Martins, E. P., and T. Garland, Jr., 1991, Phylogenetic analyses of the correlated evolution of continuous characters: A simulation study, *Evolution,* 45:534–557.

Mathur, D., and C. A. Silver, 1980, Statistical problems in studies of temperature preference of fishes, *Can. J. Fish. Aquat. Sci.,* 37:733–737.

Mautz, W. J., C. B. Daniels, and A. F. Bennett, 1992, Thermal dependence of locomotion and aggression in a xantusid lizard, *Herpetologica,* 48:271–279.

Mushinsky, H. R., 1987, Feeding ecology, in R. A. Seigel, J. T. Collins, and S. S. Novak, eds., *Snakes: Ecology and Evolutionary Biology,* McGraw-Hill, New York, pp. 302–334.

Muth, A., 1977, Thermoregulatory postures and orientation to the sun: A mechanistic evaluation for the zebra-tailed lizard, *Callisaurus draconoides, Copeia,* 1977:710–720.

Naulleau, G., 1983, The effects of temperature on digestion in *Vipera aspis*, *J. Herpetol.*, 17:166–170.
Osgood, D. W., 1970, Thermoregulation in water snakes studied by telemetry, *Copeia*, 1970:568–571.
Osgood, D. W., 1978, Effects of temperature on the development of meristic characters in *Natrix fasciata*, *Copeia*, 1978:33–47.
Peterson, C. R., 1982, Body-temperature variation in free-living garter snakes (*Thamnophis elegans vagrans*), Dissertation, Washington State Univ., Pullman, Washington.
Peterson, C. R., 1987, Daily variation in the body temperatures of free-ranging garter snakes, *Ecology*, 68:160–169.
Peterson, C. R., and S. J. Arnold, 1986, Individual variation in the thermoregulatory behavior of free-ranging garter snakes, *Thamnophis elegans*, *Amer. Zool.*, 26:112A.
Peterson, C. R., and S. J. Arnold, The effect of access to heat on growth in the garter snake *Thamnophis sirtalis*, in review.
Peterson, C. R., and M. E. Dorcas, 1992, The use of automated data acquisition techniques in monitoring amphibian and reptile populations, in D. R. McCullough and R. H. Barrett, eds., *Wildlife 2001: Populations*, Elsevier Applied Science, London, pp. 369–378.
Peterson, C. R., and M. E. Dorcas, Automated data acquisition, in W. R. Heyer, R. W. McDiarmid, M. Donnelly, and L. Hayek, eds., *Measuring and Monitoring Biological Diversity—Standard Methods for Amphibians*, Smithsonian Inst. Press, Washington, D.C., in press.
Peterson, C. R., and H. J. Fabian, 1984, *Thamnophis elegans vagrans*, coloration, Life History Notes, *Herpetol. Rev.*, 15:113.
Peterson, C. R., A. J. Vitale, and V. A. Cobb, 1989, Measuring the activity patterns of free-ranging animals with radiotelemetry, *Amer. Zool.*, 29:43A.
Plummer, M. P., 1993, Thermal ecology of arboreal green snakes (*Opheodrys aestivus*), *J. Herpetol.*, in press.
Porter, W. P., 1967, Solar radiation through the living body walls of vertebrates with emphasis on desert reptiles, *Ecol. Monogr.*, 39:245–270.
Porter, W. P., 1989, New animal models and experiments for calculating growth potential at different elevations, *Physiol. Zool.*, 62:286–313.
Porter, W. P., and C. R. Tracy, 1974, Modeling the effects of temperature changes on the ecology of the garter snake and leopard frog, in J. W. Gibbons and R. R. Sharitz, eds., *Thermal Ecology*, Atomic Energy Comm. Symp. Series, pp. 594–609.
Porter, W. P., and C. R. Tracy, 1983, Biophysical analysis of energetics, time-space utilization, and distributional limits, in R. Huey, E. Pianka, and T. Schoener, eds., *Lizard Ecology: Studies of a Model Organism*, Harvard Univ. Press, Cambridge.
Porter, W. P., J. W. Mitchell, W. A. Beckman, and C. B. DeWitt, 1973, Behavioral implications of mechanistic ecology: Thermal and behavioral modeling of desert ectotherms and their microenvironment, *Oecologia (Berlin)*, 13:1–54.
Porter, W. P., J. W. Mitchell, W. A. Beckman, and C. R. Tracy, 1975, Environmental constraints on some predator-prey interactions, in D. M. Gates and R. B. Schmerl, eds., *Perspectives of Biophysical Ecology*, Springer, New York.
Pough, F. H., 1974, Preface to *A Preliminary Study of the Thermal Requirements of Desert Reptiles*, by R. B. Cowles and C. M. Bogert, Facsimile Reprint Soc. Study Amph. Rept., pp. i–iv.
Pough, F. H., 1980, The advantages of ectothermy for tetrapods, *Amer. Nat.*, 115:92–112.
Pough, F. H., 1989, Organismal performance and Darwinian fitness: approaches and interpretations, *Physiol. Zool.*, 62:199–236.
Prosser, C. L., and J. E. Heath, 1991, Temperature, in C. L. Prosser, ed., *Environmental and Metabolic Animal Physiology*, Saunders, Philadelphia, pp. 109–166.
Regal, P. J., 1966, Thermophilic responses following feeding in certain reptiles, *Copeia*, 1966:588–590.
Regal, P. J., 1967, Voluntary hypothermia in reptiles, *Science*, 155:1551–1553.

Reinert, H. K., 1984, Habitat variation within sympatric snake populations, *Ecology,* 65:1673–1682.
Reinert, H. K., 1992, Radiotelemetric field studies of pitvipers: Data acquisition and analysis, in J. A. Campbell and E. D. Brodie III, eds., *Biology of the Pitvipers,* Selva, Tyler, Texas, pp. 185–198.
Reinert, H. K., and D. Cundall, 1982, An improved surgical implantation method for radiotracking snakes, *Copeia,* 1982:702–705.
Reinert, H. K., and R. T. Zappalorti, 1988, Timber rattlesnakes (*Crotalus horridus*) of the Pine Barrens: Their movement patterns and habitat preference, *Copeia,* 1988:964–978.
Rosen, P. C., 1991, Comparative field study of thermal preferenda in garter snakes (*Thamnophis*), *J. Herpetol.,* 25:301–312.
Saint-Girons, H., 1975, Observations preliminaires sur la thermoregulation des viperes d'Europe, *Vie Milieu,* 25:137–168.
Sanders, J. S., and J. S. Jacob, 1981, Thermal ecology of the copperhead (*Agkistrodon contortrix*), *Herpetologica,* 37:264–270.
Schieffelin, C. D., and A. de Queiroz, 1991, Temperature and defense in the common garter snake: warm snakes are more aggressive than cold snakes, *Herpetologica,* 47:230–237.
Schmidt-Nielson, K., 1983, *Animal Physiology: Adaptation and Environment,* 3rd ed., Cambridge Univ. Press, Cambridge.
Scott, J. R., 1978, Thermal biology of the wandering garter snake, Doctoral Dissertation, Colorado State Univ., Fort Collins.
Scott, J. R., and D. Pettus, 1979, Effects of seasonal acclimation on the preferred body temperature of *Thamnophis elegans vagrans, J. Thermal Biol.,* 4:307–309.
Scott, J. R., C. R. Tracy, and D. Pettus, 1982, A biophysical analysis of daily and seasonal utilization of climate space by a montane snake, *Ecology,* 63:482–493.
Secor, S. M., 1992, Activities and energetics of a sit-and-wait foraging snake, *Crotalus cerastes,* Doctoral Thesis, Univ. California, Los Angeles.
Semlitsch, R. D., 1979, The influence of temperature on ecdysis in snakes (genus *Natrix*) (Reptilia, Serpentes, Colubridae), *J. Herpetol.,* 13:212–214.
Sexton, O. J., and S. R. Hunt, 1980, Temperature relationships and movements of snakes (*Elaphe obsoleta, Coluber constrictor*) in a cave hibernaculum, *Herpetologica,* 36:20–26.
Sexton, O. J., and K. R. Marion, 1981, Experimental analysis of movements by prairie rattlesnakes, *Oecologia (Berlin),* 51:37–41.
Shine, R., 1985, The evolution of viviparity in reptiles: an ecological analysis, in *Biology of the Reptilia,* Vol. 15, edited by C. Gans and F. Billett, John Wiley and Sons, New York, pp. 605–694.
Shine, R., 1987, Intraspecific variation in thermoregulation, movements and habitat use by Australian blacksnakes, *Pseudechis porphyriacus* (Elapidae), *J. Herpetol.,* 21:165–177.
Shine, R., 1988, Parental care in reptiles, in C. Gans and R. B. Huey, eds., *Biology of the Reptilia,* Vol. 16, Alan R. Liss, New York, pp. 275–330.
Shine, R., and R. Lambeck, 1985, A radiotelemetric study of movements, thermoregulation and habitat utilization of Arafura filesnakes (Serpentes, Acrochordidae), *Herpetologica,* 41:351–361.
Shine, R., and R. Lambeck, 1990, Seasonal shifts in the thermoregulatory behavior of Australian blacksnakes, *Pseudechis porphyriacus* (Serpentes: Elapidae), *J. Therm. Biol.,* 15:301–305.
Sievert, L. M., 1989, Postprandial temperature selection in *Crotaphytus collaris, Copeia,* 1989:987–993.
Skoczylas, R., 1970, Influence of temperature on gastric digestion in the grass snakes *Natrix natrix* L., *Comp. Biochem. Physiol.,* 33:793–804.
Slip, D. J., and R. Shine, 1988, Reptilian endothermy: A field study of thermoregulation by brooding diamond pythons, *J. Zool. Lond.,* 216:367–378.
Smucny, D. A., and Gibson, A. R., Patterns of heat use by females common garter snakes in a laboratory thermal mosaic. *J. Herpetol.,* in review.

Stevenson, R. D., 1983, The ecology of body temperature control of terrestrial ectotherms, Doctoral Thesis, Univ. Washington, Seattle.
Stevenson, R. D., 1985a, Body size and limits to the daily range of body temperatures interrestrial ectotherms, *Amer. Natur.*, 125:102–117.
Stevenson, R. D., 1985b, The relative importance of behavioral and physiological adjustments controlling body temperature in terrestrial ectotherms, *Amer. Nat.*, 126:362–386.
Stevenson, R. D., C. R. Peterson, and J. S. Tsuji, 1985, The thermal dependence of locomotion, tongue flicking, digestion, and oxygen consumption in the wandering garter snake, *Physiol. Zool.*, 58:46–57.
Thomas, R. G., and F. H. Pough, 1979, The effect of rattlesnake, *Crotalus atrox*, venom on digestion of prey, *Toxicon*, 17:221–228.
Tracy, C. R., 1976, A model of the dynamic exchanges of water and energy between a terrestrial amphibian and its environment, *Ecol. Monogr.*, 46:293–326.
Tracy, C. R., 1982, Biophysical modeling in reptilian physiology and ecology, in C. Gans and F. H. Pough, eds., *Biology of the Reptilia*, Vol. 12, Academic, New York, pp. 275–321.
Tracy, C. R., and K. A. Christian, 1986, Ecological relations among space, time and thermal niche axes, *Ecology*, 67:609–615.
van Berkum, F. H., 1986, Evolutionary patterns of the thermal sensitivity of sprint speed in *Anolis* lizards, *Evolution*, 40:594–604.
Viitanen, P., 1967, Hibernation and seasonal movements of the viper, *Vipera berus berus* (L.), in southern Finland, *Ann. Zool. Fenn.*, 4:472–546.
Weatherhead, P. J., 1989, Temporal and thermal aspects of hibernation of black rat snakes (*Elaphe obsoleta*) in Ontario, *Can. J. Zool.*, 67:2332–2335.
Webb, G. J. W., and H. Heatwole, 1971, Patterns of heat distribution within the bodies of some Australian pythons, *Copeia*, 1971:209–220.
White, J. B., and R. C. Lasiewski, 1971, Rattlesnake denning: Theoretical considerations on winter temperatures, *J. Theor. Biol.*, 30:553–557.
Zurovsky, Y., T. Brain, H. Laburn, and D. Mitchell, 1987, Pyrogens fail to produce fever in the snakes *Psammophis phillipsii* and *Lamprophis fuliginosis*, *Comp. Biochem. Physiol.*, 87A:911–914.

Chapter

8

Quantitative Genetics of Snake Populations

Edmund D. Brodie III

Theodore Garland, Jr.

Introduction

Quantitative genetics is a group of techniques used to study variation in continuously distributed traits (e.g., body size, shape, many behavioral and physiological measurements). In its strongest form, it can be used to isolate the relative importance of genetic and environmental factors to the expression of phenotypic variation (Mather and Jinks, 1982; Bulmer, 1985; Falconer, 1989). The integration of quantitative genetics with evolutionary theory has yielded explicit and testable hypotheses about phenomena from life-history evolution to sexual dimorphism and sexual selection (reviewed by Lande, 1988; Arnold, 1990, 1994). Explicit methodologies for measuring natural selection and for predicting the response of multiple traits to this selection have been developed (Lande, 1979; Lande and Arnold, 1983; Endler, 1986) and, in some cases, it now may be possible to reconstruct the patterns of selection necessary to account for phenotypic differences between populations and species (Arnold, 1981c, 1988a; Price et al., 1984; Price and Grant, 1985; Grant, 1986; Lofsvold, 1988).

Although many of the origins of quantitative genetics are due to evolutionary biologists (Fisher, 1918; Wright, 1921), most of the subsequent development and application of the techniques have been in applied animal and plant breeding. Some of the most familiar statistical concepts, including correlation, regression, and analysis of variance, arose as means of dealing with evolutionary problems of continuously varying traits (see Provine, 1971). Breeders of domestic stocks recognized that many of the characters of economic importance (e.g., milk yield in dairy cattle, carcass weight of hogs, oil content in maize) varied continuously, and adopted quantitative genetics as an analytical and predictive tool. Subsequent effort was concentrated on generating breeding and selection regimens that maximized the improvement of phenotypes for commercial value. The resultant research largely concerns domestic or laboratory populations and characters that may not have obvious ecological relevance in natural populations (e.g., defecation score in open field trials with mice, bristle number in *Drosophila*). Additionally, the statistical nature and jargon of this body of literature can be quite imposing to most biologists. It is not surprising, then, that few herpetologists have studied this literature and used quantitative genetic approaches in their own investigations.

The goal of this chapter is to introduce quantitative genetics to an audience that may not yet have considered what the technique can do for their research programs. Our approach will necessarily be simple, and more detailed treatments of the methods should be consulted before embarking on quantitative genetic analyses (Turner and Young, 1969; Ehrman and Parsons, 1981; Mather and Jinks, 1982; Becker, 1992; Bulmer, 1985; Falconer, 1989; Plomin et al., 1990; Arnold, 1994; Boake, 1994). We will provide an introduction to some of the basic concepts of quantitative genetics but concentrate on pointing out the types of problems that can benefit from a quantitative genetic approach. We will give special attention to specific problems and methodologies most often encountered when applying the approach to snake populations and will point out areas of research that are particularly suited to studies of snakes. Many snake ecologists will find that, with only minor modifications in sampling regimens, they can collect the kinds of data required for quantitative genetic analyses of the traits and species they currently study.

Why Study Variation or Quantitative Genetics—Who Cares?

The study of differences among individuals is a critical step toward understanding the processes that are responsible for the evolution of characters within and among populations. The most fundamental condition necessary for natural selection to occur is individual variation in a trait (Lewontin, 1970). Selection can be defined simply as a correlation between phenotype and reproductive success (Lewontin, 1970). If a trait does not differ among individuals, then it cannot be correlated to differences in individual reproductive success and therefore cannot be subject to selection. Therefore, studies attempting to determine whether or how selection is currently acting on a character must first consider whether the necessary substrate for selection, individual variation, is present (Arnold, 1981a; Lande and Arnold, 1983; Arnold and Bennett, 1984; Bennett, 1987; Garland, 1988; Boake, 1989; van Berkum et al., 1989; Brodie, 1993a).

Natural selection acts on phenotypes and does not necessarily result in evolution (Lande and Arnold, 1983; while some authors [Endler, 1986] define "natural selection" to include evolutionary change, we find it valuable to distinguish two separate processes, selection and response to selection). Quantitative genetics comes in when considering the evolutionary response of a trait to selection (Lande, 1979; Arnold, 1981c, 1983; Grant, 1986; Falconer, 1989). In order for selection to cause an evolutionary change, relatives must resemble each other. In other words, some of the phenotypic variation in a character must be heritable. For continuously distributed traits, it is the additive genetic variance (see below) that determines the resemblance between parents and their offspring and determines the response to selection. The equation

$$R = h^2 s$$

shows how the response to selection (R) is determined by the strength of selection (s) and the heritability (h^2) of a trait (Falconer, 1989). Heritability is merely the proportion of the total phenotypic variance that is additive genetic in origin. Heritability ranges between 0 and 1, so evolutionary response is always less than or equal to the strength of selection. A regression of mean offspring on mean parent values for a trait yields a line whose slope is equal to the heritability (explained below). If selection is measured as the difference between the means of the original population and the selected parents, and response is measured as the difference between the means of all potential offspring and the offspring of the selected parents, then it is easy to show graphically how greater heritability results in greater evolutionary response to a given intensity of selection (Fig. 8.1).

It should be clear, then, that measurable selection does not necessarily mean a character will change evolutionarily. Some, possibly all, of the variation acted upon by selection may be due to nonadditive genetic invariance environmental causes. Changes in such variation due to selection will not be transmitted to the next generation. On the other hand, measurable heritability almost ensures evolutionary change, because even in the absence of selection, the effects of stochastic processes such as genetic drift will be transmitted across generations (though the changes may be in terms of variances and covariances rather than means). Quantitative genetics strives to dis-

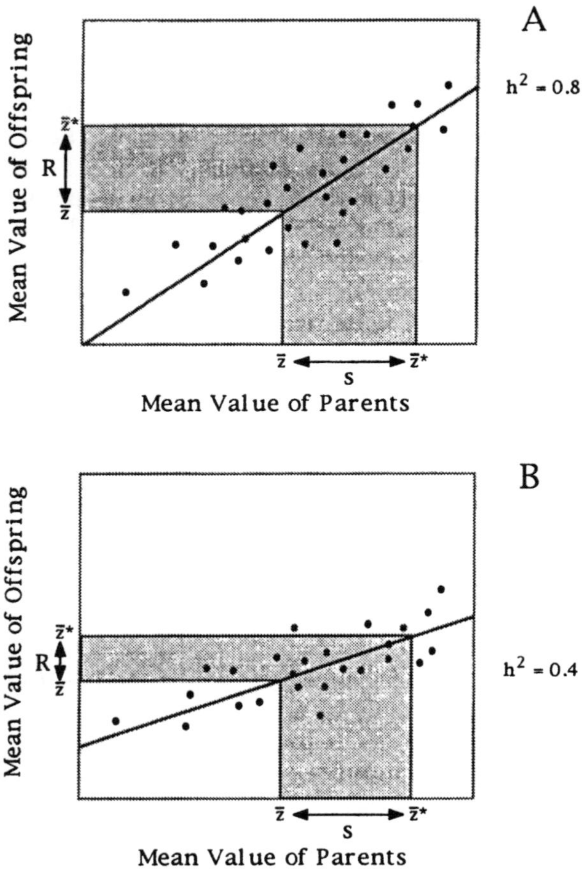

Figure 8.1 Plots of hypothetical offspring–parent data showing the effect of heritability on response to selection. Selection (s) is shown as the difference between the mean of the potential parents ($\bar{z}$) and the mean of the selected (actual) parents ($\bar{z}^*$). Evolutionary response (R) is shown as the difference between the mean of the offspring of the potential parents ($\bar{z}$) and the mean of the offspring of the selected parents ($\bar{z}^*$). The slope of the regression of offspring on parent values is equal to the narrow-sense heritability. Extrapolating from the intersection of the parental means with the heritability line to the offspring axis shows the trait mean in the next generation (shaded portion). The same strength of selection results in greater response when heritability is large (*A*) than when heritability is moderate (*B*).

cern the contributions of genetic and environmental factors to phenotypic variation, thus allowing predictions to be made regarding which traits can respond to selection, at what rate, and whether other traits will change as a result (because they are inherited together as indicated by genetic correlations [see below]).

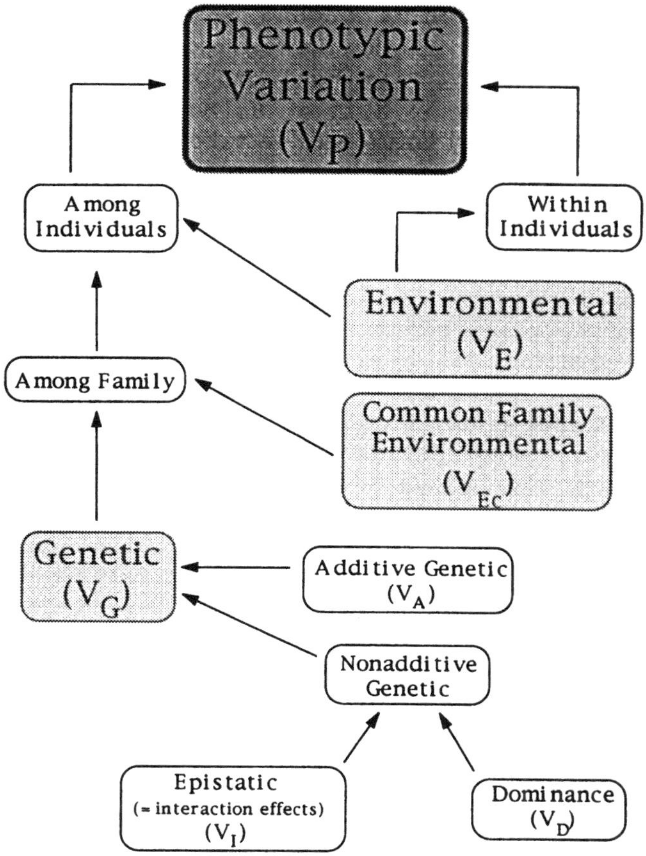

Figure 8.2 Path diagram showing components of phenotypic variation. Note that common-family environmental variance may include genetic and nongenetic maternal effects. See text for explanation of specific variance components.

Definitions

The basic goal of most quantitative genetic analyses is to separate the phenotypic variance of a trait into a number of additive elements (notation follows Falconer, 1989). Each of these *variance components* constitutes a portion of the total variance, and often itself can be further subdivided (Fig. 8.2). The calculation of variance components is often called *partitioning the variance* (the methods that are best suited to studies of snake populations will be discussed later). For instance, if one wanted to partition the variance in sprint speed in a population of snakes, the first level of analysis might be to determine how much variance is accounted for by differences between individual snakes (= *among individual component of variance*) versus variation in speed within an individual (= *within individual component of variance*) (Fig. 8.2). The ratio of the variation between individuals to the total variation is the *repeatability* and provides a measure of how consistently a trait can be measured within an individual. Variation within a single individual (except ontogenetic variation) cannot have a genetic basis, so this component may be considered *environmental variance* (V_E) (Fig. 8.2).

It will generally be of interest to subdivide the variance among individuals further. Some of the difference in speed among individuals will also be the result of environmental differences (e.g., minor differences in the testing procedure experienced by different individuals, or possibly different environments experienced by individual snakes at critical periods in development). If the variance in speed is attributable to differences among families, much of this may be *genetic variance* (V_G), which is responsible for the inherited resemblance among these relatives. It is important to realize that not all differences among families are genetic in origin (Fig. 8.2). Family members share common nongenetic factors (e.g., same maternal environment in viviparous species, the same incubation temperature during embryonic development, a common physical condition of the mother) that may also account for resemblance. This component of variance is due to the *common family environment* (V_{Ec}) and includes both genetic and nongenetic maternal effects. These nongenetic bases of among-family variation can often be controlled experimentally or isolated by comparing different groups of relatives. The genetic variance for speed may result from a variety of gene actions. The component of variance due to additive effects of genes is the *additive genetic variance* (V_A), and will determine the degree of heritable resemblance between parents and offspring. Some alleles have phenotypic effects that are not strictly additive, which lead to the *dominance variance* (V_D). Still other alleles may be expressed as different phenotypes depending on the alleles that are present at other loci. These contribute to the *epistatic* (or *interaction*) *genetic variance* (V_I) (Fig. 8.2).

If we are interested in what proportion of the total variation in speed among all of the snakes is due to genetic factors, we may calculate a heritability. A *broad-sense heritability* (h_B^2) is simply the ratio of the genetic variance to the total phenotypic variance (Table 8.1). As mentioned above, it is only the proportion of the total variance that is due to additive genetic factors, the *narrow-sense heritability* (h_N^2), that determines the response to selection by quantitative characters (Table 8.1). The narrow-sense parameters will normally be of most interest. Depending on the groups of relatives available for comparison, it may not be possible to isolate either the additive genetic variance or the total genetic variance (Table 8.1). *Full-sib heritabilities*

TABLE 8.1 Variance Components of Different Parameters

Parameter	Variance components	Techniques
Broad-sense heritability h_B^2	$\frac{V_G}{V_D} = \frac{V_A + V_D + V_I}{V_D}$	
Narrow-sense heritability h_N^2	$\frac{V_A}{V_P}$	Offspring–parent regression,[a] half-sib comparison[b]
Full-sib heritability h_{FS}^2	$\frac{V_A + \frac{1}{2}V_D + \frac{1}{2}V_I + 2V_{Ec}}{V_P}$	Full-sib comparison

[a]Offspring–parent regression will also include common family environment (e.g., "maternal") effects, especially if the parent is the mother.

[b]Maternal half-sib comparisons may include common family environment effects, but paternal half-sibs can be used to eliminate this component of variance.

(h_{FS}^2), the estimates most commonly available for natural populations of vertebrates, include some of the dominance, epistatic, and common-family environmental variance (Table 8.1). Such estimates are often mistakenly called broad-sense heritabilities (Garland, 1994).

If we expand our investigation from a single trait (e.g., speed) to multiple traits (e.g., speed and endurance), we must consider the association between these traits. The *phenotypic covariance* (COV_P) of two traits measures how closely associated they are among individuals. The *phenotypic correlation* (r_P) is merely a covariance standardized to range from −1 to 1. Quantitative genetic techniques allow us to treat covariances in the same way as variances. Throughout this chapter, comments about variances and heritabilities also hold true for covariances and genetic correlations between traits (unless otherwise noted). By partitioning the phenotypic covariance into genetic and environmental components, we can determine the degree to which certain traits are inherited together, or genetically coupled (Arnold, 1987). For comparison, genetic covariances (COV_G) are often standardized to vary between −1 and 1 and are then called *genetic correlations* (r_G). Genetic correlations among traits can arise from *pleiotropy* (where a single gene affects the expression of more than one trait) or from *linkage disequilibrium* (where traits are controlled by separate genes, but those genes tend to be inherited together, sometimes because of close physical association on a chromosome [termed *linkage*]). The full matrix of (usually additive) genetic variances and covariances for a set of traits is referred to as the *genetic variance–covariance matrix* or simply *G-matrix.*

Finally, it should be noted that phenotypic expression can vary among environments, even within an individual or among individuals sharing the same genotype. For example, it is well known that the

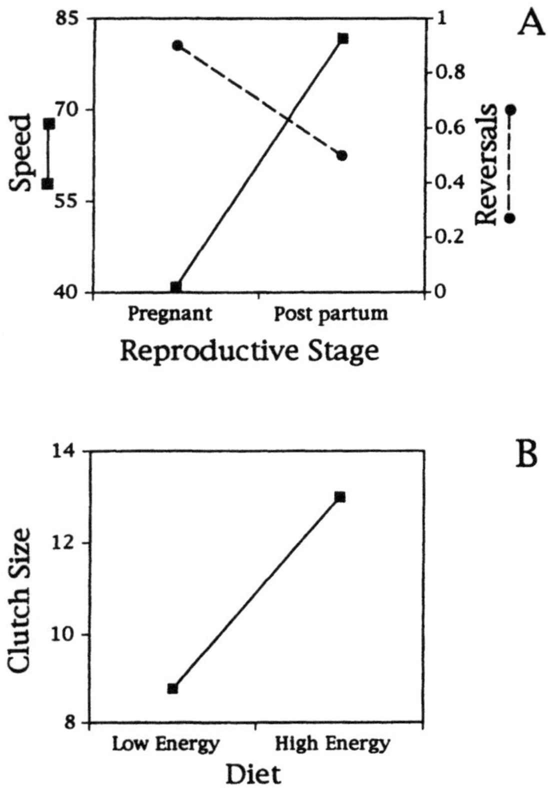

Figure 8.3 Reaction norms for phenotypes expressed in different environments. (*A*) Sprint speed and number of reversals during escape in *Thamnophis ordinoides* are shown as a function of reproductive condition (data from Brodie, 1989b). Females crawl slower and perform more reversals during pregnancy than after parturition. (*B*) Female *Thamnophis marcianus* have smaller clutch sizes when held on low-energy diets compared to high-energy diets during pregnancy (data from Ford and Seigel, 1989).

same snake crawls more slowly at 20 than at 30°C, during pregnancy compared to postparturition, or with a full stomach. The *norm of reaction* describes the phenotypic expression of a genotype across a range of possible environments (Schmalhausen, 1949; Levins, 1968; de Jong, 1990; Gomulkiewicz and Kirkpatrick, 1992) (Fig. 8.3). When the norm of reaction differs among genotypes (e.g., some snakes crawl faster at low temperatures whereas others crawl faster at high temperatures), this constitutes a *genotype-by-environment interaction,* or $G \times E$ (Via, 1984, 1987; Via and Lande, 1985; Falconer, 1989). Heritabilities and genetic correlations can also vary according to the environment in which they are measured. Because heritabilities are expressed as ratios of genetic variance to total phenotypic variance, any environment that increases the phenotypic variation of a trait inflates the denominator of the heritability (Riska et al., 1989; Schwartz and Herzog, 1993). Thus, the heritability is decreased even though the genetic control (e.g., the V_A) of the trait is unchanged. Quantitative genetic parameters are therefore properties of the specific populations and environments for which they are estimated. Extrapolations of conclusions from laboratory to field or across environments are to be made with caution (Riska et al., 1989; Bull et al., 1982; Janzen, 1992; Schwartz and Herzog, 1993).

What Data Are Required?

The key prerequisite of a data set for performing quantitative genetic analyses is some knowledge of the relationships among individuals. Techniques are available to handle almost any set of relatives, even considering complicating factors such as inbreeding and unbalanced data (Bulmer, 1985; Shaw, 1987, 1991; Falconer, 1989; Becker, 1992). All of these methods are based on the average genetic relatedness, derived from simple Mendelian principles, among individuals with a particular familial relationship (e.g., full siblings, on average, share half their alleles in common; see Falconer, 1989, Chaps. 7 and 9). The sets of relatives most commonly available to biologists working with natural systems of vertebrates are sibling groups and sometimes mother–offspring groups. The specifics of comparing these groups will be discussed below.

Sampling errors in quantitative genetics are large, so very large samples are required to estimate parameters with much precision, or to show that they differ significantly from zero. It is not so much the total number of individuals as the number and size of families measured that determines the precision. The optimal design depends on what the true heritability or genetic correlation is for the population and so can rarely be calculated before doing a study. Falconer (1989, pp. 182–183) provides equations to determine the best design if the total number of individuals that can be measured is limited. Increases in the number of families generally improve precision more than do increases in the number of individuals measured per family. In practice, the sample size and family size for studies of natural populations of snakes will usually be determined by availability of subjects.

It is also important to consider just how precise an estimate need be for the purposes of a study. If the goal is simply to determine whether a character has significant additive genetic variance or whether two characters are genetically correlated, a sample of 10–20 families may be sufficient (Arnold, 1981a, 1994; Brodie, 1989a). On the other hand, if one wishes to statistically compare parameters between populations or species, sample sizes in excess of 100 families in each group may be needed to gain the precision required to detect differences (Shaw, 1991). In a pessimistic pair of papers Klein and colleagues (Klein et al., 1973; Klein, 1974) suggested that 400 families of four offspring would have to be measured to demonstrate a significant heritability estimate of 0.20 ± 0.10. This outlook is overly gloomy for our purposes, because family sizes in snakes are usually greater than 4 (Arnold, 1994).

Efforts should be made to empirically reduce the common family environment shared by relatives, which can confound the estimated quantitative genetic parameters (Falconer, 1989). The postnatal common family environment can easily be eliminated in snakes by separating individuals at birth. However, siblings also share a common prenatal environment that is much harder to control. The condition of the female before and during pregnancy (Sinervo, 1990; Sinervo and Huey, 1990), thermoregulatory habits during gestation (Fox, 1948; Fox et al., 1961; Osgood, 1978; Arnold and Peterson, 1989), hydric and thermal features of nest sites (reviewed in Packard and Packard, 1988), and other unknown factors can all contribute to the resemblance among relatives through nongenetic effects. It may be possible to reduce prehatching common environments in oviparous species by splitting clutches as soon as they are laid (Tsuji et al., 1989). Statistical means of reducing maternal effects related to the condition of the female also may be used (explained below) but are not a substitute for experimental control.

In many cases, especially studies of behavior, it will be advantageous to measure the same trait repeatedly within individuals (Bennett, 1987; van Berkum et al., 1989; Huey et al., 1990). If there is measurement error in scoring a character, or if the character is not always expressed precisely the same by an individual (as is true for many behavioral, physiological, and performance traits), repeated measures will increase the accuracy of the estimated phenotypic variance (Arnold and Bennett, 1984; Lessels and Boag, 1987; Boake, 1989; Falconer, 1989). Differences in the trait value within an individual are attributable to the *special environmental variance* (V_{Es}), one-*n*th of which is contributed to the estimated phenotypic variance among individuals if n measurements are made:

$$V_P = V_G + V_E + V_{Es}/n.$$

The smaller the special environmental variance, the less advantage to performing multiple measurements. The optimal number of measurements depends on this component of variance and can be determined using a relationship given by Falconer (1989, p. 143). When two or more measurements are made, it is possible to calculate the repeatability, indicating how consistently a character can be measured in (or is expressed by) an individual, and then the average value of the character for an individual can be used in subsequent analyses. The heritability based on an average (or sum) of multiple measurements includes only a fraction ($1/n$) of V_{Es}, whereas the heritability (and repeatability) based on a single measure includes this whole component. Thus, if multiple measurements are used, heritabilities must be adjusted upwards to account for the reduced fraction of V_{Es} present in the denominator, or repeatabilities and heritabilities will not be of comparable scale (Brodie, 1989a, Brodie and Brodie, 1990). A table of scaling factors is provided by Becker (1992, Appendix 1) and the method is discussed in more detail by Arnold and Bennett (1984). Sometimes the highest of a series of measurements is analyzed, in an attempt to obtain measures of maximal performance abilities, free of motivational effects (see Garland, 1988; Tsuji et al., 1989; Garland and Losos, 1994).

Types of Characters Suitable for Quantitative Genetic Study

Quantitative genetics concerns itself with continuously varying characters. It is generally assumed that these characters are controlled by a large number of genes, each with small, additive effects on the phenotype (Mather and Jinks, 1982; Bulmer, 1985; Falconer, 1989). The term *quantitative* is used because individuals differ in degree (quantitatively) rather than type (qualitatively). Many of the traits of particular interest to students of evolution, ecology, and behavior exhibit this kind of variation (e.g., Boake, 1994).

The nature of quantitative inheritance (many genes, most of small effect) usually results in a phenotype that is normally distributed, at least when transformed to a suitable scale of measurement. Complex characters such as measures of performance (e.g., speed, endurance, prey-handling time), fitness components (viability, fecundity), and phenologies (date of emergence from hibernation, date of egg deposition) are especially appropriate for such analyses because they are probably comprised of many other characters and even more genes. Almost any continuously varying morphological (body size), behavioral (escape behavior, feeding preference) or physiological trait (maximal oxygen consumption) can also be examined with quantitative genetic techniques.

Traits that are controlled by one or a few genes of major effect are more appropriately studied using the techniques of populations genetics (Hartl and Clark, 1989). The presence of genes of major effect can be tested for by examining the phenotypic variances within families. If a few genes have large effects, then families with extreme phenotypes should be mostly homozygous for positive or negative alleles, whereas intermediate families should be heterozygous and therefore have higher variances (see Garland, 1988; Garland and Bennett, 1990, references therein; R. B. King, unpublished data).

Some characters are only expressible in whole-number units and so will have a necessarily discontinuous distribution. If many possible states exist, as in the number of ventral or subcaudal scales in snakes, phenotypic distributions may approach normality, and quantitative genetic theory is easily applied. Even characters with only a few discrete classes of expression (e.g., counts of head scales, number of dorsal scale rows) can be studied in a quantitative genetic context if a threshold model is assumed (Arnold, 1981b,c; Falconer, 1989; Dohm and Garland, 1993). For threshold traits, the underlying genetic and environmental variation is continuous but mapped onto fixed thresholds that determine the discrete phenotypes (Bull et al., 1982; Gianola, 1982). This interpretation of phenotypic variation allows quantitative genetic techniques to be used in the study of naturally discrete phenotypes, such as scale counts (Dohm and Garland, 1993) and color pattern elements (Beatson, 1976; Brodie, 1989a), and also of traits that can only be measured on a discrete scale, as in the case of many behaviors (Arnold, 1981b,c; Arnold and Bennett, 1984; Garland, 1988, 1994; Brodie, 1989a; Schwartz and Herzog, 1993). Additionally, discontinuous characters with only a few values may be combined into an index trait that is more continuously distributed, provided the index trait has some biological meaning (Arnold and Bennett, 1984; Garland, 1988; Brodie, 1989a, 1991, 1992, 1993b).

What Can Quantitative Genetics Do?

Most herpetologists will undoubtedly agree with the opinion of Dobzhansky (1973) that "nothing in biology makes sense except in the light of evolution." A variety of strategies is available to researchers attempting to shed this light. Traditionally, comparisons among taxa have been the most popular approach to studying phenotypic evolution in snakes and other reptiles (e.g., Seigel et al., 1987; Dunham et al., 1988). The comparative method treats taxonomic groups as "experimental" units and focuses on differences among these groups to describe patterns of evolution. This approach is historical and describes what has happened during the evolution of a particular clade. Recent advances in comparative techniques have improved the statistical rigor of this field (e.g., Harvey and Pagel, 1991; Martins and Garland, 1991; Lynch, 1991; Garland et al., 1991, 1992; Garland et al., 1993; Losos and Miles, 1994), leading to more powerful conclusions about the relative importance of selection and phylogenetic history in determining present-day patterns of phenotypic diversity.

An alternative approach is to study evolution at the level of the population. By investigating the pattern of variation and covariation in phenotype and fitness within a population, microevolutionary processes can be illuminated (Arnold, 1983, 1988b; Lande and Arnold, 1983; Grant, 1986). Quantitative genetics is one set of tools that can be used to study intrapopulational variation. This sort of microevolutionary approach was rarely employed in studies of reptiles until recently but can address a range of questions that should be of interest to many herpetologists. Aside from their obvious relevance to evolutionary ecology, results from quantitative genetic investigations may be important in systematics (Shaffer, 1986; Felsenstein, 1988; Lynch, 1989; Dohm and Garland, 1993) and conservation efforts (Dodd, Chap. 9, this volume), as well.

Nature versus nurture

The most obvious application of quantitative genetics is simply to determine the relative importance of genetic and nongenetic (environmental) factors in the expression of phenotypic variation (cf. Garland and Adolph, 1991, pp. 196–198). Besides revealing whether and how rapidly a trait will respond to selection (discussed above), this information is pertinent to a variety of other questions of interest to evolutionary ecologists.

Variation resulting from the environment is an important consideration for studies using cross-sectional data to attempt to detect selection. If phenotypic variance is regenerated each generation by the environment, selection may go undetected. Conversely, environmental changes may result in character differences across generations, even though no selection has occurred. Such scenarios seem especially important to consider in the study of certain behaviors that may be influenced by early experience. For example, feeding preferences and foraging behaviors may have high heritabilities at birth (Arnold, 1981a,c), but experience with certain prey during the first year may canalize feeding behavior in adults (see Fuchs and Burghardt, 1971; Arnold, 1977; Halloy and Burghardt, 1990). Differences in feeding behavior between generations or populations that might be interpreted as adaptive could actually be due to temporal or spatial changes in prey abundance.

Environment may affect the estimation of quantitative genetic parameters themselves. An obvious example is the difference between heritabilities estimated in a controlled laboratory environment and those estimated in the field. Field conditions typically have more environmental variation influencing the phenotype, whereas the absolute amount of genetic variation is the same. Because heritabilities are expressed as a proportion of phenotypic variance, field heritabilities are expected to be lower than those estimated under laboratory conditions (Riska et al., 1989; Bull et al., 1982; Janzen, 1992; Schwartz and Herzog, 1993). The expected response to selection based on laboratory heritabilities may overestimate the rate of evolutionary change in nature. The same holds true for different environmental conditions in the field. If the heritability of sprint speed is higher at a snake's thermal optimum than at extreme temperatures, the evolutionary response of sprint speed to selection will depend on the temperature at which selection occurs. Selection at extreme temperatures may act on environmental variation, causing little or no evolutionary change in average speed at any temperature. Conversely, selection acting at preferred temperatures should result in relatively rapid evolution of average speed, even at extreme temperatures.

Traits experiencing stronger selection are generally expected to have lower heritabilities, because selection is usually thought to reduce genetic variation at equilibrium. However, many other factors such as gene flow, population structure, and sampling effects influence heritabilities, so this interpretation of the theory is somewhat controversial (Charlesworth, 1987; Price and Schluter, 1991; Boake, 1994). Comparing the heritabilities of traits may give some clues about the relative strength of selection they have experienced (Roff and Mousseau, 1987; Mousseau and Roff, 1987; Garland et al., 1990). This approach has been applied to locomotor performance in Garter Snakes to suggest if selection is stronger on organismal level performance (e.g., speed, endurance) than suborganismal components (i.e., physiological and morphological characters such as heart mass and enzyme activities). Behavior and performance generally had higher heritabilities in this study, suggesting that physiological attributes might experience stronger selection, possibly because each trait affects many types of organismal performance (Garland et al., 1990).

Genotype-by-environment interactions (G × E)

Environmental variation can have different effects on different genotypes. This genotype-by-environment interaction can be measured as the genetic correlation of a trait in two different environments. Any nonperfect correlation ($-1 < r < 1$) indicates some $G \times E$ (Via and Lande, 1985). Genotype-by-environment interaction is one form of genetic variation for plasticity of the phenotype. The evolution of traits in multiple or variable environments (i.e., phenotypic plasticity), such as maternal investment in offspring as a function of changing resource availability (e.g., Ford and Seigel, 1989), depends on $G \times E$.

The question "Is a jack-of-all-trades a master of none?" (Huey and Hertz, 1984), can also be addressed through investigations of $G \times E$ (de Jong, 1990; Gomulkiewicz and Kirkpatrick, 1992). A genotype that performs best in one environment may not perform well in another (e.g., sprint speed on different substrates or endurance at different temperatures). High levels of $G \times E$ would indicate that one genotype is not optimal under all conditions (Arnold, 1987; Gomulkiewicz and Kirkpatrick, 1992).

Constraints

The strength and nature of evolutionary constraints on adaptation can be illuminated through quantitative genetics. The most extreme evolutionary constraint is a lack of phenotypic or genetic variation. Selection cannot act on limb number in snakes because there is no variation for this trait. Even characters that show phenotypic variation cannot respond to selection unless nonzero heritability exists.

Developmental, physiological, and genetic integrations among suites of characters are thought to constrain evolutionary response (Alberch, 1980; Gould, 1980; Clark, 1987; but see Charlesworth et al., 1982; Zeng, 1988). Genetic correlations are one way to measure such integration (Reznick, 1985; Barker and Thomas, 1987; Clark, 1987; Arnold, 1981c, 1987, 1988a; Garland, 1988; Dohm and Garland, 1993; Boake, 1994). The evolutionary response of a trait is determined not only by selection acting directly on that trait, but also by selection acting on genetically correlated traits. The pattern of genetic correlations determines both the rate and direction of short-term evolutionary change and can even result in temporary, maladaptive states in some traits (Lande, 1979; Lande and Arnold, 1983; Arnold, 1981c, 1987, 1988a, 1990). The strength of constraint is reflected in the magnitude of the genetic correlations. Congenital feeding preferences in *Thamnophis elegans* exhibit a high degree of genetic integration (Arnold, 1981a,b,c), so preferences for some prey items are not free to evolve independently. Preference for potentially hazardous prey, leeches, is presumably maintained in a coastal population of *T. elegans* because it is genetically correlated with response to the main food item, slugs (Arnold, 1981a,b,c, 1992).

Coadaptation

Multiple characters sometimes function interactively to increase performance or fitness. Traits that have been jointly selected are said to be *coadapted* (Huey and Bennett, 1987; Garland et al., 1991). Selection for coadapted suites of traits should result in genetic integration, which can in turn be recognized as genetic correlations (Cheverud, 1982, 1984, 1988; Lande, 1980, 1984; Clark, 1987). The strength of genetic correlations maintained through functional interactions depends on aspects of the breeding system and the physical relationships among genes controlling the traits (i.e., pleiotropy, physical linkage) (Lande, 1980, 1984; Endler, 1986). However, weak selection alone is expected to maintain some genetic correlation among coadapted characters (Lande, 1984). This appears to be the case for a suite of antipredator traits in some populations of the Northwestern Garter Snake, *Thamnophis ordinoides,* where color pattern and escape behavior are genetically correlated (Brodie, 1989a, 1991, 1992, 1993b).

Nonadaptive forces can also result in genetic correlations, so demonstration of genetic integration alone does not prove coadaptation (Lande, 1980, 1984; Endler, 1986; Crespi, 1990). Selection on combinations of traits (correlational selection) can be directly measured using multiple-regression techniques (Lande and Arnold, 1983; described below). Demonstration of both genetic correlation and correlational selection in the same direction can be taken as evidence that suites of traits are coadapted. Both features have been shown for combinations of color pattern and antipredator behavior in *Thamnophis ordinoides* (Brodie, 1989a, 1991, 1992, 1993b).

Systematics

Systematists can also benefit from the study of genetic variances and covariances. Character weighting in phylogeny reconstruction is usually based on phenotypic variation, assuming that this reflects genetic variation (Archie, 1985). Heritabilities directly measure genetic variation and may be more appropriate weighting parameters. Some authors would assign higher weights to characters with high heritabilities because they better reflect phylogenetic relationships (Atchley, 1983; Shaffer, 1986; Mayr and Ashlock, 1991) and are presumably less correlated with fitness (Dohm and Garland, 1993). Others would weight characters by the inverse of the heritability, reasoning that high heritabilities indicate dimensions in which genetic drift or selection could lead to the most rapid evolutionary change (Schluter, 1984; Felsenstein, 1988; Lynch, 1989). In either case, it is clear that the genetic basis of phenotypic traits is an important consideration in phylogenetic reconstruction.

Phylogeny reconstruction generally is simplified by the use of independent characters. Inclusion of nonindependent traits can inflate the confidence in a particular phylogeny because it appears to be supported by too many characters (Shaffer, 1986). Genetic correlations can be used to identify suites of evolutionarily nonindependent traits (Shaffer, 1986; Dohm and Garland, 1993), so they may be handled in a more appropriate manner (cf. Wheeler, 1986). In practice, the necessary quantitative genetic information is available for very few taxa, underscoring the need for more efforts in this arena.

Measurement of selection

One of the most exciting spin-offs of the application of quantitative genetic theory to natural populations is the development of multiple regression techniques to measure selection. It should be stressed that these techniques are purely phenotypic and require no knowledge of the genetic basis of the characters in question, but do provide measures of selection that are directly related to quantitative genetic formulas for evolutionary change (see below). Traditional methods of detecting selection look at the total change in a trait after an episode of selection (the *selection differential,* Lande and Arnold, 1983; Endler, 1986). This measure confounds selection acting directly on the trait with the indirect effects of selection acting on phenotypically correlated traits (Lande and Arnold, 1983). By considering the pattern of phenotypic correlation among traits, multiple regression measures the strength of selection directly targeting a trait (the *selection gradient*), independent of other traits in the analysis (Lande and Arnold, 1983; Endler, 1986). Gradients measuring the strength of directional (acting to change the mean of a trait), stabilizing, or disruptive (acting to change the variance in a trait), and correlational (acting to change the covariance between two traits) selection can be calculated simply by including the appropriate higher order or interaction terms in a regression model (Lande and Arnold, 1983; Endler, 1986). The technique is potentially quite powerful, but is subject to the usual problems associated with regression analysis. In particular, the gradients estimated may include selection acting indirectly through correlated characters that are not included in the analysis, and strong correlations between traits (multicollinearity) may lead to problems of estimation by ordinary multiple regression (Lande and Arnold, 1983; Slinker and Glantz, 1985; Mitchell-Olds and Shaw, 1987; Crespi, 1990).

The greatest advantage of selection gradients is that they can be combined with estimates of the genetic variance–covariance matrix to predict evolutionary change across generations (Lande, 1979; Arnold, 1981c, 1988a; Lande and Arnold, 1983; Price et al., 1984; Price and

Grant, 1985; Lofsvold, 1988). Selection on one trait will result in evolutionary change in genetically correlated traits. To predict phenotypic change over multiple generations, one must consider both selection and the pattern of genetic covariances. The equation

$$\Delta \bar{z} = G\beta$$

(where $\Delta \bar{z}$ is a vector describing the change in the mean of traits $z_i,\ldots,z_n$, G is the genetic variance–covariance matrix, and β is the vector of directional selection gradients) is the multivariate equivalent of the equation for response to selection by a single trait:

$$R = h^2 s.$$

If the amount of phenotypic divergence between two populations or taxa is known, we can use estimates of the *G*-matrix to calculate the cumulative selection necessary to account for the observed differences (Arnold, 1981c, 1988a). The reconstructed net selection gradients are the minimum amount of directional selection required to account for the phenotypic differences, assuming the most direct evolutionary trajectory. In reality, the populations may have followed more complex evolutionary paths during divergence. This method has been used to estimate the selection necessary to account for differences in vertebral number between inland and coastal populations of the Western Terrestrial Garter Snake, *Thamnophis elegans* (Arnold, 1988a,b; also for feeding preferences, Arnold, 1981c). The results suggest that nearly 50% of the divergence in number of tail vertebrae may be a correlated response to selection directly targeting the number of body vertebrate. Failure to consider genetic correlations among traits would have led to a gross overestimate of selection on tail vertebrae.

Both predictive and retrospective studies of multivariate selection assume that genetic variances and covariances remain relatively constant over the time frame in question (Lande, 1979; Lofsvold, 1986, 1988; Turelli, 1988; Barton and Turelli, 1989). Current theory is unable to make robust predictions about how long these parameters are expected to remain stable, so the problem has become an empirical one (Turelli, 1988). The few studies that have compared *G*-matrices among populations or species have produced ambiguous results (reviewed in Barton and Turelli, 1989; Wilkinson et al., 1990; Brodie, 1993b), suggesting unsurprisingly that the dynamics may vary depending on the traits and taxa being examined.

Quantitative Genetic Studies with Snakes

Several aspects of snake biology present distinct advantages for quantitative genetics, especially over other vertebrates. A variety of traits, some of special interest to herpetologists, can be examined using only a small subset of available quantitative genetic techniques. Natural populations of snakes are rarely ideal subjects for such studies, but no taxa really are. The limitations encountered with snakes usually have to do with their longevity, slow maturation, and uncertain paternity.

Why use snakes?

Subjects. Large numbers of related individuals can be obtained for many snake species. Gravid females can usually be collected early in the reproductive season, providing a researcher with two sets of relatives: sibling groups (usually assumed to be full siblings) and mother and offspring. These two sets can be used to partition different components of variance.

The large litter size of many snake species can be beneficial in several ways. Large families provide an opportunity to split litters among environmental treatments to address concepts empirically such as genotype-by-environment interactions and common family environment (including maternal) effects. Such studies have not yet been attempted with snakes. Also, the accuracy of the estimated heritabilities and genetic correlations increases as more individuals in each family are measured (Falconer, 1989). For species with very large clutches (e.g., some *Nerodia* with litters of 30–50) it may not be practical to measure all offspring, but rather to select the same number randomly from each family (Garland, 1988). Having a balanced number of individuals per family simplifies analyses and significance testing (see below).

Oviparous species present a unique opportunity to examine the importance of some specific nongenetic maternal effects. Prenatal conditions can be simply manipulated by varying parameters such as incubation temperature and hydric environment of eggs, factors known to affect postnatal phenotypes in other reptiles (Plummer and Snell, 1988; Burger, 1989, 1991). The effects of maternal condition and investment can be empirically altered by removing yolk from developing eggs as has been done in lizards (Sinervo, 1990; Sinervo and Huey, 1990).

Characters. Snakes are especially good subjects for quantitative genetic studies of behavior. Neonates are completely precocial and exhibit a variety of behaviors immediately following birth. Yolk reserves allow snakes to be tested repeatedly during the first few weeks of life, without the confounding environmental variation that accompanies feeding (or refusal to eat). The field of behavioral genetics in natural populations has already been greatly expanded by studies of feeding preferences and antipredator behavior in snakes (Arnold, 1981a,b,c, 1992; Arnold and Bennett, 1984; Garland, 1988, 1994; Herzog and Burghardt, 1988; Brodie, 1989a, 1991, 1992, 1993a,b; Herzog et al., 1989, 1992; Schwartz, 1989; Schwartz and Herzog, 1993).

Scale counts and other meristic characters can be easily scored in snakes. These traits do not change with age, so mother–offspring comparisons can be used to calculate heritabilities and genetic correlations. Because of their importance in phylogenetic reconstruction, scale counts may be particularly valuable characters to study, providing a unique opportunity to combine quantitative genetics with systematics (Dohm and Garland, 1993). Color patterns also can be easily scored, and, in some species, are ontogenetically stable (Beatson, 1976; Brodie, 1989a, 1991, 1992, 1993b; King, 1987, 1992, unpublished data; J. Barron, unpublished data). The ecological importance of coloration and its functional integration with behavior (Jackson et al., 1976; Pough, 1976; Brodie, 1989a, 1992) make it an especially relevant trait to study.

Drawbacks. The most rigorous quantitative genetic analyses require complex breeding designs, often with males mated to multiple females. The difficulty of breeding large numbers (e.g., 20 males each mated to at least 3 females) of snakes in the laboratory may prohibit these kinds of studies. Even for taxa that can be bred reliably, maturity is usually not reached until 1–2 years of age, even in the laboratory. That means a minimum of 2–3 years from scoring an individual as a neonate until scoring its offspring (all individuals should be measured at the same age to control for ontogenetic effects; see below). Additionally, the possibility of sperm storage confounds attempts to mate females to multiple males. For most traits, this means analyses will be limited to sibling comparisons, so certain components of genetic variation can never be distinguished.

One of the main reasons to use snakes in quantitative genetics may also be considered a major drawback. While some species can be obtained in huge numbers, others are rarely encountered. This factor more than any other is likely to define which snakes are suitable subjects for quantitative genetics (see Parker and Plummer, 1987, for estimates of local population sizes of many species).

Techniques used in studies of snakes

Animal and plant breeders have developed methods to analyze resemblance between virtually every set of relatives imaginable, from parents and offspring to siblings to various degrees of cousins. Crosses between populations (cf. Garland and Adolph, 1991) may be performed to test for nonadditive genetic variance and maternal effects

(Arnold, 1981b). Another strategy is to artificially impose selection and measure the response, but this method requires replicated selection lines and many generations (Falconer, 1989; Boake, 1994). Such approaches usually are impractical for studies of snake populations. In fact, either full-sib or mother–offspring comparisons account for almost all studies done to date on snakes.

Full-sib comparisons. Comparison of the resemblance between litter mates is the most widely applicable means of partitioning variances. It requires only clutches of siblings that have been scored for a trait. Because all siblings will be the same age, measurements can be made in a relatively short amount of time, and even ontogenetically variable traits can be studied. Full-sib relationships within a litter (i.e., single paternity) generally must be assumed, but even if some litters have multiple paternity, this method will yield conservative estimates (see below).

The resemblance between siblings can be measured as the *intraclass correlation, t,* or the ratio of the variance between groups (σ_B^2) to the total variance ($\sigma_B^2 + \sigma_W^2$, where σ_W^2 is the variance within groups):

$$t = \sigma_B^2/(\sigma_B^2 + \sigma_W^2).$$

What the intraclass correlation describes depends upon the average level of relatedness between members of the group (Falconer, 1989) (Table 8.1). Full-sibs, on average, share half of the additive genetic variance, and a quarter of the dominance variance. Half-sibs, on average, share only a quarter of the additive genetic variance and none of the dominance variance. To calculate a heritability, one must simply multiply the intraclass correlation from half-sib data by a factor of 4, from full-sib data by a factor of 2. The full-sib heritability will then include a fraction (½) of the dominance variance, and all interpretations of this parameter should recognize that it does not represent the pure additive genetic variance (Falconer, 1989; Garland, 1994; Arnold, 1994) (Table 8.1). Reviews of studies in a variety of taxa show very little difference between full-sib heritabilities and narrow-sense heritabilities (Roff and Mousseau, 1987; Mousseau and Roff, 1987), but the contribution of dominance effects is likely to vary depending on the trait in question (Dohm and Garland, 1993).

For full-sib data, the simplest method to derive the between- and within-family components of variance is through a one-way analysis of variance (ANOVA) with family as the main effect. The mean squares from the resulting ANOVA table can be used to calculate the necessary variance components. The error mean square is the within-family variance and the family mean square is a sum of the between-family component of variance and some function of the within-family component of variance. The exact methods depend on whether family sizes are balanced (i.e., same number of siblings in each family), and are outlined in detail in other texts (Becker, 1992; Lessels and Boag, 1987; Falconer, 1989). The technique for partitioning the variance among half-sibs is similar, but requires a slightly more complex ANOVA design (Becker, 1992; Falconer, 1989).

A useful approximation to the genetic correlation between two traits is the correlation between family (= litter) means (Arnold, 1981b; Via, 1984; Garland, 1988; Brodie, 1989a; Garland and Bennett, 1990; Garland et al., 1990) (Fig. 8.4). The numerator of the expression for a genetic correlation is the covariance among families (COV_a). The covariance of family means (COV_m) also includes a fraction of the covariance within families (COV_w):

$$COV_m = COV_a + (1/n_0)COV_w,$$

where n_0 is the average number of offspring per family corrected for differences in family size (Arnold, 1981b; Via, 1984). Thus, the family-mean correlation includes a fraction of the within-family covariance, but this factor becomes diminishingly small as family size increases.

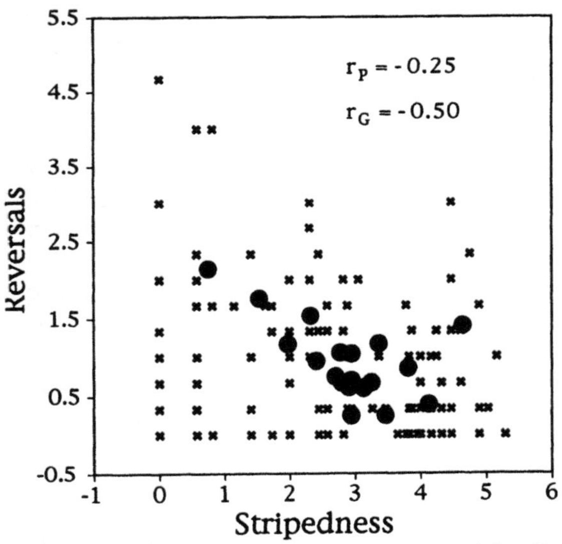

Figure 8.4 Negative phenotypic (crosses) and family-mean (dots) correlations between reversals and stripedness (square-root transformed) in *T. ordinoides* (Alsea population, see Brodie, 1989a). Note that the genetic correlation (r_G) approximated by the litter-mean correlation is greater in magnitude than the phenotypic correlation (r_P).

For large families, litter-mean correlations may be reasonable alternative estimates of genetic correlations (Fig. 8.4). An additional advantage of this estimate is that significance testing is clear-cut even for unbalanced designs, because it is a standard product–moment correlation (Via, 1984).

Multiple paternity has been detected whenever examined in snakes (*Thamnophis sirtalis,* Gibson and Falls, 1975; Schwartz et al., 1989; *Nerodia sipedon,* Barry et al., 1992; *Vipera berus,* Stille et al., 1986; Madsen et al., 1992; *Lampropeltis getula,* Zweifel and Dessauer, 1983). This phenomenon acts to reduce the average relatedness of litter mates below one-half (if all litter mates had a different father, they would all be half-sibs and the average relatedness would be ¼). Ideally, the average relatedness within litters could be calculated based on paternity analysis (cf. Schwartz et al., 1989), but this usually will not be possible. In the absence of such information, a full-sib assumption renders quantitative genetic parameters underestimates (the intraclass correlation is multiplied by 2 for full-sibs and 4 for half-sibs; partial multiple paternity would be somewhere intermediate). Full-sib parameters are also assumed to include some dominance variance; the actual contribution of this confounding factor to the estimate becomes less important as multiple paternity increases.

Mother–offspring comparisons. A somewhat more powerful method of partitioning variances is to compare the resemblance between mother and offspring. This is done by regressing the average value of the trait in the offspring on the mother's value (Bulmer, 1985; Falconer, 1989; Becker, 1992). The slope of the regression line equals one-half of the additive genetic variance (Fig. 8.5). This estimate is not confounded by nonadditive genetic variance components, but does not include maternal effects. Genetic covariances can be obtained similarly by comparing one trait in the offspring with another in the parents.

The use of the parent–offspring regression technique is limited to traits that do not change with age or to traits measured at the same age and environment in both generations. Characters that change ontogenetically (and, in fact, some that do not) may have different genetic control at different ages (Arnold, 1990). In such circumstances, a parent–offspring heritability would really describe the genetic covariance between the trait at two ages (which may be considered two different traits), rather than the genetic variance of the trait at a particular age (Falconer 1989; Arnold, 1990).

Population crosses. Crossing individuals from two divergent populations can reveal several components of variance that may not be distinguishable in analyses of either population alone (Falconer, 1989). If phenotypes of the F_1 hybrids are not intermediate between those of

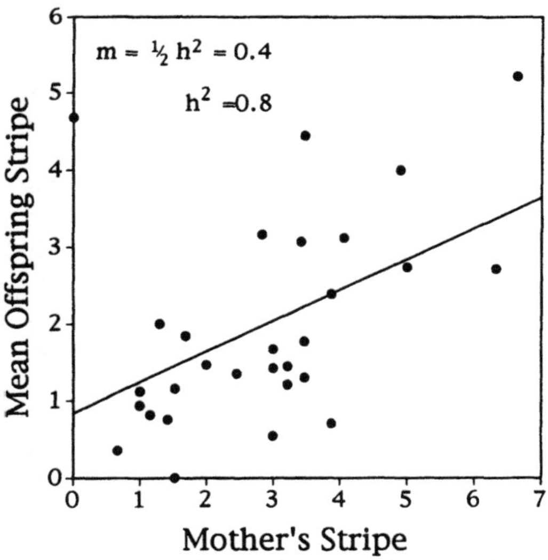

Figure 8.5 Offspring–mother plots for stripedness (square-root transformed) in *Thamnophis ordinoides* (McGribble population, see Brodie, 1989a). The slope of the regression of mean stripedness of offspring on the stripedness of their mother is, in the absence of maternal effects, equal to one-half the narrow-sense heritability (the estimate of heritability reported here is lower than the full-sib heritability reported by Brodie [1989a] for the same data and may indicate some contribution of dominance variance to the full-sib heritability).

the parent populations, dominance or maternal effects would be suspected. If F_1 progeny from reciprocal crosses more closely resemble their maternal phenotype, then maternal effects are likely to be important determinants.

Although this method requires laboratory breeding, relatively few crosses are required to get useful results. One study of *T. elegans* has employed this technique (Arnold, 1981b), and it stands as the only direct investigation of dominance or maternal effects on any trait in snakes (see below).

Common family environment. Estimates of quantitative genetic parameters from either full-sib or mother–offspring comparisons will be confounded by common family environment, including maternal effects (Table 8.1, Fig. 8.2). The importance of splitting families as soon as possible was discussed above, and, especially with oviparous species, several methods of empirically reducing or investigating the effects of common family environments can be employed.

In spite of these efforts, it will be impossible to reduce some prenatal maternal effects empirically. Statistical methods may be useful in further eliminating this source of confounding environmental variance. Many nongenetic maternal effects are likely to be mediated through the condition of the female during energy allocation of the ova and gestation. Statistical covariation with measures of female condition (female mass, mass-to-length ratios, etc.) can be eliminated by regressing the trait in question on such measures (cf. Garland, 1988, 1994; Tsuji et al., 1989; Brodie, 1989a, 1991, 1993b; Garland and Bennett, 1990; Garland et al., 1990; Brodie and Brodie, 1990; Dohm and Garland, 1993). Residuals from this regression will reflect variation in the trait independent of female condition. These residuals may then be substituted for the original trait in the usual analyses. This technique will eliminate nongenetic maternal effects that are associated with these condition measures, but will also eliminate genetic covariances between female condition and the trait in question. This conservative approach has been used previously, especially in studies of locomotor performance (Garland 1988, 1994; Brodie 1989a, 1991, 1993b; Garland and Bennett, 1990; Garland et al., 1990; Brodie and Brodie, 1990). No empirical validation of the technique has yet been attempted, but would be extremely valuable.

Statistical considerations. In addition to the ANOVA and regression methods outlined above, maximum likelihood estimation (MLE) may be used to calculate quantitative genetic parameters, using any set of relatives (e.g., Shaw, 1987, 1991). MLE has several advantages over standard techniques, including better handling of unbalanced family sizes, the ability to use data from multiple sets of relatives, and more direct significance testing, but is somewhat difficult to implement in practice (but see Garland, 1988; Garland and Bennett, 1990; Dohm and Garland, 1993 for some successes). The computer resources needed to analyze even modest data sets with MLE have prevented many researchers from utilizing the technique. But, as computational facilities improve and algorithms for MLE are refined, this promises to become a more practical tool for quantitative genetic analyses.

Significance testing of heritabilities and genetic correlations can be problematic. The expected distributions for quantitative genetic parameters are poorly understood, and unbalanced family sizes further complicate matters. In the past, most researchers have calculated standard errors for their estimates and used t-tests for approximate significance testing. The popularization of computer-intensive resampling testing procedures allows more rigorous tests to be performed (Efron and Tibshirani, 1991; Crowley, 1992). Jackknifing has been demonstrated to be effective for many types of data distributions (Sokal and Rohlf, 1981; Arveson and Schmitz, 1970; Knapp et al., 1989; Mitchell-Olds and Bergelson, 1990). This method creates a t-distributed set of parameter estimates by resampling the original data without replacement. The standard errors from this distribution can be used to test a variety of hypotheses by means of a t-test (Knapp et al., 1989). Bootstrapping resamples the data with replacement, creating a distribution of the parameter based on each data set. Parameter estimates can then be tested against this expected distribution for the probability of obtaining such an estimate by chance (Efron and Tibshirani, 1991). Randomization tests are another nonparametric method of significance testing (Sokal and Rohlf, 1981; Mitchell-Olds, 1987; Jayne and Bennett, 1990b) and may be applied to quantitative genetic data.

What has been done with snakes

As compared with other wild vertebrates, snakes have been the subject of a relatively large number of quantitative genetic analyses (reviews in Arnold, 1981c, 1983; Ehrman and Parsons, 1981; Mousseau and Roff, 1987; Weir et al., 1988; Plomin et al., 1990; Boake, 1994; Garland, 1994). A few species of lizards have been studied (reviews in Bennett and Huey, 1990; Garland and Losos, 1994). Very few quantitative genetic studies of chelonians (Bull et al., 1982; Janzen, 1992, 1993), and apparently none of crocodilians have been attempted, although their large clutches, relative ease of incubation and husbandry (e.g., in commercial crocodile farms), and conservation status would seem to make them good models.

Scale counts. Some of the earliest studies of quantitative variation among related individuals in nonhuman vertebrates involved scale counts of snakes (Dunn, 1915, 1942; Inger, 1943; Endler, 1986); however, these early studies were concerned with demonstrating natural selection and did not attempt to estimate heritabilities. Snake scale counts are obvious targets for quantitative genetic analyses, because (1) they are easy to score on both living and preserved specimens; (2) they do not change ontogenetically, thus allowing immediate offspring–parent comparisons; and (3) they show quasicontinuous variation (Arnold, 1988a; Dohm and Garland, 1993).

Numbers of ventral and subcaudal scutes, the scale counts studied most commonly, show a 1-to-1 correspondence with numbers of body and tail vertebrae, respectively (references in Arnold, 1988a; Dohm and Garland, 1993). The first quantitative genetic study of any trait in snakes estimated a full-sib heritability of 0.75 for ventral scales in *Nerodia sipedon* (Beatson, 1976). Using offspring-on-dam regressions, narrow-sense heritabilities of 0.65–0.79 for ventrals and 0.46–0.64 for subcaudals were estimated in *Thamnophis elegans* (Arnold, 1988a). Corresponding values of 0.3 and 0.4, respectively, were calculated for

T. sirtalis (Dohm and Garland, 1993) using both conventional and restricted maximum likelihood (Shaw, 1987, 1991). Schwartz (1989) reports full-sib heritabilities of 0.6–0.7 and 0.8 for ventrals and subcaudals, respectively, in a Michigan and a Wisconsin population of *T. sirtalis*. Numbers of ventrals and subcaudals showed a significant positive additive genetic correlation in each of these populations: 0.3 in *T. elegans* (Arnold, 1988a), 0.7 in California *T. sirtalis* (Dohm and Garland, 1993), and full-sib correlations of 0.5–1.0 in Michigan and Wisconsin *T. sirtalis* (Schwartz, 1989).

Other scale counts, especially head scales, are less variable and have been studied as threshold characters. Dohm and Garland (1993), again using maximum likelihood analyses of full-sib plus mother data, showed a significant narrow-sense heritability for the number of temporal scales in *Thamnophis sirtalis*. Three other head-scale counts (supralabials, infralabials, and postoculars) and two derived characters involving the position of the umbilical scar were not significantly heritable. Schwartz (1989) reported full-sib heritabilities for preoculars, postoculars, supralabials, infralabials, and pretemporals (all analyzed as threshold characters), but did not test significance. Dohm and Garland (1993) also found that some of the genetic correlations between body and head scales were significant. Thus, the meristic characters they studied do not represent eight evolutionarily independent traits. Arnold (1988a, personal communication) is continuing a large-scale study of the evolutionary stability of the genetic variance–covariance matrix in *Thamnophis* and *Nerodia*, using both full-sib and parent–offspring analyses.

Color patterns. Brodie (1989a, 1991, 1993b) showed significant full-sib heritabilities for a composite index of color pattern termed "stripedness" that describes the overall linearity of color pattern in *Thamnophis ordinoides*. This index trait was analyzed after preliminary analysis detected high levels of genetic correlation among individual pattern components including presence and contrast of dorsal and lateral stripes and spot rows. All four populations of *T. ordinoides* studied to date have full-sib heritabilities in the range 0.5–1.0 (Brodie, 1989a, 1991, 1993b).

The number of dorsal and lateral blotches, size of lateral blotches, and the extent of ventral pigmentation are all significantly heritable in Northern Water Snakes (*Nerodia sipedon*) (Beatson, 1976; R. B. King, personal communication; J. Barron, personal communication). Both full-sib and mother–offspring comparisons reveal moderate to high heritabilities (0.3–1.0) and positive genetic correlations (0.24–0.82) among most of the pattern components measured (R. B. King, unpublished data). Genes with major effects also appear to influence some color pattern components of *N. sipedon* from Lake Erie (R. B. King, unpublished data).

Antipredator behavior. In the fall of 1981, S. J. Arnold and A. F. Bennett began studying locomotor performance of Garter Snakes. Their initial protocol was to chase newborn snakes around a rectangular track and record the time to cover the initial 1.0 m ("burst speed"), the time to cover from 1.0 to 2.5 m ("mid-distance speed"), and the total distance and time crawled until the snake stopped locomoting (Arnold and Bennett, 1988). They discovered that snakes typically adopted some type of "antipredator display" when they stopped crawling. Based on measures of whole-body lactic acid concentrations, snakes seemed to have reached their anaerobic capacities when they assumed antipredator displays, although this may not be true for other species (see Jayne and Bennett, 1990b, p. 1223; Brodie, 1991, 1992, 1993a,b). Arnold and Bennett (1984) argued that the antipredator display could be scored on a quasicontinuous scale of 0–6 and hence analyzed by standard quantitative genetic techniques. The behavioral score of an individual *Thamnophis radix* was repeatable across trial days (and even across different testing temperatures), and showed full-sib heritabilities of 0.37 for a single trial and 0.45 for the average of two trials. This was the first study of heritability of antipredator behavior in any terrestrial vertebrate.

Antipredator displays of *Thamnophis sirtalis* crawling to exhaustion on a motorized treadmill were studied by Garland (1988). He used an expanded scoring scale (0–9.9) to reflect the exhibition of more offensive behavioral components, such as striking and biting, in this species. The displays of *T. sirtalis* under these conditions were also repeatable across trial days, and also showed a full-sib heritability of about 0.4.

Brodie (1989a,b, 1991, 1992, 1993a,b) has studied antipredator behavior in several populations of *Thamnophis ordinoides*. When chased around a circular track, this species performed stereotyped changes in direction called "reversals." The number of reversals during a trial was taken as a measure of evasiveness during flight and is thought to be important in causing predators to lose sight of the prey. Four different populations of *T. ordinoides* showed significant full-sib heritabilities (0.33–0.65) for reversals, and in two populations reversals were negatively genetically correlated with stripedness (Brodie, 1989a, 1991, 1993b). Antipredator display (cf. Arnold and Bennett, 1984) was also scored in three populations and showed significant but low full-sib heritability (phenotypic variation in the display was much less in *T. ordinoides* than in *T. radix* or *T. sirtalis*; Brodie, 1989a).

Antipredator responses of snakes (numbers of strikes) to moving and nonmoving stimuli (a human index finger) have been studied by

Schwartz and colleagues (Schwartz, 1989; Herzog and Schwartz, 1990; Schwartz and Herzog, 1993; references therein). Schwartz and Herzog (1993) showed significant full-sib heritabilities for *T. sirtalis* (Michigan and Wisconsin), *T. butleri,* and *T. melanogaster*. While not formally employing quantitative genetic methods, several other studies have detected significant family differences in antipredator behavior and rate of habituation to threat stimuli (Herzog and Burghardt, 1988; Herzog et al., 1989, 1992). Litter differences among neonates appear to be stable for at least one year (Herzog and Burghardt, 1988).

Schwartz (1989) measured aggregative tendencies as the nearest-neighbor distance of a snake with any of its litter mates when placed into a 1-m-diameter arena. No population differences were found between animals from Michigan and Wisconsin, but a significant full-sib heritability of 0.26 was estimated for the Michigan population.

Locomotor performance and physiology. Motivated by the recent emphasis in physiological ecology and functional morphology on direct measurement of whole-animal performance abilities (reviews in Bennett and Huey, 1990; Garland and Losos, 1993), several workers have examined the genetics of locomotor abilities in snakes, in all cases using full-sib data. California *T. sirtalis* showed significant full-sib heritabilities for sprint speed over 0.5 m on a photocell-timed racetrack (0.58) and for treadmill endurance at 0.4 km/h (0.70) (in both cases, size-corrected values were analyzed) (Garland, 1988). Contrary to expectations based on exercise physiology, the two performances showed a positive phenotypic (0.36) and genetic (0.59) correlation. Using the same techniques (although analyzing mean rather than best performances) and the same population, Jayne and Bennett (1990a) also reported a positive phenotypic correlation between speed and endurance, as well as similar heritability estimates. Four separate populations of *T. ordinoides* from Oregon exhibited significant heritabilities for both speed over 0.5 m and distance crawled in a circular track, as well as positive phenotypic and genetic correlations (Brodie, 1989a, 1991, 1993b). Garland (1988, 1993) discusses the possible biological basis for the surprising lack of a genetic trade-off between speed and stamina (see also Garland and Losos, 1994).

Locomotor performance was used as a bioassay of resistance to tetrodotoxin in *Thamnophis sirtalis* by Brodie and Brodie (1990, 1991). A full-sib heritability of 0.72 for resistance (measured as a relative reduction in sprint speed after injection with tetrodotoxin) was detected in a population of *T. sirtalis* that feeds on the toxic Roughskin Newt, *Taricha granulosa* (Brodie and Brodie, 1990). This study was the first to establish the ability of predators to respond evolutionarily to defensive adaptations of prey, providing empirical support for the arms-race view of predator–prey coevolution.

Garland and Bennett (1990) reported significant full-sib heritabilities for maximal oxygen consumption (0.89), blood hemoglobin content (0.63), and relative heart (ventricle) mass (0.41). Treadmill endurance (from Garland, 1988) showed a significant positive phenotypic (0.18) but not a genetic correlation with maximal oxygen consumption. In turn, maximal oxygen consumption was significantly positively correlated with ventricle mass both phenotypically (0.27) and genetically (0.64), presumably reflecting the effect of heart size on cardiac output and hence oxygen delivery. In general, low full-sib heritabilities were found for enzyme activities measured as maximal in vitro catalytic rates of citrate synthase (a key aerobic enzyme) and pyruvate kinase (an anaerobic enzyme) in liver, ventricle, and skeletal muscle samples from these same *Thamnophis sirtalis* (Garland et al., 1990).

Schwartz (1989) has shown significant full-sib heritability of 0.6–0.8 for resistance to cold temperatures (critical thermal minimum) in *T. sirtalis*.

Feeding behavior. Garter Snakes show individual and geographic variation in responses to prey odors, in their propensity to eat pieces of slugs in the laboratory, and in their diet in nature (Burghardt, 1970, 1975; Arnold, 1977, 1981a,b,c, 1992; references therein). In a path-breaking series of papers that included the first estimates of behavioral heritabilities for any natural population of vertebrates, Arnold (1977, 1981a,b,c, 1992; Ayres and Arnold, 1983) has studied the quantitative genetics of these behaviors. Slug-eating by naive, newborn snakes was bimodally distributed within each of several populations of *Thamnophis elegans,* although coastal and inland populations showed significant differences in the frequencies of slug eaters (Arnold, 1977, 1981b,c, 1992). Full-sib heritabilities for slug eating were about 0.3–0.5, and probably not different, in two populations (Arnold, 1981a,c; Ayres and Arnold, 1983). Litter differences in response to slug, fish, anuran, and salamander odors were also detected in each of three populations of *T. elegans* and *T. sirtalis* (Arnold, 1992). In a one-of-a-kind study for snake quantitative genetics, Arnold (1981b) crossed populations and demonstrated general intermediacy of slug-eating scores in the F_1, but with some directional dominance for slug-refusing (see also Ayres and Arnold, 1983), and no indication of maternal effects.

Chemoreception responses of *T. elegans* to 10 different prey odors on cotton-tipped swabs were continuously distributed, and most showed significant full-sib heritabilities (Arnold, 1981a,c). Estimated genetic correlations between leech and slug chemoreceptive scores,

variance–covariance matrices of two of the populations show no significant differences (Brodie, 1991, 1993b).

Schwartz (1989) studied scale counts, antipredator responses, thermal tolerance, responses to prey extracts, and aggregative tendencies in two populations of *Thamnophis sirtalis*. Interestingly, he reports several significant full-sib correlations between different types of traits, e.g., ventrals with CTMin, antipredator response with tongue-flick response to minnow extract. The biological significance of such correlations is unclear, and some may be inflated by uncontrolled maternal effects.

Studies of selection in the wild. As discussed above, studies of inheritance comprise one-half of the equation for describing microevolution; studies of natural (or sexual) selection, drift, and gene flow form the other half. The combination of quantitative genetics with selection estimation enables ecologists to predict (at least micro-) evolutionary change (Lande and Arnold, 1983). To date, virtually no study has been able to measure anything near true fitness (e.g., lifetime reproductive success) of individual snakes in nature (but see Madsen and Shine, 1993). Rather, cross-sectional comparisons (Lande and Arnold, 1983; Endler, 1986) of juveniles and adults have been used to infer selection (Dunn, 1942; Inger, 1943; Beatson, 1976; King, 1987, 1992), or survivorship or growth rate has been correlated with some phenotypic trait(s).

Lake Erie Water Snakes (*Nerodia sipedon insularum*) have received continued study by herpetologists and evolutionary biologists (Camin et al., 1954; Camin and Ehrlich, 1958; Ehrlich and Camin, 1960). Most recently, King (1992, personal communication) has found "support for the hypothesis that unbanded morphs are favoured by natural selection in island populations" (King, 1992, p. 115).

Natural selection in a Kansas population of Northern Water Snakes favored increases in both dorsal and lateral blotches (Beatson, 1976). Genetic correlations with ventral scutes, which experienced stabilizing selection, apparently retarded the evolutionary increase of the number of dorsal and lateral blotches.

In a longitudinal study, Arnold (1988a) found some correlations between the combination of ventral and caudal scutes and growth rate in *Thamnophis elegans*. He further employed estimates of the *G*-matrix for scale counts to estimate the strength of selection that would have been necessary to account for phenotypic differences in scale counts between inland and coastal populations of *T. elegans* (Arnold, 1988a). His results suggest that both body and tail vertebrae experienced direct positive selection, but that nearly half of the divergence in tail vertebrae was due to a correlated response to selection acting directly on the number of body vertebrae.

Jayne and Bennett (1990b) scored three kinds of locomotor performance on several hundred *T. sirtalis,* marked the animals by scale-clipping, then released them back into nature. Of 275 cohort snakes tested within eight days of birth in 1985, 79 were recaptured in subsequent years; only birth snout–vent length was a statistically significant (positive) predictor of their probability of survival from 1985 to 1986. However, separate tests of yearling and older snakes did show significant predictive value for both speed and distance crawled (with effects of body sized removed by computing residuals), but apparently not for residual endurance.

Brodie (1992) documented significant correlational selection on stripedness and the tendency to reverse direction in the Tenmile population of *Thamnophis ordinoides.* The direction of this correlated selection is the same as the genetic correlation between these two traits, suggesting that previous selection may have shaped the genetic architecture.

Correlating phenotypic and genetic correlations. As stressed elsewhere in this chapter, studies of heritabilities and genetic correlations can be crucial to microevolutionary analyses (see also Arnold, 1981c, 1983, 1987, 1988a, 1990; Grant, 1986; Garland et al., 1990; Garland, 1994; Brodie, 1992; Boake, 1994). Since phenotypic correlations are much easier to estimate than genetic correlations, it would expedite evolutionary analyses if we could use the former to approximate the latter. Cheverud (1988) reviewed the available data and concluded that phenotypic correlations were generally similar to genetic correlations and "may be substituted for their genetic counterparts in models of phenotypic evolution" (p. 966). However, problems with the data and the statistical analyses prompted Willis et al. (1991) to disagree with this assertion. What do the available data for snakes indicate?

For the behavioral and physiological traits measured in *Thamnophis sirtalis* (Garland, 1988, 1994; Garland and Bennett, 1990; Garland et al., 1990) phenotypic correlations are similar to genetic correlations, but their relationship is not exceptionally tight (Fig. 8.6; $r = 0.76$, $n = 76$). Moreover, the phenotypic correlations tend to underestimate genetic correlations; the reduced major axis (RMA) slope is 1.23 (Fig. 8.6). A statistical comparison of phenotypic and genetic correlations estimated from the same data is problematical (estimates are nonindependent), but for a general comparison of their relative magnitude, consider the plot in Fig. 8.6. While the bulk of the genetic correlations are of similar sign and greater magnitude than the corresponding phenotypic correlations (as evidenced by the positive RMA slope), some values (18 of 76 or 24%; all those in the upper left and lower right quadrants of Fig. 8.6) are of different sign.

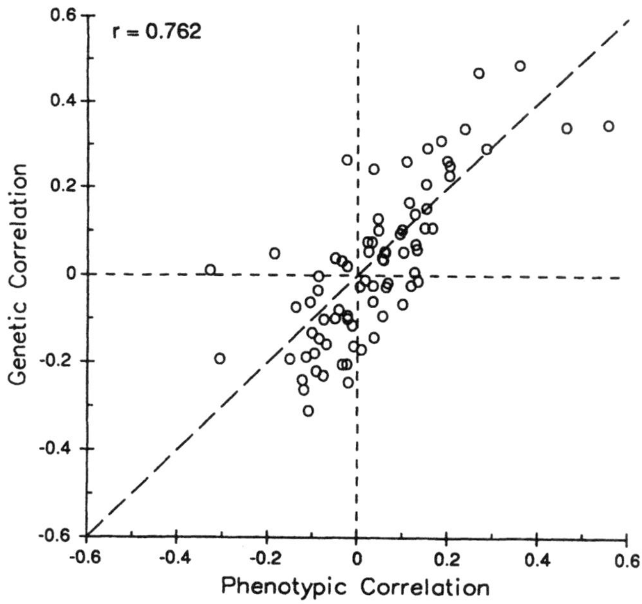

Figure 8.6 Relationship between phenotypic and genetic (estimated as correlations of weighted litter means) correlations (n = 78) for behavioral and physiological traits in *Thamnophis sirtalis* (from Garland, 1988; Garland and Bennett, 1990; Garland et al., 1990). Points falling on the dashed line at a 45° angle correspond perfectly in both sign and magnitude. For correlations falling in the positive–positive and negative–negative quadrants, the genetic correlations tend to be stronger.

For the eight scale characters measured in *Thamnophis sirtalis,* phenotypic and additive genetic correlations (estimated by maximum likelihood) show a considerably weaker relationship (r = 0.40, n = 28), and the underestimation of genetic by phenotypic correlations is even more extreme (RMA slope = 3.79; data from Dohm and Garland, 1993).

The phenotypic and genetic correlations for stripedness, reversals, speed, and distance crawled in two populations of *Thamnophis ordinoides* reported by Brodie (1991, 1993b) also show rather strong correlations (n = 6, r = 0.97 for the Tenmile population and r = 0.87 for the CCQ population). Again, genetic correlations tend to be greater (RMA slope = 1.24 and 1.58, respectively).

Arnold (1981a) presents phenotypic and genetic (weighted-litter-mean) correlations for tongue-flick scores in response to 10 prey odors and two controls for both of two populations of *T. elegans* (his Tables 1 and 2). The correlations between these two estimates (n = 66 correlations for each) are 0.62 and 0.70 for coastal and inland populations, respectively. In both cases, phenotypic correlations greatly underestimate genetic correlations; reduced major axis slopes are 2.24 and 2.45, respectively.

In general then, the data from studies of snakes suggest that phenotypic correlations underestimate genetic correlations, and the correlation between the two is positive (0.4–0.97), but not always strong. Some of the discrepancies may be related to small sample sizes (Cheverud [1988] found genetic and phenotypic correlations tend to be more similar when effective sample sizes are large), but real differences in the two parameters cannot be ruled out. The use of phenotypic correlations when genetic estimates are unavailable will likely lead to flawed conclusions about phenotypic evolution (see Willis et al., 1991).

Summary and Future Research

Quantitative genetics and related techniques can be profitable tools for studying a variety of problems in snake ecology. Understanding the sources and patterns of variation at different levels is integral to any study of evolutionary processes. The specific problems that can be addressed with quantitative genetics are virtually limitless (see above for some examples) and the techniques are relatively simple. The necessary requisites for a quantitative genetic analysis amount to little more than individuals of known relationship, large samples, and some basic statistics.

Although snakes have been the subjects of a large proportion of previous quantitative genetic studies of vertebrates, there has actually been very little work in this area. Two perspectives may guide future quantitative genetics studies of snake populations: (1) what can this approach tell us about snake ecology and evolution; and (2) what general problems can be particularly well addressed using snakes?

Examples of problems in snake biology that can benefit from quantitative genetics may be found in this chapter and throughout this volume. Mating systems, thermoregulatory ecology, endocrinology, and habitat preferences are all unstudied from a quantitative genetic perspective (see Duvall, Schuett, and Arnold, Chap. 5; Ford and Burghardt, Chap. 4; Peterson, Gibson, and Dorcas, Chap. 7; and Reinert, Chap. 6, all this volume). Perhaps the most important direction is to study different taxa. With the exception of two studies on *Nerodia,* all quantitative genetic studies to date have been conducted on *Thamnophis* (and even these two genera are closely related; Table 8.2). Other taxa are common enough to obtain the large samples needed for quantitative genetics (see Parker and Plummer, 1987, for population estimates) and some may even be bred in captivity. It is time such taxa are examined.

TABLE 8.2 Quantitative Genetic Studies of Snake Populations

Traits	Species	Locality	Number of families[a]	Parameters	Method[b]	Study
Scale counts						
Ventrals	*N. sipedon*	Kansas	14	h^2_{FS}	1	Beatson (1976)
Ventrals, subcaudals	*T. elegans*	California	100	h^2_N	2	Arnold (1986)
		Inland, coastal California	84–159	h^2_N, r_G	2	Arnold (1988a)
	T. sirtalis	Wisconsin	≤5	h^2_{FS}, r_G	1	Schwartz (1989)
		Michigan	≤23	h^2_{FS}, r_G	1	Schwartz (1989)
Ventrals, subcaudals, head scales	*T. sirtalis*	California	47	h^2_N, h^2_{FS}, r_G	1,2,3,4	Dohm and Garland (1993)
Head scales	*T. sirtalis*	Wisconsin and Michigan	28	h^2_{FS}	1	Schwartz (1989)
Color pattern						
Number of blotches	*N. sipedon*	Kansas	14	h^2_{FS}, r_G	1,2	Beatson (1976)
		Indiana	58	h^2_N	2	J. Barron (unpublished data)
Number and size of blotches	*N. sipedon*	Lake Erie, Michigan	38	h^2_{FS}, h^2_N, r_G	1,2	R. B. King (unpublished data)
Stripedness	*T. ordinoides*	Oregon	19–126	h^2_{FS}	1	Brodie (1989a, 1991, 1993b)
Antipredator behavior						
Display	*T. radix*	Illinois	15	h^2_{FS}	1	Arnold and Bennett (1984)
	T. sirtalis	California	46	h^2_{FS}	1,3,4	Garland (1988)
Display, reversals	*T. ordinoides*	Oregon	19–77	h^2_{FS}, r_G	1,4	Brodie (1989a)
Reversals	*T. ordinoides*	Oregon	77, 126	h^2_{FS}	1	Brodie (1991, 1993b)
Number of strikes	*T. butleri*	Michigan	7	h^2_{FS}	1	Schwartz and Herzog (1993)
	T. melanogaster	Jalisco, Mexico	9	h^2_{FS}	1	Schwartz and Herzog (1993)
	T. sirtalis	Wisconsin	5–20	h^2_{FS}	1	Schwartz and Herzog (1993)
		Michigan	3–27	h^2_{FS}	1	Schwartz and Herzog (1993)

(Continued)

TABLE 8.2 Quantitative Genetic Studies of Snake Populations (Continued)

Traits	Species	Locality	Number of families[a]	Parameters	Method[b]	Study
Number of strikes (various stimuli)	*T. sirtalis*	Wisconsin	≤5	h^2_{FS}	1	Schwartz (1989)
		Michigan	≤23	h^2_{FS}	1	Schwartz (1989)
Aggregative tendency	*T. sirtalis*	Wisconsin	≤5	h^2_{FS}	1	Schwartz (1989)
		Michigan	≤23	h^2_{FS}	1	Schwartz (1989)
Locomotor performance						
Speed, endurance	*T. sirtalis*	California	46	h^2_{FS}, r_G	1,3,4	Garland (1988)
		California	34	h^2_{FS}	1	Jayne and Bennett (1990a)
Speed, distance	*T. ordinoides*	Oregon	19–126	h^2_{FS}, r_G	1,4	Brodie (1989a, 1991, 1993b)
Physiology						
Cold resistance	*T. sirtalis*	Wisconsin	≤5	h^2_{FS}	1	Schwartz (1989)
		Michigan	≤23	h^2_{FS}	1	Schwartz (1989)
Tetrodotoxin resistance	*T. sirtalis*	Oregon	23	h^2_{FS}	1	Brodie and Brodie (1990)
Enzyme activities	*T. sirtalis*	California	45	h^2_{FS}, r_G	1,4	Garland et al. (1990)
Maximum oxygen consumption, blood hemoglobin, relative heart and liver mass	*T. sirtalis*	California	45	h^2_{FS}, r_G	1,3,4	Garland and Bennett (1990); Garland et al. (1990)
Feeding behavior						
Slug-eating tendency	*T. elegans*	Inland, coastal California	56, 68	h^2_{FS}	1	Arnold (1981c); Ayres and Arnold (1983)
			8–68	Dominance, maternal effects	5	Arnold (1981b)

Feeding preferences (tongue-flick score)	*T. elegans*	Inland, coastal California	10,19	h^2_{FS}, r_G	1,4	Arnold (1981a,c)
	T. sirtalis	Wisconsin	≤5	h^2_{FS}	1	Schwartz (1989)
		Michigan	≤23	h^2_{FS}	1	Schwartz (1989)
Integrative studies						
Scale counts, color pattern	*N. sipedon*	Kansas	...	h^2_{FS}, r_G, selection	1,2,6	Beatson (1976)
Scale counts, locomotor performance	*T. elegans*, *T. radix*	California, Illinois	...	h^2_N, r_G, selection	2,7	Arnold (1988a); Arnold and Bennett (1988)
Antipredator behavior, color pattern, locomotor performance	*T. ordinoides*	Oregon	...	h^2_{FS}, r_G, selection	1,4,7	Brodie (1989a, 1991, 1992, 1993a,b)
Scale counts, antipredator behavior, locomotor performance, physiology	*T. sirtalis*	California	...	h^2_N, h^2_{FS}, r_G, selection	1,2,3,4,7	Garland (1988); Garland and Bennett (1990); Garland et al. (1990); Jayne and Bennett (1990a,b); Garland (1993); Dohm and Garland (1993)
Scale counts, antipredator behavior, physiology, feeding behavior	*T. sirtalis*	Wisconsin, Michigan	...	h^2_{FS}, r_G	1	Schwartz (1989)

[a]Number of families differs according to the population and specific trait in some studies.
[b]1 = full-sib ANOVA; 2 = mother–offspring regression; 3 = maximum likelihood estimation; 4 = family mean correlation; 5 = population crosses; 6 = intensity of selection; 7 = selection gradient.

Ontogeny has been virtually unstudied from a quantitative genetic perspective in any taxon (Cheverud et al., 1983; Atchley, 1984; Riska et al., 1984; Price and Grant, 1985; Arnold, 1990). Techniques and theoretical frameworks for applying quantitative genetics to an ontogenetic dimension are rapidly becoming available (see references in Kirkpatrick et al., 1990a,b). Antipredator behavior of snakes is a model system to begin this avenue of research. A large body of literature on the genetics of neonate behavior already exists, as well as studies of the phenotypic development of behavior, and predictions about the dynamics of the *G*-matrix for behaviors and color patterns have been offered (Brodie, 1993a).

Maternal effects on phenotype are also largely unknown. Again, theory is outpacing empirical investigation. Oviparous reptiles are ideal subjects for studies of maternal effects, but only recently have studies of this kind been performed (Sinervo, 1990; Sinervo and Huey, 1990; for viviparous taxa, see Osgood, 1978; Arnold and Peterson, 1989; Ford and Seigel, 1989). Eggs allow manipulation of investment (i.e., yolk), temperature, and other prenatal environmental components, and many species of snakes have large clutch sizes that allow for families to be split among several treatments. Empirical tests of the efficacy of statistical reduction of maternal effects through regression would be especially valuable.

The approach of studying genotype-by-environment interactions is especially well suited to ectotherms, so many of whose traits depend on microenvironment (e.g., behavior and physiology as a function of temperature). A quantitative genetic approach promises to increase our understanding not only of snake ecology, but also of the evolution of reaction norms and phenotypic plasticity in general (de Jong, 1990; Gomulkiewicz and Kirkpatrick, 1992). Again, split-family designs could be incorporated to test for environmental affects, while providing partial genetic controls.

Finally, some of the greatest advances may come from the integration of macro- and microevolutionary approaches. Comparative studies may provide clues to interesting systems to study at the population level and vice versa (Brodie's [1989a, 1991, 1992, 1993a,b] study of the integration of color pattern and antipredator behavior was motivated by the species comparisons of Jackson et al. [1976]). Investigations of microevolutionary processes may help to explain patterns of diversity among taxa (Price et al., 1984; Emerson and Arnold, 1989; Garland and Losos, 1994). Even hypotheses of genetic drift and selection as differentiating forces among populations or taxa can now be tested through quantitative genetic studies of multiple groups (Lofsvold, 1988).

Acknowledgments

We thank Brent Charland, Harry Greene, Fred Janzen, and an anonymous reviewer for comments on earlier versions of the manuscript. S. J. Arnold, J. Barron, R. B. King, and J. Schwartz graciously provided unpublished data and/or manuscripts. Supported in part by N.S.F. grant BSR-9157268 to T.G. and a Miller Fellowhsip to EDB III.

Literature Cited

Alberch, P., 1980, Ontogenesis and morphological diversification, *Amer. Zool.*, 20:653–667.

Archie, J. W., 1985, Methods for coding variable morphological features for numerical taxonomic analysis, *Syst. Zool.*, 34:326–345.

Arnold, S. J., 1977, Polymorphism and geographic variation in the feeding behavior of the garter snake *Thamnophis elegans, Science,* 197:676–678.

Arnold, S. J., 1981a, Behavioral variation in natural populations, I. Phenotypic, genetic and environmental correlations between chemoreceptive responses to prey in the garter snake, *Thamnophis elegans, Evolution,* 35:489–509.

Arnold, S. J., 1981b, Behavioral variation in natural populations, II. The inheritance of a feeding response in crosses between geographic races of the garter snake, *Thamnophis elegans, Evolution,* 35:510–515.

Arnold, S. J., 1981c, The microevolution of feeding behavior, in A. Kamil and T. Sargent, eds., *Foraging Behavior: Ecological, Ethological and Psychological Approaches,* Garland, New York, pp. 409–453.

Arnold, S. J., 1983, Morphology, performance and fitness, *Amer. Zool.,* 23:347–361.

Arnold, S. J., 1986, Laboratory and field approaches to the study of adaptation, in M. E. Feder and G. V. Lauder, eds., *Predator–Prey Relationships: Perspectives and Approaches from the Study of Lower Vertebrates,* Univ. Chicago Press, Chicago, pp 157–179.

Arnold, S. J., 1987, Genetic correlation and the evolution of physiology, in M. E. Feder, A. F. Bennett, W. W. Burggren, and R. B. Huey, eds., *New Directions in Ecological Physiology,* Cambridge Univ. Press, Cambridge, pp. 189–215.

Arnold, S. J., 1988a, Quantitative genetics and selection in natural populations: Microevolution of vertebral numbers in the garter snake *Thamnophis elegans,* in B. S. Weir, E. J. Eisen, M. J. Goodman, and G. Namkoong, eds., *Proceedings of the Second International Conference on Quantitative Genetics,* Sinauer, Sunderland, Massachusetts, pp. 619–636.

Arnold, S. J., 1988b, Behavior, energy and fitness, *Amer. Zool.,* 28:815–827.

Arnold, S. J., 1990, Inheritance and the evolution of behavioral ontogenies, in M. E. Hahn, J. K. Hewitt, N. D. Henderson, and R. H. Benno, eds., *Developmental Behavior Genetics: Neural, Biometrical, and Evolutionary Approaches,* Oxford Univ. Press, New York, pp. 167–189.

Arnold, S. J., 1992, Behavioural variation in natural populations, VI. Prey responses by two species of garter snakes in three regions of sympatry, *Anim. Behav.,* 44:705–719.

Arnold, S. J., 1994, Multivariate inheritance and evolution: A review of concepts, in C. R. Boake, ed., *Quantitative Genetic Analyses of the Evolution of Behavior,* Univ. Chicago Press, Chicago, in press.

Arnold, S. J., and A. F. Bennett, 1984, Behavioural variation in natural populations, III. Antipredator displays in the garter snake *Thamnophis radix, Anim. Behav.,* 32:1108–1118.

Arnold, S. J., and A. F. Bennett, 1988, Behavioral variation in natural populations, V. Morphological correlates of locomotion in the garter snake *Thamnophis radix, Biol. J. Linn. Soc.,* 34:175–190.

Arnold, S. J., and C. R. Peterson, 1989, A test for temperature effects on the ontogeny of shape in the garter snake *Thamnophis sirtalis, Physiol. Zool.,* 62:1316–1333.

Arveson, J. N., and T. H. Schmitz, 1970, Robust procedures for variance component problems using the jackknife, *Biometrics,* 26:677–686.

Atchley, W. R., 1983, Some genetic aspects of morphometric variation, in J. Felsenstein, ed., *Numerical Taxonomy,* NATO ASI Series, Vol. G1, Springer, Berlin, pp. 346–363.

Atchley, W. R., 1984, Ontogeny, timing of development, and genetic variance-covariance structure, *Amer. Nat.,* 123:519–540.

Ayres, F. A., and S. J. Arnold, 1983, Behavioural variation in natural populations, IV. Mendelian models and heritability of a feeding response in the garter snake, *Thamnophis elegans, Heredity,* 51:405–413.

Barker, J. S. F., and R. H. Thomas, 1987, A quantitative genetic perspective on adaptive evolution, in V. Loeschcke, ed., *Genetic Constraints on Adaptive Evolution,* Springer, Berlin, pp. 3–23.

Barry, F. E., P. J. Weatherhead, and D. E. Phillipp, 1992, Multiple paternity in a wild population of northern water snakes, *Nerodia sipedon, Behav. Ecol. Sociobiol.,* 30:193–199.

Barton, N. H., and M. Turelli, 1989, Evolutionary quantitative genetics: How little do we know? *Ann. Rev. Genet.,* 23:337–370.

Beatson, R. R., 1976, Environmental and genetical correlates of disruptive coloration in the water snake, *Natrix s. sipedon, Evolution,* 30:241–252.

Becker, W. A., 1992, *Manual of Quantitative Genetics,* Academic Enterprises, Pullman, Washington.

Bennett, A. F., 1987, Inter-individual variability: An underutilized resource, in M. E. Feder, A. F. Bennett, W. W. Burggren, and R. B. Huey, eds., *New Directions in Ecological Physiology,* Cambridge Univ. Press, Cambridge, pp. 147–169.

Bennett, A. F., and R. B. Huey, 1990, Studying the evolution of physiological performance, *Oxford Surv. Evol. Biol.,* 7:251–284.

Boake, C. R. B., 1989, Repeatability: its role in evolutionary studies of mating behavior, *Evol. Ecol.,* 3:173–182.

Boake, C. R. B., 1994, *Quantitative Genetic Analyses of the Evolution of Behavior,* Univ. Chicago Press, Chicago, in press.

Brodie, E. D., III, 1989a, Genetic correlations between morphology and antipredator behaviour in natural populations of the garter snake *Thamnophis ordinoides, Nature,* 342:542–543.

Brodie, E. D., III, 1989b, Behavioral modification as a means of reducing the cost of reproduction, *Amer. Nat.,* 134:225–238.

Brodie, E. D., III, 1991, Functional and genetic integration of color pattern and antipredator behavior in the garter snake *Thamnophis ordinoides,* Doctoral Dissertation, Univ. Chicago.

Brodie, E. D., III, 1992, Correlational selection for color pattern and antipredator behavior in the garter snake *Thamnophis ordinoides, Evolution,* 46:1284–1298.

Brodie, E. D., III, 1993a, Consistency of individulal differences in antipredator behaviour and colur patterns in the garter snake *Thamnophis ordinoides, Anim. Behav.,* in press.

Brodie, E. D., III, 1993b, Homogeneity of the genetic variance-covariance matrix for antipredator traits in two natural populations of the garter snake *Thamnophis ordinoides, Evolution,* in press.

Brodie, E. D., III, and E. D. Brodie, Jr., 1990, Tetrodotoxin resistance in garter snakes: An evolutionary response of predators to dangerous prey, *Evolution,* 44:651–659.

Brodie, E. D., III, and E. D. Brodie, Jr., 1991, Evolutionary response of predators to dangerous prey: Reduction of toxicity of newts and resistance of garter snakes in island populations, *Evolution,* 45:221–224.

Bull, J. J., R. C. Vogt, and M. G. Bulmer, 1982, Heritability of sex ratio in turtles with environmental sex determination, *Evolution,* 36:333–341.

Bulmer, M. G., 1985, *The Mathematical Theory of Quantitative Genetics,* Clarendon, Oxford.

Burger, J., 1989, Incubation has long-term effects on behaviour of young pine snakes (*Pituophis melanoleucus*), *Behav. Ecol. Sociobiol.*, 24:201–207.
Burger, J., 1991, Effects of incubation temperature on behavior of hatchling pine snakes: Implications for reptilian distribution, *Behav. Ecol. Sociobiol.*, 28:297–303.
Burghardt, G. M., 1970, Intraspecific geographical variation in chemical food cue preferences of newborn garter snakes (*Thamnophis sirtalis*), *Behaviour,* 36:246–257.
Burghardt, G. M., 1975, Chemical preference polymorphism in newborn garter snakes *Thamnophis sirtalis, Behaviour,* 52:202–225.
Camin, J., and P. Ehrlich, 1958, Natural selection in water snakes (*Natrix sipedon* L.) on islands in Lake Erie, *Evolution,* 12:504–511.
Camin, J., C. Triplehorn, and H. Walter, 1954, Some indications of survival value in the type 'A' pattern of the island water snakes in Lake Erie, *Nat. Hist. Misc.*, 131:1–3.
Charlesworth, B., 1987, The heritability of fitness, in J. W. Bradley and M. B. Andersson, eds., *Sexual Selection: Testing the Alternatives,* Wiley, Chichester, pp. 21–40.
Charlesworth, B., R. Lande, and M. Slatkin, 1982, A neo-Darwinian commentary on macroevolution, *Evolution,* 36:474–498.
Cheverud, J. M., 1982, Phenotypic, genetic, and environmental morphological integration in the cranium, *Evolution,* 36:499–516.
Cheverud, J. M., 1984, Quantitative genetics and developmental constraints on evolution by selection, *J. Theoret. Biol.*, 110:155–171.
Cheverud, J. M., 1988, A comparison of genetic and phenotypic correlations, *Evolution,* 42:958–968.
Cheverud, J. M., J. J. Rutledge, and W. R. Atchley, 1983, Quantitative genetics of development: Genetic correlations among age-specific trait values and the evolution of ontogeny, *Evolution,* 37:895–905.
Cheverud, J. M., G. P. Wagner, and M. M. Dow, 1989, Methods for the comparative analysis of variation patterns, *Syst. Zool.*, 38:201–213.
Clark, A. G., 1987, Genetic correlations: The quantitative genetics of evolutionary constraints, in V. Loeschcke, ed., *Genetic Constraints on Adaptive Evolution,* Springer, Berlin, pp. 25–45.
Crespi, B. J., 1990, Measuring the effect of natural selection of phenotypic interaction systems, *Amer. Nat.*, 135:32–47.
Crowley, P. H., 1992, Resampling methods for computation-intensive data analysis in ecology and evolution, *Annu. Rev. Ecol. Syst.*, 23:405–447.
de Jong, G., 1990, Quantitative genetics of reaction norms, *J. Evol. Biol.*, 3:447–468.
Dobzhansky, Th., 1973, Nothing in biology makes sense except in the light of evolution, *Amer. Biol. Teach.*, March 1973, pp. 125–129.
Dohm, M. R., and T. Garland, Jr., 1993, Quantitative genetics of scale counts in the garter snake *Thamnophis sirtalis, Copeia,* in press.
Dunham, A. E., D. B. Miles, and D. N. Reznick, 1988, Life history patterns in squamate reptiles, in C. Gans and R. B. Huey, eds., *Biology of the Reptilia, Ecology B: Defense and Life History,* Vol. 16, Alan R. Liss, New York, pp. 441–522.
Dunn, E. R., 1915, The variations of a brood of watersnakes, *Proc. Biol. Soc. Washington,* 28:61–68.
Dunn, E. R., 1942, Survival value of varietal characters in snakes, *Amer. Nat.*, 76:104–109.
Efron, B., and R. Tibshirani, 1991, Statistical data analysis in the computer age, *Science,* 253:390–395.
Ehrlich, P., and J. Camin, 1960, Natural selection in Middle Island water snakes (*Natrix sipedon* L.), *Evolution,* 14:136.
Ehrman, L., and P. A. Parsons, 1981, *Behavior Genetics and Evolution,* McGraw-Hill, New York.
Emerson, S. B., and S. J. Arnold, 1989, Intra- and interspecific relationships between morphology, performance, and fitness, in D. B. Wake and G. Roth, eds., *Complex Organismal Functions: Integration and Evolution in Vertebrates,* Wiley, Chichester, pp. 295–314.
Endler, J. A., 1986, *Natural Selection in the Wild,* Princeton Univ. Press., Princeton, New Jersey.

Falconer, D. S., 1989, *Introduction to Quantitative Genetics,* 3rd ed., Longman, London.
Felsenstein, J., 1988, Phylogenies and quantitative characters, *Annu. Rev. Ecol. Syst.,* 19:445–471.
Fisher, R. A., 1918, The correlations between relatives on the supposition of Mendelian inheritance, *Trans. R. Soc. Edinb.,* 52:399–433.
Ford, N. B., and R. A. Seigel, 1989, Phenotypic plasticity in reproductive traits: Evidence from a viviparous snake, *Ecology,* 70:1768–1774.
Fox, W., 1948, Effect of temperature on development of scutellation in the garter snake, *Thamnophis elegans atratus, Copeia,* 1948:252–262.
Fox, W., C. Gordon, and M. H. Fox, 1961, Morphological effects of low temperatures during the embryonic development of the garter snake *Thamnophis elegans, Zoologica,* 46:57–71.
Fuchs, J. L., and G. M. Burghardt, 1971, Effects of early feeding experience on the response of garter snakes to food chemicals, *Learn. Motiv.,* 2:271–279
Garland, T., Jr., 1988, Genetic basis of activity metabolism, I. Inheritance of speed, stamina, and antipredator displays in the garter snake *Thamnophis sirtalis, Evolution,* 42:335–350.
Garland, T., Jr., 1994, Quantitative genetics of locomotor behavior and physiology in a garter snake, in C. R. B. Boake, ed., *Quantitative Genetic Analyses of the Evolution of Behavior,* Univ. Chicago Press, Chicago, in press.
Garland, T., Jr., and S. C. Adolph, 1991, Physiological differentiation of vertebrate populations, *Annu. Rev. Ecol. Syst.,* 22:193–228.
Garland, T., Jr., and A. F. Bennett, 1990, Quantitative genetics of maximal oxygen consumption in a garter snake, *Amer. J. Physiol.,* 259 (*Regulatory Integrative Comp. Physiol.,* 28):R986–R992.
Garland, T., Jr., and J. B. Losos, 1994, Ecological morphology of locomotor performance in squamate reptiles, in P. C. Wainwright and S. M. Reilly, eds., *Ecological Morphology: Integrative Organismal Biology,* Univ. Chicago Press, Chicago, in press.
Garland, T., Jr., A. F. Bennett, and C. B. Daniels, 1990, Heritability of locomotor performance and its correlates in a natural population, *Experientia,* 46:530–533.
Garland, T., Jr., A. W. Dickerman, C. M. Janis, and J. A. Jones, 1993, Phylogenetic analysis of covariance by computer simulation, *Syst. Biol.,* in press.
Garland, T. Jr., R. B. Huey, and A. F. Bennett, 1991, Phylogeny and coadaptation of thermal physiology in lizards: A reanalysis, *Evolution,* 45:1969–1975.
Garland, T. Jr., P. H. Harvey, and A. R. Ives, 1992, Procedures for the analysis of comparative data using phylogenetically independent contrasts, *Syst. Biol.,* 41:18–32.
Gianola, D., 1982, Theory and analysis of threshold characters, *J. Anim. Sci.,* 54:1079–1096.
Gibson, A. R., and J. B. Falls, 1975, Evidence for multiple insemination in the common garter snake, *Thamnophis sirtalis, Can. J. Zool.,* 53:1362–1368.
Gomulkiewicz, R., and M. Kirkpatrick, 1992, Quantitative genetics and the evolution of reaction norms, *Evolution,* 46:390–411.
Gould, S. J., 1980, Is a new and general theory of evolution emerging? *Paleobiology,* 6:119–130.
Grant, P. R., 1986, *Ecology and Evolution of Darwin's Finches,* Princeton Univ. Press, Princeton, New Jersey.
Halloy, M., and G. M. Burghardt, 1990, Ontogeny of fish capture and ingestion in four species of garter snakes (*Thamnophis*), *Behaviour,* 112:299–317.
Hartl, D. L., and A. G. Clark, 1989, *Principles of Population Genetics,* Sinauer, Sunderland, Massachusetts.
Harvey, P. H., and M. D. Pagel, 1991, *The Comparative Method in Evolutionary Biology,* Oxford Univ. Press, Oxford.
Herzog, H. A., Jr., and G. M. Burghardt, 1988, Development of antipredator responses in snakes, III. Stability of individual and litter differences over the first year of life, *Ethology,* 77:250–258.
Herzog, H. A., Jr., and J. Schwartz, 1990, Geographical variation in the antipredator behaviour of neonate garter snakes, *Thamnophis sirtalis, Anim. Behav.,* 40:597–598.
Herzog, H. A., Jr., B. B. Bowers, and G. M. Burghardt, 1989, Stimulus control of

antipredator behavior in newborn and juvenile garter snakes (*Thamnophis*), *J. Comp. Psych.*, 103:233–242.

Herzog, H. A., Jr., B. B. Bowers, and G. M. Burghardt, 1992, Development of antipredator response of snakes: V. Species differences in ontogenetic trajectories, *Develop. Psycholbiol.*, 25:199–211.

Huey, R. B., and A. F. Bennett, 1987, Phylogenetic studies of coadaptation: Preferred temperatures versus optimal performance temperatures of lizards, *Evolution*, 41:1098–1115.

Huey, R. B., and P. E. Hertz, 1984, Is a jack-of-all-temperatures a master of none? *Evolution*, 38:441–444.

Huey, R. B., A. E. Dunham, K. L. Overall, and R. A. Newman, 1990, Variation in locomotor performance in demographically known populations of the lizard *Sceloporus merriami*, *Physiol. Zool.*, 63:845–872.

Inger, R. F., 1943, Further notes on differential selection of variant juvenile snakes, *Amer. Nat.*, 77:87–90.

Jackson, J. F., W. Ingram, III, and H. W. Campbell, 1976, The dorsal pigmentation pattern of snakes as an antipredator strategy: A multivariate approach, *Amer. Nat.*, 110:1020–1053.

Janzen, F. J., 1992, Heritable variation for sex ratio under environmental sex determination in the common snapping turtle (*Chelydra serpentina*), *Genetics*, 131:155–161.

Janzen, F. J., 1993, An experimental analysis of natural selection on body size of hatchling turtles, *Ecology*, in press.

Jayne, B. C., and A. F. Bennett, 1989, The effect of tail morphology on locomotor performance of snakes: A comparison of experimental and correlative methods, *J. Exper. Zool.*, 252:126–133.

Jayne, B. C., and A. F. Bennett, 1990a, Scaling of speed and endurance in garter snakes: A comparison of cross-sectional and longitudinal allometries, *J. Zool., Lond.*, 220:257–277.

Jayne, B. C., and A. F. Bennett, 1990b, Selection on locomotor performance capacity in a natural population of garter snakes, *Evolution*, 44:1204–1229.

King, R. B., 1987, Color pattern polymorphism in the Lake Erie water snake, *Nerodia sipedon insularum*, *Evolution*, 41:241–255.

King, R. B., 1992, Lake Erie water snakes revisited: Morph- and age-specific variation in relative crypsis, *Evol. Ecol.*, 6:115–124.

Kirpatrick, M., D. Lofsvold, and M. Bulmer, 1990a, Analysis of the inheritance of growth trajectories and other complex quantitative characters, *J. Math. Biol.*, 27:429–450.

Kirpatrick, M., D. Lofsvold, and M. Bulmer, 1990b, Analysis of the inheritance, selection and evolution of growth trajectories, *Genetics*, 124:979–993.

Klein, T. W., 1974, Heritability and genetic correlation: Statistical power, population comparisons, and sample size, *Behav. Genet.*, 4:171–189.

Klein, T. W., J. C. DeFries, and C. T. Finkbeiner, 1973, Heritability and genetic correlation: standard errors of estimates and sample size, *Behav. Genet.*, 3:355–364.

Knapp, S. J., W. C. Bridges, Jr., and M.-H. Yang, 1989, Nonparametric confidence interval estimators for heritability and expected selection response, *Genetics*, 121:891–898.

Lande, R., 1979, Quantitative genetic analysis of multivariate evolution, applied to brain: Body size allometry, *Evolution*, 33:402–426.

Lande, R., 1980, The genetic covariance between characters maintained by pleiotropic mutations, *Genetics*, 94:203–215.

Lande, R., 1984, The genetic correlation between characters maintained by selection, linkage and inbreeding, *Genet. Res.*, 44:309–320.

Lande, R., 1988, Quantitative genetics and evolutionary theory, in B. Weir, E. Eisen, M. Goodman, and G. Namkoong, eds., *Proceedings of the Second International Conference on Quantitative Genetics*, Sinauer, Sunderland, Massachusetts, pp. 71–84.

Lande, R., and S. J. Arnold, 1983, The measurement of selection on correlated characters, *Evolution*, 37:1210–1226.

Lessels, C. M., and P. T. Boag, 1987, Unrepeatable repeatabilities: A common mistake, *Auk*, 104:116–121.

Levins, R., 1968, *Evolution in Changing Environments,* Princeton Univ. Press., Princeton, New Jersey.

Lewontin, R. C., 1970, The units of selection, *Ann. Rev. Ecol. Syst.,* 1:1–18.

Lofsvold, D., 1986, Quantitative genetics of morphological differentiation in *Peromyscus,* I. Tests of the homogeneity of genetic covariance structure among species and subspecies, *Evolution,* 40:559–573.

Lofsvold, D., 1988, Quantitative genetics of morphological differentiation in *Peromyscus,* II. Analysis of selection and drift, *Evolution,* 42:54–67.

Losos, J. B., and D. B. Miles, 1994, Adaptation, constraint, and the comparative method: Phylogenetic issues and methods, in P. C. Wainwright and S. M. Reilly, eds., *Ecological Morphology: Integrative Organismal Biology,* Univ. Chicago Press, Chicago, in press.

Lynch, M., 1989, Phylogenetic hypotheses under the assumption of neutral quantitative-genetic variation, *Evolution,* 43:1–17.

Lynch, M., 1991, Methods for the analysis of comparative data in evolutionary biology, *Evolution,* 45:1065–1080.

Madsen, T., and R. Shine, 1993, Determinants of reproductive success in female adders, *Vipera berus, Oecologia (Berlin),* in press.

Madsen, T., Shine, R., Loman., J., and Hakansson, T., 1992, Why do female adders copulate so frequently? *Nature,* 335:440–441.

Martins, E. P., and T. Garland, Jr., 1991, Phylogenetic analyses of the correlated evolution of continuous characters: A simulation study, *Evolution,* 45:534–557.

Mather, K., and J. L. Jinks, 1982, *Biometrical Genetics: The Study of Continuous Variation,* 3rd ed., Chapman and Hall, London.

Mayr, E., and P. D. Ashlock, 1991, *Principles of Systematic Zoology,* McGraw-Hill, New York.

Mitchell-Olds, T., 1987, Analysis of local variation in plant size, *Ecology,* 68:82–87.

Mitchell-Olds, T., and J. Bergelson, 1990, Statistical genetics of an annual plant, *Impatiens capensis.* I. Genetic basis of quantitative variation, *Genetics,* 124:407–415.

Mitchell-Olds, T., and R. G. Shaw, 1987, Regression analysis of natural selection: Statistical inference and biological interpretation, *Evolution,* 41:1149–1161.

Mousseau, T. A., and D. A. Roff, 1987, Natural selection and the heritability of fitness components, *Heredity,* 59:181–187.

Osgood, D. W., 1978, Effect of temperature on the development of meristic characters in *Natrix fasciata, Copeia,* 1978:33–37.

Packard, G. C., and M. J. Packard, 1988, The physiological ecology of reptilian eggs and embryos, in C. Gans and R. B. Huey, eds., *Biology of the Reptilia, Ecology B: Defense and Life History,* Vol. 16, Alan R. Liss, New York, pp. 523–605.

Parker, W. S., and M. V. Plummer, 1987, Population ecology, in R. A. Seigel, J. T. Collins, S. S. Novak, eds., *Snakes: Ecology and Evolutionary Biology,* McGraw-Hill, New York, pp. 253–301.

Plomin, R., J. C. DeFries, and G. E. McClearn, 1990, *Behavioral Genetics: A Primer,* W. H. Freeman, New York.

Plummer, M. V., and H. L. Snell, 1988, Nest site selection and water relations of eggs in the snake, *Opheodrys aestivus, Copeia,* 1988:58–63.

Pough, F. H., 1976, Multiple cryptic effects of cross-banded and ringed patterns of snakes, *Copeia,* 1976:834–836.

Price, T. D., and P. R. Grant, 1985, The evolution of ontogeny in Darwin's finches: A quantitative genetic approach, *Amer. Nat.,* 125:169–188.

Price, T. D., and D. Schluter, 1991, On the low heritability of life-history traits, *Evolution,* 45:853–861.

Price, T. D., P. R. Grant, H. L. Gibbs, and P. T. Boag, 1984, Recurrent patterns of natural selection in a population of Darwin's finches, *Nature,* 309:787–789.

Provine, W. B., 1971, *The Origins of Theoretical Population Genetics,* Univ. Chicago Press, Chicago.

Reznick, D., 1985, Costs of reproduction: An evaluation of the empirical evidence, *Oikos,* 44:257–267.

Riska, B., W. R. Atchley, and J. J. Rutledge, 1984, A genetic analysis of targeted growth in mice, *Genetics,* 107:79–101.
Riska, B., T. Prout, and M. Turelli, 1989, Laboratory estimates of heritabilities and genetic correlations in nature, *Genetics,* 123:865–871.
Roff, D. A., and T. A. Mousseau, 1987, Quantitative genetics and fitness: Lessons from *Drosophila, Heredity,* 58:103–118.
Schluter, D., 1984, Morphological and phylogenetic relations among Darwin's finches, *Evolution,* 38:921–930.
Schmalhausen, I. I., 1949, *The Factors of Evolution,* Blakiston, Philadelphia.
Schwartz, J. M., 1989, Multiple paternity and offspring variability in wild populations of the garter snake *Thamnophis sirtalis* (Colubridae), Doctoral Dissertation, Univ. Tennessee, Knoxville.
Schwartz, J. M., and H. A. Herzog, Jr., 1993, Estimates of heritability of antipredator behavior in three garter snake species (*Thamnophis butleri, T. melanogaster,* and *T. sirtalis*) in nature, *Behav. Genetics,* in press.
Schwartz, J. M., G. F. McCracken, and G. M. Burghardt, 1989, Multiple paternity in wild populations of the garter snake, *Thamnophis sirtalis, Behav. Ecol. Sociobiol.,* 25:269–273.
Seigel, R. A., J. T. Collins, and S. S. Novak, 1987, *Snakes: Ecology and Evolutionary Biology,* McGraw-Hill, New York.
Shaffer, B., 1986, Utility of quantitative genetic parameters in character weighting, *Syst. Zool.,* 35:124–134.
Shaw, R. G., 1987, Maximum likelihood approaches applied to quantitative genetics of natural populations, *Evolution,* 41:812–826.
Shaw, R. G., 1991, The comparison of quantitative genetic parameters between populations, *Evolution,* 45:143–151.
Sinervo, B., 1990, The evolution of maternal investment in lizards: An experimental and comparative analysis of egg size and its effect on offspring performance, *Evolution,* 44:279–294.
Sinervo, B., and R. B. Huey, 1990, Allometric engineering: An experimental test of the causes of interpopulational differences in performance, *Science,* 248:1106–1109.
Slinker, B. K., and S. A. Glantz, 1985, Multiple regression for physiological data analysis: The problem of multicollinearity, *Amer. J. Physiol.,* 249 (*Regulatory Integrative Comp. Physiol.,* 18):R1–R12.
Sokal, R. R., and F. J. Rohlf, 1981, *Biometry,* W. H. Freeman, New York.
Stille, B., T. Madsen, and M. Niklasson, 1986, Multiple paternity in the adder, *Vipera berus, Oikos,* 47:173–175.
Tsuji, J. S., R. B. Huey, F. H. van Berkum, T. Garland, Jr., and R. G. Shaw, 1989, Locomotor performance of hatchling fence lizards (*Sceloporus occidentalis*): Quantitative genetics and morphometric correlates, *Funct. Ecol.,* 3:240–252.
Turelli, M., 1988, Phenotypic evolution, constant covariances, and the maintenance of additive variance, *Evolution,* 42:1342–1347.
Turner, H. N., and S. Y. Young, 1969, *Quantitative Genetics in Sheep Breeding,* Macmillan, Melbourne, Australia.
van Berkum, F. H., R. B. Huey, J. S. Tsuji, and T. Garland, Jr., 1989, Repeatability of individual differences in locomotor performance and body size during early ontogeny of the lizard *Sceloporus occidentalis* (Baird & Girard), *Funct. Ecol.,* 3:97–105.
Via, S., 1984, The quantitative genetics of polyphagy in an insect herbivore, II. Genetic correlations in larval performance within and among host plants, *Evolution,* 38:896–905.
Via, S., 1987, Genetic constraints on the evolution of phenotypic plasticity, in V. Loeschcke, ed., *Genetic Constraints on Adaptive Evolution,* Springer, Berlin, pp. 46–71.
Via, S., and R. Lande, 1985, Genotype-environment interaction and the evolution of phenotypic plasticity, *Evolution,* 39:505–522.
Weir, B., E. Eisen, M. Goodman, and G. Namkoong, 1988, *Proceedings of the Second International Conference on Quantitative Genetics,* Sinauer, Sunderland, Massachusetts.

Wheeler, Q. D., 1986, Character weighting and cladistic analysis, *Syst. Zool.*, 35:102–109.
Wilkinson, G. S., K. Fowler, and L. D. Partridge, 1990, Resistance of genetic correlation structure to directional selection in *Drosophila melanogaster*, *Evolution*, 44:1990–2003.
Willis, J. H., J. A. Coyne, and K. Kirkpatrick, 1991, Can one predict the evolution of quantitative characters without genetics? *Evolution*, 45:441–444.
Wright, S., 1921, Systems of mating, *Genetics*, 6:111–173.
Zeng, Z.-B., 1988, Long-term correlated response, interpopulation covariation and interspecific allometry, *Evolution*, 42:363–374.
Zweifel, R. G., and H. C. Dessauer, 1983, Multiple insemination demonstrated experimentally in the kingsnake (*Lampropeltis getulus*), *Experientia*, 39:317–319.

Chapter

9

Strategies for Snake Conservation

C. Kenneth Dodd, Jr.

"It is not enough to understand the natural world; the point is to defend and preserve it"
EDWARD ABBEY (1989)

Introduction

Snakes generally do not attract human attention unless a person is bitten or a snake decimates an endangered bird community (Savidge, 1987; Fritts, 1988). When it comes to bad things happening to snakes, however, few supportive resources are available. The loss of the earth's biological diversity has attracted much attention (Wilson, 1985, 1988), yet discussions of biodiversity in the mainstream conservation literature rarely mention amphibians and reptiles, much less snakes. When conservation funds are handed out, amphibians and reptiles are generally overlooked, and funds for snake conservation are extremely scarce. Even among conservation biologists and herpetologists, research on status assessment or the development of recovery or management programs for snakes has been directed at relatively few species.

Consistent methods of status determination and the development of management strategies for declining snake populations have, in general, not been developed. More than 180 species or populations of snakes were identified as declining, rare, or in otherwise need of assessment or management in the biological literature through the mid-1980s (Dodd, 1987); few studies have been published since then. Much status data are anecdotal, and conservation criteria are not often explained. On the other hand, there has been an explosion in the conservation biology literature on a great variety of topics that are germane to both single-species and community conservation.

In this paper, I review problems related to the assessment of status, the development of conservation programs, opportunities and limitations at the individual, population, and community level, and research needs. I purposely try to avoid the species versus community approach because both are necessary. Because few data exist on these problems directly relating to snakes, I draw heavily from the literature on conservation biology. More detailed discussions are found in primary source books (e.g., Soulé and Wilcox, 1980; Frankel and Soulé, 1981; Schonewald-Cox et al., 1983; Soulé, 1986, 1987; Western and Pearl, 1989; Shafer, 1990) and journals, particularly *Biological Conservation, Conservation Biology,* and *The Natural Areas Journal.*

Status Identification

How is status determined?

The ways that species can be identified as in need of conservation and management are numerous. However, there are four general categories through which conservation problems involving snakes have been identified: personal experience, extensive trade, rarity, and a specific identifiable cause, such as habitat destruction, affecting an individual species or community.

Historically, the best sources of information on snake populations were from people who, for one reason or another, were interested in the species in question. Personal experience is often anecdotal but may result from directed research. Biologists may notice that certain snakes are not as plentiful as they used to be, or that populations appear to have declined. For example, R. H. Mount (personal communication) reports that the number of snakes killed on highways between Auburn, Alabama and Gainesville, Florida declined drastically from the 1960s to the present. Mount suspects that traffic mortality combined with predation of ground-dwelling snakes by Fire Ants (*Solenopsis invicta*) can explain the declines (Mount, 1981). Since the 1970s, Common Kingsnakes (*Lampropeltis getula*) virtually disappeared at Paynes Prairie, Alachua County, Florida, where they were once quite common (R. Franz, personal communication). Franz also notes that Cottonmouths (*Agkistrodon piscivorus*) increased in abundance between 1977 and 1981, but attributes the kingsnake decline to collection, altered water flow regimes on the prairie, and habitat changes along U.S. Highway 441. While often subjective, personal experience may be the only hint that declines have occurred or are occurring. Anecdotal information may and should lead to a more careful examination of status, but it rarely can be used to influence legislative or management decisions effectively.

In a few fortunate cases, such as Spellerberg's (1988) work on *Coronella austriaca* in Britain and Brown's (1993) work on *Crotalus horridus* in New York, research into one facet of a snake's life history leads to information that contributes substantially to the species' conservation. Radiotracking Timber Rattlesnakes in New York showed the extent of habitat use, differential patterns of dispersion between the sexes, and the location of important den sites (Brown et al., 1982; Brown, 1991). These data then were used to develop a management plan and suggest reserve design and spatial requirements (Brown, 1993).

A large volume of trade in live snakes or snake products suggests that conservation efforts need to be instituted. The snake trade continues to be substantial from many regions, particularly Asia and South America (Dodd, 1987; Andrews and Birkinshaw, 1988; Luxmoore et al., 1988; Fitzgerald, 1989). Although population data are unavailable for most species, a large amount of trade coupled with known life-history characteristics (e.g., long life-span, large habitat requirements) have led to national (e.g., *Drymarchon corais couperi* in the United States) or international (e.g., the listing of many large Indian species in Appendices of the Convention on Trade in Endangered Species of Wild Fauna and Flora [CITES]) protection. Extensive trade focuses the need for management and population assessment, but the effects of trade on long-term survival remains unknown. Wild populations of species heavily exploited by commercial trade should be managed.

A species may be rare and therefore may be subject to protection or at least may attract the interest of regulatory agencies. For example, many snakes in India and Sri Lanka are known from single or a few specimens and are thus included on "rare" species lists (Dodd, 1987). However, an assessment of rarity is often a subjective judgment that reflects a lack of collecting effort or familiarity with appropriate habitats. In some cases, a species may be rare in one geopolitical jurisdiction but common elsewhere (Dodd, 1987; Branch, 1988). Rarity per se may not be a cause for concern. Unfortunately, most of the literature on "rare" snakes does not give information on collection effort or other pertinent information such as available habitat or real (or even potential) threats. Emphasis on rarity also places undue attention to numbers rather than problems. Protecting "rare" species without adequate assessment may impede research and conservation efforts.

Finally, conservation and management programs for individual species, and sometimes even their habitats, have been instituted in response to a known threat, particularly habitat destruction (e.g.,

Hoplocephalus bungaroides in Australia [Shine and Fitzgerald, 1989], *Nerodia harteri* in Texas [Mathews, 1989; Scott et al., 1989], and many species of European snakes [Corbett, 1989]; also see Dodd, 1987). Species with small habitat ranges or populations at the periphery of a wider geographic range are easily recognized as meriting conservation or management. Snakes are sometimes specifically recognized in calls for the conservation of an imperiled reptile community (e.g., Corke, 1987).

Only in very rare cases is a snake community or assemblage recognized as threatened (Henderson and Sajdak, 1986; Sajdak and Henderson, 1991). Island species warrant special attention because of restricted ranges coupled with immediate threats, particularly habitat destruction and the introduction of exotic species (Dodd, 1987; Henderson, 1992; Lillywhite and Henderson, Chap. 1, this volume). A class of threat also may result in studies designed to monitor the threat or assess its effects. For example, concern about the effects of rattlesnake roundups in the United States has prompted both increased public scrutiny (Williams, 1990; Weir, 1992) and biological investigations (Campbell et al., 1989; Reinert, 1990; Warwick, 1990).

How should status be determined?

While the four categories discussed above will continue to serve as a foundation for future status assessment, especially in regions without resident biological expertise or resources, a more comprehensive and systematic set of procedures needs to be developed for future planning. I suggest that piecemeal status assessment will prove inadequate in providing for the long-term survival of snake populations, and that a regional approach using a variety of criteria is necessary (see Noss, 1990). Research on life history and distribution is vital to the success of assessment programs to ensure that measures to conserve snake communities are initiated prior to the dubious attainment of officially recognized endangered or threatened status.

When feasible, status assessment of snake populations should be undertaken on all species to help identify actual and potential problems, not just on those species a priori presumed to need management. Early assessment of the status of even common species may avoid future conflicts and help track status through time. Species may be targeted for monitoring without detailed knowledge of population size and trends. At minimum, general data on life-history and habitat requirements are necessary. For example, species that possess certain morphological or life-history characteristics, such as conspicuousness, large body size, a long life-span, delayed sexual maturity, low reproductive potential or effort, or specialized diet or physiological requirements, may be prone to environmental stress and have particular management requirements (Scott and Seigel, 1992).

Habitat monitoring and management should be required for those species that have specialized needs for foraging, dens or other retreats from unfavorable habitat conditions, or large home ranges. Examples include Timber Rattlesnakes (*Crotalus horridus*) in cooler climates where denning is common, and the large communal dens of Canadian Red-Sided Garter Snakes (*Thamnophis sirtalis parietalis*) (Gregory, 1977). In many regions, land-use planning can provide insights to a snake community's future prospects. For example, each Florida county is required by state law to develop a comprehensive land-use plan through the year 2000. Examination of these plans reveals that little habitat will remain undeveloped in some counties if the plans are implemented. Knowledge of snake microdistribution is essential to ensuring that planners are aware of unique species and communities before conflicts arise.

Various ranking indices have been developed to assess status, some with more success than others. Schlauch (1978) evaluated five indices using both qualitative and quantitative status categories. He concluded that, while quantitative assessments are valuable to planning, numerical indices would probably not prove sufficiently rigorous to serve as the primary criteria in determining conservation priorities. Common sense is necessary. Nevertheless, ranking indices can be a valuable tool in recognizing future problems.

A comprehensive regionwide status ranking assessment for future conservation planning was recently carried out by the Nongame Wildlife Program of the Florida Game and Fresh Water Fish Commission (Millsap et al., 1990). All of Florida's 668 vertebrate species and subspecies were ranked on biological (population size, population trends, range size, distribution trends, population concentration, reproductive potential for recovery, ecological specialization), action (knowledge of distribution, monitoring population trends, population limitations, ongoing management activities), and supplemental (systematic significance, percentage of the range that occurs in Florida, trend in Florida's population, period of occurrence in Florida, harvest) variables. Criteria and related discussions were published (see Tables 1–3 in Millsap et al., 1990).

Combined variable scores were used to identify candidate taxa and set priorities for further survey, monitoring, research, and management. Eighteen major Florida ecosystems then were ranked with regard to all taxa to identify those habitats most in need of protection. For reptiles, a combined geographic point score identified three major regions for special attention: the southeastern coast, the Central Florida Ridge, and the Apalachicola Basin. The Florida approach provides a sound biological basis for planning future activities. With minor modifications, it could be used for both small- and large-scale conservation assessment of snake species and communities.

Making decisions

One of the most difficult factors in assessing status is making decisions. Some authors have suggested that status assessment and conservation decisions should be based on complex decision trees (e.g., Thibodeau, 1983), population viability analysis (Soulé, 1987), or mathematical formulas for extinction factors (e.g., Mace and Lande, 1991). The "neatness" of such proposals revolves around attempts to take the subjectivity out of status assessment. Data on snake populations are not generally available to plug into formulas, although the Mace and Lande (1991) model provides a working basis to define categories (Table 9.1). In an imperfect world, conservation biologists must be willing to use the best data available to make judgments and decisions. Failing to make a decision because one does not have all the data one might like to have is in reality making a decision that could lead to irreversible decline.

Development of a Conservation Program

Goals and objectives

Conservation programs should begin with a goal or series of goals and set of objectives to achieve those goals. The goals should be clearly stated and attainable within a specified time period. A lack of goals and criteria for the determination of the effectiveness of a conservation program is a clear indication that the program is ill-conceived. Biologists must have a clear understanding about what they are trying to do; that is, is the immediate goal conservation or preservation (Frankel and Soulé, 1981)? Conservation generally implies a long-term program that attempts to maintain evolutionary potential, whereas preservation implies an immediate attempt to avert extinction that is but one choice in a broad range of conservation options. The procedures used to affect conservation or preservation may be quite different; for example, acquisition of reserves versus captive breeding programs. Investigators also should be clear as to the time scale of concern—weeks, years, or decades (Frankel and Soulé, 1981).

Goals may be far-reaching, but they should be task-oriented and related to the biological requirements of the species. Once goals are identified and objectives defined, specific tasks can be assigned to fulfill the objectives. Conservation programs should conserve the species or community in question rather than have the removal of regulatory control as a primary goal.

TABLE 9.1 Proposed Status Categories[a]

Population trait	Critical	Endangered	Vulnerable
Probability of extinction	50% within five years or two generations, whichever is longer or any two of following criteria	20% within 20 years or 10 generations, whichever is longer or any two of following criteria or any one critical criterion	10% within 100 years or any two of following criteria or any one endangered criterion
Effective population	$N_e < 50$	$N_e < 500$	$N_e < 2000$
Total population	$N < 250$	$N < 2000$	$N < 10,000$
Subpopulations	≤ 2 with $N_e > 25$, $N > 125$ with immigration <1 per generation	≤ 5 with $N_e > 100$, $N > 500$ or ≤ 2 with $N_e > 250$, $N > 1250$ with immigration <1 per generation	≤ 5 with $N_e > 500$, $N > 2500$ or ≤ 2 with $N_e > 1000$, $N > 5000$ with immigration <1 per generation
Population decline	> 20%/yr for last two years or > 50% in last generation	>5%/yr for last five years or >10% per generation for last two generations	> 1%/yr for last 10 years
Catastrophe: rate and effect	> 50% decline per 5–10 yr or 2–4 generations; subpopulations highly correlated	>20% decline per 5–10/yr or 2–4 generations; > 50% decline per 10–20 yr, 5–10 generations with subpopulations correlated	>10% decline per 5–10 yr, >20% decline/10–20 yr, or >50% decline/50 yr, with subpopulations correlated
or			
Habitat change	Results in above effects	Results in above effects	Results in above effects
or			
Commercial exploitation or interaction, or introduced taxa	Results in above effects	Results in above effects	Results in above effects

[a]Based on criteria oulined in Mace and Lande (1991) and originally presented in tabular form by Foose (in press). (Reprinted with permission.)

For example, a goal might be set to protect habitat for a snake population. A stepwise progression of specific objectives could then be developed to attain the goal, such as research on habitat requirements, spatial use, population structure, and minimum-population-size analysis. A second set of objectives would use this information to address questions concerning reserve design, reserve acquisition, immediate management needs, and long-term monitoring. Once specific objectives are stated, the details needed to obtain the objectives can be developed as finances and other nonbiological considerations make options available.

At any stage in a conservation plan, criteria should be available to assess the effectiveness of ongoing activity. Of course, many activities can take place concurrently depending on need. Evaluation is critical. How do you know that the plan is successful? One might choose very specific criteria. In the example above, an objective might be stated as the protection of a minimum population size of 2500 breeding snake pairs on a 5000-ha reserve over a period of 30 years, but the actual success will be measured only by monitoring the population for those 30 years. Attainment of conservation goals, and subsequent declarations of success, will require a long-term monitoring commitment to ensure that the goals are actually met. Conservation programs without the means or will to carry them out are exercises in futility and a waste of biological expertise.

Constraints to conservation

The greatest constraint in conservation planning for either individual species or entire snake assemblages is the fundamental lack of basic biological information on most species. For most species, data on population size and spatial limits, genetic variability, social structure and spacing, dispersal, habitat requirements of juveniles, and survivorship are limited (but see Parker and Plummer, 1987; Brodie and Garland, Chap. 8, this volume; Reinert, Chap. 6, this volume). As a result, examples of the ways biological constraints should be considered when preparing snake conservation programs often are hypothetical. Few detailed conservation programs have been developed for snakes.

Conservation programs must be undertaken within both biological and human-related constraints. Of these, the biological constraints are most important because they set the options, limits, and probabilities that goals and objectives can be achieved. Conservation programs undertaken without regard to or in ignorance of a species' biological constraints have little chance of success (Dodd and Seigel, 1991). Therefore, the first question to ask when developing a conservation program is "What do we know about the species' biology?" Biological constraints fall into a number of major categories, including habitat, demographic, biophysical, social or behavioral (both intra- and interspecific), disease and parasite, and genetic considerations. It might seem obvious that these requirements must be met but, unfortunately, conservation programs are replete with examples of failure to adequately consider a species' life-history requirements (e.g., Dodd and Seigel, 1991).

Biological constraints

Habitat constraints involve the physical factors that relate to a species' ability to maintain an effective population size. At both the individual and population levels, they include enough physical space for the acquisition of food and mates, dispersal, and protective cover. Prey should be available for all size classes of a snake population because diet can change during ontogeny (Mushinsky, 1987). Threats to a prey species should be regarded as a threat to the managed species. Structural habitat diversity may be important for some species, especially on a seasonal or ontogenetic basis (Gregory et al., 1987). Specialized shelter and habitat needs, such as friable soils for burrowing species, must be taken into consideration (Reinert, Chap. 6, this volume). At the population or even metapopulation (Pickett and Thompson, 1978) level, sufficient space should be available for a population size large enough to minimize the effects of environmental fluctuation and demographic stochasticity (Soulé, 1983). If required habitat components are missing or if the area is too small to support a minimum viable population, the conservation program will fail. Habitats should contain protected den sites when appropriate. If dens and summer foraging areas are widely separated (Gregory, 1984), planning will need to provide for movement corridors.

Demographic, or life-history, constraints include those factors affecting a population's size and age–class structure, sex ratio (either primary, secondary, or operational), clutch size and frequency, survivorship, and growth (Parker and Plummer, 1987). Very little is known about how demographic changes actually affect snake populations, but changes in demography as a result of human disturbance have resulted in population declines in other vertebrates (e.g., Gerrodette and Gilmartin, 1990). Certain demographic traits, such as long life span, few offspring, and high adult survivorship, might indicate a low recruitment rate. A species with these characteristics that is heavily involved in trade or experiencing massive habitat destruction would be more prone to population decline than a short-lived highly fecund species. Conservation would require a longer time to effect as the population could not rebound quickly, if ever. Similarly, if collecting were directed toward one sex because of physical appearance or behavior, sex ratios could be altered, thus adversely affecting social structure, ability to find mates, and reproduction.

Because they are ectotherms, reptiles have biophysical requirements that are different from the biophysical requirements of endotherms (Pough, 1983; Scott and Seigel, 1992). The main biophysical requirement of snakes is the need for raising the body temperature to a preferred optimum. Snakes thermoregulate for a variety of reasons, including digestion, reproduction, and the optimum maintenance of physiological processes (Lillywhite, 1987). This is done behaviorally by basking. Conservation programs need to ensure that basking sites appropriate to the species are available and that minimal disturbance occurs. Disturbance may be more important at some times of the year than others. For instance, snakes basking at den sites in winter are more exposed to predation or human interference than in the summer because reaction times are slower due to reduced ambient temperatures. Likewise, gravid females need to bask prior to oviposition for successful embryonic development, yet have reduced locomotor capacity because of the added weight of the developing eggs or young (Seigel et al., 1987).

Knowledge of snake social behavior is still minimal for most species, leading Gillingham (1987) to refer to the "social behavior in a solitary reptile." However, snakes are not entirely solitary, and their social behavior consists of a complex set of interactions relating to reproduction, habitat spacing, and communal aggregation (Gillingham, 1987; Ford and Burghardt, Chap. 4, this volume). Snakes are very olfactory-oriented animals; they rely on chemoreception for species, mate, and prey capture. Conservation programs must take olfactory-based social interactions into account.

Chemoreception might be used in territorial spacing, which in turn could affect the overall number of animals within an area. For example, male Florida Pine Snakes (*Pituophis melanoleucus mugitus*) have nonoverlapping home ranges whereas females do not (R. Franz, personal communication). Several females occupy the home range of a single male. This spatial arrangement could imply a polygynous spatially based mating system. As such, the habitat required for the long-term conservation of Florida Pine Snakes will have to contain a much larger population of snakes than it might if the effective population size was based solely on an adult 1:1 sex ratio with overlapping home ranges.

Snake reproduction may be solitary or carried out in a group. Canadian Red-Sided Garter Snakes (*Thamnophis sirtalis parietalis*) mate in large aggregations immediately upon emergence from hibernation (Gregory, 1984). Literally hundreds of snakes mass in snake "balls" as frantic males court females (Aleksiuk and Gregory, 1974). At this time, the snakes are very susceptible to collection. If large numbers of animals were removed during mating, there might be adverse effects on the snake's mating system. A conservation program would ensure that the snakes are free to complete reproduction with a minimum of disturbance.

Conservation programs should attempt to maintain genetic diversity in wild populations and to minimize the loss of heterozygosity. Extensive considerations for the genetic management of wild populations are provided in many papers in Schonewald-Cox et al. (1983). One of the first considerations in the development of a conservation plan might be to determine the level of genetic variation in the population under question. Nondestructive blood- and tissue-sampling methods are now available for electrophoretic, mitochondrial DNA, or DNA sequencing techniques.

Over a short time scale, genetic considerations may have minimal impact on a population's survival provided the population is large. Small populations are more susceptible to the long-term loss of fitness, however. One of the many problems in snake conservation is that the size of the population is often not known. Computer simulations suggest that in populations of less than 100 individuals, genetic drift is far more important than either selection or mutation in the loss of genetic diversity (Lacy, 1987). Even the immigration of only one or two migrants per generation is sufficient to counter the effects of drift. The loss of heterozygosity is particularly important because it can diminish a population's response to a changing environment, especially in small populations (Lande, 1988). Therefore, the long-term management of small fragmented snake populations might consider the occasional introduction of individuals from genetically diverse populations.

On the other hand, theoretical genetic considerations may not be critical to the immediate conservation of a snake population. Lande (1988) suggested that "primary" demographic factors, defined as social structure, life-history variation caused by environmental fluctuation, dispersal in spatially heterogeneous environments, local extinction and colonization, are more important than population genetics in determining minimum viable population sizes (also see Woodruff, 1989). For example, genetic considerations might argue for a minimum viable population size of 500 individuals, whereas stochastic events might render such a population extinct long before the adverse effects of inbreeding depression and the loss of genetic diversity would be felt. Conservation programs that rely solely on population genetics criteria and ignore demographic factors risk species extinction. If competition arises for funds, conservation programs might focus more effectively on life-history rather than genetic questions.

Minimum viable population (MVP) models have received much attention as tools for conservation planning (Soulé and Wilcox, 1980; Frankel and Soulé, 1981; Soulé, 1987). MVP models and simulations attempt to predict the time to extinction given a set of variables reflecting a species' life-history characteristics, genetics, and spatial requirements. As such, the goal of MVP models is the prevention of extinction and the maintenance of genetic variability (Shaffer, 1981; Samson, 1983; Shaffer and Samson, 1985; Samson et al., 1985). MVP models take extinction factors (habitat alteration, predation, competition, random population fluctuation) into consideration to build theoretical models that incorporate genetic diversity to determine an estimation of species-specific extinction probability (Samson, 1983). Such models are dependent on both habitat size and patchiness, and are thus used for guidance in the development of habitat reserves.

An important point, often misunderstood by administrators, is that a minimum viable population is not the maximum number of animals that can exist within the smallest possible reserve, even over an extended period. Instead, it should be viewed as "the smallest isolated population having a 99% chance of remaining extant for 1000 years despite the foreseeable effects of demographic, environmental, and genetic stochasticity, and natural catastrophes" (Shaffer, 1981). The objective of establishing MVPs is not to determine the smallest possible reserve to mitigate conflicts, but to assist in conservation planning and reserve establishment.

In the past, biologists seized on several "magic" numbers (usually 50 or 500) as general estimates of theoretical MVP size, but it is now recognized that there is no universal MVP size on which conservation programs should be based (Woodruff, 1989). Current approaches focus on models that are a modification of the MVP, population viability models (PVA) (Gilpin and Soulé, 1986; Gilpin, 1987; Soulé, 1987). PVA models also attempt to integrate life-history, genetics, and stochastic events and to put them into a time-scale framework of extinction. Such models have generally led to the conclusion that very large populations requiring large amounts of habitat will be needed to conserve species (Woodruff, 1989). In this regard, it is crucial to understand the relationship between overall preserve area and the amount of habitat available to the species. There have been no MVP or PVA analyses published for any snake, although a viability analysis and risk assessment were prepared for *Nerodia harteri* when it was proposed for federal protection (M. E. Soulé, personal communication).

Human-related constraints

Snakes are difficult to conserve because the general public does not particularly like snakes (Morris and Morris, 1965; Steinhart, 1984; Bruno and Maugeri, 1990). Education approaches featuring mere exposure or information do not seem to improve attitudes concerning snakes, at least among elementary students in eastern North America (Morgan and Gramann, 1989). Although it would likely be denied by those involved, negative reactions to serpents probably carry over into conservation-related decisions by administrators in both public and private agencies. Prejudice against snakes may be a major impediment in their conservation (Dodd, 1987). For example, prejudice against amphibians and reptiles, including poisonous snakes, led to reservations when Spain acceded to the Berne Convention (Corbett, 1989).

Few funds are available for life-history studies on any species (Tangley, 1988), much less for snake conservation. By and large, national and international conservation organizations have ignored reptiles, except for sea turtles. Most nongovernmental conservation organizations contribute well under 4% of their research funds toward herpetofaunal research (Mittermeier and Carr, 1993; Mittermeier et al., 1992). Much attention has been directed at commercially important crocodilians whereas commercially valuable snakes receive no funding for population or assessment studies. Governmental agencies probably do better in funding, although virtually no published funding breakdowns are available. Dodd (1980) reported that 23% of the money available for herpetofaunal research grants in the U.S. Endangered Species Program went for snake projects in 1979–80. State cooperative agreements allocated 5.4% of money spent on amphibians and reptiles on snake projects from 1976 to 1980 (Dodd, 1980). These funds were a small fraction of the total money available, however, and snakes have probably not fared better in the intervening years. Clearly, lack of funding is a major constraint to snake conservation.

Many regional, national, and international laws protect reptiles, including snakes. Some of these laws, such as the U.S. Endangered Species Act of 1973, have been effective in addressing certain facets of snake conservation (e.g., controlling trade). Species, subspecies, or populations may be protected, such as the northeastern population of the Timber Rattlesnake (*Crotalus horridus*) currently under consideration for federal protection. In other geopolitical units, however, laws are not enforced because of the lack of resources or political will, or appropriate legislation is nonexistent (see Corbett, 1989). Even with the best of intentions, conservation legislation is sometimes misused or does not address critical issues such as habitat protection (Yaffee, 1982; Kohm, 1991; Rohlf, 1991). Laws designed to protect one aspect of conservation are confused by impressions that they protect other aspects of a species' status. For example, the Convention on International Trade in Endangered Species of Wild Fauna and Flora (CITES) is a trade convention, not an endangered species convention. Statutory protection is provided to species listed in Appendix I of the Convention, whereas Appendix II species can be freely traded as long as the country of origin issues a permit. Endangered or threatened species are not eligible for listing unless they are in trade. Appendix I is not equivalent to endangered status and Appendix II is not equivalent to threatened status.

The general public and scientific community is often confused about differences between legislation and regulation. This distinction is important because legislation sets the limits within which conservation activities take place within a government agency. If legislative authority protects a species but not its habitat, the regulatory agency cannot set out to create protection for which authority does not exist. Biologists may want to change a law and make it more reflective of biological reality, but until that occurs, they must work within the existing legislative framework.

On the other hand, regulations are adopted by an agency to enforce generally broad-based legislative mandates. They should provide flexibility to accomplish legislative goals. The scientific community needs to provide input when regulations are proposed. If regulations are not sufficient to accomplish their goals or if they are contrary to the purpose of the legislation, biologists should be prepared to lobby for change or lend their expertise to legal challenges. Biology without some form of public involvement will not conserve snake populations.

Finally, logistic problems may limit conservation. I include both physical factors (lack of equipment, roads, transportation) as well as human factors (lack of personnel, human encroachment on reserves, dense human populations). Conservation research at times may suffer from a lack of optimum statistical rigor because of logistic limitations. This is a cost of field biology that may be offset, in part, by carefully planned laboratory experiments. In some cases, innovative approaches incorporating local inhabitants in snake conservation programs have overcome human-related logistic problems to the benefit of both communities. Perhaps the best example of such a program is the development of the Irula Snake Cooperative by the Madras Crocodile Bank Trust in India (Z. Whitaker, 1979).

Methods to Conserve Snakes

Community-based options: Habitat protection

If the goal of a conservation program is to ensure the long-term survival of a snake species, its habitat (sensu Harris and Kangas, 1988) will have to be conserved. Conservation and recovery plans should focus on the ecosystem or community rather than a single species, except under special circumstances. Although goals may focus on the conservation of a particular species, single-species management is not always effective and may jeopardize other members of the biological community. When funding constraints restrict options, efforts should be focused on the animal in its native habitat rather than experimental programs such as relocation or captive propagation.

Habitats may be protected by establishing parks, reserves (wildlife refuges, management areas, watershed protection zones, etc.) or conservation easements with private landowners. Many snake communities are protected within national or regional parks (e.g., in Europe [Corbett, 1989]), but few reserves have been established specifically for them (albino *Elaphe climacophora* in Japan [Tokunaga et al., 1991]; *Elaphe longissima* in Poland, *Natrix tessellata* in Czechoslovakia, and *Vipera* sp. in Hungary [Corbett, 1989]). In Western nations, parks are perceived as the pinnacle of habitat conservation. They usually protect beautiful scenery or unique geological attributes rather than biological communities per se. An exception is Australia where national parks have been established throughout the country to preserve representative vegetative and animal communities. Parks are generally large, but not always so. In the past, parks were often seen as places where nature could "take its course" without very much human manipulation. With fragmentation, global environmental concerns, and increasing human population pressure, however, this attitude is changing (White and Bratton, 1980; Hales, 1989; McNeely, 1989). In Florida, for example, state parks are managed with a goal of recreating environmental conditions prior to European settlement. All protected areas need some form of management.

Reserves may be established for individual species or in connection with other specific land-use practices. Indeed, reserves can be integrated into a regional landscape that allows a multiplicity of uses compatible with small vertebrate conservation (Western, 1989). Various models are available, from sanctuaries held inviolate from human disturbance to large wildlife management areas created for hunting game species. Consumptive use of reserves, such as those permitted in the United States on National Wildlife Refuges, often would be compatible with snake conservation as long as efforts are made to avoid malicious vandalism, incidental take, and the destruction of crucial microenvironments such as dens and basking sites. No snake in a conservation reserve, park, or publicly owned refuge should be the target of organized directed take, such as rattlesnake roundups or eradication programs.

Once it is decided that a reserve or reserves should be created, various theoretical factors might be considered (Shafer, 1990), including size (Soulé and Simberloff, 1986), optimal shape (Blouin and Connor, 1985), the effects of boundaries and edges on reserve viability (Schonewald-Cox and Bayless, 1986; Schonewald-Cox, 1988; Harris, 1988), and number of reserves (Gilpin and Diamond, 1980). Shape and areal extent may be inadequate considerations by themselves because reserves are generally not self-contained ecosystems (Kushlan, 1979). In any case, it is unlikely that conservation biologists will have the luxury of taking all these factors into consideration to design an "optimum" reserve for a snake community, especially in long-settled parts of the world. Biologists should be cognizant of the effects that design might have on the species to be managed, and attempt to incorporate life-history information into plans for the creation and management of the reserve.

Conserving wildlife on private lands has proved difficult because of concerns for private property rights. Vast land tracts are owned by private companies. For example, millions of hectares are owned by timber companies in the Southeastern United States. Such lands are often the only large tracts of undeveloped land left for wildlife. Government agencies with mandates different from conservation or wildlife management also own millions of hectares. At the same time, it is unlikely that even a fraction of the earth's lands needed to maintain some semblance of biotic diversity will be held in public and private conservation-based ownership.

One way to protect large tracts of land without buying it is to develop conservation easements. Landowners, both private and public, often are willing to conserve wildlife as long as they retain land ownership. An agreement to maintain species or communities can be placed into a formal arrangement as a conservation easement, as long as land-use practices that do not harm the target species are maintained. In return, the landowner gains expertise, assistance in land management, good public relations, and most importantly, tax incentives to maintain natural habitats. Millions of hectares of tropical forest in economically depressed countries have been set aside in exchange for release from international loan debts. Inasmuch as herpetofaunal communities are preserved, the use of such strategies should be encouraged and perhaps expanded to include private debts in more developed countries.

Whether natural habitats are protected in parks, reserves, or in large private holdings, habitat fragmentation is becoming an increasingly important management problem (Wilcox and Murphy, 1985) because isolation increases the chances of extinction among habitat patches or islands (e.g., Pickett and Thompson, 1978; Fahrig and

Merriam, 1985; Wright, 1985). One proposed solution on a landscape scale is to plan for movement corridors between habitat patches. The concept has been widely embraced (Mackintosh, 1989), particularly in Florida (Harris, 1985; Harris and Gallagher, 1989; Lines and Harris, 1989), but is not without valid criticism (Simberloff and Cox, 1987). Much corridor planning has emphasized riparian regions. Certain snake species use riparian habitats (e.g., Rudolph and Dickson, 1990), but it is by no means clear that riparian strips can be used by snakes as movement corridors. In addition, snakes in other habitats, such as surrounding uplands or deserts, will not likely abandon preferred habitats to move between isolated habitat patches (e.g., Szaro and Belfit, 1986). If corridors are to be considered in planning, different habitat types will have to be contained within them.

On a more localized scale, one of the most serious barriers to wildlife dispersal is the increasingly complex system of roads and highways that surround or crisscross nearly all natural habitats (Mader, 1984). The genetic structure of a European frog (*Rana temporaria*) even varies between nearby habitat patches separated by roads (Reh and Seitz, 1990), showing the potential effect of restricted gene flow on a small vertebrate. Roads bisecting snake habitats have resulted in significant mortality (Klauber, 1939; Seigel, 1986; Dodd et al., 1989; R. Franz, personal communication). One way to minimize road mortality is to construct tunnels or culverts under highways to allow passage between habitats. A second way would be to prevent snakes from crawling onto roads through the use of baffles or barriers, such as "snake pit" and modified three-beam guardrail barriers (Fig. 9.1) (Southall, 1991).

Individual-based options

Conservation strategies directed toward individual animals (and thus only indirectly at communities) fall into three areas: restrictions, movement of individuals, and captive propagation. Regulations restricting ownership and collection are the most common forms of wildlife protection. Most such regulations focus on the individual as if the habitat does not exist. For this reason, individual-based regulations may be effective in stopping trade or organized exploitation, such as rattlesnake roundups, but are not effective in protecting ecosystems on which species depend. Misguided regulations can restrict research on species at the same time whole ecosystems are lost to development and massive commercial trade continues unabated (e.g., Conniff, 1990). Restrictions on ownership and research should be based on threats to the species rather than the ease of enforcement or confusion about "animal rights."

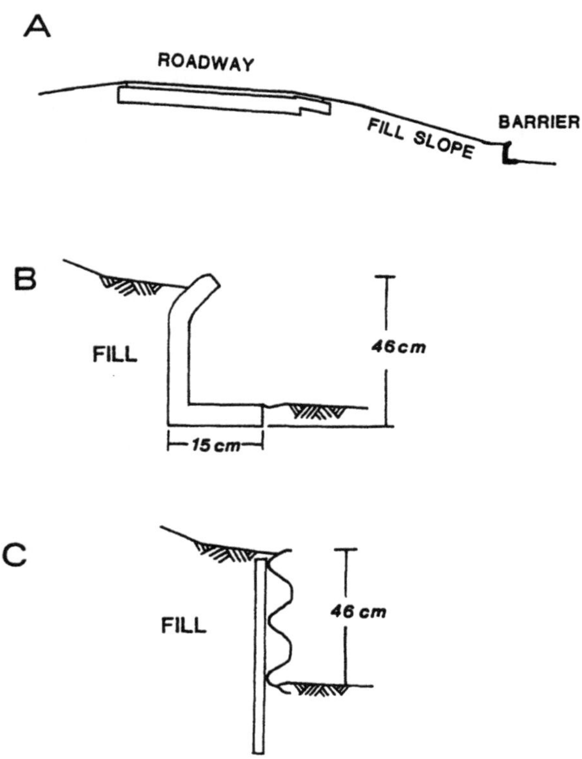

Figure 9.1 Types of wildlife barriers that might prove useful in deterring snakes from crossing a roadway. *A:* Barrier location on roadway. *B:* Snake pit-type barrier. *C:* Modified three-beam guardrail barrier. (Designs reprinted from Southall, 1991, with permission.)

Regulations for endangered species protection may include some provisions for habitat protection, such as land acquisition or ensuring that government agencies do not undertake actions that result in species loss (Yaffee, 1982). However, biologists interested in snake conservation also should become familiar with broad-based regulations, both national and local, that affect reptile communities. Examples include regulations involving the distribution and handling of toxic chemicals and pesticides, wetlands protection, zoning or other forms of land use planning, and water and air quality.

Moving animals from one location to another (relocation, repatriation, translocation, augmentation; see Dodd and Seigel, 1991; Reinert, 1991) may serve to establish or augment populations or remove individuals from harm's way. The only snake that has ever been repatriated is the Eastern Indigo Snake (*Drymarchon corais couperi*) in the southeastern United States. Eastern Indigo Snakes were headstarted, that is, raised from hatchlings to a large size and released into suitable habitat. Details of the project are scant (Dodd and Seigel, 1991) and whether the releases were successful is unknown. Many biological questions that should be considered prior to undertaking movement and headstarting programs often are not adequately considered (Dodd and Seigel, 1991). Moving animals is best considered unproved and experimental. Unless a long-term commitment can be ensured, other conservation measures are more appropriate.

Captive propagation of snakes has been used rather infrequently in connection with habitat-related conservation plans. In a few cases, captive propagation may be critical to ensuring the survival of the species. The best example is the captive breeding program for the Round Island Boa (*Casarea dussumieri*) at the Jersey Wildlife Preservation Trust (Bloxam and Tonge, 1987). Zoos and many private individuals are increasingly taking an organized and systematic approach to snake captive-propagation programs (Hutchins and Wiese, 1991). For example, the American Association of Zoological Parks and Aquariums (AAZPA) maintains a 47-member snake taxon advisory group (STAG) that makes recommendations on action plans and proposals for studbooks. Species survival plans are available for *Acrantophis dumerili* and *Crotalus unicolor*. Several studbooks have been approved (*Acrantophis madagascariensis, Pituophis melanoleucus lodingi, P. m. ruthveni, Ophiophagus hannah, Lachesis muta*) and proposals for four more are pending (*Corallus annulatus, Sanzinia madagascariensis, Bothriechis aurifer, Thamnophis sirtalis tetrataenia*). Other AAZPA STAG committees are examining the status of Mexican, Brazilian, and Caribbean faunas, pit vipers, and large boids (H. Quinn, personal communication).

Captive propagation works as a conservation tool only if provisions are made to protect and conserve species and their habitats in the wild. It has not been demonstrated that amateur herpetoculture, although an interesting and enjoyable pastime, contributes to lessening collection pressure on wild populations. However, there are many responsible breeders and collectors that will purchase only captive-bred animals and keep detailed records on reproduction and behavior. This trend should be encouraged. On the other hand, the vast production of albino, unusually colored, or genetically aberrant individuals does not contribute to conservation.

Education

Education may assist in the future conservation of snake species and communities, but human fear is a powerful force to overcome.

Education is likely to be more effective when aimed at children, because children are less prejudicial than adults who may have ingrained negative feelings toward snakes. On the other hand, remaining snake communities will vanish if it is necessary to wait 20 or 30 years for conservation messages to be effective. Conservation education must be aimed at all segments of society.

Education should take a hands-on approach when dealing with snake conservation. Indeed, one of the biggest problems in teaching environmental awareness and responsibility is the detachment of many people, especially in urban environments, from natural habitats and associated wildlife (Williams, 1988). The reasons various educational exposures do or do not work are complex (Morgan and Gramann, 1989). These authors demonstrated that attitudes toward snakes among elementary students improved significantly when modeling programs were used, especially modeling with exposure to a snake. Direct contact by itself had no effect, possibly because repeated exposures are necessary to make a nonsignificant object (in terms of daily experience) take on a positive connotation. Providing information alone did not improve attitudes although knowledge increased. These results have important implications for snake conservation.

Zoos have increasingly taken on an educational as well as a behind-the-scenes conservation role. Educational materials provided at zoos must be accurate, amenable to relatively quick dissemination, and sensitive to local needs. Zoos, snake parks, and other forms of animal exhibits may be very important as places where people can interact with native wildlife that rarely would be seen otherwise. Conservation of native habitats likely begins with an acquaintance with the local fauna. Local programs can be developed that are effective by integrating both cultural and conservation objectives (e.g., R. Whitaker, 1979; Z. Whitaker, 1985; Whitaker and Andrews, 1992).

Conservation Needs

Identification of priority species and habitats

In 1987, I suggested that several categories of snake species deserved special attention to their conservation and management (Dodd, 1987). I included island species and species with limited geographic ranges, species known to have a small population size, and species that were rare based on good sampling efforts. Species on the geographic limits of their ranges also are prone to extirpation (Dodd, 1987) as are communally denning species, and all species in trade need data on wild populations. The situation in 1992 remains the same, that is, island species are most prone to extinction while most species identified as declining or in need of management fall into one of the other categories. As in 1987, there remains a paucity of studies on wild populations of snakes in commercial trade (e.g., Bhupathy and Vijayan, 1989). Based on my review of snake status, the following should be priority for international concern: (1) island species, particularly in the West Indies (Lillywhite and Henderson, Chap. 1, this volume), Sri Lanka, and Madagascar; (2) European *Vipera*; (3) species in trade, particularly the large boids of Asia and South America; and (4) narrow-ranged endemics.

The highest priority region for snake conservation concern today is the West Indies. The Caribbean and the Bahamas include hundreds of islands, many of which contain unique species and subspecies (Schwartz and Henderson, 1991). Henderson (1992) noted the following extinctions or extirpations of snakes in the West Indies: *Alsophis antiguae* (Antigua), *A. antillensis* (Marie-Galante), *A. melanichnus* (Hispaniola), *A. portoricensis* (St. Thomas), *A. rufiventris* (St. Christopher, Nevis), *A. sanctaecrucis* (St. Croix), *Liophis cursor* (Martinique), *L. juliae* (Marie-Galante), *L. ornatus* (St. Lucia). Many other species are declining or now confined to extremely small islands. Small, diurnal, ground-dwelling, active foraging, oviparous species seem most prone to extirpation. Mongooses and habitat destruction are implicated in declines and extinction (Henderson and Sajdak, 1986; Sajdak and Henderson, 1991; Henderson, 1992).

If West Indian snakes are in trouble, it suggests that other small, heavily populated, islands need conservation assessment and management of their snake communities. Islands in the Mediterranean (e.g., *Vipera schweizeri* on Milos; Corbett, 1989), the Mexican Gulf of California [especially endemic species and subspecies of *Crotalus* sp.; Campbell and Lamar, 1989], Indian Ocean (Andamans [particularly the Andaman Krait, *Bungarus andamanensis*], Sunderbans, Nicobars, Mauritius), Southeast Asia, and Oceania should be surveyed as to the status of their snakes. Large islands (Sri Lanka, Madagascar, Java, the Philippines) also merit special attention because of high species diversity.

On a local level, priority species and communities need to be identified. The analysis might be aided through the use of a species-priority ranking program (Millsap et al., 1990) based on the best available data, particularly museum collections. Subjective assessments should be supplemented by "ground-truthing," employing a wide variety of sampling techniques. Data can then be entered into a geographic information system (GIS) to highlight regions of high diversity (e.g., Millsap et al., 1990). GIS systems facilitate assessment because a variety of data can be stored and manipulated to project impacts once distributions are known (Scott et al., 1987). A GIS system recently has been used to analyze worldwide regions of turtle diversity (A. Kiester, personal communication).

Research priorities

Research should be addressed at questions pertaining to deficiencies in knowledge of snake biology, the effects of various land-use practices on snakes (e.g., Reynolds, 1979), the effects of exotic animals (e.g., Fire Ants, Cane Toads, Cattle Egrets, Armadillos), and at what conservation or management practices will benefit snakes given their life-history requirements. Even the most basic information is needed on distribution, habitat use, environmental physiology, movement patterns, and population size and sex–class structure for many species, especially in the tropics. National and regional biological surveys, such as that undertaken by the Canadian National Museum of Natural Sciences, are urgently needed (Kim and Knutson, 1986). Much of the required information is not based on complex hypothesis testing, but rather on gathering data that may someday lead to applied applications. In short, we need good data on the systematics and natural history of snakes and snake communities (Bartholomew, 1986; Greene and Losos, 1988).

Once basic life-history data are available, conservation applications may become apparent. Brown (1993) studied the population structure, denning, and movement patterns of the Timber Rattlesnake in New York state. Based on the results, den sites were selected as focal points for immediate conservation. However, summer activity areas often were located some distance from the dens. By using radiotelemetry, Brown was able to determine summer foraging areas, migratory routes, and sexual differences in activity and habitat use. These data then were used to develop a management plan suggesting reserve design and the amount of habitat needed to conserve the species (Brown, 1993). Zappalorti and Burger (1985) showed that New Jersey Pine Snakes (*Pituophis melanoleucus*) preferred disturbed sites. A management plan for Pine Snakes in New Jersey should ensure that disturbed sites are created within preserves. Detailed knowledge of distribution and life history is helping to plan conservation strategies for Grass Snakes (*Natrix natrix*) in the Netherlands (Smit and Zuiderwijk, 1991).

There is an urgent need to evaluate what are rapidly becoming "accepted" sampling, conservation, and management techniques. Examples include conservation corridors (will snakes use them?, how wide do they need to be?, in what habitats?), tunnels under roads (what design and size?, over what distance?), headstarting, repatriation, sampling techniques (biases?, trap shyness?, design?), reserve design (shape?, large or small?), restoration (e.g., Humphrey et al., 1985), etc. Do these practices work for snakes, or indeed any species of reptile or amphibian? Do other techniques work better? Now that rabbits have been removed from Round Island in Mauritius (Tonge, 1989) and rats from offshore New Zealand islands (Veitch and Bell, 1990), can similar techniques be used to exterminate introduced predators elsewhere? Carefully planned observations and experiments should be able to answer such questions that unfortunately seem to have been overlooked. Techniques of reducing human–snake conflicts, such as using glueboards to remove snakes from dwellings (Knight, 1986), need to be devised.

Only a small amount of research has been directed to assessing the effects of environmental toxicants on snake communities. Indeed, there is very little information in the literature on the levels of toxicants in snakes (Dodd, 1987). Long-term research needs to monitor changes in populations known to be contaminated by toxicants (e.g., Bagshaw and Brisbin, 1984) and to assess the impacts on snake communities suspected of being affected (e.g., Fleet et al., 1972).

Finally, there is very little information available on the genetics of most snake populations. While electrophoretic and other molecular techniques have been increasingly used to answer systematic questions, most work has focused on broad topics, such as generic relationships. Questions involving genetic variation, the persistence of small populations, particularly on islands, fragmentation effects, and hybridization might be relevant to snake conservation. In addition, genetic information is becoming increasingly important in forensic science as it relates to the wildlife trade (Brazaitis, 1986).

The tie that binds research on life-history and conservation applications together is the need for long-term data. Long-term research is needed to both answer questions and direct future research. It is particularly appropriate for studies that involve processes that are slow, rare or episodic, contain high variability, or are subtle or complex (Franklin, 1989; Magnusson, 1990). These are all factors that affect the conservation of snakes and snake communities. A system of long-term study sites needs to be developed internationally, nationally, and regionally. For example, the U.S. National Science Foundation's Long-Term Ecological Research Program (Swanson and Sparks, 1990) should include herpetofaunal monitoring. Other sites with good baseline data, such as the U.S. Department of Energy's Savannah River Site, the University of Kansas Natural History Reservation, the Katharine Ordway Preserve and Archbold Biological Station, both in Florida, and areas of focused university-based research (e.g., Madsen, 1987; Ford et al., 1991) should be considered for long-term studies. In the Neotropics, studies at Barro Colorado Island (Panama), La Selva

(Costa Rica), Manaus (Brazil), and Cocha Cashu (Peru) should be continued (see Gentry, 1990). Long-term community studies, such as those now under way at Korup National Park in Cameroon (B. Powell, personal communication), are urgently needed in Africa and Asia.

Communication

Once studies are completed and techniques worked out, biologists have the responsibility to communicate their results to both professional peers and the general public. Failure to publish results leads to unnecessary duplication and in many cases advocacy of techniques that are untested and unproved (e.g., Dodd and Seigel, 1991). Popular articles and informal talks are major forms of communication that should not be overlooked.

Status Update

Since 1987, reviews of the status of snakes have appeared for various parts of the world, including Europe (Corbett, 1989; Bruno and Maugeri, 1990), The Netherlands (Zuiderwijk and Smit, 1991), South Africa (Branch, 1988), the West Indies (Henderson and Sajdak, 1986; Sajdak and Henderson, 1991; Henderson, 1992) and Virginia, U.S. (Mitchell and Pague, 1987; Mitchell, 1991). Revised versions of the Red Data Books of Russia (Bannikov and Darevskii, 1985) and the former USSR (Bannikov and Darevskii, 1984) are now available. Darevskii and Orlov (1988) provide a general worldwide review, while Shine (1991) discusses snake conservation in Australia. Conant (1992) reviewed the status of snakes in the *Agkistrodon* complex and Lillywhite (1991) reviewed *Acrochordus*. The most serious news is the declining status of West Indian snakes, although previous suggestions that *Chironius vincenti* was extinct (Dodd, 1987) have proven premature (Henderson et al., 1988, 1992). Henderson et al. (1992) also report that *Alsophis antillensis* may still occur on the Grand Terre portion of Guadeloupe. Unless conservation actions are initiated immediately, many West Indian snakes will probably not survive into the 21st century.

Summary

The status of snakes is generally poorly known despite the large amount of recent interest and research in conservation biology. Concern for snake species results from a combination of personal experience, from perceptions of the volume of trade or rarity of a few species, and from specific easily recognized causes such as habitat destruction. Research contributes directly to conservation by providing baseline data on the biological constraints on which management programs depend. More research is necessary to determine if the techniques advocated for the conservation of other species and habitat reserves will work for snakes. Biologists should question the advocacy and adoption of conservation techniques to ensure that programs are carefully planned and appropriate to the species. A clear definition of goals and a strong commitment to long-term research and reserve management is required.

Very little research has been conducted on population viability analysis of snake species. In part, this results from a lack of available data. Snakes are difficult to study in the field, and relatively few investigators have attempted to obtain the types of data required for viability analysis. However, there now appears to be sufficient data for at least preliminary models involving species such as *Vipera berus, Thamnophis sirtalis,* and others. With the publication of this book and its predecessor, much renewed interest has been and hopefully will continue to be expressed in the study of snakes' life histories. Researchers are now in an exciting position to extend their studies beyond the presentation of field results.

A more serious problem, perhaps, is that most snake researchers are unfamiliar with the advances in population viability analysis undertaken on a variety of species within the recent past. For that matter, few herpetologists have delved into this expanding field. However, new software is available that will help biologists relate their field results to the practical application of species' conservation. The intellectual and professional challenges are just as rewarding as traditional population modeling. We need to change our attitudes concerning conservation and stimulate our interest into its advances and challenges. Conservation is not "soft" science.

This chapter, perhaps in contrast to the others in this book, seems to emphasize what we do not know about snakes. While that is true, we know a great deal more about the biology of snakes than we did a few years ago. Conservation will benefit as a result of the basic research on snakes discussed in each of the previous chapters. I hope that this chapter will point to questions that need attention and to stimulate additional work and interest, especially among graduate students. I also hope that biologists will come to view past results in a broader context and to reexamine their results in terms of their applicability to conservation and management problems.

Acknowledgments

I thank W. S. Brown, J. L. Carr, R. Franz, J. W. Gibbons, R. W. Henderson, K. B. Jones, H. Quinn, R. Sajdak, R. Shine, and R. A.

Seigel for information on snake conservation. S. S. Novak and T. Novikov translated Russian language information. D. Auth, R. Franz, R. Henderson, H. Lillywhite, R. Noss, and R. A. Seigel provided comments on early versions of the manuscript.

Literature Cited

Abbey, E., 1989, *A Voice Crying in the Wilderness (Vox Clamantis in Deserto),* St. Martin's, New York.

Aleksiuk, M., and P. T. Gregory, 1974, Regulation of seasonal mating behavior in *Thamnophis sirtalis parietalis, Copeia,* 1974:681–689.

Andrews, J., and C. Birkinshaw, 1988, India's snakeskin trade, *IUCN Traffic Bull.,* 9:66–77.

Bagshaw, C., and I. L. Brisbin, Jr., 1984, Long-term declines in radiocesium of two sympatric snake populations, *J. Appl. Ecol.,* 21:407–413.

Bannikov, A. S., and I. S. Darevskii, 1984, Amphibians and reptiles, in *Red Book of the USSR, Rare and Endangered Species of Animals and Plants,* Vol. 1, 2nd ed., Lesnaja promish Lennost, Moscow, pp. 171–201 [in Russian].

Bannikov, A. S., and I. S. Darevskii, 1985, Reptiles, in *Red Book of the Russian Federal Socialist Republic, Animals,* Rosselkhozizdat, Moscow, pp. 351–366 [in Russian].

Bartholomew, G. A., 1986, The role of natural history in contemporary biology, *Bioscience,* 36:324–329.

Bhupathy, S., and V. S. Vijayan, 1989, Status, distribution and general ecology of the Indian python *Python molurus molurus* Linn. in Keoladeo National Park, Bharatpur, Rajasthan, *J. Bombay Nat. Hist. Soc.,* 86:381–387.

Blouin, M. S., and E. F. Connor, 1985, Is there a best shape for nature reserves?, *Biol. Conserv.,* 32:277–288.

Bloxam, Q. M. C., and S. J. Tonge, 1987, The Round Island boa, *Casarea dussumieri,* breeding programme at the Jersey Wildlife Preservation Trust, *Dodo, J. Jersey Wildl. Preserv. Trust,* 23:101–107.

Branch, W. R. (ed.), 1988, *South African Red Data Book—Amphibians and Reptiles,* South African Nat. Sci. Prog. Rep. No. 151.

Brazaitis, P., 1986, Reptile leather trade: The forensic science examiner's role in litigation and wildlife law enforcement, *J. Forensic Sci.,* 31:621–629.

Brown, W. S., 1991, Female reproductive ecology in a northern population of the timber rattlesnake, *Crotalus horridus, Herpetologica,* 47:101–115.

Brown, W. S., 1993, Biology, status, and management of the timber rattlesnake (*Crotalus horridus*): A guide for conservation, *SSAR Herpetol. Circular* 22:1–78.

Brown, W. S., D. W. Pyle, K. R. Greene, and J. B. Friedlaender, 1982, Movements and temperature relationships of timber rattlesnakes (*Crotalus horridus*) in northeastern New York, *J. Herpetol.,* 16:151–161.

Bruno, S., and S. Maugeri, 1990, *Serpenti D'Italia e D'Europa,* Editoriale Giorgio Mondadori, Milano.

Campbell, J. A., and W. W. Lamar, 1989, *The Venomous Reptiles of Latin America,* Comstock, Ithaca, New York.

Campbell, J. A., D. R. Formanowicz, Jr., and E. D. Brodie, Jr., 1989, Potential impact of rattlesnake roundups on natural populations, *Texas J. Sci.,* 41:301–317.

Conant, R., 1992, Comments on the survival status of members of the *Agkistrodon* complex, in P. D. Strimple and J. L. Strimple, eds., *Contributions in Herpetology,* Greater Cincinnati Herp. Soc., Cincinnati, Ohio, pp. 29–33.

Conniff, R., 1990, Fuzzy-wuzzy thinking about animal rights, *Audubon* (Nov):120–122, 129–133.

Corbett, K., 1989, *Conservation of European Reptiles and Amphibians,* Christopher Helm, London.

Corke, D., 1987, Reptile conservation on the Maria Islands (St. Lucia, West Indies), *Biol. Conserv.,* 40:263–279.

Darevskii, I. S., and N. L. Orlov, 1988, *Rare and Vanishing Animals: Amphibians and Reptiles,* Vyshaia Shkola, Moscow [in Russian].
Dodd, C. K., Jr., 1980, Money for research in the Federal Endangered Species Program, *Herpetol. Rev.,* 11:70–72.
Dodd, C. K., Jr., 1987, Status, conservation, and management, in R. A. Seigel, J. T. Collins, and S. S. Novak, eds., *Snakes: Ecology and Evolutionary Biology,* McGraw-Hill, New York, pp. 478–513.
Dodd, C. K., Jr., and R. A. Seigel, 1991, Relocation, repatriation, and translocation of amphibians and reptiles: Are they conservation strategies that work?, *Herpetologica,* 47:336–350.
Dodd, C. K., Jr., K. M. Enge, and J. N. Stuart, 1989, Reptiles on highways in north-central Alabama, USA, *J. Herpetol.,* 23:197–200.
Fahrig, L., and G. Merriam, 1985, Habitat patch connectivity and population survival, *Ecology,* 66:1762–1768.
Fitzgerald, S., 1989, *International Wildlife Trade: Whose Business Is It?,* World Wildlife Fund, Washington, D.C.
Fleet, R. R., D. R. Clark, Jr., and F. W. Plapp, Jr., 1972, Residues of DDT and dieldrin in snakes from two Texas agro-systems, *Bioscience,* 22:664–665.
Foose, T. J., 1993, Genetic and demographic management of small populations, in J. B. Murphy, J. T. Collins, and K. Adler, eds., *Captive Management and Conservation of Amphibians and Reptiles—A Tribute to Roger Conant,* SSAR Contrib. Herpetol., in press.
Ford, N. B., V. A. Cobb, and J. Stout, 1991, Species diversity and seasonal abundance of snakes in a mixed pine-hardwood forest of eastern Texas, *Southwest. Nat.,* 36:171–177.
Frankel, O. H., and M. E. Soulé, 1981, *Conservation and Evolution,* Cambridge Univ. Press, Cambridge.
Franklin, J. F., 1989, Importance and justification of long-term studies in ecology, in G. E. Likens, ed., *Long-term Studies in Ecology,* Springer, New York, pp. 3–19.
Fritts, T. H., 1988, The brown tree snake, *Boiga irregularis,* a threat to Pacific islands, *U.S. Fish Wildl. Serv. Biol. Rep.,* 88(31):1–36.
Gentry, A. H. (ed.), 1990, *Four Neotropical Rainforests,* Yale Univ. Press, New Haven, Connecticut.
Gerrodette, T., and W. G. Gilmartin, 1990, Demographic consequences of changed pupping and hauling sites of the Hawaiian monk seal, *Conserv. Biol.,* 4:423–430.
Gillingham, J. C., 1987, Social behavior, in R. A. Seigel, J. T. Collins, and S. S. Novak, eds., *Snakes: Ecology and Evolutionary Biology,* McGraw-Hill, New York, pp. 184–209.
Gilpin, M. E., 1987, Spatial structure and population vulnerability, in M. E. Soulé, ed., *Viable Populations for Conservation,* Cambridge Univ. Press, Cambridge, pp. 125–139.
Gilpin, M. E., and J. M. Diamond, 1980, Subdivision of nature reserves and the maintenance of species diversity, *Nature,* 285:567–568.
Gilpin, M. E., and M. E. Soulé, 1986, Minimum viable populations: Processes of species extinction, in M. E. Soulé, ed., *Conservation Biology: The Science of Scarcity and Diversity,* Sinauer, Sunderland, Massachusetts, pp. 19–34.
Greene, H. W., and J. B. Losos, 1988, Systematics, natural history, and conservation, *Bioscience,* 38:458–462.
Gregory, P. T., 1977, Life-history parameters of the red-sided garter snake (*Thamnophis sirtalis parietalis*) in an extreme environment in the Interlake Region of Manitoba, *Nat. Mus. Canada Publ. Zool.,* 13:1–44.
Gregory, P. T., 1984, Communal denning in snakes, in R. A. Seigel, L. E. Hunt, J. L. Knight, L. Malaret and N. L. Zuschlag, eds., *Vertebrate Ecology and Systematics—A Tribute to Henry S. Fitch,* Univ. Kansas Spec. Publ. No. 10, pp. 57–75.
Gregory, P. T., J. M. MacCartney, and K. W. Larsen, 1987, Spatial patterns and movement, in R. A. Seigel, J. T. Collins, and S. S. Novak, eds., *Snakes: Ecology and Evolutionary Biology,* McGraw-Hill, New York, pp. 366–395.
Hales, D., 1989, Changing concepts of national parks, in D. Western and M. Pearl, eds.,

Conservation for the Twenty-first Century, Oxford Univ. Press, New York, pp. 139–144.

Harris, L. D., 1985, Conservation corridors: A highway system for wildlife, *ENFO (Florida Conserv. Found.),* 85-5:1–10.

Harris, L. D., 1988, Edge effects and the conservation of biodiversity, *Conserv. Biol.,* 2:330–332.

Harris, L. D., and P. B. Gallagher, 1989, New initiatives for wildlife conservation: The need for movement corridors, in G. Mackintosh, ed., *In Defense of Wildlife: Preserving Communities and Corridors,* Defenders of Wildlife, Washington, D.C., pp. 11–34.

Harris, L. D., and P. Kangas, 1988, Reconsideration of the habitat concept, *Trans. 53rd N. A. Wildl. Nat. Res. Conf.,* pp. 137–144.

Henderson, R. W., 1992, Consequences of predator introductions and habitat destruction on amphibians and reptiles in the post-Columbus West Indies, *Carib. J. Sci.,* 28:1–10.

Henderson, R. W., and R. A. Sajdak, 1986, West Indian racers: A disappearing act or a second chance? *Lore,* 36(3):13–18.

Henderson, R. W., R. A. Sajdak, and R. M. Henderson, 1988, The rediscovery of the West Indian colubrid snake *Chironius vincenti, Amphibia-Reptilia,* 9:415–416.

Henderson, R. W., J. Daudin, G. T. Haas, and T. J. McCarthy, 1992, Significant distribution records for some amphibians and reptiles in the Lesser Antilles, *Carib. J. Sci.,* 28:101–103.

Humphrey, S. H., J. F. Eisenberg, and R. Franz, 1985, Possibilities for restoring wildlife of a longleaf pine savanna in an abandoned citrus grove, *Wildl. Soc. Bull.,* 13:487–496.

Hutchins, M., and R. J. Wiese, 1991, Beyond genetic and demographic management: The future of the Species Survival Plan and related AAZPA conservation efforts, *Zoo Biol.,* 10:285–292.

Kim, K. C., and L. Knutson (eds.), 1986, *Foundations for a National Biological Survey,* Assoc. Systematic Collections, Lawrence, Kansas.

Klauber, L. M., 1939, Studies of reptile life in the arid southwest, Part 1. Night collecting on the desert with ecological statistics, *Bull. Zool. Soc. San Diego,* 14:7–64.

Knight, J. E., 1986, A humane method for removing snakes from dwellings, *Wildl. Soc. Bull.,* 14:301–303.

Kohm, K. A. (ed.), 1991, *Balancing on the Brink of Extinction: The Endangered Species Act and Lessons for the Future,* Island, Washington, D.C.

Kushlan, J. A., 1979, Design and management of continental wildlife reserves: Lessons from the Everglades, *Biol. Conserv.,* 15:281–290.

Lacy, R. C., 1987, Loss of genetic diversity from managed populations: Interacting effects of drift, mutation, immigration, selection, and population subdivision, *Conserv. Biol.,* 1:143–158.

Lande, R., 1988, Genetics and demography in biological conservation, *Science,* 241:1455–1460.

Lillywhite, H. B., 1987, Temperature, energetics, and physiological ecology, in R. A. Seigel, J. T. Collins, and S. S. Novak, eds., *Snakes: Ecology and Evolutionary Biology,* McGraw-Hill, New York, pp. 422–477.

Lillywhite, H. B., 1991, The biology and conservation of acrochordid snakes, *Hamadryad,* 16:1–9.

Lines, L. G., Jr., and L. D. Harris, 1989, Isolation of nature reserves in north Florida: Measuring linkage exposure, *Trans. 54th N. A. Wild. Nat. Res. Conf.,* pp. 113–120.

Luxmoore, R., B. Groombridge, and S. Broad (eds.), 1988, *Significant Trade in Wildlife: A Review of Selected Species in CITES Appendix II. Reptiles and Invertebrates,* Vol. 2, IUCN Conservation Monitoring Centre, Cambridge.

Mace, G. M., and R. Lande, 1991, Assessing extinction threats: Toward a reevaluation of IUCN threatened species categories, *Conserv. Biol.,* 5:148–157.

Mackintosh, G. (ed.), 1989, *In Defense of Wildlife: Preserving Communities and Corridors,* Defenders of Wildlife, Washington, D.C.

Mader, H.-J., 1984, Animal habitat isolation by roads and agricultural fields, *Biol. Conserv.,* 29:81–96.

Madsen, T., 1987, Natural and sexual selection in grass snakes, *Natrix natrix,* and adders, *Vipera berus,* Doctoral Thesis, Univ. Lund, Sweden.

Magnusson, J. J., 1990, Long-term research and the invisible present, *Bioscience,* 40:495–501.

Mathews, A. E., 1989, Conflict, controversy, and compromise: The Concho water snake (*Nerodia harteri paucimaculata*) versus the Stacy Dam and Reservoir, *Environ. Manage.,* 13:297–307.

McNeely, J. A., 1989, Protected areas and human ecology: How national parks can contribute to sustaining societies in the twenty-first century, in D. Western and M. Pearl, eds., *Conservation for the Twenty-First Century,* Oxford Univ. Press, New York, pp. 150–157.

Millsap, B. A., J. A. Gore, D. E. Runde, and S. I. Cerulean, 1990, Setting priorities for the conservation of fish and wildlife species in Florida, *Wildl. Monogr.,* 111:1–57.

Mitchell, J. C., 1991, Amphibians and reptiles, in *Virginia's Endangered Species,* McDonald and Woodward, Blacksburg, Virginia, pp. 411–476.

Mitchell, J. C., and C. A. Pague, 1987, A review of reptiles of special concern in Virginia, *Virginia J. Sci.,* 38:319–328.

Mittermeier, R. A., and J. L. Carr, 1993, Conservation of reptiles and amphibians: A global perspective, in J. B. Murphy, J. T. Collins, and K. Adler, eds., *Captive Management and Conservation of Amphibians and Reptiles—A Tribute to Roger Conant,* SSAR Contrib. Herpetol., in press.

Mittermeier, R. A., J. L. Carr, I. R. Swingland, T. B. Werner, and R. B. Mast, 1992, Conservation of amphibians and reptiles, in K. Adler, ed., *Herpetology: Current Research on the Biology of Amphibians and Reptiles,* SSAR Contrib. Herpetol., 9:59–80.

Morgan, J. M., and J. H. Gramann, 1989, Predicting effectiveness of wildlife education programs: A study of students' attitudes and knowledge toward snakes, *Wildl. Soc. Bull.,* 17:501–509.

Morris, R., and D. Morris, 1965, *Men and Snakes,* Hutchinson, London.

Mount, R. H., 1981, The red imported fire ant, *Solenopsis invicta* (Hymenoptera: Formicidae), as a possible serious predator on some native Southeastern vertebrates: Direct observations and subjective impressions, *J. Alabama Acad. Sci.,* 52:71–78.

Mushinsky, H. R., 1987, Foraging ecology, in R. A. Seigel, J. T. Collins, and S. S. Novak, eds., *Snakes: Ecology and Evolutionary Biology,* McGraw-Hill, New York, pp. 302–334.

Noss, R. F., 1990, Indicators for monitoring biodiversity: A hierarchical approach, *Conserv. Biol.,* 4:355–364.

Parker, W. S., and M. V. Plummer, 1987, Population ecology, in R. A. Seigel, J. T. Collins, and S. S. Novak, eds., *Snakes: Ecology and Evolutionary Biology,* McGraw-Hill, New York, pp. 253–301.

Pickett, S. T. A., and J. N. Thompson, 1978, Patch dynamics and the design of nature reserves, *Biol. Conserv.,* 13:27–37.

Pough, F. H., 1983, Amphibians and reptiles as low-energy systems, in W. P. Aspey and S. I. Lustick, eds., *Behavioral Energetics: The Cost of Survival in Vertebrates,* Ohio State Univ. Press, Columbus, pp. 141–188.

Reh, W., and A. Seitz, 1990, The influence of land use on the genetic structure of populations of the common frog *Rana temporaria, Biol. Conserv.,* 54:239–249.

Reinert, H. K., 1990, A profile and impact assessment of organized rattlesnake hunts in Pennsylvania, *J. Pennsylvania Acad. Sci.,* 64:136–144.

Reinert, H. K., 1991, Translocation as a conservation strategy for amphibians and reptiles: Some comments, concerns, and observations, *Herpetologica,* 47:357–363.

Reynolds, T. D., 1979, Response of reptile populations to different land management practices on the Idaho National Engineering Laboratory Site, *Great Basin Nat.,* 39:255–262.

Rohlf, D. J., 1991, Six biological reasons why the Endangered Species Act doesn't work—and what to do about it, *Conserv. Biol.,* 5:273–282.

Rudolph, D.C., and J. G. Dickson, 1990, Streamside zone width and amphibian and reptile abundance, *Southwest. Nat.,* 35:472–476.

Sajdak, R. A., and R. W. Henderson, 1991, Status of West Indian racers in the Lesser Antilles, *Oryx,* 25:33–38.
Samson, F. B., 1983, Minimum viable populations—a review, *Nat. Areas J.,* 3:15–23.
Samson, F. B., F. Perez-Trejo, H. Salwasser, L. F. Ruggiero, and M. L. Shaffer, 1985, On determining and managing minimum population size, *Wildl. Soc. Bull.,* 13:425–433.
Savidge, J. A., 1987, Extinction of an island forest avifauna by an introduced snake, *Ecology,* 68:660–668.
Schlauch, F. C., 1978, New methodologies for measuring species status and their application to the herpetofauna of a suburban region, *Englehardtia,* 6:30–41.
Schonewald-Cox, C., 1988, Boundaries in the protection of nature reserves, *Bioscience,* 38:480–486.
Schonewald-Cox, C., and J. W. Bayless, 1986, The boundary model: A geographical analysis of design and conservation of nature reserves, *Biol. Conserv.,* 38:305–322.
Schonewald-Cox, C. M., S. M. Chambers, B. MacBryde, and L. Thomas (eds.), 1983, *Genetics and Conservation: A Reference for Managing Wild Animal and Plant Populations,* Benjamin/Cummings, Menlo Park, California.
Schwartz, A., and R. W. Henderson, 1991, *Amphibians and Reptiles of the West Indies,* Univ. Florida Press, Gainesville, Florida.
Scott, J. M., B. Csuti, J. D. Jacobi, and J. E. Estes, 1987, Species richness, *Bioscience,* 37:782–788.
Scott, N. J., Jr., and R. A. Seigel, 1992, The management of amphibian and reptile populations: Species priorities and methodological and theoretical constraints, in D. R. McCullough and R. H. Barrett, eds., *Wildlife 2001: Populations,* Elsevier Applied Science, London, pp. 343–368.
Scott, N. J., Jr., T. C. Maxwell, O. W. Thornton, Jr., L. A. Fitzgerald, and J. W. Flury, 1989, Distribution, habitat, and future of Harter's water snake, *Nerodia harteri,* in Texas, *J. Herpetol.,* 23:373–389.
Seigel, R. A., 1986, Ecology and conservation of an endangered rattlesnake, *Sistrurus catenatus,* in Missouri, U.S.A., *Biol. Conserv.,* 35:333–346.
Seigel, R. A., M. M. Huggins, and N. B. Ford, 1987, Reduction in locomotor ability as a cost of reproduction in gravid snakes, *Oecologia (Berlin),* 73:481–485.
Shafer, C. L., 1990, *Nature Reserves: Island Theory and Conservation Practice,* Smithsonian Inst. Press, Washington, D.C.
Shaffer, M. L., 1981, Minimum population sizes for species conservation, *Bioscience,* 31:131–134.
Shaffer, M. L., and F. B. Samson, 1985, Population size and extinction: A note on determining critical population sizes, *Amer. Nat.,* 125:144–152.
Shine, R., 1991, *Australian Snakes, A Natural History,* Cornell Univ. Press, Ithaca, New York.
Shine, R., and M. Fitzgerald, 1989, Conservation and reproduction of an endangered species: The broad-headed snake, *Hoplocephalus bungaroides* (Elapidae), *Aust. Zool.,* 25:65–67.
Simberloff, D., and J. Cox, 1987, Consequences and costs of conservation corridors, *Conserv. Biol.,* 1:63–71.
Smit, G., and A. Zuiderwijk, 1991, Nieuwland voor de Ringslang, *De Levende Natuur,* 1991(6):212–222.
Soulé, M. E., 1983, What do we really know about extinction?, in C. M. Schonewald-Cox, S. M. Chambers, B. MacBryde, and L. Thomas, eds., *Genetics and Conservation: A Reference for Managing Wild Animal and Plant Populations,* Benjamin/Cummings, Menlo Park, California, pp. 111–124.
Soulé, M. E. (ed.), 1986, *Conservation Biology: The Science of Scarcity and Diversity,* Sinauer, Sunderland, Massachusetts.
Soulé, M. E. (ed.), 1987, *Viable Populations for Conservation,* Cambridge Univ. Press, New York.
Soulé, M. E., and D. Simberloff, 1986, What do genetics and ecology tell us about the design of nature reserves? *Biol. Conserv.,* 35:19–40.
Soulé, M. E., and B. A. Wilcox (eds.), 1980, *Conservation Biology,* Sinauer, Sunderland, Massachusetts.

Southall, P. D., 1991, *The Relationship Between Wildlife and Highways in the Paynes Prairie Basin,* Florida Dept. Transportation, District 2, Lake City, Florida.

Spellerberg, I. F., 1988, Ecology and management of reptile populations in forests, *Quart. J. For.,* 82:99–109.

Steinhart, P., 1984, Fear of snakes, *Audubon,* 86:2, 8–9.

Swanson, F. J., and R. E. Sparks, 1990, Long-term ecological research and the invisible place, *Bioscience,* 40:502–508.

Szaro, R. C., and S. C. Belfit, 1986, Herpetofaunal use of a desert riparian island and its adjacent scrub habitat, *J. Wildl. Manage.,* 50:752–761.

Tangley, L., 1988, Research priorities for conservation, *Bioscience,* 38:444–448.

Thibodeau, F. R., 1983, Endangered species: Deciding which species to save, *Environ. Manage.,* 7:101–107.

Tokunaga, S., Y. Ono, and S. Akagishi, 1991, The Iwakuni Shirohebis, a group of albino *Elaphe climacophora, Herpetol. Rev.,* 22:120–121.

Tonge, S., 1989, A preliminary account of changes in reptile populations on Round Island following the eradication of rabbits, *Dodo, J. Jersey Wildl. Preserv. Trust,* 26:8–17.

Veitch, C. R., and B. D. Bell, 1990, Eradication of introduced animals from the islands of New Zealand, in D. R. Towns, C. H. Daugherty, and I. A. E. Atkinson, eds., *Ecological Restoration of New Zealand Islands,* Conserv. Sci. Publ., 2, Dept. Conservation, Wellington, pp. 137–146.

Warwick, C., 1990, Disturbance of natural habitats arising from rattlesnake round-ups, *Environ. Conserv.,* 17:172–174.

Weir, J., 1992, The Sweetwater rattlesnake round-up: A case study in environmental ethics, *Conserv. Biol.,* 6:116–127.

Western, D., 1989, Conservation without parks: Wildlife in the rural landscape, in D. Western and M. Pearl, eds., *Conservation for the Twenty-First Century,* Oxford Univ. Press, New York, pp. 158–165.

Western, D., and M. Pearl (eds.), 1989, *Conservation for the Twenty-First Century,* Oxford Univ. Press, New York.

Whitaker, R., 1979, The Madras Snake Park: Its role in public education and reptile research, *Int. Zoo Yb.,* 19:31–38.

Whitaker, R., and H. Andrews, 1992, The Madras Crocodile Bank, in P. D. Strimple and J. L. Strimple, eds., *Contributions in Herpetology,* Greater Cincinnati Herp. Soc., Cincinnati, Ohio, pp. 77–83.

Whitaker, Z., 1979, Artful catchers, deadly prey, *Int. Wildl.,* (Mar/Apr):26–33.

Whitaker, Z., 1985, A snake park as a conservation centre, *Oryx,* 19:17–21.

White, P. S., and S. P. Bratton, 1980, After preservation: Philosophical and practical problems of change, *Biol. Conserv.,* 18:241–255.

Wilcox, B. A., and D. D. Murphy, 1985, Conservation strategy: The effects of fragmentation on extinction, *Amer. Nat.,* 125:879–887.

Williams, T., 1988, Why Johnny shoots stop signs, *Audubon,* (Sep):112, 114, 116–121.

Williams, T., 1990, Driving out the dread serpent, *Audubon,* (Dec):26–28, 30–32.

Wilson, E. O., 1985, The biological diversity crisis, *Bioscience,* 35:700–706.

Wilson, E. O. (ed.), 1988, *Biodiversity,* National Academy Press, Washington, D.C.

Woodruff, D. S., 1989, The problems of conserving genes and species, in D. Western and M. Pearl, eds., *Conservation for the Twenty-First Century,* Oxford Univ. Press, New York, pp. 76–88.

Wright, S. J., 1985, How isolation affects rates of turnover of species on islands, *Oikos,* 44:331–340.

Yaffee, S. L., 1982, *Prohibitive Policy: Implementing the Federal Endangered Species Act,* MIT Press, Cambridge, Massachusetts.

Zappalorti, R. T., and J. Burger, 1985, On the importance of disturbed sites to habitat selection by pine snakes in the Pine Barrens of New Jersey, *Environ. Conserv.,* 12:358–361.

Zuiderwijk, A., and G. Smit, 1991, De Nederlandse slangen in de jaren tachtig, *Lacerta,* 49:43–60.

Chapter

10

Summary: Future Research on Snakes, or How to Combat "Lizard Envy"

Richard A. Seigel

Introduction

During the American Society of Ichthyologists and Herpetologists meetings at the University of Victoria in 1986, I was asked to give the summary address at the first-ever symposium on snake ecology and behavior. This excellent symposium, organized by Neil Ford and Patrick Gregory, brought together an impressive group of researchers, including Richard Shine, Patrick Gregory, Neil Ford, Henry Fitch, William Brown, Hubert Saint Girons, Charles Peterson, Thomas Madsen, and many others. I recall vividly listening to their talks, and wondering what I could possibly say that would add to what these distinguished biologists had already discussed.

Just a few hours before I was scheduled to speak, it finally struck me that I had heard a recurring theme throughout many of the papers. A frequent comment was "well, in comparison to lizards my data are meager, but...." This apologetic tone led me to the main theme of my talk in 1986, and the main message for this chapter, i.e., how do we combat what I termed "lizard envy"?

Roots of Lizard Envy

The idea that snakes are poor research animals in comparison to so-called model organisms (Huey et al., 1983) is probably a long-standing tradition, but recent ideas may stem from a frequently cited paper by Turner (1977), which reviewed studies of reproduction and demography of lizards, snakes, crocodilians, and tuataras. The low recapture rate among most of the studies on snakes (especially for juveniles), combined with high variability in density and survival estimates, led Turner to conclude that "one is left with distinct reservations about the suitability of snake populations for this sort of ecological endeavor" (Turner, 1977, p. 228).

Later authors seconded Turner's view. For example, Parker and Plummer (1987) noted that "no single study on snake populations measures up to numerous studies on lizards or mammals." Vitt (1987) suggested that despite considerable effort, studies on snakes have contributed little to our understanding of community ecology. Considering that both of these papers were in the original *Snakes: Ecology and Evolutionary Biology* volume (coedited by Joseph T. Collins, Susan S. Novak, and myself), it seems that even specialists who work with snakes have accepted Turner's conclusions.

I do not dispute the specific problems identified by these workers; snakes can (and often are) difficult to work with. However, this is not the issue; as I argue below, all species are difficult to work with for some kinds of studies but are ideally suited for others. The real problem with lizard envy stems from the mindset that such statements engender. For example, when I was a new graduate student at the University of Kansas, Henry Fitch encouraged me to make the transition from studying turtle ecology to working with snakes. Having just read Turner's (1977) paper, I asked Fitch "can you really do good science with snakes?" Obviously, Fitch convinced me that you can (rather quickly, I might add), but I wonder how many young biologists have been turned away from imaginative research on snakes by the negative attitude described above.

Combating Lizard Envy

Here, I argue that lizard envy results from (1) not recognizing the limitations of snakes for certain kinds of studies, (2) not using different and innovative techniques when it is apparent that traditional techniques are inadequate, (3) not properly matching question, study animal, and technique, and (4) not focusing on the aspects of snake biology that make them "model" organisms for certain kinds of research.

Recognizing the limitations of studying snakes

The main problems that authors associate with working with snakes are as follows: (1) snakes exist in apparently low densities, making obtaining adequate sample sizes difficult; (2) snakes are difficult to observe in the field, making behavioral studies difficult, (3) snakes are prone to long periods of inactivity, and (4) snakes feed infrequently, reducing sample sizes for studies on foraging and community ecology (Turner, 1977; Parker and Plummer, 1987; Vitt, 1987). Are these problems real? Yes. Do they preclude high-quality science? No. The perception that snakes are "intractable" study animals (Parker and Plummer, 1987) may be largely a function of not properly matching the question you are seeking to answer with the appropriate species and/or group. For example, the difficulty in observing snakes in the field is quite real for many, if not most species. What this tells me is that if one is interested in observing behavior in the field, lizards are (in general) a better choice. What I suspect happens is that students interested in field behavior (as an example) attempt to conduct such studies using locally available species. Because a randomly selected snake is probably not suitable for such efforts, the result is that the data are poor or inadequate, reinforcing the contention that snakes are intractable study animals. Similar difficulties exist for studies of survivorship and demography if species are not selected carefully (see below).

Using new and innovative techniques

I suggest that many of the apparent difficulties in working with snakes result not as much from snakes being "intractable" as from the use of traditional techniques that are sometimes inefficient or inappropriate for the particular species under study. For example, demographic studies of snakes frequently suffer from a low recapture rate (Parker and Plummer, 1987). In many cases, this may be the result of the use of hand captures as a source of specimens rather than through the use of traps for adults and juveniles (Fitch, 1987). The same reliance on hand captures may also explain the low apparent density (and hence, sample size) of snakes for field studies. Quantitative data are lacking on these ideas (few, if any, studies have compared recapture rates between techniques), but two anecdotal examples from the Savannah River Ecology Laboratory (SREL), South Carolina, illustrate my point.

Researchers at SREL have long placed heavy reliance on traps and drift fences rather than hand captures (Gibbons and Semlitsch, 1978). Use of these techniques have provided a very different picture on snake abundance and recapture rates than achieved by collecting snakes by hand. For example, use of drift fences and aquatic minnow traps at Ellenton Bay between 1983 and 1992 resulted in the capture of almost 1000 aquatic snakes (*Farancia abacura, F. erytrogramma, Nerodia fasciata, N. floridana,* and *Seminatrix pygaea*), species almost never captured by hand during these years, despite large numbers of person-hours spent at the study area (Seigel and Gibbons, unpublished). Use of funnel traps along drift fences also resulted not only in large numbers of Racers (*Coluber constrictor*) between 1983 and 1987, but also in a recapture rate that approached 100% in some years (Seigel, unpublished). In comparison, Parker and Plummer (1987) found that most recapture rates for snakes leveled off at about 15%.

I do not suggest that the high recapture rates at SREL were the result of any special talent on our part. Indeed, I found the recapture rate among hand-collected Ribbon and Garter Snakes (*Thamnophis proximus, T. sirtalis,* and *T. radix*) in Missouri was less than 10% after four years of effort. Instead, the high recapture rates at SREL reflected the fact that Racers continually reused portions of Ellenton Bay as foraging sites and were easily recaptured by the drift fence as they approached the bay to forage (M. Plummer, personal communication). Nor do I suggest that drift fences are a panacea for all the difficulties encountered in working with snakes nor that traps are the only technique that works well for demographic studies. Instead, I argue that the use of traps and drift fences illustrate an important lesson; if one technique does not work, try another. If we have learned that, in many cases, hand captures are an inefficient means of sampling for demographic studies of snakes, then we should reduce the emphasis on hand-collecting and emphasize other techniques. However, the message that actually has been sent seems to be that snakes are not suitable for this research (Turner, 1977). I would argue that such a conclusion is both premature and misleading.

Synthesis: Matching question, technique, and species

From the previous discussion, it is clear that what we need is a better match between the question, the technique, and the study animal. For example, although I noted above that snakes are not generally suitable for observing behavior in the wild, not all snakes are difficult to observe under field conditions, and some offer spectacular opportunities. The large mating aggregations of the Red-Sided Garter Snake (*Thamnophis sirtalis parietalis*) are a good illustration; D. Crews and associates have gathered excellent data on the behavior of this species by matching the right question to the right animal. Other populations of snakes offer similar opportunities for behavioral studies, including *Vipera berus* in Sweden (Shine, Chap. 2, this volume), marine snakes in Fiji (R. Shine, personal communication), and several populations of North American viperids (Duvall, Schuett, and Arnold, Chap. 5, this volume).

Another example of matching question, technique, and animal concerns demography. As noted previously, we were able to achieve high recapture rates for Racers using the appropriate method (drift fences and funnel traps). However, even hand captures can be used as a technique for studying demography under the proper conditions. J. L. Knight (unpublished) was able to mark 100% of a population of *Tretanorhinus nigroluteus* in Belize by finding a discrete population inhabiting a stream, where snakes were easily captured and recaptured. Madsen's recent studies on *Vipera berus* (see Shine, Chap. 2, this volume, for review) also show how, under the proper conditions, even hand captures provide high-quality data on demography, reproduction, and behavior. I suggest that when question, technique, and study animal are matched carefully, studies on snakes can produce excellent results. This contention is supported by the fact that at least 20 papers on snakes were published between 1988 and 1992 in such prestigious journals as *Nature, American Naturalist, Ecology, Evolution,* and *Oecologia* (personal observation).

Focusing on areas for which snakes are well-suited

Finally, we need to recognize that there are some topics for which studies on snakes are particularly well-suited, and perhaps these areas deserve special attention. What follows is a brief (and highly incomplete) list of some of the areas for which I see studies on snakes as having high potential.

Experimental studies using captive breeding populations. Perhaps an area of research where snakes are truly model organisms concerns experimental studies based on captive breeding programs. Many (if not most) groups of snakes are bred and maintained in captivity with relative ease, including most colubrids, viperids, boids, and some elapids, and the literature on this topic is enormous (see Murphy and Campbell, 1987, for review). In addition, gravid female snakes captured in nature can usually be induced to produce offspring in captivity. Thus, researchers are able to maintain and produce large numbers of snakes for experimental studies.

This ease of captive maintenance has led to some of the best studies on the ecology and behavior of snakes, including Arnold's classic studies on genetics and behavior (Arnold, 1977; Arnold and Bennett, 1984), and Burghardt's studies on feeding behavior (see Ford and Burghardt, Chap. 4, this volume). Captive populations have also led to important studies on effects of temperature on behavior (Burger et al., 1987; Burger and Zappalorti, 1988), and effects of energy on life-history traits (Ford and Seigel, 1989). This is a fertile area for future research on topics ranging from neurobiology to population genetics (Brodie and Garland, Chap. 8, this volume).

Thermal ecology and habitat selection using telemetry. Although snakes were once difficult subjects for radio telemetric studies, the development of implantable transmitters (Reinert and Cundall, 1982) has radically altered the suitability of snakes for such research. This has led to a rapid increase in research in two areas, thermal ecology (Peterson, Gibson, and Dorcas, Chap. 7, this volume), and habitat selection and utilization (Reinert, Chap. 6, this volume). Because transmitters are easily implanted in the body cavity and because temperature-sensitive transmitters are widely available, snakes are a natural subject for studies of thermal ecology, especially when such studies incorporate biophysical models (Peterson, Gibson, and Dorcas, Chap. 7, this volume). The availability of relatively small transmitters also makes snakes highly suitable for studies of habitat selection, especially since snakes frequently move long distances over the course of the activity season (Reinert, Chap. 6, this volume). Transmitters can also be used to study behavioral interactions among snakes during the breeding season (Madsen and Shine, in press).

Field studies of reproduction and feeding. Snakes possess two traits that make them well-suited for studies on feeding and reproduction in the field. First, snakes swallow their prey whole and can be easily induced to regurgitate this prey in the field (Fitch, 1987). Thus, both identification and measurement of prey items are simpler for snakes than for most other organisms, and such studies can be conducted with little or no harm to the individual. Arnold (Chap. 3, this volume) reviews studies of foraging ecology in snakes. Second, the clutch size of most snakes can be counted accurately under field conditions (Fitch, 1987), and in most cases, offspring can be produced in the lab (see above). These characteristics mean that field studies of snakes can obtain data on feeding and reproduction without killing the individuals, an important consideration in today's "politically correct" climate.

Sexual selection, sexual dimorphism, and reproductive success. As Shine (Chap. 2, this volume) notes, snakes have a number of advantages for studies of sexual selection, sexual dimorphism, and reproductive success. For example, male body size seems uncorrelated with reproductive success in many snakes (Shine, Chap. 2, this volume), and snakes apparently cannot rape (Shine, in litt.), so two potential confounding factors in determining reproductive success are removed. In addition, the fact that combat behavior in snakes does not involve biting means that the widespread sexual differences in head size in snakes cannot be attributed to sexual selection on males, a situation quite different from lizards (Shine, Chap. 2, this volume). Finally, the fact that courtship and copulation are so prolonged in many snakes means that (under the proper circumstances) it is possible for an observer to record all the successful matings in a population.

Summary

Despite gloomy discussions about the intractable nature of snakes, the chapters in this and the preceding volume show that high-quality science has been and continues to be done with snakes. The notion that data on snakes are less rigorous and interesting than those for lizards (lizard envy) may be largely a function of (1) not asking appropriate questions about snakes, (2) not using new and innovative techniques, (3) not properly matching question, technique, and study animal, and (4) not focusing on areas of research for which snakes are well-suited. This perception will only be altered by studies of the kinds suggested in the chapters in this text and its predecessor.

Acknowledgments

I thank Neil Ford and Patrick Gregory for inviting me to present my summary address in 1986, upon which this paper is based. This paper benefited from the lively and enjoyable discussions at the Snake Ecology Group meeting at Turtle Cove in 1990, and from the comments of Neil Ford, Michael Plummer, and Richard Shine.

Literature Cited

Arnold, S. J., 1977, Polymorphism and geographic variation in the feeding behavior of the garter snake *Thamnophis elegans*, *Science*, 197:676–678.

Arnold, S. J., and A. F. Bennett, 1984, Behavioural variation in natural populations, III: Antipredator displays in the garter snake *Thamnophis radix*, *Anim. Behav.*, 32:1108–1118.

Burger, J., and R. T. Zappalorti, 1988, Effect of incubation temperature on sex ratios in pine snakes: Differential vulnerability of males and females, *Am. Nat.*, 132:492–505.

Burger, J., R. T. Zappalorti, and M. Gochfeld, 1987, Developmental effects of incubation temperature on hatchling pine snakes *Pituophis melanoleucus*, *Comp. Biochem. Physiol.*, 87A:727–732.

Fitch, H. S., 1987, Collecting techniques, in R. A. Seigel, J. T. Collins, and S. S. Novak, eds., *Snakes: Ecology and Evolutionary Biology*, McGraw-Hill, New York, pp. 143–164.

Ford, N. B., and R. A. Seigel, 1989, Phenotypic plasticity in reproductive traits: Evidence from a viviparous snake, *Ecology*, 70:1768–1774.

Gibbons, J. W., and R. D. Semlitsch, 1978, Terrestrial drift fences with pitfall traps: An effective technique for quantitative sampling of animal populations, *Brimleyana*, 7:1–16.

Huey, R. B., E. R. Pianka, and T. W. Schoener (eds.), 1983, *Lizard Ecology: Studies of a Model Organism,* Harvard Univ. Press, Cambridge, Massachusetts.
Madsen, T., and R. Shine, A rapid, sexually-selected shift in mean body size in a population of snakes, *Evolution,* in press.
Murphy, J. B., and J. A. Campbell, 1987, Captive maintenance, in R. A. Seigel, J. T. Collins, and S. S. Novak, eds., *Snakes: Ecology and Evolutionary Biology,* McGraw-Hill, New York, pp. 165–181.
Parker, W. S., and Plummer, M. V., 1987, Population ecology, in R. A. Seigel, J. T. Collins, and S. S. Novak, eds., *Snakes: Ecology and Evolutionary Biology,* McGraw-Hill, New York, pp. 253–301.
Reinert, H. K., and D. Cundall, 1982, An improved surgical implantation method for radio-tracking snakes, *Copeia,* 1982:702–705.
Turner, F. B., 1977, The dynamics of populations of squamates, crocodilians, and rhynchocephalians, in C. Gans and D. W. Tinkle, eds., *Biology of the Reptilia,* Vol. 7, Academic, New York, pp. 157–264.
Vitt, L. J., 1987, Communities, in R. A. Seigel, J. T. Collins, and S. S. Novak, eds., *Snakes: Ecology and Evolutionary Biology,* McGraw-Hill, New York, pp. 335–365.

Index

ABOUT THE EDITORS

RICHARD A. SEIGEL is an Associate Professor in the Department of Biological Sciences at Southeastern Louisiana University. He is also the editor of *Vertebrate Ecology and Systematics: A Tribute to Henry S. Fitch* (1984) and *Snakes: Ecology and Evolutionary Biology* (1987).

JOSEPH T. COLLINS is a Vertebrate Zoologist in the Museum of Natural History at The University of Kansas. He has written and edited numerous books, including *Reproductive Biology and Diseases of Captive Reptiles* (1980), *Snakes: Ecology and Evolutionary Biology* (1987), *Peterson Field Guide to Reptiles and Amphibians of Eastern and Central North America* (1991), and *Amphibians and Reptiles in Kansas*, Third Edition (1993).

www.ingramcontent.com/pod-product-compliance
Ingram Content Group UK Ltd.
Pitfield, Milton Keynes, MK11 3LW, UK
UKHW020128290926
15471UKWH00003B/348

9 781930 66514